3.2 THE DIFFERENCE OF TWO SQUARES; THE SUM AND DIFFERENCE OF TWO CUBES

$$x^2 - y^2 = (x + y)(x - y)$$
$$x^3 + y^3 = (x + y)(x^2 - xy + y^2)$$
$$x^3 - y^3 = (x - y)(x^2 + xy + y^2)$$

3.3 FACTORING TRINOMIALS

$$x^2 + 2xy + y^2 = (x + y)^2$$
$$x^2 - 2xy + y^2 = (x - y)^2$$

4.1 LINEAR EQUATIONS AND THEIR SOLUTIONS

If a, b, and c are real numbers, if there are no divisions by 0, and if $a = b$, then

$$a + c = b + c \qquad a - c = b - c$$
$$ac = bc \qquad \frac{a}{c} = \frac{b}{c}$$

4.4 SOLVING EQUATIONS BY FACTORING

If $ab = 0$, then $a = 0$ or $b = 0$.

4.5 ABSOLUTE VALUE EQUATIONS

$|x| = k$ is equivalent to $x = k$ or $x = -k$.

4.6 LINEAR INEQUALITIES

If a, b, and c are real numbers, if there are no divisions by 0, and if $a < b$, then

$$a + c < b + c \qquad a - c < b - c$$
$$ac < bc \quad (c > 0) \qquad ac > bc \ (c < 0)$$
$$\frac{a}{c} < \frac{b}{c} \quad (c > 0) \qquad \frac{a}{c} > \frac{b}{c} \ (c < 0)$$

$c < x < d$ is equivalent to $c < x$ and $x < d$.

4.7 INEQUALITIES CONTAINING ABSOLUTE VALUES

$|x| < k$ is equivalent to $-k < x < k$.
$|x| > k$ is equivalent to $x < -k$ or $x > k$.

6.1 RATIONAL EXPONENTS

If all expressions represent real numbers, then

$$(a^{1/n})^n = a$$
$$a^{m/n} = (a^{1/n})^m = (a^m)^{1/n}$$
$$a^{-m/n} = \frac{1}{a^{m/n}}$$

6.2 RADICALS

If all expressions represent real numbers, then

$$a^{1/n} = \sqrt[n]{a} \qquad \left(\sqrt[n]{a}\right)^n = a$$
$$x^{m/n} = \sqrt[n]{x^m} = \left(\sqrt[n]{x}\right)^m$$
$$\sqrt[n]{a^n} = |a| \quad \text{if } n \text{ is even}$$
$$\sqrt[n]{a^n} = a \quad \text{if } n \text{ is odd}$$
$$\sqrt[n]{ab} = \sqrt[n]{a}\,\sqrt[n]{b} \qquad \sqrt[n]{\frac{a}{b}} = \frac{\sqrt[n]{a}}{\sqrt[n]{b}}$$

6.6 RADICAL EQUATIONS

If a, b, and c are real numbers and if $a = b$, then
$$a^n = b^n$$

7.1 THE RECTANGULAR COORDINATE SYSTEM

If $P(x_1, y_1)$ and $Q(x_2, y_2)$ are two points on a line, then the distance between P and Q is

$$d = \sqrt{(x_2 - x_1)^2 + (y_2 - y_1)^2} \quad \text{distance formula}$$

The coordinates of the midpoint of line segment PQ are

$$\left(\frac{x_1 + x_2}{2}, \frac{y_1 + y_2}{2}\right) \quad \text{midpoint formula}$$

7.2 SLOPE OF A NONVERTICAL LINE

The slope of a nonvertical line passing through $P(x_1, y_1)$ and $Q(x_2, y_2)$ is given by

$$m = \frac{y_2 - y_1}{x_2 - x_1} \quad (x_2 \neq x_1)$$

7.3 EQUATIONS OF LINES

$y - y_1 = m(x - x_1)$ point–slope form of a line
$y = mx + b$ slope–intercept form of a line
$Ax + By = C$ general form of a line
$y = k$ a horizontal line
$x = k$ a vertical line

Books in the Gustafson and Frisk Series

BEGINNING ALGEBRA, Second Edition
INTERMEDIATE ALGEBRA, Second Edition
ALGEBRA FOR COLLEGE STUDENTS, Second Edition
COLLEGE ALGEBRA, Third Edition
PLANE TRIGONOMETRY, Second Edition
COLLEGE ALGEBRA AND TRIGONOMETRY, Second Edition
FUNCTIONS AND GRAPHS

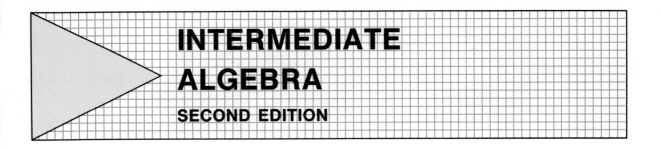

INTERMEDIATE ALGEBRA

SECOND EDITION

R. David Gustafson
Rock Valley College

Peter D. Frisk
Rock Valley College

697-3498

Brooks/Cole Publishing Company
Pacific Grove, California

To
Harold and Monie,
Harder and Evelyn,
with love and affection

Brooks/Cole Publishing Company
A Division of Wadsworth, Inc.

© 1988, 1984 by Wadsworth, Inc., Belmont, California, 94002.

Printed in the United States of America

10   9   8   7   6   5   4

Library of Congress Cataloging-in-Publication Data

Gustafson, R. David (Roy David), [date]
    Intermediate algebra.
    Includes index.
    1. Algebra.   I. Frisk, Peter D., [date]
II. Title.
QA154.2.G874   1987   512.9   87-6657
ISBN 0-534-08388-9

Sponsoring Editor: *Jeremy Hayhurst*
Editorial Assistant: *Maxine Westby*
Production Editors: *Steven Bailey & Suzanne Ewing*
Production Assistant: *Linda Loba*
Manuscript Editor: *Patricia E. Cain*
Permissions Editor: *Carline Haga*
Interior and Cover Design: *Roy R. Neuhaus*
Cover Photo: *Lee Hocker*
Art Coordinator: *Lisa Torri*
Interior Illustration: *Lori Heckelman*
Typesetting: *Syntax International, Ltd., Singapore*
Cover Printing: *Phoenix Color Corp., Long Island City, New York*
Printing and Binding: *R. R. Donnelley & Sons Co., Crawfordsville, Indiana*

Much of the discussion on CAREERS AND MATHEMATICS has been adapted from the
Occupational Outlook Handbook, 1987–1988 edition, Bulletin 2250, published by
    U.S. Department of Labor
    Bureau of Labor Statistics
For more information on these and other careers, please consult that publication.

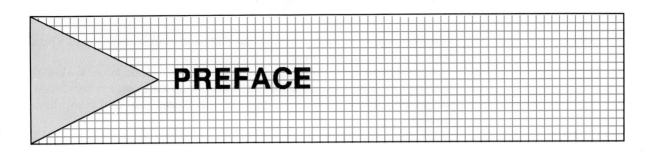

PREFACE

TO THE INSTRUCTOR

In most colleges, the mathematical backgrounds of intermediate algebra students are quite varied. Some students have recently completed a beginning course in algebra, whereas others have not studied algebra for years. In this mixture, many students grasp algebra quickly whereas others find the concepts to be difficult. To minimize the problems caused by this heterogeneous group, we have prepared a text that meets the needs of a variety of students in many different teaching situations. For example, in *Intermediate Algebra, Second Edition*, students with less mathematics background will appreciate the non-technical writing style, the authors' notes in the worked examples, and the review of basic algebra. Students with more skill will appreciate the comprehensive coverage of topics and the challenging problems that appear at the end of most exercise sets.

Because several chapters are independent from the others, the material in the text can be sequenced to meet the needs of any curriculum. Depending on the topics chosen, the text can support a three-, four-, or five-hour course.

This second edition of *Intermediate Algebra* retains the basic philosophy of the highly successful first edition. However, we have made some significant improvements, including the following:

1. The chapter on factoring has been reorganized and expanded to cover the **key number method** of factoring trinomials. A new section that summarizes the factoring techniques has been added.
2. The chapter on solving equations now precedes the chapter on fractions. This ordering allows students to get to applications more quickly. Because equations containing fractions remain in Chapter 5, the work with equations is now spread over two chapters, and word problems are distributed more evenly throughout the text.
3. The chapter on radicals has been rewritten using simpler and more concise language.
4. The chapters on functions and graphing have been reorganized. The material on coordinate systems, graphing linear equations, and writing equations of lines now precedes the discussion of functions. This organization provides students with preparatory material before they encounter the more formal aspects of the function.

5. The treatment of conic sections has been expanded to include conics whose centers are not at the origin.

6. The chapter on logarithms has been expanded to include more work with base-10 and base-*e* exponential functions, more work with natural logarithms, and many more applications of the exponential and logarithmic functions.

7. Many more word problems have been included throughout the text.

Our first edition users responded well to the following features:

Review   The text includes a thorough and continuing review of basic topics. Skills taught in the early chapters are used throughout the text. This review provides a built-in redundancy that allows students several opportunities to review, or even relearn, the material. This constant review helps improve student confidence and helps reduce student attrition.

Many Exercises   The text includes nearly 4000 exercises. Each exercise set is carefully graded and contains an ample supply of both drill and challenging problems. Answers to the odd-numbered exercises are provided in Appendix I.

Comprehensiveness   The text covers all the topics that are essential to providing a strong background for work in college algebra, finite mathematics, or statistics.

Mathematical Honesty   The mathematical developments preserve the integrity of the mathematics, but they are not so rigorous as to confuse students.

Relevance   The text includes many applications and discusses occupations requiring a background in mathematics.

Teacher Support   A test manual containing three tests for each chapter is available in hard copy format. For compiling individualized examinations, a computer-based test bank of test items with EXP-TEST®, a full-featured test-generating system for the IBM-PC, is available to those who adopt the text. Also available is a teacher's manual that gives answers to the problems that are not answered in the Appendix or in the Student's Solutions Manual.

Our students like the text because of the following features:

Informal Writing   The text is written for students to read. The writing is informal rather than technical and is at the tenth-grade reading level on the Fry Readability Test.

Worked Examples   There are over 350 worked examples, many of which include authors' notes to explain each step in the solution process.

Functional Use of Second Color   The text uses second color effectively—not just to highlight important definitions and theorems, but to "point" to terms and expressions that the instructor would point to in a classroom discussion.

Review Exercises   Each chapter concludes with a chapter summary, review exercises, and a sample chapter test. Cumulative review exercises appear regularly throughout the text.

Applications   Careers in mathematics are discussed after every chapter. Applications appear throughout.

Summary of Information   Key formulas and ideas from the text are listed inside the front and back covers for easy reference.

Student Support   The STUDENT'S SOLUTIONS MANUAL provides complete solutions to the even-numbered exercises that are not multiples of four. A student STUDY GUIDE is also available.

ORGANIZATION AND COVERAGE

Several of the chapters are independent from the others to allow for flexibility in the sequencing of topics. The following diagram shows how the chapters are interrelated.

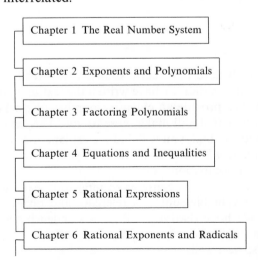

Chapter 1 The Real Number System

Chapter 2 Exponents and Polynomials

Chapter 3 Factoring Polynomials

Chapter 4 Equations and Inequalities

Chapter 5 Rational Expressions

Chapter 6 Rational Exponents and Radicals

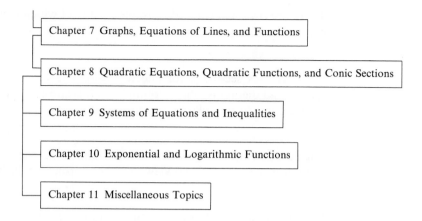

CALCULATORS

We encourage the use of calculators. We believe that students should learn calculator skills in the mathematics classroom. They will then be able to use a calculator for science and business classes and for nonacademic purposes, as well.

ACCURACY

Dozens of mathematics teachers have reviewed either all or part of the text. We are grateful for their many constructive criticisms and helpful suggestions. Both authors have independently worked all the exercises, as have two professional problem checkers.

TO THE STUDENT

Because we believe that many students who read this text do not intend to major in mathematics, we have written the text in an informal rather than technical way. We provide an extensive number of worked examples and present them in a way that will make sense to you. This text has been written for *you* to read, and we think you will find the explanations clear and instructive. What you learn here will be of great value to you, both in other course work and in your chosen occupation.

We suggest that you consider keeping your text after completing this course. It will be a valuable source of reference and review for later courses such as calculus. We hope that you will gain confidence and take more mathematics courses in the future.

We wish you well.

ACKNOWLEDGMENTS

We are grateful to the following people who reviewed the entire text at various levels of its development.

Dale Boye
Schoolcraft College

Lee R. Clancy
Golden West College

Elias Deeba
University of Houston—Downtown

Robert B. Eicken
Illinois Central College

Paul Finster
El Paso Community College

George Grisham
Bradley University

David W. Hansen
Monterey Peninsula College

Steven Hatfield
Marshall University

John Hooker
Southern Illinois University

William A. Hutchings
Diablo Valley College

Herbert Kasube
Bradley University

Diane Koenig
Rock Valley College

Thomas McCready
California State University—Chico

Kenneth Shabell
Riverside Community College

Ray Tebbetts
San Antonio College

Steve Thomassin
Ventura County Community College

Jerry Wilkerson
Missouri Western State College

Our thanks go to William Hinrichs, Darrell Ropp, David Hinde, Jerry Frang, Gary Schultz, and James Yarwood for their helpful comments and suggestions. We thank Archie Strole for his suggested inclusion of the key number method for factoring trinomials.

We also thank Jeremy Hayhurst, Steven Bailey, Trisha Cain, Roy Neuhaus, Lee Hocker, Maxine Westby, Lisa Torri, Sue Ewing, Diane Koenig, and Jeanne Wyatt for their valuable assistance in the creation of this text.

R. David Gustafson
Peter D. Frisk

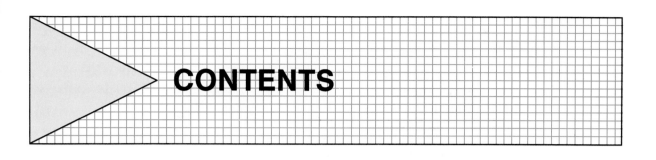

CONTENTS

1

The Real Number System

Algebra is the result of contributions from many cultures over thousands of years. Possibly some algebra was known to the ancient Babylonians, although its first recorded traces are found in the writings of an Egyptian priest named Ahmes, who lived before 1700 B.C. Diophantus, a Greek who lived around A.D. 300, was the first to use special symbols to represent unknown quantities. He is famous for his work with equations. The first known woman mathematician, a Greek scholar named Hypatia, studied and explained the work of Diophantus. Within one hundred years of her death in A.D. 415, the Arabians were using + and − signs and were working with fractions. The symbols we now use to write numbers were developed by the Hindu and Arabic cultures. They were introduced to the West by the thirteenth century Italian merchant Leonardo of Pisa, also known as Fibonacci. The name *algebra* comes from the title of a book written by the Arabian mathematician Al-Khowarazmi around A.D. 800. Its title, *Ihm al-jabr wa'l muqabalah*, means restoration and reduction, a process then used to solve equations.

During the middle ages there was little further development of algebra. Not until the sixteenth century did scholars again become interested in mathematics. At that time, France and Spain were at war. François Viète, a French lawyer with an interest in mathematics, devised a system to break the codes the Spaniards used to send secret messages. His system developed the algebraic notation that we still use today.

Because the concept of *number* is basic in algebra, we begin by discussing the various sets of real numbers.

1.1 SETS OF NUMBERS

A **set** is a collection of objects. To denote a set, we often use braces to enclose a list of its **members** or **elements**. For example, the notation

$$\{a, b, c\}$$

denotes the set with elements a, b, and c. To indicate that b is an element of this set, we write

$$b \in \{a, b, c\} \qquad \text{Read } \in \text{ as "is an element of."}$$

1

The expression

$$d \notin \{a, b, c\}$$

indicates that d is not an element of $\{a, b, c\}$.

Capital letters are used to name sets. For example, the expression

$$A = \{a, e, i, o, u\}$$

means that A is the set containing the vowels a, e, i, o, and u.

In **set-builder notation** a rule is given that establishes membership in a set. The set of vowels in the English alphabet, for example, can be denoted as

$$V = \{x : x \text{ is a vowel of the English alphabet.}\}$$

The statement above is read as "V is the set of all letters x such that x represents a vowel of the English alphabet." Because x can represent many different elements of the set, x is called a **variable**.

When two sets such as A and V have exactly the same elements, we say that they are equal, and we write $A = V$.

If $B = \{a, c, e\}$ and $A = \{a, b, c, d, e\}$, each element of B is also an element of A. When this is so, we say that B is a **subset** of A. In symbols, we write

$$B \subseteq A \qquad \text{Read as "B is a subset of A."}$$

Because every element in set A is an element in set A, set A is a subset of itself. In symbols,

$$A \subseteq A$$

In general, any set is a subset of itself. The expression

$$A \nsubseteq B$$

indicates that A is not a subset of B.

A set with no elements is called the **empty set**, or **null set**, and is denoted as $\varnothing$. Thus,

$$\varnothing = \{ \quad \}$$

The empty set is considered to be a subset of every set.

If the elements of some set A are joined with the elements of some set B, the **union** of set A and set B is formed. The union of set A and set B is denoted as

$$A \cup B \qquad \text{Read as "the union of set A and set B."}$$

The elements of $A \cup B$ are *either* elements of set A, *or* elements of set B, *or both*.

The set of elements that are common to set A and set B is called the **intersection** of set A and set B. The intersection of set A and set B is denoted as

$$A \cap B \qquad \text{Read as "the intersection of set A and set B."}$$

The elements of $A \cap B$ are elements of *both* set A *and* set B. If A and B have no elements in common, then $A \cap B = \varnothing$, and the sets A and B are said to be **disjoint**.

Example 1 If $A = \{a, e, i, o, u\}$ and $B = \{a, b, c, d, e\}$, find **a.** $A \cup B$, **b.** $A \cap B$, and **c.** $(A \cap \varnothing) \cup B$.

Solution **a.** $A \cup B = \{a, b, c, d, e, i, o, u\}$

b. $A \cap B = \{a, e\}$

c. $(A \cap \varnothing) \cup B = \varnothing \cup B$ Do the work in parentheses first.

$= B$

$= \{a, b, c, d, e\}$ ■

Sets of Numbers

The set of numbers that we use for counting is called the set of **natural numbers** or the set of **counting numbers**.

Definition. The **natural numbers** are the numbers

$1, 2, 3, 4, 5, 6, 7, 8, 9, \ldots$

The three dots used in the previous definition, called the **ellipsis**, indicate that the list of natural numbers continues endlessly. If a set, such as the set of natural numbers, has an unlimited number of elements, it is called an **infinite set**. If a set has a limited number of elements, it is called a **finite set**.

Two important subsets of the natural numbers are the **prime numbers** and the **composite numbers**.

Definition. A **prime number** is any natural number greater than 1 that is divisible without a remainder only by itself and by 1.

A **composite number** is any natural number greater than 1 that is not a prime number.

The prime numbers less than 20 are

2, 3, 5, 7, 11, 13, 17, and 19

and the composite numbers less than 20 are

4, 6, 8, 9, 10, 12, 14, 15, 16, and 18

If we join 0 to the set of natural numbers, we have the set of **whole numbers**.

Definition. The **whole numbers** are the numbers

$0, 1, 2, 3, 4, 5, 6, 7, 8, 9, \ldots$

It is often necessary to use numbers that indicate direction as well as quantity—for example, profit or loss, temperatures above or below 0, and gains or losses in the stock market. To do so, we must extend the set of whole numbers to include the negatives of the natural numbers. These negatives are denoted with − signs. For example, the negative of 7 is written as −7 and is read as "negative 7." The union of the set of whole numbers and the set of negatives of the natural numbers forms the set of **integers**.

> **Definition.**   The **integers** are the numbers
>
> $$\ldots, -7, -6, -5, -4, -3, -2, -1, 0, 1, 2, 3, 4, 5, 6, 7, \ldots$$

Integers that are divisible by 2 are called **even integers**, and integers that are not divisible by 2 are called **odd integers**. Because $\frac{0}{2} = 0$, 0 is an even integer. The even integers from −10 to 10 are

$$-10, -8, -6, -4, -2, 0, 2, 4, 6, 8, 10$$

and the odd integers between −10 and 10 are

$$-9, -7, -5, -3, -1, 1, 3, 5, 7, 9$$

If two integers are added, subtracted, or multiplied, the result is always another integer. However, the result obtained when two integers are divided is not always another integer. For example, when 8 is divided by 5, we obtain the fraction $\frac{8}{5}$. When an integer is divided by a nonzero integer, the result is called a **rational number**. The numbers

$$\frac{2}{3}, \quad -\frac{44}{23}, \quad 16, \quad \text{and} \quad -0.25$$

are examples of rational numbers. The number 16 is rational because it can be written as the fractions $\frac{16}{1}, \frac{32}{2}, \frac{48}{3}$, and so on. The number -0.25 is rational because it can be written as the fraction $-\frac{1}{4}$.

We note that $\frac{8}{4} = 2$ because $4(2) = 8$, that $\frac{24}{8} = 3$ because $8(3) = 24$, and that $\frac{0}{9} = 0$ because $9(0) = 0$. However, the fraction $\frac{5}{0}$ is undefined because there is no number that when multiplied by 0 gives 5. The fraction $\frac{0}{0}$ is undefined also because *all* numbers when multiplied by 0 give 0. Thus, it is understood that *the denominator of a fraction can never be* 0. We emphasize this important fact in the following definition.

> **Definition.**   A **rational number** is any number that can be written in the form a/b, where a and b are integers and b is not 0.

Example 2   **a.**   The fraction $\frac{5}{3}$ is a rational number because it is the quotient of two integers and the denominator is not 0.

b.   The number -7 is rational because it can be written in the form $-\frac{7}{1}$, $-\frac{14}{2}$, $-\frac{21}{3}$, and so on.

c.   The number 0.125 is rational because it can be written in the form $\frac{1}{8}$.

d.   The number $-0.666\ldots$ is rational because it can be written as $-\frac{2}{3}$.   ■

Every rational number written in fractional form can be written in decimal form. For example, to change $\frac{3}{4}$ to a decimal fraction, we divide 3 by 4 to obtain 0.75.

$$
\begin{array}{r}
0.75 \\
4\,\overline{)\,3.00} \\
2\,8 \\
\overline{20} \\
20 \\
\overline{0}
\end{array}
$$

Because the division leaves a remainder of 0, the division stops and the quotient 0.75 is called a **terminating decimal**. If we change a fraction such as $\frac{421}{990}$ to a decimal fraction, we obtain $0.4252525\ldots$, called a **repeating decimal**, in which the block of digits "25" repeats forever.

$$
\begin{array}{r}
0.42525\ldots \\
990\,\overline{)\,421.0000} \\
3960 \\
\overline{2500} \\
1980 \\
\overline{5200} \\
4950 \\
\overline{250}
\end{array}
$$

The repeating decimal $0.4252525\ldots$ is often written as $0.4\overline{25}$, where the overbar indicates the repeating block of digits.

It can be shown that all decimal forms of rational numbers are either terminating or repeating decimals.

It is easy to write a terminating decimal as a rational number in fractional form. For example, to write 0.25 as a fraction we note that 0.25 means $\frac{25}{100}$ and that $\frac{25}{100}$ can be simplified to obtain $\frac{1}{4}$.

$$0.25 = \frac{25}{100} = \frac{25 \cdot 1}{25 \cdot 4} = \frac{25 \cdot 1}{25 \cdot 4} = \frac{1}{4} \qquad \text{Read } 25 \cdot 1 \text{ as "25 times 1."}$$

In Chapter 4 we will show that any repeating decimal can be written as a rational number in fractional form.

Because it is possible both to write rational numbers in fractional form as terminating or repeating decimals and to write terminating and repeating decimals as rational numbers in fractional form, these two sets of numbers are one and the same. Thus, the set of rational numbers is equal to the set of all decimals that either terminate or repeat.

Numbers whose decimal representations neither terminate nor repeat are called **irrational numbers**. For example, the decimal

$$0.31\ 331\ 3331\ldots$$

follows a pattern, but it will never have a repeating block of digits. Thus, it represents an irrational number. Other examples of irrational numbers are $\sqrt{3}$ (the square root of 3) and $-\pi$ (negative pi).

$$\sqrt{3} = 1.7320508075 \quad \text{and} \quad -\pi = -3.141592653\ldots$$

If **R** is used to represent the set of rational numbers and **H** is used to represent the set of irrational numbers, the union of set **R** and set **H** is the set of all decimals. This set, denoted as $\mathscr{R}$, is called the set of **real numbers**. In symbols, we write

$$\mathbf{R} \cup \mathbf{H} = \mathscr{R}$$

Since there are no real numbers that are both rational and irrational, sets **R** and **H** are disjoint. Thus,

$$\mathbf{R} \cap \mathbf{H} = \varnothing$$

The relationship of the sets of numbers developed thus far is shown in Figure 1-1. Each set of numbers in the figure is a subset of those sets that precede it.

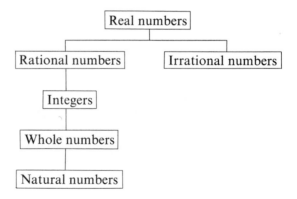

Figure 1-1

■ EXERCISE 1.1

In Exercises 1–6, $A = \{1, 3, 5, 9\}$, $B = \{3, 5, 9, 11\}$, and $C = \{x : x \text{ is an odd integer.}\}$. Insert either an $\in$ or a $\subseteq$ symbol to make a true statement.

1. $A \underline{\ \subseteq\ } C$
2. $3 \underline{\ \in\ } A$
3. $9 \underline{\ \in\ } B$
4. $\{3, 5\} \underline{\ \subseteq\ } C$
5. $\varnothing \underline{\ \subseteq\ } B$
6. $A \underline{\ \subseteq\ } A$

In Exercises 7–10, list the elements in each set, if possible.

7. $\{x : x$ is a prime number less than 10.$\}$ **8.** $\{x : x$ is a composite number between 3 and 7.$\}$

9. $\{x : x$ is a number that is both rational and irrational.$\}$

10. $\{x : x$ is the name of a state in the United States beginning with the letter I$\}$

In Exercises 11–18, $A = \{2, 3, 5, 7\}$, $B = \{2, 4, 6, 8, 10\}$, and $C = \{1, 3, 5, 7, 9\}$. Find each set.

11. $A \cup B$ **12.** $A \cap C$ **13.** $B \cap C$ **14.** $A \cup C$

15. $(C \cap \varnothing) \cap B$ **16.** $B \cup (\varnothing \cap C)$ **17.** $(A \cup B) \cap C$ **18.** $(A \cap B) \cup C$

In Exercises 19–24, tell whether each set is a finite or an infinite set.

19. $\{x : x$ is a prime number.$\}$ **20.** $\{1, 2, 3, 4, 5, 6, 7, 8\}$

21. $\{x : x$ is a natural number that is neither prime nor composite.$\}$

22. $\{x : x$ is an even integer.$\}$

23. $\{x : x$ is a rational number.$\}$ **24.** $\{x : x$ is a state in the United States.$\}$

In Exercises 25–36, simplify each expression, if necessary. Then classify each number as a natural number, a whole number, an integer, a rational number, an irrational number, and/or a real number. Most numbers will be in many classifications.

25. 2   **26.** -2 **27.** 0   **28.** $\dfrac{2}{3}$

29. 0.75   **30.** $0.333\ldots$ **31.** $-\dfrac{8}{9}$ **32.** $-3.\overline{15}$

33. $\dfrac{20}{4}$ **34.** $\dfrac{4}{8}$ **35.** $0.232232223\ldots$ **36.** $-2.373373337\ldots$

In Exercises 37–48, simplify each expression, if necessary, and classify each result as an even integer, an odd integer, a prime number, and/or a composite number. Some numbers will be in more than one classification.

37. 8   **38.** -5 **39.** -9 **40.** 4   **41.** $\dfrac{12}{6}$ **42.** $8 - 5$

43. 0   **44.** 1   **45.** $3(5)$ **46.** -16 **47.** $\dfrac{0}{4}$ **48.** 2

In Exercises 49–68, identify each statement as true or false. If a statement is true, give an example to confirm that it is true. If a statement is false, give an example to show that it is false. If a statement is sometimes true and sometimes false, consider it to be a false statement.

49. The product of two natural numbers is a natural number.

50. The sum of two natural numbers is a natural number.

51. All of the whole numbers are natural numbers.

52. All of the natural numbers are integers.

53. The sum of two prime numbers is a prime number.

54. The product of two prime numbers is a prime number.

55. The sum of two even integers is even.

56. The product of two even integers is even.

57. The sum of two odd integers is odd.

58. The product of two odd integers is odd.

59. The sum of two composite numbers is a composite number.

60. The product of two composite numbers is composite.

61. The product of two prime numbers is composite.

62. The sum of two prime numbers is composite.

63. The sum of 0 and any natural number is a natural number.

64. The product of 0 and any natural number is a natural number.

65. The only even prime number is 2.

66. The negative of a prime number is also prime.

67. Every integer is a rational number.

68. No rational number is an integer.

In Exercises 69–72, write each rational number as a decimal. Then classify the decimal as a terminating or a repeating decimal.

69. $\dfrac{7}{8}$ **70.** $\dfrac{7}{3}$ **71.** $-\dfrac{11}{15}$ **72.** $-\dfrac{19}{16}$

In Exercises 73–76, write each terminating decimal as a rational number in fractional form.

73. 0.5   **74.** 0.2   **75.** 0.75   **76.** 0.125

1.2 EQUALITY AND PROPERTIES OF REAL NUMBERS

If two variables such as a and b represent the same number, we say that a and b are equal, and we write $a = b$. To show that a and b are not equal, we write $a \neq b$, where the symbol $\neq$ means "is not equal to." There are several properties of equality that we shall use throughout this book.

Properties of Equality.   If a, b, and c are real numbers, then

$a = a$	The reflexive property
If $a = b$, then $b = a$.	The symmetric property
If $a = b$ and $b = c$, then $a = c$.	The transitive property

---

The reflexive property of equality states that any number is equal to itself. The symmetric property states that if one number is equal to a second, then the second number is equal to the first. The transitive property states that if one number is equal to a second and the second number is equal to a third, then the first number is equal to the third.

Another property of equality enables us to substitute a quantity for its equal in any mathematical expression without changing the meaning of the expression.

The Substitution Property.   If a and b are real numbers and $a = b$, then b can be substituted for a in any mathematical expression to obtain an equivalent expression.

Example 1   Each statement is true because of the given reason.

a. $x - 4 = x - 4$ The reflexive property

b.   If $5x = 3y$, then $3y = 5x$.   The symmetric property

c.   If $3x = 8$ and $8 = 2y$, then $3x = 2y$.   The transitive property

d.   If $x + 3 = xy$ and $x = 9$, then $9 + 3 = 9y$.   The substitution property   ∎

There are many properties involving operations on the real numbers.

The Closure Properties.   If a, b, and c are real numbers, then

$a + b$ (the sum of a and b) is a real number.

$a - b$ (the difference of a and b) is a real number.

$a \cdot b$ (the product of a and b) is a real number.

$\dfrac{a}{b}$ (the quotient of a and b) is a real number, provided that $b \neq 0$.

The product of a and b is often written as $a(b)$, $(a)(b)$, or just ab.

Because of the closure properties, the sum, difference, product, and quotient of any two real numbers is, again, a real number (provided there are no divisions by 0).

The Associative Properties.   If a, b, and c are real numbers, then

$(a + b) + c = a + (b + c)$ The associative property of addition

$(ab)c = a(bc)$ The associative property of multiplication

The associative properties enable us to group, or associate, the numbers in a sum, or the numbers in a product, in any way that we wish and still be assured of the same answer. For example,

$$(2 + 3) + 4 = 5 + 4 \qquad \text{and} \qquad 2 + (3 + 4) = 2 + 7$$
$$= 9 \qquad\qquad\qquad\qquad = 9$$

The answer is 9 regardless of how we group the numbers in this sum. Likewise,

$$(2 \cdot 3) \cdot 4 = 6 \cdot 4 \qquad \text{and} \qquad 2 \cdot (3 \cdot 4) = 2 \cdot 12$$
$$= 24 \qquad\qquad\qquad\qquad = 24$$

The product is 24 regardless of how we group the numbers in this product.

Example 2   Use an associative property to simplify the expression $3 + (2 + a)$.

Solution $3 + (2 + a) = (3 + 2) + a$ Use the associative property of addition.

$\qquad\qquad\qquad = 5 + a$ ∎

The Commutative Properties.   If a and b are real numbers, then

$\qquad a + b = b + a$ The commutative property of addition

$\qquad ab = ba$ The commutative property of multiplication

The commutative properties enable us to add or multiply two numbers in either order. We can add the first number to the second, or the second number to the first. The results are the same. Likewise, we can multiply the first number by the second, or the second number by the first, and the results are the same. For example,

$2 + 3$ and $3 + 2$ are both 5.

$7 \cdot 9$ and $9 \cdot 7$ are both 63.

Example 3   Use a commutative and an associative property to simplify the expression $(3a)4$.

Solution $(3a)4 = 4(3a)$ Use the commutative property of multiplication.

$\qquad\qquad = (4 \cdot 3)a$ Use the associative property of multiplication.

$\qquad\qquad = 12a$ ∎

The Distributive Property.   If a, b, and c are real numbers, then

$\qquad a(b + c) = ab + ac$

Because of the distributive property, there are two ways of evaluating certain expressions that involve both a multiplication and an addition: We can either add first and then multiply, or multiply first and then add. Either way, the answer is the same. For example, $2(3 + 7)$ can be computed in two ways. One way is to perform the indicated addition and then the multiplication.

$$2(3 + 7) = 2 \cdot 10$$
$$= 20$$

The other way is to distribute the multiplication by 2 by first multiplying each number within the parentheses by 2 and then adding.

$$2(3 + 7) = 2 \cdot 3 + 2 \cdot 7$$
$$= 6 + 14$$
$$= 20$$

Either way, the answer is 20.

A more general form of the distributive property is called the **extended distributive property**:

$$a(b + c + d + e + \cdots) = ab + ac + ad + ae + \cdots$$

Example 4 Use the extended distributive property to simplify the expression $2(x + y + 7)$.

Solution
$$2(x + y + 7) = 2x + 2y + 2(7)$$
$$= 2x + 2y + 14 \qquad \blacksquare$$

Because adding 0 to a number leaves that number identically the same, 0 is called the **identity for addition**.

The Identity Element for Addition. There is a unique number 0, called the **additive identity**, such that

$$0 + a = a + 0 = a$$

Another property of 0, called the **multiplication property of zero**, states that the product of any number and 0 is 0.

$$a \cdot 0 = 0 \cdot a = 0$$

Because multiplying a number by 1 leaves that number identically the same, 1 is called the **identity for multiplication**.

The Identity Element for Multiplication. There is a unique number 1, called the **multiplicative identity**, such that

$$1 \cdot a = a \cdot 1 = a$$

If the sum of two numbers is 0, the numbers are called **additive inverses** or **negatives** of each other.

The Additive Inverse Elements.   For each real number a, there is a single number $-a$ such that

$$a + (-a) = -a + a = 0$$

The number $-a$ is called the **additive inverse** or the **negative** of a. Also, a is called the additive inverse or the negative of $-a$.

Because of the previous definition, the sum of a number and its negative is 0. For example,

$$5 + (-5) = 0 \qquad \text{and} \qquad -7 + 7 = 0$$

The symbol $-(-6)$ is read as "the negative of negative 6." Because the sum of two numbers that are negatives is 0, we have

$$-6 + [-(-6)] = 0$$

but

$$-6 + \quad 6 \quad = 0$$

Because -6 has only one additive inverse, it follows that

$$-(-6) = 6$$

In general, we have the following rule:

The Double Negative Rule.   If a represents any number, then

$$-(-a) = a$$

If the product of two numbers is 1, the numbers are called **multiplicative inverses** or **reciprocals** of each other.

The Multiplicative Inverse Elements.   For every nonzero real number a, there exists a single real number $\dfrac{1}{a}$ such that

$$a \cdot \frac{1}{a} = \frac{1}{a} \cdot a = 1$$

The number $\dfrac{1}{a}$ is called the **multiplicative inverse**, or the **reciprocal**, of a.

Also, a is called the multiplicative inverse, or the reciprocal, of $\dfrac{1}{a}$.

The existence of multiplicative inverse elements guarantees that whatever nonzero real number we start with, another real number can be found so that their product is 1. For example, the reciprocal, or multiplicative inverse, of 5 is $\frac{1}{5}$ because $5 \cdot \frac{1}{5} = 1$. The reciprocal of $-\frac{3}{2}$ is $-\frac{2}{3}$ because $(-\frac{3}{2})(-\frac{2}{3}) = 1$. A reciprocal for 0 does not exist because $\frac{1}{0}$ is an undefined mathematical expression.

Example 5   The statements in the left column are true because of the properties listed in the right column.

$2 + 7$ is a real number	The closure property for addition
$2(7)$ is a real number	The closure property for multiplication
$9 + 3 = 3 + 9$	The commutative property for addition
$8 \cdot 3 = 3 \cdot 8$	The commutative property for multiplication
$9 + (2 + 3) = (9 + 2) + 3$	The associative property for addition
$2(xy) = (2x)y$	The associative property for multiplication
$2(x + 3) = 2x + 2 \cdot 3$	The distributive property
$(a + b) + c = c + (a + b)$	The commutative property for addition
$37 + 0 = 37$	The identity property for addition
$17 \cdot 1 = 17$	The identity property for multiplication
$\frac{3}{7} + \left(-\frac{3}{7}\right) = 0$	The additive inverse property
$\frac{4}{5} \cdot \frac{5}{4} = 1$	The multiplicative inverse property

■ EXERCISE 1.2

In Exercises 1–4, insert either an $=$ or an $\neq$ symbol to make a true statement.

1. $3 \_\_ 2 + 1$ **2.** $\frac{2}{3} \_\_ \frac{3}{4}$ **3.** $\pi \_\_ 3$ **4.** $0.375 \_\_ \frac{3}{8}$

In Exercises 5–12, tell which property of equality justifies each statement.

5. If $a = b + c$, then $b + c = a$.

6. If $x = y + z$ and $z = 3$, then $x = y + 3$.

7. $a + b + c = a + b + c$

8. If $a = 37$ and $37 = b$, then $a = b$.

9. If $x = y + z$ and $y + z = 10$, then $x = 10$.

10. If $x + y = c + d$, then $c + d = x + y$.

11. If $3x = 3y$ and $y = 4$, then $3x = 3(4)$.

12. $(a + b) + c = (a + b) + c$

In Exercises 13–16, use an associative property to help simplify each expression.

13. $5 + (2 + x)$ **14.** $(a + 3) + 4$ **15.** $5(3b)$ **16.** $3(2x)$

In Exercises 17–20, use a commutative property and then an associative property to help simplify each expression.

17. $(3 + b) + 7$ **18.** $7 + (a + 3)$ **19.** $(3y)2$ **20.** $(5z)3$

In Exercises 21–24, use the distributive property to remove parentheses and then simplify, if possible.

21. $3(x + 2)$ **22.** $2(3y + 4)$ **23.** $5(x + y + 4)$ **24.** $9(3 + a + b)$

In Exercises 25–40, tell which property of the real numbers justifies each statement.

25. $3(4)$ is a real number.   **26.** $5 + 5$ is a real number.

27. $3 + 7 = 7 + 3$ **28.** $2(9 \cdot 13) = (2 \cdot 9)13$

29. $3(2 + 5) = 3 \cdot 2 + 3 \cdot 5$ **30.** $1 \cdot 3 = 3 \cdot 1$

31. $81 + 0 = 0 + 81$ **32.** $3(9 + 2) = 3 \cdot 9 + 3 \cdot 2$

33. $81 + 0 = 81$ **34.** $3 + (9 + 0) = (9 + 0) + 3$

35. $5 \cdot \dfrac{1}{5} = 1$ **36.** $a + (3 + y) = (a + 3) + y$

37. $2 + (7 + 8) = (2 + 7) + 8$ **38.** $1 \cdot 3 = 3$

39. $(2 \cdot 3)4 = 4(2 \cdot 3)$ **40.** $8 + (-8) = 0$

*In Exercises 41–52, find the **additive inverse** of each number. If necessary, simplify the expression first, and then find the additive inverse of the result.*

41. 1 **42.** 3 **43.** -8 **44.** -7 **45.** 0 **46.** $\dfrac{1}{2}$

47. π **48.** a **49.** $2 + 8$ **50.** $5 - 3$ **51.** $-(7 - 4)$ **52.** $-\dfrac{8}{2}$

*In Exercises 53–60, find the **multiplicative inverse** (the reciprocal) of each number, provided one exists.*

53. 1 **54.** 3 **55.** $\dfrac{1}{2}$ **56.** $-\dfrac{7}{5}$

57. -0.25 **58.** $0.333\ldots$ **59.** 0 **60.** 1.25

In Exercises 61–64, give a reason for each step in each proof.

61. Prove that $(a + b) + c = a + (c + b)$.

$(a + b) + c = a + (b + c)$ _____

$= a + (c + b)$ _____

62. Prove that $a(b + c) = ca + ba$.

$a(b + c) = ab + ac$ _____

$= ac + ab$ _____

$= ca + ab$ _____

$= ca + ba$ _____

63. Prove that $(b + c)a = ba + ca$.

$(b + c)a = a(b + c)$ _____

$= ab + ac$ _____

$= ba + ac$ _____

$= ba + ca$ _____

64. Prove that $(ab)(cd) = (ad)(bc)$.

$(ab)(cd) = (cd)(ab)$ _____

$= c[d(ab)]$ _____

$= c[(da)b]$ _____

$= c[(ad)b]$ _____

$= [(ad)b]c$ _____

$= (ad)(bc)$ _____

1.3 INEQUALITIES AND GRAPHS OF SETS OF REAL NUMBERS

Sets of numbers can be pictured, or graphed, on a number line. To do so, we construct a number line by choosing some point on a line (called the **origin**) and giving it a number value (a **coordinate**) of 0. We then locate points that are equal distances to the right and to the left of 0, and label them with coordinates, as shown in Figure 1-2. The point on the number line corresponding to the real number -4 is the point with coordinate -4. The point corresponding to the real number $\frac{13}{2}$ is the point midway between the points with coordinates 6 and 7. The point corresponding to the real number π is the point with coordinate 3.14159 . . . . To every real number there corresponds exactly one point on the number line, called its **graph**, and to each point there corresponds exactly one real number, which is its coordinate.

Figure 1-2

Real numbers such as 5 and 25.347 that are to the right of 0 are called **positive numbers**. Sometimes these numbers are preceded by a $+$ sign:

$$+5 = 5 \qquad +25.347 = 25.347 \qquad +\pi = \pi$$

Real numbers such as -4 and $-\frac{17}{2}$ that are to the left of 0 are called **negative numbers**. The number 0 is neither positive nor negative.

Example 1 Graph the set of even integers between -5 and 5.

Solution The graph of the set of even integers between -5 and 5 includes the points with coordinates -4, -2, 0, 2, and 4. The graph is shown in Figure 1-3.

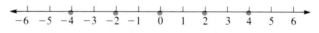

Figure 1-3   ■

If a point lies to the right of a second point on the number line, its coordinate is the greater. For example, on the number line the point with coordinate 4 lies to the right of the point with coordinate -2. Thus,

$$4 > -2 \qquad \text{Read as "4 is greater than negative 2."}$$

If a point on the number line is to the left of another, its coordinate is the smaller. The point with coordinate -5, for example, lies to the left of the point with coordinate -1. Thus,

$$-5 < -1 \qquad \text{Read as "negative 5 is less than negative 1."}$$

Two other common inequality symbols are

$\leq$ Read as "is less than or equal to."

and

$\geq$ Read as "is greater than or equal to."

Example 2 **a.** $-7 > -10$ because -7 is to the right of -10 on the number line.

b. $5 < 8$ because   5 is to the left of 8 on the number line.

c. $15 \leq 15$ because $15 = 15$.

d. $20 \geq -19$ because $20 > -19$.   ■

Inequality statements can be written so that the inequality symbol points in the opposite direction. For example, the inequality

$-3 \leq 9$ can be written as $9 \geq -3$.

To say that a number is not less than 0, we write $x \not< 0$. This is equivalent to saying that $x \geq 0$. Likewise,

$x \not> 0$ is equivalent to $x \leq 0$.

$x \not\geq 0$ is equivalent to $x < 0$.

$x \not\leq 0$ is equivalent to $x > 0$.

$x \neq 0$ is equivalent to $x < 0$ or $x > 0$.

If a and b are two numbers, either a and b are equal or they are not. If they are not equal, then one or the other must be the larger. The possibilities are summed up in the following property.

> **The Trichotomy Property.**   If a and b are real numbers, then exactly one of the following statements is true:
>
> $a < b$ or $a = b$ or $a > b$

Many of the inequality relationships have a transitive property.

> **The Transitive Property of Inequality.**   If a, b, and c are real numbers, then
>
> If $a < b$ and $b < c$, then $a < c$.
>
> A similar statement is true for the $>$, $\leq$, and $\geq$ symbols.

Example 3 **a.**   By the trichotomy property, if x is a real number, then

$$x < 5 \quad \text{or} \quad x = 5 \quad \text{or} \quad x > 5$$

b.   By the transitive property,

if $x > 12$ and $12 > 5$, then $x > 5$.   ■

Graphs of sets of real numbers are often portions of the number line, called **intervals**. For example, Figure 1-4 shows the graph of all real numbers greater than 3. The open circle at 3 indicates that 3 is *not* included. This interval, which includes no endpoints, is called an **open interval** and is denoted by the inequality

$$x > 3$$

In **interval notation** this interval is denoted as $(3, \infty)$, where the symbol ∞ is read as "infinity." In this notation the parentheses indicate that neither endpoint is included.

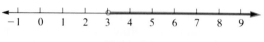

Figure 1-4

The interval shown in Figure 1-5 includes all real numbers x such that $x \le -3$. The solid circle at -3 indicates that -3 is included. Because exactly one endpoint is included in the graph, this interval is called a **half-open interval**. In interval notation, it is denoted as $(-\infty, -3]$. The bracket indicates that the endpoint with coordinate -3 is included.

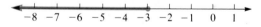

Figure 1-5

To graph the set of real numbers between -4 and 2 (see Figure 1-6), we graph the open interval denoted by the inequalities

$$x > -4 \quad \text{and} \quad x < 2$$

or, more briefly, by the double inequality

$$-4 < x < 2$$ Read as "-4 is less than x, which, in turn, is less than 2."

In interval notation this open interval is expressed as $(-4, 2)$.

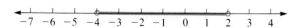

Figure 1-6

Figure 1-7

The interval shown in Figure 1-7, called a **closed interval**, includes two end-points. It is denoted by the double inequality $-3 \le x \le 1$, or in interval notation by $[-3, 1]$. The brackets indicate that both endpoints are included.

The half-open interval shown in Figure 1-8 is denoted by the double inequality $0 \le x < 5$, or in interval notation by $[0, 5)$.

Figure 1-8

Example 4   If $A = (-2, 4)$ and $B = [1, 5)$, find the graph of

a. $A \cup B$ and   **b.** $A \cap B$.

Solution   **a.**   The union of intervals A and B is the set of all real numbers that are elements of either set A or set B or both. Numbers between -2 and 4 are in set A, and numbers between 1 and 5 (including 1) are in set B. Numbers between -2 and 5 are in at least one of these sets. To see this, refer to Figure 1.9(a). Thus,

$$A \cup B = (-2, 4) \cup [1, 5) = (-2, 5)$$

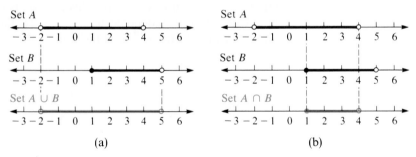

Figure 1-9

b.   The intersection of intervals A and B is the set of all real numbers that are elements of both set A and set B. The numbers that are in both of these sets are those numbers between 1 and 4 (including 1). To see this, refer to Figure 1-9(b). Thus,

$$A \cap B = (-2, 4) \cap [1, 5) = [1, 4)$$ ■

The **absolute value** of any real number a, denoted as $|a|$, is the distance on the number line between the origin and the point with coordinate a. Because the points shown in Figure 1-10 with coordinates of 3 and -3 both lie 3 units

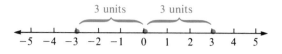

Figure 1-10

from 0,

$$|3| = |-3| = 3$$

Similarly, because the points with coordinates -15 and 15 both lie 15 units from 0,

$$|-15| = |15| = 15$$

In general, for any real number a,

$$|a| = |-a|$$

The absolute value of a number can be defined more formally as follows:

Definition.   If $x \geq 0$, then $|x| = x$.

If $x < 0$, then $|x| = -x$.

The previous definition indicates that if x is a positive number or 0, then x is its own absolute value. However, if x is a negative number, then $-x$ (which is a positive number) is the absolute value of x. Thus, $|x|$ always represents a nonnegative number. In general, for all x

$$|x| \geq 0$$

Example 5   **a.** $|8| = 8$ **b.** $|-4| = -(-4) = 4$

c. $|0| = 0$ **d.** $-|7| = -7$

e. $-|-7| = -(7)$ **f.** $|7 - 2| - (-3) = |5| + 3$

$\qquad\qquad\quad = -7$ $\qquad\qquad\qquad = 5 + 3$

$\qquad\qquad\qquad\qquad\qquad\qquad = 8$ ■

■ **EXERCISE 1.3** ▬▬▬▬▬▬▬▬▬▬▬▬▬▬▬▬▬▬▬▬▬▬▬▬▬▬▬

In Exercises 1–4, graph each set on the number line.

1.   The set of prime numbers less than 10.

2.   The set of negative integers between -10 and 0.

3.   The set of odd integers between 10 and 20.

4.   The set of composite numbers less than 10.

In Exercises 5–10, insert one of the symbols $<$ or $>$ to make a true statement.

5.   5 __ 9   **6.**   9 __ 0   **7.** -5 __ -10 **8.** -3 __ 10

9. -7 __ 7   **10.**   6 __ -6

In Exercises 11–16, rewrite each statement with the inequality symbol pointing in the opposite direction.

11. $19 > 12$ **12.** $-3 \geq -5$ **13.** $-6 \leq -5$ **14.** $-10 < 13$

15. $5 \geq -3$ **16.** $-10 \leq 0$

In Exercises 17–20, rewrite each statement using one of the symbols $<$ or $>$.

17. $x \nleq 3$ **18.** $y \ngeq 4$ **19.** $z \ngeq 4$ **20.** $t \nleq -2$

In Exercises 21–24, rewrite each expression using one of the symbols $\leq$ or $\geq$.

21. $x \nless 7$ **22.** $x \ngtr 3$ **23.** $x \ngtr -3$ **24.** $x \nless -7$

In Exercises 25–36, graph each interval on the number line.

25. $x > 3$ **26.** $x < 0$ **27.** $-3 < x < 2$ **28.** $-5 \leq x < 2$

29. $0 < x \leq 5$ **30.** $-4 \leq x \leq -2$ **31.** $(-2, \infty)$ **32.** $(-\infty, 4]$

33. $[-6, 9]$ **34.** $(-1, 3)$ **35.** $(2, 4]$ **36.** $[-5, 2]$

In Exercises 37–42, A, B, and C are intervals with $A = [-4, 4]$, $B = (0, 6)$, and $C = [2, 8)$. Graph each set.

37. $A \cap C$ **38.** $B \cup C$ **39.** $A \cup C$ **40.** $A \cap B$

41. $A \cap B \cap C$ **42.** $A \cup B \cup C$

In Exercises 43–52, write each expression without using absolute value symbols.

43. $|20|$ **44.** $|-20|$ **45.** $-|-6|$ **46.** $-|8|$

47. $|-5| + |0|$ **48.** $-|0| + |-4|$

49. $|-15| - |10| - (-2)$ **50.** $|-4| - |-4| - (-4)$

51. $|5 - 3| - (-|-1|)$ **52.** $|18 - 5| - (-|-18|)$

53. How many integers have an absolute value that is less than 50?

54. How many odd integers have an absolute value between 20 and 40?

55. What numbers are equal to their own absolute values?

56. What numbers when added to their own absolute values give a sum of 0?

57. What numbers must x and y be if $|x| + |y| = 0$?

58. What numbers must x and y be if $|x + y| = 0$?

59. Does the absolute value of the product of two numbers equal the product of their absolute values? Explain.

60. Does the absolute value of the sum of two numbers equal the sum of their absolute values? Explain.

61. If $|x| = 3$, what numbers could x be?

62. If $|x| = 7$, what numbers could x be?

1.4 ARITHMETIC OF REAL NUMBERS

If x and y are added, the result (denoted by $x + y$) is called their **sum**. Each of the numbers x and y is called a **term** of that sum.

Suppose we wish to add the numbers $+2$ and $+3$. Because the positive direction on the number line is to the right, we can represent $+2$ with an arrow

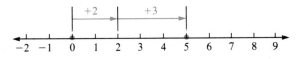

Figure 1-11

of length 2 pointing to the right. Likewise, we can represent $+3$ with an arrow of length 3 also pointing to the right. To find the sum $(+2) + (+3)$, we start at the origin and place the arrows end to end as in Figure 1-11. The endpoint of the second arrow is the point with coordinate $+5$. Thus,

$$+2 + (+3) = +5$$

We can represent the numbers in the addition problem

$$(-2) + (-3)$$

with arrows as in Figure 1-12. We represent -2 with an arrow of length 2 that begins at the origin and points to the left. Then using -2 as a starting point, we represent -3 with an arrow of length 3 that also points to the left. Because the endpoint of the final arrow has coordinate -5,

$$(-2) + (-3) = -5$$

Figure 1-12

Because two numbers with like signs are represented by arrows pointing in the same direction, we have the following rule:

Adding Real Numbers with Like Signs.   If two real numbers a and b have the same sign, their sum is found by adding their absolute values and using their common sign.

Example 1   **a.** $+4 + (+6) = +(|+4| + |+6|)$ **b.** $-4 + (-6) = -(|-4| + |-6|)$
$= +(4 + 6)$ $= -(4 + 6)$
$= +10$ $= -10$ ■

Two real numbers with unlike signs can be represented by arrows that point in opposite directions. For example, to add -6 and $+2$, we refer to the number line in Figure 1-13 where the number -6 is represented by an arrow of length 6 that begins at the origin and points to the left. The arrow representing $+2$

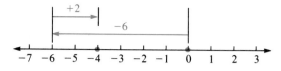

Figure 1-13

begins at -6, has length 2, and points to the right. The endpoint of this final arrow is the point with coordinate -4. Thus,

$$(-6) + (+2) = -4$$

The arrows in Figure 1-14 represent the numbers in the addition problem

$$(+7) + (-4)$$

The first arrow begins at the origin, has length 7, and points to the right. The second arrow begins at point 7, has length 4, and points to the left. The endpoint of the final arrow is the point with coordinate $+3$. Thus,

$$(+7) + (-4) = +3$$

Figure 1-14

Because two real numbers with unlike signs are represented by arrows pointing in opposite directions, we have the following rule:

Adding Real Numbers with Unlike Signs. If two real numbers a and b have unlike signs, their sum is found by subtracting their absolute values (the smaller from the larger) and using the sign of the number with the greater absolute value.

Example 2

a.
$$+6 + (-5) = +(|+6| - |-5|)$$
$$= +(6 - 5)$$
$$= +1$$

b.
$$-2 + (+5) = +(|+5| - |-2|)$$
$$= +(5 - 2)$$
$$= +3$$

c.
$$6 + (-9) = -(|-9| - |6|)$$
$$= -(9 - 6)$$
$$= -3$$

d.
$$-10 + (4) = -(|-10| - |4|)$$
$$= -(10 - 4)$$
$$= -6$$
■

It is always possible to express a subtraction problem as an equivalent addition problem. For example, we can think of the subtraction problem

$$7 - 4$$

as the addition problem

$$7 + (-4)$$

because they have the same answer:

$$7 - 4 = 3 \quad \text{and} \quad 7 + (-4) = 3$$

We use this idea to define the **difference** (the answer to a subtraction problem) when b is to be subtracted from a.

> **Subtracting Real Numbers.**   If a and b are real numbers, then
>
> $$a - b = a + (-b)$$

Example 3   Evaluate   **a.** $12 - 4$,   **b.** $-13 - 5$, and   **c.** $-14 - (-6)$.

Solution   Use the rule for finding the difference of two real numbers:

a. $12 - 4 = 12 + (-4)$ **b.** $-13 - 5 = -13 + (-5)$
$\qquad\quad = 8$ $\qquad\qquad\qquad\qquad\quad = -18$

c. $-14 - (-6) = -14 + [-(-6)]$
$\qquad\qquad\quad = -14 + 6$ Use the double negative rule.
$\qquad\qquad\quad = -8$ ∎

The result of multiplying two numbers x and y is called the **product** of x and y. Each of the numbers x and y is called a **factor** of that product. Multiplication by a positive number can be thought of as repeated addition. The expression $5(4)$, for example, means that 4 is to be used as a term in an indicated sum five times. Thus,

$$5(4) = 4 + 4 + 4 + 4 + 4$$
$$= 20$$

Likewise, the expression $5(-4)$ means that -4 is to be used as a term in a sum five times. Thus,

$$5(-4) = (-4) + (-4) + (-4) + (-4) + (-4)$$
$$= -20$$

If multiplication by a positive number means repeated addition, then it is reasonable to assume that multiplication by a negative number means repeated subtraction. The expression $-5(4)$ means that 4 is to be used as a term in a repeated subtraction five times. Thus,

$$-5(4) = -4 - 4 - 4 - 4 - 4$$
$$= -4 + (-4) + (-4) + (-4) + (-4)$$
$$= -20$$

Likewise, the expression $-5(-4)$ means that -4 is to be used as a term in a repeated subtraction five times. Thus,

$$-5(-4) = -(-4) - (-4) - (-4) - (-4) - (-4)$$
$$= 4 + [-(-4)] + [-(-4)] + [-(-4)] + [-(-4)]$$
$$= 4 + 4 + 4 + 4 + 4$$
$$= 20$$

Note that the products $5(4)$ and $-5(-4)$ both equal $+20$, and that the products $5(-4)$ and $-5(4)$ both equal -20. These results suggest the following rule:

Multiplying Real Numbers.

1. The product of two real numbers with like signs is the positive product of their absolute values.
2. The product of two real numbers with unlike signs is the negative of the product of their absolute values.

Note that, if x is any real number, then $x \cdot 0 = 0 \cdot x = 0$.

Example 4 Use the rules for multiplying real numbers to find each product:

    a. $4(-7)$,   **b.** $-5(-6)$,   **c.** $-7(6)$, and   **d.** $8(6)$.

Solution **a.** $4(-7) = -(|4| \cdot |-7|)$ **b.** $-5(-6) = +(|-5| \cdot |-6|)$
$$= -(4 \cdot 7)$$ $$= +(5 \cdot 6)$$
$$= -28$$ $$= +30$$

  c. $-7(6) = -(|-7| \cdot |6|)$ **d.** $8(6) = +(|8| \cdot |6|)$
$$= -(7 \cdot 6)$$ $$= +(8 \cdot 6)$$
$$= -42$$ $$= +48$$ ∎

Just as subtraction is defined in terms of addition, division is defined in terms of multiplication. When x is divided by a nonzero number y, the result is called the **quotient** of x and y, and we can write

$$\frac{x}{y} = q$$

The quotient q is that number which, when multiplied by y, gives x.

$$yq = x$$

This special relationship can be used to develop the rules for dividing one real number by a nonzero real number. We consider four divisions:

$$\frac{+10}{+2} = +5 \quad \text{because} \quad +2(+5) = +10.$$

$$\frac{-10}{-2} = +5 \quad \text{because} \quad -2(+5) = -10.$$

$$\frac{+10}{-2} = -5 \quad \text{because} \quad -2(-5) = +10.$$

$$\frac{-10}{+2} = -5 \quad \text{because} \quad +2(-5) = -10.$$

The results of the previous examples suggest that the rules for dividing real numbers are very similar to the rules for multiplying real numbers.

Dividing Real Numbers.

1. The quotient of two real numbers with like signs is the positive quotient of their absolute values.
2. The quotient of two real numbers with unlike signs is the negative of the quotient of their absolute values.
3. Division by 0 is undefined.

Note that, if x is any nonzero number, then $\frac{0}{x} = 0$.

Example 5 Use the rules for dividing real numbers to find each quotient:

a. $\frac{36}{18}$, b. $\frac{-44}{11}$, c. $\frac{27}{-9}$, and d. $\frac{-64}{-8}$

Solution a. $\frac{36}{18} = +\frac{|36|}{|18|}$ b. $\frac{-44}{11} = -\frac{|-44|}{|11|}$

$= \frac{36}{18}$ $= -\frac{44}{11}$

$= 2$ $= -4$

c. $\frac{27}{-9} = -\frac{|27|}{|-9|}$ d. $\frac{-64}{-8} = +\frac{|-64|}{|-8|}$

$= -\frac{27}{9}$ $= \frac{64}{8}$

$= -3$ $= 8$

Order of Operations

The expression $2 \cdot 3 + 4$ seems to have two answers depending upon whether the addition or the multiplication is done first. If we were to add first, the answer would be $2(3 + 4) = 2 \cdot 7 = 14$. If we were to multiply first, the answer would be $(2 \cdot 3) + 4 = 6 + 4 = 10$. Unless there is agreement on which operation is to be done first, there is no way of knowing which of these answers is correct. To remove this ambiguity, mathematicians agree on the following order of operations.

Order of Operations.   Unless grouping symbols indicate otherwise, perform the multiplications and/or divisions, in order from left to right. After that, perform the additions and/or subtractions, in order from left to right.

Example 6   Calculate: **a.** $2 \cdot 3 + 4$,   **b.** $2(3 + 4)$,   **c.** $10 \div 5 \cdot 2$, and   **d.** $10 \div (5 \cdot 2)$.

Solution   **a.**   Because parentheses do not indicate otherwise, the multiplication is done first, followed by the addition.

$$2 \cdot 3 + 4 = 6 + 4$$
$$= 10$$

b.   Here the parentheses indicate that the addition must be done first.

$$2(3 + 4) = 2 \cdot 7$$
$$= 14$$

c.   With the absence of grouping symbols to indicate otherwise, the multiplications and/or divisions are done in the order in which they are encountered (from left to right).

$$10 \div 5 \cdot 2 = (10 \div 5)2$$
$$= 2 \cdot 2$$
$$= 4$$

d.   The parentheses indicate that the multiplication is to be done first. Hence,

$$10 \div (5 \cdot 2) = 10 \div 10$$
$$= 1$$ ■

Example 7   Perform the following calculations: **a.** $5 \cdot 3 - 6 \div 3 + 1$,   **b.** $5(3 - 6) \div 3 + 1$, and   **c.** $5(3 - 6 \div 3) + 1$.

Solution   **a.**   Do the multiplications and divisions first, and then the additions and subtractions, from left to right.

$$5 \cdot 3 - 6 \div 3 + 1 = 15 - 2 + 1$$
$$= 13 + 1$$
$$= 14$$

b.   The parentheses indicate that a subtraction must be done first, then the multiplications and divisions in order from left to right. The addition is done last.

$$5(3 - 6) \div 3 + 1 = 5(-3) \div 3 + 1$$
$$= -15 \div 3 + 1$$
$$= -5 + 1$$
$$= -4$$

c.   The operations within the parentheses must be done first; and within the parentheses, the division has priority over the subtraction.

$$5(3 - 6 \div 3) + 1 = 5(3 - 2) + 1$$
$$= 5 \cdot 1 + 1$$
$$= 5 + 1$$
$$= 6$$ ∎

Often, we will encounter arithmetic that needs to be done in both the numerator and denominator of a fraction. To simplify such a fraction, we carry out all indicated operations in the numerator and the denominator separately. Then we simplify the fraction, if possible.

Example 8   If $a = 2$, $b = -3$, and $c = -5$, evaluate $\dfrac{ab + 3c}{b(c - a)}$.

Solution   Substitute 2 for a, -3 for b, -5 for c, do the work in the numerator and denominator separately, and then simplify.

$$\frac{ab + 3c}{b(c - a)} = \frac{2(-3) + 3(-5)}{-3(-5 - 2)}$$
$$= \frac{-6 + (-15)}{-3(-7)}$$
$$= \frac{-21}{21}$$
$$= -1$$ ∎

Properties of Fractions

In algebra we will often encounter fractions. We summarize the properties of arithmetic fractions here and will discuss algebraic fractions in detail in Chapter 5.

Properties of Fractions.   If no denominators are 0, then

1. $\dfrac{a}{b} = \dfrac{c}{d}$ if and only if $ad = bc$

2. $\dfrac{a}{1} = a$ and   if $a \neq 0$, $\dfrac{a}{a} = 1$

3. If $k \neq 0$, $\dfrac{a}{b} = \dfrac{ak}{bk}$

4. $\dfrac{a}{b} \cdot \dfrac{c}{d} = \dfrac{ac}{bd}$ and $\dfrac{a}{b} \div \dfrac{c}{d} = \dfrac{a}{b} \cdot \dfrac{d}{c} = \dfrac{ad}{bc}$

5. $\dfrac{a}{b} + \dfrac{c}{b} = \dfrac{a+c}{b}$ and $\dfrac{a}{b} - \dfrac{c}{b} = \dfrac{a-c}{b}$

6. $-\dfrac{a}{b} = \dfrac{-a}{b} = \dfrac{a}{-b}$

Property 1 points out that fractions are equal if the same product occurs when "cross multiplying." For example,

$$\frac{2}{3} = \frac{8}{12} \quad \text{because} \quad \begin{array}{c} 2(12) = 3(8) \\ 24 = 24 \end{array}$$

Property 2 points out that any number divided by 1 is left unchanged and that any nonzero number divided by itself is 1.

Property 3 is used to simplify fractions because it enables us to simplify fractions containing common factors in both the numerator and the denominator. For example, to simplify the fraction $\frac{27}{90}$ we proceed as follows:

$$\frac{27}{90} = \frac{9 \cdot 3}{9 \cdot 10} = \frac{\cancel{9} \cdot 3}{\cancel{9} \cdot 10} = \frac{3}{10}$$

Property 3 is also used to build fractions. For example, to write the fraction $\frac{3}{4}$ as a fraction with a denominator of 20 we use Property 3 and multiply both the numerator and the denominator of $\frac{3}{4}$ by 5:

$$\frac{3}{4} = \frac{3 \cdot 5}{4 \cdot 5} = \frac{15}{20}$$

Property 4 asserts that the product of two fractions is found by multiplying their numerators and multiplying their denominators, and that the quotient of two fractions is found by inverting the **divisor** (the fraction following the $\div$ symbol) and multiplying. For example,

$$\frac{1}{3} \cdot \frac{2}{5} = \frac{1 \cdot 2}{3 \cdot 5} = \frac{2}{15}$$

$$\frac{3}{7} \cdot \frac{21}{5} = \frac{3 \cdot 21}{7 \cdot 5} = \frac{3 \cdot 3 \cdot \cancel{7}}{\cancel{7} \cdot 5} = \frac{3 \cdot 3}{5} = \frac{9}{5}$$

and

$$\frac{3}{7} \div \left(-\frac{5}{4}\right) = \frac{3}{7} \cdot \left(-\frac{4}{5}\right) = -\frac{3 \cdot 4}{7 \cdot 5} = -\frac{12}{35}$$

Property 5 enables us to add or subtract fractions with common denominators. To add or subtract fractions with unlike denominators, we must use Property 3 to build the fractions into fractions with common denominators. For example,

$$\frac{6}{7} - \frac{2}{7} = \frac{6}{7} + \left(-\frac{2}{7}\right) = \frac{6 + (-2)}{7} = \frac{4}{7}$$

$$\frac{3}{5} + \frac{4}{7} = \frac{3 \cdot 7}{5 \cdot 7} + \frac{4 \cdot 5}{7 \cdot 5} = \frac{21}{35} + \frac{20}{35} = \frac{41}{35}$$

$$\frac{5}{6} - \frac{3}{4} = \frac{5 \cdot 2}{6 \cdot 2} - \frac{3 \cdot 3}{4 \cdot 3} = \frac{10}{12} - \frac{9}{12} = \frac{1}{12}$$

We need not write a fraction such as $\frac{41}{35}$ as the mixed number $1\frac{6}{35}$. In algebra, **improper fractions** (fractions with numerators greater than their denominators) are preferred.

Property 6 points out that a $-$ sign placed in front of a fraction can be written in either the numerator or the denominator of the fraction. For example,

$$-\frac{2}{3} = \frac{-2}{3} = \frac{2}{-3}$$

■ EXERCISE 1.4

In Exercises 1–50, perform the indicated operations.

1. $-3 + (-5)$	**2.** $2 + (+8)$	**3.** $-7 + (-2)$	**4.** $3 + (-5)$
5. $2 + (-8)$	**6.** $-5 + 3$	**7.** $-7 + 2$	**8.** $3 + (-7)$
9. $8 + (-2)$	**10.** $-17 + (-8)$	**11.** $0 - (-17)$	**12.** $0 - 21$
13. $0 - 93$	**14.** $0 - (-57)$	**15.** $3 - 5$	**16.** $3 - (-4)$
17. $-3 - 4$	**18.** $-11 - (-17)$	**19.** $-33 - (-33)$	**20.** $14 - (-13)$
21. $-2(6)$	**22.** $3(-5)$	**23.** $-3(-7)$	**24.** $-2(-5)$
25. $\dfrac{-8}{4}$	**26.** $\dfrac{25}{-5}$	**27.** $\dfrac{-16}{-4}$	**28.** $\dfrac{2-6}{-8-(-12)}$
29. $\dfrac{9-5}{5-9}$	**30.** $\dfrac{8+2}{3-8}$	**31.** $3 - 2 - 1$	**32.** $5 - 3 - 1$

33. $3 - (2 - 1)$ **34.** $5 - (3 - 1)$ **35.** $2 - 3 \cdot 5$ **36.** $6 + 4 \cdot 7$

37. $8 \div 4 \div 2$ **38.** $50 \div 10 \div 5$ **39.** $8 \div (4 \div 2)$ **40.** $50 \div (10 \div 5)$

41. $2 + 6 \div 3 - 5$ **42.** $6 - 8 \div 4 - 2$ **43.** $(2 + 6) \div (3 - 5)$ **44.** $(6 - 8) \div (4 - 2)$

45. $\dfrac{3(8 + 4)}{2 \cdot 3 - 9}$ **46.** $\dfrac{5(4 - 1)}{3 \cdot 2 + 5 \cdot 3}$ **47.** $\dfrac{100(2 - 4)}{1000 \div 10 \div 10}$ **48.** $\dfrac{8(3) - 4(6)}{5(3) + 3(-7)}$

49. $\dfrac{5 \div (2 - 3)}{5(2 - 3)}$ **50.** $\dfrac{4(3 - 7)}{4(3) - 7}$

In Exercises 51–58, a = 3, b = −2, c = −1, and d = 2. Evaluate each expression.

51. $ab + cd$ **52.** $ad + bc$ **53.** $a(b + c)$ **54.** $d(b + a)$

55. $\dfrac{ad + c}{cd + b}$ **56.** $\dfrac{ab + d}{bd + a}$ **57.** $\dfrac{ac - bd}{cd - ad}$ **58.** $\dfrac{bc - ad}{bd + ac}$

In Exercises 59–64, a = −1, b = 3, c = −3, and d = −2. Evaluate each expression.

59. $ad - d$ **60.** $db - d$ **61.** $\dfrac{ad + dc - bd}{(a - d)(b + d)}$ **62.** $\dfrac{ab - bc + cd}{(a - c)(b - d)}$

63. $\dfrac{(a + c)(b - d)}{(a + b \cdot d)(-c)}$ **64.** $\dfrac{(c \cdot d - b + a)(a - b \cdot c + d)}{(c - d \cdot c)(-b + c)(c - d)}$

In Exercises 65–74, use signed numbers to solve each problem.

65. One day Scott earned $22.17 mowing lawns and $39.56 painting a picnic bench. How much did he earn?

66. Wendy lost 3 pounds after an illness. She then dieted and lost 11 pounds. How much weight did Wendy lose?

67. The temperature rose 7 degrees in 1 hour. It then dropped 3 degrees in the next hour. What was the net change in the temperature?

68. An army retreated 2300 meters. After regrouping, they moved forward 1750 meters. The next day they gained another 1875 meters. What was the army's net gain (or loss)?

69. Sally had $437 in a bank account. One month, she had deposits of $25, $37, and $45. That same month, she had withdrawals of $17, $83, and $22. How much was in her account at the end of the month?

70. If the temperature is dropping 4 degrees each hour, how much warmer was it 3 hours ago?

71. In Las Vegas, Harry lost $30 per hour playing the slot machines. How much did he lose after gambling for 15 hours?

72. The flow of water from a pipe is filling a pool at the rate of 23 gallons per minute. How much less water was in the pool 5 hours ago?

73. If a drain is emptying a pool at the rate of 12 gallons per minute, how much more water was in the pool 2 hours ago?

74. John worked all day mowing lawns. He was paid $8 per hour. If he had $94 in his pocket at the end of an 8-hour day, how much money did he have before he started working?

In Exercises 75–76, tell whether the given fractions are equal.

75. $\dfrac{14}{15}, \dfrac{42}{45}$ **76.** $\dfrac{11}{13}, \dfrac{133}{169}$

In Exercises 77–80, simplify each fraction.

77. $\dfrac{12}{15}$ 78. $-\dfrac{18}{36}$ 79. $-\dfrac{48}{72}$ 80. $\dfrac{539}{637}$

In Exercises 81–92, perform the indicated operations and simplify the answer, if possible.

81. $\dfrac{1}{4} \cdot \dfrac{3}{5}$ 82. $-\dfrac{3}{5}\left(\dfrac{20}{27}\right)$ 83. $-\dfrac{2}{3} \div \left(-\dfrac{3}{7}\right)$ 84. $\dfrac{1}{4} \div \dfrac{3}{5}$

85. $-\dfrac{3}{5} \div \dfrac{9}{15}\left(-\dfrac{5}{27}\right)$ 86. $\dfrac{1}{3} \div \dfrac{3}{4} \cdot \dfrac{5}{3} \div \dfrac{1}{2}$ 87. $\dfrac{2}{3} + \dfrac{4}{3}$ 88. $\dfrac{9}{14} - \dfrac{3}{14}$

89. $\dfrac{2}{5} - \dfrac{7}{9}$ 90. $\dfrac{2}{3} + \dfrac{4}{5}$ 91. $-\dfrac{3}{5}\left(\dfrac{1}{7} + \dfrac{2}{3}\right)$ 92. $\dfrac{2}{3}\left(-\dfrac{3}{4} + \dfrac{1}{3}\right)$

93. Jim baked 4 dozen cookies, and took two-thirds of them to a bake sale. How many were left for his family to eat?

94. One dessert recipe calls for 8 eggs, while another recipe requires only three-fourths of that number. How many eggs will Jim need to make both desserts?

95. Sandy lost 15 pounds on a recent diet. Following the same diet, Jane lost one-third as much as Sandy, and Sue lost two-fifths as much as Sandy. How many pounds did Sandy, Jane, and Sue lose combined?

96. In January, Stu earned $3000 in commissions. In February, his commissions dropped to five-sixths of January's earnings, and during March, commissions were only two-fifths of January's. How much did Stu earn in commissions during the first quarter of that year?

CHAPTER SUMMARY

Key words

absolute value   (1.3)
closed interval   (1.3)
coordinate   (1.3)
difference   (1.4)
disjoint sets   (1.1)
element of a set   (1.1)
ellipsis   (1.1)
empty set   (1.1)
factor   (1.4)
finite set   (1.1)
graph   (1.3)
half-open interval   (1.3)
improper fraction   (1.4)
infinite set   (1.1)
intersection of two sets   (1.1)
interval   (1.3)
negative numbers   (1.3)

null set   (1.1)
open interval   (1.3)
origin   (1.3)
positive numbers   (1.3)
product   (1.4)
quotient   (1.4)
reciprocal   (1.2)
repeating decimal   (1.1)
set   (1.1)
set-builder notation   (1.1)
subset   (1.1)
sum   (1.4)
term   (1.4)
terminating decimal   (1.1)
union of two sets   (1.1)
variable   (1.1)

Key Ideas

(1.1)   Two sets are equal if they have the same elements.

The **natural numbers** are the numbers 1, 2, 3, 4, 5, . . .

A **prime number** is a natural number greater than 1 that is divisible without remainder only by itself and 1.

A **composite number** is a natural number greater than 1 that is not a prime number.

The **whole numbers** are the numbers 0, 1, 2, 3, 4, 5, . . .

The **integers** are the numbers . . . , -4, -3, -2, -1, 0, 1, 2, 3, 4, . . .

If an integer n is divisible by 2, then n is an **even integer**.

If an integer n is not divisible by 2, then n is an **odd integer**.

A **rational number** is any number that can be written as a fraction with an integer numerator and a nonzero integer denominator. Any rational number can be written in decimal form as either a terminating or a repeating decimal.

An **irrational number** is any number that can be written in decimal form as a nonterminating, nonrepeating decimal.

A **real number** is any number that can be written in decimal form.

(1.2)   The **reflexive property**: $a = a$.

The **symmetric property**:   If $a = b$, then $b = a$.

The **transitive property**:   If $a = b$ and $b = c$, then $a = c$.

The **substitution property**:   In any expression a quantity can be substituted for its equal without changing the meaning of the expression.

The **closure properties**: $a + b$ is a real number.

   $a - b$ is a real number.

   ab is a real number.

   $\dfrac{a}{b}$ is a real number, provided that $b \neq 0$.

The **associative properties**: $(a + b) + c = a + (b + c)$, and $(ab)c = a(bc)$.

The **commutative properties**: $a + b = b + a$, and $ab = ba$.

The **distributive property**: $a(b + c) = ab + ac$.

0 is the **additive identity**: $a + 0 = 0 + a = a$.

1 is the **multiplicative identity**: $a \cdot 1 = 1 \cdot a = a$.

$-a$ is the **negative** (or **additive inverse**) of a: $a + (-a) = 0$.

The **double negative rule**: $-(-a) = a$.

$\dfrac{1}{a}$ is the **multiplicative inverse** (or **reciprocal**) of a: $a\left(\dfrac{1}{a}\right) = 1$, provided $a \neq 0$.

(1.3) The **trichotomy property**: $a < b$ or $a = b$ or $a > b$.

The **transitive property** for $<$:   If $a < b$ and $b < c$, then $a < c$. A similar statement is true for $>$, $\leq$, and $\geq$.

$|x| = x$ if $x \geq 0$ and $|x| = -x$ if $x < 0$.

$|x| \geq 0$.

(1.4) To add two numbers with like signs, add their absolute values and use their common sign.

To add two numbers with unlike signs, find the difference of their absolute values and use the sign of the number with the greater absolute value.

The difference $a - b$ is equivalent to $a + (-b)$.

The product of two numbers with like signs is the positive product of their absolute values.

The product of two numbers with unlike signs is the negative of the product of their absolute values.

The quotient of two numbers with like signs is the positive quotient of their absolute values.

The quotient of two numbers with unlike signs is the negative of the quotient of their absolute values.

Division by 0 is undefined.

Unless parentheses indicate otherwise, do multiplications and/or divisions first, in order from left to right. Then do the additions and/or subtractions, from left to right. In a fraction, perform the operations within the numerator and denominator separately. Then simplify the fraction, if possible.

Properties of fractions: If no denominators are 0, then

$$\frac{a}{b} = \frac{c}{d} \quad \text{if and only if} \quad ad = bc.$$

$$\frac{a}{1} = a \qquad \text{and} \qquad \text{if } a \neq 0, \quad \frac{a}{a} = 1.$$

If $k \neq 0$, $\dfrac{a}{b} = \dfrac{ak}{bk}$.

$$\dfrac{a}{b} \cdot \dfrac{c}{d} = \dfrac{ac}{bd} \quad \text{and} \quad \dfrac{a}{b} \div \dfrac{c}{d} = \dfrac{a}{b} \cdot \dfrac{d}{c} = \dfrac{ad}{bc}.$$

$$\dfrac{a}{b} + \dfrac{c}{b} = \dfrac{a+c}{b} \quad \text{and} \quad \dfrac{a}{b} - \dfrac{c}{b} = \dfrac{a-c}{b}.$$

$$-\dfrac{a}{b} = \dfrac{-a}{b} = \dfrac{a}{-b}.$$

============ REVIEW EXERCISES ============

In Review Exercises 1–6, $A = \{1, 2, 4, 6, 8, 9\}$, $B = \{1, 2, 4, 9\}$, *and* $C = \{3, 5, 7, 9\}$. *Tell whether each statement is true. If a statement is false, change the symbol between the letters to make it true.*

1. $4 \in A$ **2.** $B \subseteq C$ **3.** $B \in A$ **4.** $\{3, 7\} \subseteq C$
5. $\varnothing \subseteq B$ **6.** $\varnothing \in C$

In Review Exercises 7–14, simplify each expression, if necessary, and classify the result as a natural number, whole number, integer, rational number, irrational number, and/or real number. Most numbers will be in several classifications.

7. -10 **8.** $\dfrac{4}{2}$ **9.** π **10.** $\sqrt{7}$

11. $-\dfrac{8}{9}$ **12.** 17 **13.** $5 - 5$ **14.** $6 - 8$

In Review Exercises 15–22, classify each integer as an even, odd, prime, or composite number.

15. 10 **16.** -8 **17.** -11 **18.** 7
19. 1 **20.** 0 **21.** -1 **22.** 2

In Review Exercises 23–34, state the property of equality or the property of real numbers that justifies each statement.

23. $3(4 + 2) = 3 \cdot 4 + 3 \cdot 2$ **24.** If $3 = 2 + 1$, then $2 + 1 = 3$.
25. $3 + (x + 7) = (x + 7) + 3$ **26.** $3 + (x + 7) = (3 + x) + 7$
27. $3 + 0 = 3$ **28.** $3 + (-3) = 0$
29. $xy = xy$ **30.** $5(3) = 3(5)$
31. $3(xy) = (3x)y$ **32.** $3x \cdot 1 = 3x$
33. $a\left(\dfrac{1}{a}\right) = 1 \quad (a \neq 0)$ **34.** If $x = 7$ and $7 = y$, then $x = y$.

In Review Exercises 35–38, find the additive inverse and the multiplicative inverse (the reciprocal), if any, of each number.

35. 1 **36.** -3 **37.** 0 **38.** $\dfrac{1}{3}$

In Review Exercises 39–40, write each expression using < and > symbols.

39. $a \not\geq 4$

40. $a \neq b$

In Review Exercises 41–42, write each expression so that the inequality symbol points in the opposite direction.

41. $3 \leq 10$

42. $-4 > -8$

43. Graph the set of prime numbers between 20 and 30.

44. Graph the set of composite numbers between 5 and 13.

In Review Exercises 45–50, graph each interval.

45. $x \geq -4$

46. $-2 < x \leq 6$

47. $(-2, 3)$

48. $[2, 6]$

49. $(2, \infty)$

50. $(-\infty, -1)$

In Review Exercises 51–54, write each expression without using absolute value symbols.

51. $|0|$

52. $|-1|$

53. $-|-8|$

54. $|3 - 8|$

In Review Exercises 55–70, perform the indicated operations and simplify, if possible.

55. $3 + (-5)$

56. $5 - 3$

57. $-2 + 5$

58. $-3 - 5$

59. $-8 - (-3)$

60. $7 - (-9)$

61. $4(-3)$

62. $-3(8)$

63. $-4(3 - 6)$

64. $3[8 - (-1)]$

65. $\dfrac{-8}{2}$

66. $\dfrac{8}{-4}$

67. $\dfrac{-16}{-4}$

68. $\dfrac{-25}{-5}$

69. $\dfrac{3 - 8}{10 - 5}$

70. $\dfrac{|-32 - 8|}{6 - 16}$

In Review Exercises 71–78, $a = 5$, $b = -2$, $c = -3$, and $d = 2$. Simplify each expression.

71. $\dfrac{3a - 2b}{cd}$

72. $\dfrac{3b + 2d}{ac}$

73. $\dfrac{ab + cd}{c(b - d)}$

74. $\dfrac{ac - bd}{a(d + c)}$

75. $\dfrac{a(b + d) + c}{ad - bc}$

76. $\dfrac{b(c - d) - a}{a + c(b - d)}$

77. $\dfrac{ac}{-4b} + \dfrac{bc}{-2d}$

78. $\dfrac{ab}{c} - \dfrac{-5b}{a - d}$

CHAPTER ONE TEST

In Problems 1–4, $A = \{1, 2, 3, 4\}$, $B = \{3, 4, 5, 6\}$, and $C = \{5, 6, 7, 8\}$. Insert either an $\in$ or a $\subseteq$ symbol to make a true statement.

1. $3 \underline{\in} A$

2. $\emptyset \underline{\subseteq} B$

3. $A \cap B \underline{\subseteq} B$

4. $\{7, 8\} \underline{\in} C$

In Problems 5–8, $A = \{1, 2, 3, 4\}$, $B = \{3, 4, 5, 6\}$, and $C = \{5, 6, 7, 8\}$. Find each set.

5. $A \cup B$

6. $A \cap B \cap C$

7. $(A \cup \emptyset) \cap C$

8. $A \cup B \cup C$

In Problems 9–12, let $A = \{-2, 0, 1, 2, \frac{6}{5}, 5, \sqrt{7}\}$.

9. List the elements in A that are natural numbers.

10. List the elements in A that are integers.

11. List the elements in A that are rational numbers.

12. List the elements in A that are irrational numbers.

13. Express the fraction $\dfrac{7}{9}$ as a decimal. , $7\overline{7}7$

In Problems 14–18, tell which property of equality or property of real numbers justifies each statement.

14. $3 = 3$

15. $3 + 5 = 5 + 3$

16. $a(b + c) = ab + ac$

17. $7 + (4 + 3) = (7 + 4) + 3$

18. $(4 \cdot 3) \cdot 5 = 5 \cdot (4 \cdot 3)$

In Problems 19–20, graph each set on the number line.

19. The set of odd integers between -4 and 6.

20. The set of prime numbers less than 12.

In Problems 21–22, graph each interval on the number line.

21. $-2 \leq x < 4$

22. $[-1, 3)$

In Problems 23–24, write each expression in simplest form without using absolute value symbols.

23. $-|8 - 5|$

24. $|-5| - |2| + |0|$

In Problems 25–30, $a = 2$, $b = -3$, and $c = 4$. Evaluate each expression.

25. ab

26. $a + bc$

27. $ab - bc$

28. $b(a + c)$

29. $\dfrac{-3b + a}{ac - b}$

30. $\dfrac{(4a + b)(b + 2c)}{-4c - ab}$

COMPUTER SYSTEMS ANALYST

Computer system analysts help businesses and scientific research organizations develop analytical systems to process and interpret data. Using such techniques as cost accounting, sampling, and mathematical model building, they analyze information and often present results graphically by charts and diagrams. They may also prepare cost–benefit analyses to help management decide whether proposed solutions are satisfactory.

Once the system is accepted, systems analysts adapt its logical requirements to the capabilities of computer machinery. They work closely with programmers to debug possible errors in the system.

Systems analysts solve a wide range of problems in many different industries. Because the work is so varied and complex, analysts usually specialize in either business or scientific and engineering applications.

Qualifications

Businesses prefer a college degree in accounting, business management, or economics. Scientific organizations prefer a college degree in the physical sciences, mathematics, or engineering. Advanced degrees and degrees in computer science and information science are becoming more important in both employment areas.

Job outlook

The demand for systems analysts is rising because of the expansion of computer usage and computer capability for problem solving in computer service firms, accounting firms, and organizations engaged in research and development. Systems analysts also will be needed by computer manufacturers to design software packages.

Example application

The process of arranging records into a sequential order, called **sorting**, is a common and important task in electronic data processing. In any sorting operation, records must be compared with other records to determine which one should precede the other. One sorting technique, called a **selection sort**, requires C comparisons to sort N records into their proper order, where C and N are related by the following formula.

$$C = \frac{N(N-1)}{2}$$

How many comparisons are required to **a.** sort 20 records? **b.** sort 10,000 records?

Solution

a. Substitute 20 for N in the formula

$$C = \frac{N(N-1)}{2}$$

and calculate C.

$$C = \frac{N(N-1)}{2}$$

$$C = \frac{20(20-1)}{2}$$

$$C = \frac{20 \cdot 19}{2}$$

$$C = 190$$

Sorting 20 records requires 190 comparisons.

b.   Substitute 10,000 for N in the formula

$$C = \frac{N(N-1)}{2}$$

and calculate C.

$$C = \frac{N(N-1)}{2}$$

$$C = \frac{10,000(10,000-1)}{2}$$

$$C = 49,995,000$$

Sorting 10,000 records requires almost 50 million comparisons. A selection sort is not efficient for large lists.

Exercises

1.   How many comparisons are required to sort 500 records?

2.   If the computer time required to sort 20 records were 0.5 second, how long would it take to sort 10,000 records?

3.   Another important data processing task is that of finding a particular entry in a large list. In a **sequential search**, an average of C comparisons is required to find an entry in a list of N items. C and N are related by the formula

$$C = \frac{(N+1)}{2}$$

How many comparisons are needed (on the average) to search a list of 25 items?

4.   How many comparisons are needed (on the average) to search 10,000 items?

(Answers: **1.**   124,750   **2.**   36.5 hrs   **3.**   13   **4.**   5000)

Exponents and Polynomials

One of the most common expressions in algebra is the polynomial. Because polynomials often involve exponents, we begin this chapter by discussing the properties of exponents.

2.1 EXPONENTS

Multiplication indicates repeated addition. For example, $4x$ means $x + x + x + x$ and $3y$ means $y + y + y$. The number 4, called the **numerical coefficient**, or simply **coefficient**, of the expression $4x$, indicates that x is to be used as a term in an addition four times. The coefficient 3 of the expression $3y$ indicates that y is to be used as a term three times.

Exponents are used to indicate repeated multiplication. For example,

$$y^2 = y \cdot y \qquad \text{Read } y^2 \text{ as "}y\text{ to the second power" or "}y\text{ squared."}$$

$$z^3 = z \cdot z \cdot z \qquad \text{Read } z^3 \text{ as "}z\text{ to the third power" or "}z\text{ cubed."}$$

$$x^4 = x \cdot x \cdot x \cdot x \qquad \text{Read } x^4 \text{ as "}x\text{ to the fourth power."}$$

These examples suggest the following definition:

Definition. If n is a natural number, then

$$x^n = \overbrace{x \cdot x \cdot x \cdots \cdots x}^{n \text{ factors of } x}$$

The exponential expression x^n is called a **power of x**. In this expression, x is called the **base** and n is the exponent. A natural number exponent tells how many times the base of an exponential expression is to be used as a factor in a product.

Example 1   **a.** $2^5 = 2 \cdot 2 \cdot 2 \cdot 2 \cdot 2$ **b.** $(-2)^5 = (-2)(-2)(-2)(-2)(-2)$
                      $= 32$ $= -32$

c. $\quad -4^4 = -(4^4)$

$\quad\quad\quad = -(4 \cdot 4 \cdot 4 \cdot 4)$

$\quad\quad\quad = -256$

d. $\quad (-4)^4 = (-4)(-4)(-4)(-4)$

$\quad\quad\quad\quad = 256$

e. $\quad 8a^2 = 8(a^2)$

$\quad\quad\quad = 8aa$

f. $\quad (8a)^2 = (8a)(8a)$

$\quad\quad\quad\quad = 64a^2$ ◼

It is important to note the difference between $-x^n$ and $(-x)^n$, and between ax^n and $(ax)^n$:

$$-x^n = -\overbrace{(x \cdot x \cdot x \cdots x)}^{n \text{ factors of } x}$$

$$(-x)^n = \overbrace{(-x)(-x)(-x) \cdots (-x)}^{n \text{ factors of } -x}$$

$$ax^n = a \cdot \overbrace{x \cdot x \cdot x \cdots x}^{n \text{ factors of } x}$$

$$(ax)^n = \overbrace{(ax)(ax)(ax) \cdots (ax)}^{n \text{ factors of } ax}$$

Several properties of exponents follow directly from the definition of *exponent*. Because x^5 means that x is to be used as a factor five times, and x^3 means that x is to be used as a factor three times, the product $x^5 \cdot x^3$ indicates that x is to be used as a factor eight times:

$$x^5 x^3 = \overbrace{x \cdot x \cdot x \cdot x \cdot x}^{5 \text{ factors of } x} \cdot \overbrace{x \cdot x \cdot x}^{3 \text{ factors of } x} = \overbrace{x \cdot x \cdot x \cdot x \cdot x \cdot x \cdot x \cdot x}^{8 \text{ factors of } x}$$

In general,

$$x^m x^n = \overbrace{x \cdot x \cdot x \cdots x}^{m \text{ factors of } x} \cdot \overbrace{x \cdot x \cdots x}^{n \text{ factors of } x} = \overbrace{x \cdot x \cdot x \cdot x \cdots x}^{m + n \text{ factors of } x}$$

Thus, to multiply exponential expressions with the same base, we keep the same base and add the exponents.

> **The Product Rule of Exponents.** If m and n are natural numbers, then
>
> $$x^m x^n = x^{m+n}$$

Note that the product rule of exponents applies only to exponential expressions with the same base. The expression $x^5 y^3$, for example, cannot be simplified because the bases of the exponential expressions are different.

Example 2 **a.** $x^{11}x^5 = x^{11+5}$ **b.** $a^5a^4a^3 = (a^5a^4)a^3$
$\qquad\qquad\quad = x^{16}$ $\qquad\qquad\quad = a^9a^3$
$\qquad\qquad\qquad\qquad\qquad\qquad\qquad = a^{12}$

c. $a^2b^3a^3b^2 = a^2a^3b^3b^2$ **d.** $8x^4x^4 = 8x^{4+4}$
$\qquad\qquad\quad = a^5b^5$ $\qquad\qquad = 8x^8$ ∎

To find another property of exponents, we simplify the expression $(x^4)^3$. This expression means "x^4 cubed" or $x^4 \cdot x^4 \cdot x^4$, which can be written as x^{12} because it is the product of 12 factors of x. Thus,

$$(x^4)^3 = x^{4 \cdot 3} = x^{12}$$

In general,

$$(x^m)^n = \overbrace{x^m \cdot x^m \cdot x^m \cdots \cdot x^m}^{n \text{ factors of } x^m} = \overbrace{x \cdot x \cdot x \cdot x \cdot x \cdots \cdot x}^{mn \text{ factors of } x} = x^{mn}$$

Thus, to raise an exponential expression to a power, we keep the same base and multiply the exponents.

To raise a product to a power, we raise each factor to that power.

$$(xy)^n = \overbrace{(xy)(xy)(xy) \cdots \cdot (xy)}^{n \text{ factors of } xy} = \overbrace{xxx \cdots \cdot x}^{n \text{ factors of } x} \cdot \overbrace{yyy \cdots \cdot y}^{n \text{ factors of } y} = x^ny^n$$

To raise a fraction to a power, we raise both the numerator and the denominator to that power:

$$\left(\frac{x}{y}\right)^n = \overbrace{\left(\frac{x}{y}\right)\left(\frac{x}{y}\right)\left(\frac{x}{y}\right) \cdots \cdot \left(\frac{x}{y}\right)}^{n \text{ factors of } \frac{x}{y}}$$

$$= \frac{\overbrace{xxx \cdots \cdot x}^{n \text{ factors of } x}}{\underbrace{yyy \cdots \cdot y}_{n \text{ factors of } y}}$$

Recall that we multiply fractions by multiplying the numerators and multiplying the denominators.

$$= \frac{x^n}{y^n}$$

The previous three results are often called the **power rules of exponents**.

The Power Rules of Exponents. If m and n are natural numbers, then

$$(x^m)^n = x^{mn} \qquad (xy)^n = x^ny^n \qquad \left(\frac{x}{y}\right)^n = \frac{x^n}{y^n} \quad (y \neq 0)$$

Example 3   **a.** $(3^2)^3 = 3^{2 \cdot 3}$ **b.** $(x^{11})^5 = x^{11 \cdot 5}$
$= 3^6$ $= x^{55}$
$= 729$

c. $(x^2 x^3)^6 = (x^5)^6$ **d.** $(x^2)^4(x^3)^2 = x^8 x^6$
$= x^{30}$ $= x^{14}$

Example 4   **a.** $(x^2 y)^3 = (x^2)^3 y^3$ **b.** $(x^3 y^4)^4 = (x^3)^4 (y^4)^4$

$= x^6 y^3$ $= x^{12} y^{16}$

c. $\left(\dfrac{x}{y^2}\right)^4 = \dfrac{x^4}{(y^2)^4}$ **d.** $\left(\dfrac{x^3}{y^4}\right)^2 = \dfrac{(x^3)^2}{(y^4)^2}$

$= \dfrac{x^4}{y^8}$ $= \dfrac{x^6}{y^8}$

Until now we have defined only natural-number exponents. We can, how-ever, extend the definition to include other exponents. For example, if we assume that the rules for natural-number exponents hold for exponents of 0, we can write

$$x^0 x^n = x^{0+n} = x^n = 1 x^n$$

Because $x^0 x^n = 1 x^n$, it follows that if $x \neq 0$, then $x^0 = 1$. Thus, we make the following definition:

Definition.   If $x \neq 0$, then

$$x^0 = 1$$

Furthermore, if we assume that the rules for natural-number exponents hold for exponents that are negative integers, then if $x \neq 0$, we can write

$$x^{-n} x^n = x^{-n+n} = x^0 = 1$$

However, because

$$\frac{1}{x^n} \cdot x^n = 1$$

we make the following definition:

Definition.   If n is an integer and $x \neq 0$, then

$$x^{-n} = \frac{1}{x^n} \qquad \text{and} \qquad \frac{1}{x^{-n}} = x^n$$

Because of the two previous definitions, all of the rules for natural-number exponents hold for integer exponents.

Example 5   **a.** $(3x)^0 = 1$ **b.** $3x^0 = 3(x^0)$
$$= 3(1)$$
$$= 3$$

c. $x^{-5} = \dfrac{1}{x^5}$ **d.** $\dfrac{1}{x^{-6}} = x^6$

e. $x^{-5}x^3 = x^{-5+3}$ **f.** $(x^{-3})^{-2} = x^{(-3)(-2)}$
$$= x^{-2} \qquad\qquad\qquad = x^6$$
$$= \dfrac{1}{x^2}$$

■

To develop the quotient rule for exponents, we proceed as follows:

$$\frac{x^m}{x^n} = x^m\left(\frac{1}{x^n}\right) = x^m x^{-n} = x^{m+(-n)} = x^{m-n}$$

Thus, to divide two exponential expressions with the same nonzero base, we keep the same base and subtract the exponents.

The Quotient Rule for Exponents.   If m and n are integers and $x \neq 0$, then

$$\frac{x^m}{x^n} = x^{m-n}$$

Example 6   **a.** $\dfrac{a^5}{a^3} = a^{5-3}$ **b.** $\dfrac{x^{-5}}{x^{11}} = x^{-5-11}$
$$= a^2 \qquad\qquad\qquad = x^{-16}$$
$$= \dfrac{1}{x^{16}}$$

c. $\dfrac{x^4 x^3}{x^{-5}} = \dfrac{x^7}{x^{-5}}$ **d.** $\dfrac{(x^2)^3}{(x^3)^2} = \dfrac{x^6}{x^6}$
$$= x^{7-(-5)} \qquad\qquad = x^{6-6}$$
$$= x^{12} \qquad\qquad\qquad = x^0$$
$$= 1$$

e. $\dfrac{x^2y^3}{xy^4} = x^{2-1}y^{3-4}$

$= xy^{-1}$

$= x\left(\dfrac{1}{y}\right)$

$= \dfrac{x}{y}$

f. $\left(\dfrac{a^{-2}b^3}{a^2a^3b^4}\right)^3 = \left(\dfrac{a^{-2}b^3}{a^5b^4}\right)^3$

$= (a^{-2-5}b^{3-4})^3$

$= (a^{-7}b^{-1})^3$

$= a^{-21}b^{-3}$

$= \dfrac{1}{a^{21}b^3}$ ■

Note that part a of Example 6 could be simplified by using Property 3 of fractions:

$$\frac{a^5}{a^3} = \frac{a \cdot a \cdot a \cdot a \cdot a}{a \cdot a \cdot a} = \frac{\not a \cdot \not a \cdot \not a \cdot a \cdot a}{\not a \cdot \not a \cdot \not a} = a^2$$

One final property often is used to simplify exponential expressions.

Theorem.   If n is an integer, $x \neq 0$, and $y \neq 0$, then

$$\left(\frac{x}{y}\right)^{-n} = \left(\frac{y}{x}\right)^{n}$$

This theorem states that a fraction, when raised to the power $-n$ is equal to the reciprocal of that fraction raised to the power n.

To prove this theorem, we proceed as follows:

$\left(\dfrac{x}{y}\right)^{-n} = \dfrac{x^{-n}}{y^{-n}}$

$= \dfrac{x^{-n}x^ny^n}{y^{-n}x^ny^n}$ Multiply both numerator and denominator by x^ny^n.

$= \dfrac{x^0y^n}{y^0x^n}$

$= \dfrac{y^n}{x^n}$

$= \left(\dfrac{y}{x}\right)^n$ □

Example 7   **a.** $\left(\dfrac{2}{3}\right)^{-4} = \left(\dfrac{3}{2}\right)^4$

$= \dfrac{81}{16}$

b. $\left(\dfrac{y^2}{x^3}\right)^{-3} = \left(\dfrac{x^3}{y^2}\right)^3$

$= \dfrac{x^9}{y^6}$

c. $\left(\dfrac{4x^{-2}}{3y^3}\right)^{-4} = \left(\dfrac{3y^3}{4x^{-2}}\right)^4$ **d.** $\left(\dfrac{a^{-2}b^3}{a^2a^3b^4}\right)^{-3} = \left(\dfrac{a^2a^3b^4}{a^{-2}b^3}\right)^3$

$\qquad\qquad = \dfrac{81y^{12}}{256x^{-8}} \qquad\qquad\qquad\qquad = \left(\dfrac{a^5b^4}{a^{-2}b^3}\right)^3$

$\qquad\qquad = \dfrac{81y^{12}x^8}{256x^{-8}x^8} \qquad\qquad\qquad\quad = (a^{5-(-2)}b^{4-3})^3$

$\qquad\qquad = \dfrac{81y^{12}x^8}{256x^0} \qquad\qquad\qquad\quad = (a^7b)^3$

$\qquad\qquad = \dfrac{81y^{12}x^8}{256} \qquad\qquad\qquad\qquad = a^{21}b^3$

We summarize the rules of exponents as follows:

Rules for Exponents.

$\qquad\qquad\qquad\qquad\qquad\qquad\qquad$ n factors of x

If n is a natural number, then $x^n = \overbrace{x \cdot x \cdot x \cdot \cdots \cdot x}$.

If $x \neq 0$, then $x^0 = 1$.

For all integers n, $x^{-n} = \dfrac{1}{x^n}$ and $\dfrac{1}{x^{-n}} = x^n$, provided $x \neq 0$.

If there are no divisions by zero, then for all integers m and n,

$\qquad x^m x^n = x^{m+n} \qquad\qquad (x^m)^n = x^{mn}$

$\qquad (xy)^n = x^n y^n \qquad\qquad \left(\dfrac{x}{y}\right)^n = \dfrac{x^n}{y^n}$

$\qquad \dfrac{x^m}{x^n} = x^{m-n} \qquad\qquad \left(\dfrac{x}{y}\right)^{-n} = \left(\dfrac{y}{x}\right)^n$

In Chapter 1 we agreed that multiplications and divisions take priority over additions and subtractions. For example, $3 \cdot 4 + 5$ means $(3 \cdot 4) + 5$, rather than $3(4 + 5)$. To avoid any confusion when simplifying expressions containing exponents, we will follow this convention: *Unless parentheses indicate otherwise, we find the power of each base first, and then follow the priority rules given in Section 1.4.* Thus, $5 \cdot 2^3$ means $5(2^3)$ or $5(8) = 40$. However, $(5 \cdot 2)^3$ means 10^3, or 1000, because parentheses indicate that the multiplication is to be done first.

Example 8 Evaluate **a.** $3x^2y^3$, **b.** $(3x)^2y^3$, **c.** $-z^4$, and **d.** $(-z)^4$ for $x = 2$, $y = -1$, and $z = 3$.

Solution **a.** $3x^2y^3 = 3(2)^2(-1)^3 = 3(4)(-1) = -12$

b. $(3x)^2y^3 = (3 \cdot 2)^2(-1)^3 = 6^2(-1) = -36$

c. $-z^4 = -(3)^4 = -81$

d. $(-z)^4 = (-3)^4 = 81$ ∎

Example 9   Simplify each expression. Assume that neither a nor x is zero.

a. $\dfrac{a^n a}{a^2} = \dfrac{a^{n+1}}{a^2} = a^{n+1-2} = a^{n-1}$

b. $\dfrac{x^3 x^2}{x^n} = \dfrac{x^5}{x^n} = x^{5-n}$

c. $\left(\dfrac{x^n}{x^2}\right)^2 = \dfrac{x^{2n}}{x^4} = x^{2n-4}$ ∎

■ EXERCISE 2.1

In Exercises 1–78, simplify each expression and write all answers without using negative exponents. Assume that no denominators are 0.

1. 3^2	**2.** 3^4	**3.** -3^2	**4.** -3^4
5. $(-3)^2$	**6.** $(-3)^3$	**7.** $(-2x)^5$	**8.** $(-3a)^3$
9. $-(2x)^7$	**10.** $(-2a)^4$	**11.** $(-2x)^6$	**12.** $(-3y)^5$
13. 5^{-2}	**14.** 5^{-4}	**15.** -5^{-2}	**16.** -5^{-4}
17. $(-5)^{-2}$	**18.** $(-5)^{-4}$	**19.** 8^0	**20.** 9^0
21. -8^0	**22.** -9^0	**23.** $(-8)^0$	**24.** $(-9)^0$
25. $x^2 x^3$	**26.** $y^3 y^4$	**27.** $k^0 k^7$	**28.** $x^8 x^{11}$
29. $x^2 x^3 x^5$	**30.** $y^3 y^7 y^2$	**31.** $p^9 p p^0$	**32.** $z^7 z^0 z$
33. $aba^3 b^4$	**34.** $x^2 y^3 x^3 y^2$	**35.** $(-x)^2 y^4 x^3$	**36.** $-x^2 y^7 z^3 y^3 x^{-2}$
37. $(x^4)^7$	**38.** $(y^7)^5$	**39.** $(b^{-8})^9$	**40.** $(z^{12})^2$
41. $(x^3 y^2)^4$	**42.** $(x^2 y^5)^2$	**43.** $(r^{-3} s)^3$	**44.** $(m^5 n^2)^{-3}$
45. $(a^2 a^3)^4$	**46.** $(bb^2 b^3)^4$	**47.** $(-d^2)^3 (d^{-3})^3$	**48.** $(c^3)^2 (c^4)^{-2}$
49. $(x^{-2} y x^3 y^4)^2$	**50.** $(-a^2 b^{-4} a^3 b^2)^3$	**51.** $\left(\dfrac{a^3}{b^2}\right)^5$	**52.** $\left(\dfrac{a^2}{b^3}\right)^4$
53. $\left(\dfrac{a^{-3}}{b^{-2}}\right)^{-2}$	**54.** $\left(\dfrac{k^{-3}}{k^{-4}}\right)^{-1}$	**55.** $\dfrac{a^4 a^4}{a^3}$	**56.** $\dfrac{c^3 c^4}{c^2}$
57. $\dfrac{c^{12} c^5}{(c^5)^2}$	**58.** $\dfrac{(a^3)^{11}}{a^2 a^3}$	**59.** $\dfrac{m^9 m^{-2}}{(m^2)^3}$	**60.** $\dfrac{a^{10} a^{-3}}{a^5 a^{-2}}$
61. $\dfrac{1}{a^{-4}}$	**62.** $\dfrac{3}{b^{-5}}$	**63.** $\dfrac{m^5 m^{-7}}{m^2 m^{-5}}$	**64.** $\dfrac{(a^{-2})^3}{a^3 a^{-4}}$
65. $\left(\dfrac{4a^{-2} b}{3ab^{-3}}\right)^3$	**66.** $\left(\dfrac{2ab^{-3}}{3a^{-2} b^2}\right)^2$	**67.** $\left(\dfrac{3a^{-2} b^2}{17a^2 b^3}\right)^0$	**68.** $\left(\dfrac{-3x^{-2} y^5}{7xy^4}\right)^0$

69. $\left(\dfrac{-2a^4b}{a^{-3}b^2}\right)^{-3}$ 　　**70.** $\left(\dfrac{-3x^{-5}y^2}{-9x^5y^{-2}}\right)^{-2}$ 　　**71.** $\left(\dfrac{2a^3b^2}{3a^{-3}b^2}\right)^{-3}$ 　　**72.** $\left(\dfrac{3x^5y^{-2}}{2x^5y^{-2}}\right)^{-4}$

73. $\dfrac{(3x^2)^{-2}}{x^3x^{-4}x^0}$ 　　**74.** $\dfrac{y^{-3}y^{-4}y^0}{(2y^{-2})^3}$ 　　**75.** $\dfrac{-3x^{-2}y^2}{(-2x^{-3})^0}$ 　　**76.** $\dfrac{-4x^{-2}x^2(y^0)^2}{(-4x^2y^{-4})^0}$

77. $\dfrac{(4m^{-2}n^{-3})^{-2}(m^{-4}n^{-3})^2}{-3m^{-4}n^{-3}}$ 　　**78.** $\dfrac{(3t^2s^{-2})^{-1}(2^{-3}ts^{-4})^{-2}}{(6t^2s^{-3})^{-2}}$

In Exercises 79–90, evaluate each expression if $x = -2$ and $y = 3$.

79. x^2y^3 　　**80.** x^3y^2 　　**81.** x^y 　　**82.** y^x

83. $\dfrac{x^{-3}}{y^3}$ 　　**84.** $\dfrac{x^2}{y^{-3}}$ 　　**85.** $(xy^2)^{-2}$ 　　**86.** $(x^3y)^{-3}$

87. $-y^3x^{-1}$ 　　**88.** $-y^3x^{-2}$ 　　**89.** $(-yx^{-1})^3$ 　　**90.** $(-y)^3x^{-2}$

In Exercises 91–98, simplify each expression.

91. $\dfrac{a^na^3}{a^4}$ 　　**92.** $\dfrac{b^9b^7}{b^n}$ 　　**93.** $\left(\dfrac{b^n}{b^3}\right)^3$ 　　**94.** $\left(\dfrac{a^2}{a^n}\right)^4$

95. $\dfrac{a^{-n}a^2}{a^3}$ 　　**96.** $\dfrac{a^na^{-2}}{a^4}$ 　　**97.** $\dfrac{a^{-n}a^{-2}}{a^{-4}}$ 　　**98.** $\dfrac{a^n}{a^{-3}a^5}$

99. Show that, if m is a natural number, $(xy)^m = x^my^m$.

100. Show that, if m is a natural number and if $y \neq 0$, then $\left(\dfrac{x}{y}\right)^m = \dfrac{x^m}{y^m}$.

101. Construct an example to show that $x^mx^n = x^{m+n}$ when m is a negative integer and n is a positive integer.

102. Construct an example to show that $x^mx^n = x^{m+n}$ when both m and n are negative integers.

103. Construct an example to show that $(x^m)^n = x^{mn}$ when m is a negative integer and n is a positive integer.

104. Construct an example to show that $(x^m)^n = x^{mn}$ when both m and n are negative integers.

105. Construct an example to show that $\dfrac{x^m}{x^n} = x^{m-n}$ when both m and n are negative integers and x is not zero.

106. Construct an example to show that $\left(\dfrac{y}{x}\right)^m = \dfrac{y^m}{x^m}$ when m is a negative integer and x is not zero.

107. Construct an example using numbers to show that $x^m + x^n$ is *not* equal to x^{m+n}.

108. Construct an example using numbers to show that $x^m + y^m$ is *not* equal to $(x + y)^m$.

In Exercises 109–116, use a calculator to verify that each statement is true.

109. $(3.68)^0 = 1$ 　　**110.** $(2.1)^4(2.1)^3 = (2.1)^7$

111. $(7.2)^2(2.7)^2 = [(7.2)(2.7)]^2$ 　　**112.** $(3.7)^2 + (4.8)^2 \neq (3.7 + 4.8)^2$

113. $(3.2)^2(3.2)^{-2} = 1$ 　　**114.** $[(5.9)^3]^2 = (5.9)^6$

115. $(7.23)^{-3} = \dfrac{1}{(7.23)^3}$ 　　**116.** $\left(\dfrac{5.4}{2.7}\right)^{-4} = \left(\dfrac{2.7}{5.4}\right)^4$

2.2 SCIENTIFIC NOTATION

Scientists often deal with very large and with very small numbers. For example, the speed of light is approximately 29,980,000,000 centimeters per second, and the mass of a hydrogen atom is 0.0000000000000000000001673 gram. The large number of zeros in these numbers makes them difficult to read and hard to remember. However, the use of exponents makes it possible for scientists to write such numbers more compactly in a form called **scientific notation**.

> **Definition.**   A number is written in **scientific notation** if it is written as the product of a number between 1 (including 1) and 10, and an appropriate power of 10.

Example 1   Change   **a.**   29,980,000,000 and   **b.**   0.0000000000000000000001673 to scientific notation.

Solution   **a.**   To write the number 29,980,000,000 in scientific notation, you must express that number as a product of a number between 1 and 10, and some power of 10. The number 2.998 lies between 1 and 10. To get the number 29,980,000,000 the decimal point in 2.998 must be moved 10 places to the right. This is accomplished by multiplying 2.998 by 10^{10}. Hence, the number 29,980,000,000 written in scientific notation is 2.998×10^{10}.

b.   To write the number 0.0000000000000000000001673 in scientific notation, you must express that number as a product of a number between 1 and 10, and some power of 10. To get the number 0.0000000000000000000001673, the decimal point in 1.673 must be moved 24 places to the left. This is accomplished by dividing 1.673 by 10^{24}. This is equivalent to multiplying 1.673 by $1/10^{24}$, or by 10^{-24}. Hence, the number 0.0000000000000000000001673 written in scientific notation is 1.673×10^{-24}.   ■

Example 2   Change **a.** 93,000,000 and **b.** 0.0000000667 to scientific notation.

Solution   **a.** $93,000,000 = 9.3 \times 10^7$ because 9.3 is a number between 1 and 10, and $9.3 \times 10^7 = 93,000,000$.

b. $0.0000000667 = 6.67 \times 10^{-8}$ because 6.67 is between 1 and 10, and

$$6.67 \times 10^{-8} = 6.67 \times \frac{1}{100,000,000} = 0.0000000667$$ ■

Example 3   Change   **a.** 3.7×10^5 and   **b.** 1.1×10^{-3} to standard notation.

Solution   **a.**   Because $10^5 = 100,000$, it follows that $3.7 \times 10^5 = 370,000$.

b.   Because $10^{-3} = \dfrac{1}{1000}$, it follows that $1.1 \times 10^{-3} = 0.0011$.   ■

Study each of the following numbers written in both scientific and standard notation. In each case, note that the exponent gives the number of places that the decimal point moves, and the sign of the exponent indicates the direction that it moves:

$$5.32 \times 10^4 = 5\,3\,2\,0\,0$$

4 places to the right

$$6.45 \times 10^7 = 6\,4\,5\,0\,0\,0\,0\,0$$

7 places to the right

$$5.37 \times 10^{-4} = 0.\,0\,0\,0\,5\,3\,7$$

4 places to the left

$$5.234 \times 10^{-2} = 0.\,0\,5\,2\,3\,4$$

2 places to the left

$$5.89 \times 10^0 = 5.\,8\,9$$

no movement of the decimal point

Example 4   Change   **a.** 47.2×10^{-3} and   **b.** 0.043×10^{-2} to scientific notation.

Solution   Neither number is in scientific notation because the first factors are not between 1 and 10. You can, however, change these numbers to scientific notation as follows:

a. $47.2 \times 10^{-3} = (4.72 \times 10^1) \times 10^{-3} = 4.72 \times (10^1 \times 10^{-3}) = 4.72 \times 10^{-2}$

b. $0.043 \times 10^{-2} = (4.3 \times 10^{-2}) \times 10^{-2} = 4.3 \times (10^{-2} \times 10^{-2}) = 4.3 \times 10^{-4}$

Another advantage of scientific notation becomes evident when we must multiply and divide combinations of very large and very small numbers. For example, the fraction

$$\frac{(0.0000064)(24,000,000,000)}{0.000000048}$$

can be evaluated by ordinary arithmetic. However, there is an easier way: change each number to scientific notation. Next do the arithmetic on the numbers and on the exponential expressions separately. Finally, write the answer in standard notation, if desired.

$$\frac{(0.0000064)(24,000,000,000)}{0.000000048} = \frac{(6.4 \times 10^{-6})(2.4 \times 10^{10})}{4.8 \times 10^{-8}}$$

$$= \frac{(6.4)(2.4)}{4.8} \cdot \frac{10^{-6}\,10^{10}}{10^{-8}}$$

$$= 3.2 \times 10^{12}$$

$$= 3,200,000,000,000$$

Example 5   Evaluate $\dfrac{(1,920,000)(0.0015)}{(0.000032)(45,000)}$.

Solution   First, express all numbers in scientific notation to get

$$\frac{(1.92 \times 10^6)(1.5 \times 10^{-3})}{(3.2 \times 10^{-5})(4.5 \times 10^4)}$$

Then do the arithmetic on the numbers and exponents separately.

$$\frac{(1.92)(1.5)}{(3.2)(4.5)} \cdot \frac{10^6}{10^{-5}} \frac{10^{-3}}{10^4} = 0.2 \cdot \frac{10^3}{10^{-1}}$$

$$= 0.2 \cdot 10^4$$

$$= 2.0 \cdot 10^{-1} \cdot 10^4 \qquad 0.2 = 2.0 \times 10^{-1}$$

$$= 2.0 \cdot 10^3$$

$$= 2000 \qquad \blacksquare$$

■ EXERCISE 2.2

In Exercises 1–20, change each expression to scientific notation.

1.	3,900	**2.**	1,700	**3.**	0.0078	**4.**	0.068
5.	17,600,000	**6.**	89,800,000	**7.**	0.0000096	**8.**	0.000000467
9.	323×10^5	**10.**	689×10^9	**11.**	$6,000 \times 10^{-7}$	**12.**	$76,543 \times 10^{-5}$
13.	0.0527×10^5	**14.**	0.0298×10^3	**15.**	0.0317×10^{-2}	**16.**	0.0012×10^{-3}
17.	731.0×10^4	**18.**	817.6×10^3	**19.**	$9,137 \times 10^{-2}$	**20.**	1000×10^{-3}

In Exercises 21–32, change each expression to standard notation.

21.	2.7×10^2	**22.**	7.2×10^3	**23.**	3.23×10^{-3}	**24.**	6.48×10^{-2}
25.	7.96×10^5	**26.**	9.67×10^6	**27.**	3.7×10^{-4}	**28.**	4.12×10^{-5}
29.	5.23×10^0	**30.**	8.67×10^0	**31.**	23.65×10^9	**32.**	75.62×10^{-9}

In Exercises 33–40, use the method of Example 5 to simplify each expression. Give all answers in standard notation.

33. $\dfrac{(0.006)(0.008)}{0.0012}$

34. $\dfrac{(600)(80,000)}{120,000}$

35. $\dfrac{(640,000)(2,700,000)}{120,000}$

36. $\dfrac{(0.0000013)(0.000090)}{0.00039}$

37. $\dfrac{(220,000)(0.000009)}{0.00033}$

38. $\dfrac{(0.00024)(96,000,000)}{640,000,000}$

39. $\dfrac{(320,000)^2(0.0009)}{(12,000)^2}$

40. $\dfrac{(0.000012)^2(49,000)^2}{0.021}$

41. The speed of sound (in air) is 3.31×10^4 centimeters per second. Use scientific notation to compute the speed of sound in centimeters per hour.

42. Calculate the volume of a tank that has dimensions of 3000 by 7000 by 4000 millimeters. Use scientific notation to perform the calculation, and express the answer in scientific notation.

43. The mass of a proton is 0.00000000000000000000000167248 gram. Express the mass of 1 million protons using scientific notation.

44. The speed of light (in a vacuum) is approximately 30,000,000,000 centimeters per second. Find the speed of light in miles per hour, and express your answer in scientific notation. There are approximately 160,000 centimeters in 1 mile.

45. The moon is approximately 235,000 miles from the earth. Express this distance in inches using scientific notation.

46. One *angstrom* is 0.0000001 millimeter. One inch is 25.4 millimeters. Use scientific notation to express the number of angstroms in an inch.

47. One *astronomical unit* (AU) is the distance from the earth to the sun—93,000,000 miles. Halley's comet ranges from 0.6 to 18 AU from the sun. Use scientific notation to express the comet's range in miles.

48. Light travels 300,000,000 meters per second. A *light-year* is the distance that light would travel in one year. Use scientific notation to express the number of meters in 1 light-year.

49. Light travels 186,000 miles per second. A *parsec* is 3.26 light-years. The star Alpha Centauri is 1.3 parsecs from the earth. Use scientific notation to express this distance in miles.

50. The moon is approximately 378,196 kilometers from the earth. Use scientific notation to express this distance in feet. (*Hint:* 1 mile $\approx$ 1.61 kilometers. Read "$\approx$" as "is approximately equal to.")

Most calculators have either 8- or 10-digit displays. When doing arithmetic involving very large or very small numbers, it is common to get answers requiring more than 8 or 10 digits. To solve this problem, many scientific calculators give answers to such problems in scientific notation. In Exercises 51–56, evaluate each expression using a calculator. Note that each answer is given in scientific notation.

51. $(23,437)^3$

52. $(0.00034)^4$

53. $(63,480)(893,322)$

54. $(0.0000413)(0.0000049)^2$

55. $\dfrac{(69.4)^8(73.1)^2}{(0.0043)^3}$

56. $\dfrac{(0.0031)^4(0.0012)^5}{(0.0456)^{-7}}$

2.3 ADDING AND SUBTRACTING POLYNOMIALS

A fundamental expression in algebra is the **polynomial**.

> **Definition.** A **polynomial in one variable**, say x, is the sum of one or more terms of the form ax^n, where a is a real number and n is a whole number.

The expressions

$$3x^2 + 2x + 3$$

$$\frac{3}{2}x^5 - \frac{7}{3}x^4 - \frac{8}{3}x^3$$

$$19x^{20} - 22.2x^{15} + \sqrt{3}x^{14} + 4.5x^{11} - 17x^2$$

are polynomials in x.

> **Definition.**   A **polynomial in several variables**, say x, y, and z, is the sum of one or more terms of the form $ax^m y^n z^p$, where a is a real number and m, n, and p are whole numbers.

The expressions

$$3xy$$
$$5x^2 y + 2yz^3 - 3xz$$
$$u^2 v^2 w^2 + x^3 y^3 + 1$$

are polynomials in several variables.

If a polynomial has a single term, it is called a **monomial**. If it has two terms, it is called a **binomial**. If it has three terms, it is called a **trinomial**. Here are some examples of each:

Monomials	*Binomials*	*Trinomials*
$2x^3$	$2x^4 + 5$	$2x^3 + 4x^2 + 3$
$x^2 y$	$-17t^{45} - 3xy$	$3xy^3 - x^2 y^3 + 7y$
$3x^3 y^5 z^2$	$32x^{13} y^5 + 47x^3 yz$	$-12x^5 y^2 + 13x^4 y^3 - 7x^3 y^3$

> **Definition.**   If $a \neq 0$, the **degree of the monomial** ax^n is n. The degree of a monomial containing several variables is the sum of the exponents of those variables.

By the previous definition,

$3x^4$ is a monomial of degree 4.

$4^7 x^2 y^3$ is a monomial of degree 5.   The sum of the exponents of x and y is 5.

$-18x^3 y^2 z^{12}$ is a monomial of degree 17.

3 is a monomial of degree 0. $3 = 3x^0$

The monomial 0 is of undefined degree because, by definition of the degree of a monomial, a cannot be zero.

> **Definition.**   The **degree of a polynomial** is the degree of the term with largest degree that is contained within the polynomial.

By the previous definition,

$3x^5 + 4x^2 + 7$ is a trinomial of degree 5.

$7x^2 y^8 - 3xy$ is a binomial of degree 10.

$3x + 2y - xy$ is a trinomial of degree 2.

$18x^2y^3 - 12x^7y^2 + 3x^9y^3 - 3$ is a polynomial of degree 12.

Polynomials in one variable can be denoted by symbols such as

$P(x)$ Read $P(x)$ as "P of x."

where the letter within the parentheses represents the variable of the polynomial. As we read each of the polynomials

$$P(x) = 3x^2 + 4x + 5$$
$$Q(y) = 5y^4 + 3y^3 + 4y^2 - 3y - 4$$
$$R(z) = 2z^3 - 3z^2 + 7z + 1$$

from left to right, the degrees of successive terms get smaller. When this is so, we say the polynomials are written in descending powers of their variables. In the polynomial

$$3 - 2x + 3x^2 + 4x^3 - 7x^4$$

the exponents of x increase. When this is so, we say the polynomial is written in ascending powers of x. In the polynomial

$$x^5 - 5x^4y + 10x^3y^2 - 10x^2y^3 + 5xy^4 - y^5$$

the exponents of x decrease and the exponents of y increase. We say the polynomial is written in descending powers of x and in ascending powers of y.

The symbol $P(x)$ gives a convenient way to indicate the value of a polynomial at different values of x. For example, $P(0)$ represents the value of the polynomial $P(x)$ when $x = 0$. Likewise $P(-5)$ represents the value of $P(x)$ when $x = -5$.

Example 1   Consider the trinomial $P(x) = 3x^2 - 2x + 7$. Find   **a.** $P(0)$,   **b.** $P(4)$, **c.** $P(-5)$, and   **d.** $P(-x)$.

Solution   **a.** $P(0) = 3(0)^2 - 2(0) + 7 = 7$

b. $P(4) = 3(4)^2 - 2(4) + 7 = 47$

c. $P(-5) = 3(-5)^2 - 2(-5) + 7 = 92$

d. $P(-x) = 3(-x)^2 - 2(-x) + 7 = 3x^2 + 2x + 7$ ∎

Example 2   Consider the polynomial in x and y: $4x^2y - 5xy^3$. Find the value of the polynomial when $x = 3$ and $y = -2$.

Solution   Substitute 3 for x and -2 for y in the polynomial and simplify:

$$\begin{aligned} 4x^2y - 5xy^3 &= 4(3)^2(-2) - 5(3)(-2)^3 \\ &= 4(9)(-2) - 5(3)(-8) \\ &= -72 + 120 \\ &= 48 \end{aligned}$$ ∎

Adding and Subtracting Polynomials

If two terms have the same variables with the same exponents, they are called **like** or **similar terms**. Like terms can differ only in their numerical coefficients.

$3x^2$, $5x^2$, and $7x^2$ are like terms.

$5x^3y^2$, $17x^3y^2$, and $103x^3y^2$ are like terms.

$4x^4y^2$, $12xy^5$, and $98x^7y^9$ are unlike terms.

Recall that multiplication distributes over addition and that this fact is expressed by the formula

$$a(b + c) = ab + ac$$

Because multiplication is commutative, we can write the distributive property in the form

$$ba + ca = (b + c)a$$

This form of the distributive property enables us to combine like terms. For example,

$$3x + 7x = (3 + 7)x = 10x$$
$$5x^2y^3 + 22x^2y^3 = (5 + 22)x^2y^3 = 27x^2y^3$$
$$9xy^4 + 11xy^4 + 20xy^4 = (9 + 11 + 20)xy^4 = 40xy^4$$

However, the terms in the binomials

$$3x^2 - 5y^2, \qquad -2a^2 + 3a^3, \qquad \text{and} \qquad 5y^2 + 17xy$$

cannot be combined because they are not like terms.

The previous results suggest that to combine like terms we simply add their numerical coefficients and keep the same variables with the same exponents.

Example 3   **a.** $12x^2z + 13x^2z = 25x^2z$

b. $-28x^5y^2 + 11x^5y^2 = -17x^5y^2$

c. $22x^2y^3 - 7x^2y^3 = 15x^2y^3$ ∎

To subtract one monomial from another, we add the negative of the monomial that is to be subtracted.

Example 4   **a.** $8x^2 - 3x^2 = 8x^2 + (-3x^2) = 8x^2 + (-3)x^2 = 5x^2$

b. $3x^2y - 9x^2y = 3x^2y + (-9x^2y) = 3x^2y + (-9)x^2y = -6x^2y$

c. $-5x^5y^3z^2 - 3x^5y^3z^2 = -5x^5y^3z^2 + (-3x^5y^3z^2)$
$$= -8x^5y^3z^2$$ ∎

Because of the distributive property, we can remove parentheses enclosing several terms when the sign preceding the parentheses is a + sign. We simply drop the parentheses:

$$+(a + b - c) = +1(a + b - c) = 1a + 1b - 1c = a + b - c$$

Polynomials are added by removing parentheses, if necessary, and combining any like terms that are contained within the polynomials.

Example 5
$$(3x^2 - 2x + 4) + (2x^2 + 4x - 3) = 3x^2 - 2x + 4 + 2x^2 + 4x - 3$$
$$= 3x^2 + 2x^2 - 2x + 4x + 4 - 3$$
$$= 5x^2 + 2x + 1 \qquad \blacksquare$$

For the sake of convenience, problems like the one in Example 5 are sometimes written with the terms aligned vertically to facilitate addition:

$$3x^2 - 2x + 4$$
$$\underline{2x^2 + 4x - 3}$$
$$5x^2 + 2x + 1$$

Because of the distributive property, we can also remove parentheses enclosing several terms when the sign preceding the parentheses is a $-$ sign. We simply drop the $-$ sign and the parentheses and *change the sign of each term within the parentheses.*

$$-(a + b - c) = -1(a + b - c)$$
$$= -1a + (-1)b - (-1)c$$
$$= -a - b + c$$

This suggests that the way to subtract polynomials is to remove parentheses and combine like terms.

Example 6   **a.** $(8x^3y + 2x^2y) - (2x^3y - 3x^2y) = 8x^3y + 2x^2y - 2x^3y + 3x^2y$
$$= 6x^3y + 5x^2y$$

b. $(3rt^2 + 4r^2t^2) - (8rt^2 - 4r^2t^2 + r^3t^2) = 3rt^2 + 4r^2t^2 - 8rt^2 + 4r^2t^2 - r^3t^2$
$$= -5rt^2 + 8r^2t^2 - r^3t^2 \qquad \blacksquare$$

To subtract polynomials in vertical form, we add the negative of the **subtrahend** (the bottom polynomial) to the **minuend** (the top polynomial).

$$-\;\begin{matrix} 8x^3y + 2x^2y \\ \underline{2x^3y - 3x^2y} \end{matrix} \qquad \Rightarrow \qquad +\;\begin{matrix} 8x^3y + 2x^2y \\ \underline{-2x^3y + 3x^2y} \\ 6x^3y + 5x^2y \end{matrix}$$

Example 7   Subtract $8rt^2 - 4r^2t^2 + r^3t^2$ from $3rt^2 + 4r^2t^2$.

Solution
$$-\;\begin{matrix} 3rt^2 + 4r^2t^2 \\ \underline{8rt^2 - 4r^2t^2 + r^3t^2} \end{matrix} \qquad \Rightarrow \qquad +\;\begin{matrix} 3rt^2 + 4r^2t^2 \\ \underline{-8rt^2 + 4r^2t^2 - r^3t^2} \\ -5rt^2 + 8r^2t^2 - r^3t^2 \end{matrix} \qquad \blacksquare$$

Because of the distributive property, we can remove parentheses enclosing several terms that are multiplied by a constant. We simply multiply each term

within the parentheses by the constant. For example,
$$3(2x^2 + 4x - 7) = 3(2x^2) + 3(4x) + 3(-7)$$
$$= 6x^2 + 12x - 21$$

Thus, to add multiples of one polynomial to another, or to subtract multiples of one polynomial from another, we proceed as in Example 8.

Example 8 Simplify $3(2x^2 + 4x - 7) - 2(3x^2 - 4x - 5)$.

Solution $3(2x^2 + 4x - 7) - 2(3x^2 - 4x - 5) = 6x^2 + 12x - 21 - 6x^2 + 8x + 10$
$$= 20x - 11$$ ■

■ EXERCISE 2.3 ■

In Exercises 1–8, classify each polynomial as a monomial, binomial, trinomial, or none of these.

1. $3x^2$ **2.** $2y^3 + 4y^2$ **3.** $3x^2y - 2x + 3y$ **4.** $a^2 + b^2$

5. $x^2 - y^2$ **6.** $\dfrac{17}{2}x^3 + 3x^2 - x - 4$ **7.** 5 **8.** $8x^3y^5$

In Exercises 9–20, give the degree of each polynomial.

9. $3x^2 + 2$ **10.** x^{17} **11.** $4x^8 + 3x^2y^4$ **12.** $19x^2y^4 - y^{10} + x^4$

13. $4x^2 - 5y^3z^3t^4$ **14.** $7x$ **15.** $121y$ **16.** $x^2y^3z^4 + z^{12}$

17. 77 **18.** 43 **19.** $x + y + xy$ **20.** $5xy - x$

In Exercises 21–26, rewrite each polynomial in descending powers of x.

21. $2x + 3x^3 - 4x^2 + 7$ **22.** $3x^3 + 5x - 3x^2 - 8$

23. $5xy^3 - 3x^2y^2 + 2x^3y - x^4 + y^4$ **24.** $x^2y + y^2x^{10} - x^{13}y + x^5 - y$

25. $3x^3z - 4x^6y + 3 - 4x$ **26.** $-6x^2yz + 7x^3y^2z - 8x^4yz^2 - 10$

In Exercises 27–30, consider the polynomial $P(x) = 2x^2 + x + 2$. Find the indicated value.

27. $P(0)$ **28.** $P(1)$ **29.** $P(4)$ **30.** $P(-3)$

In Exercises 31–44, consider the polynomial $P(x) = -3x^2 + 4x - 3$. Find the indicated value.

31. $P(1)$ **32.** $P(0)$ **33.** $P(-2)$ **34.** $P(3)$

35. $P(t)$ **36.** $P(z)$ **37.** $P(-x)$ **38.** $P(-r)$

39. $P(2x)$ **40.** $P(3x)$ **41.** $P(P(0))$ **42.** $P(P(2))$

43. $P(P(-1))$ **44.** $P(P(-2))$

In Exercises 45–58, find the value of the given expression if $x = 2$ and $y = -3$.

45. $2x^8 - 3x^2 - 4x + 2$ **46.** $3y^3 + 4y^2 - 2y - 4$

47. $x^2 + y^2$ **48.** $x^3 + y^3$ **49.** $x^3 - y^3$ **50.** $x^2 - y^2$

51. $3x^2y + xy^3$ **52.** $8xy - xy^2$ **53.** $\dfrac{9x^3}{y} - \dfrac{8y^3}{x}$ **54.** $\dfrac{27x^2}{y} + \dfrac{16y^4}{x}$

55. $\dfrac{3x^3 - 2y^2}{2x^2 + y^3}$ **56.** $\dfrac{2x^4 + 3y^3}{2x^3 - y}$ **57.** $\dfrac{3x^3 + 6y^2}{-2x - 3y^3}$ **58.** $\dfrac{4x^4 - 3y}{-2x^2 - y}$

In Exercises 59–64, use a calculator to find the value of each expression given that $x = 3.7$, $y = -2.5$, and $z = 8.9$.

59. x^2y

60. xyz^2

61. $\dfrac{x^2}{z^2}$

62. $\dfrac{z^3}{y^2}$

63. $\dfrac{x + y + z}{xyz}$

64. $\dfrac{x + yz}{xy + z}$

In Exercises 65–72, simplify each expression, if necessary. Tell whether the terms are like or unlike terms. If they are like terms, find their sum.

65. $3x, 7x$

66. $-8x, 3y$

67. $7x, 7y$

68. $3r^2t^3, -8r^2t^3$

69. $9, 3x^0$

70. $9u^2v, x^0u^2v$

71. $(3x)^2, 3x^2$

72. $(3x^2)^3, (2x^2)^3$

In Exercises 73–80, simplify each expression.

73. $8x + 4x$

74. $-2y + 16y$

75. $5x^3y^2z - 3x^3y^2z$

76. $8wxy - 12wxy$

77. $-2x^2y^3 + 3xy^4 - 5x^2y^3$

78. $3ab^4 - 4a^2b^2 - 2ab^4 + 2a^2b^2$

79. $(3x^2y)^2 + 2x^4y^2 - x^4y^2$

80. $(5x^2y^4)^3 - (5x^3y^6)^2$

In Exercises 81–100, perform the indicated operations.

81. $(3x^2 + 2x + 1) + (-2x^2 - 7x + 5)$

82. $(-2a^2 - 5a - 7) + (-3a^2 + 7a + 1)$

83. $(-a^2 + 2a + 3) - (4a^2 - 2a - 1)$

84. $(x^2 - 3x + 8) - (3x^2 + x + 3)$

85. $(7y^3 + 4y^2 + y + 3) + (-8y^3 - y + 3)$

86. $(6x^3 + 3x - 2) - (2x^3 + 3x^2 + 5)$

87. $(3x^2 + 4x - 3) + (2x^2 - 3x - 1) - (x^2 + x + 7)$

88. $(-2x^2 + 6x + 5) - (-4x^2 - 7x + 2) - (4x^2 + 10x + 5)$

89. $(3x^3 - 2x + 3) + (4x^3 + 3x^2 - 2) + (-4x^3 - 3x^2 + x + 12)$

90. $(x^4 - 3x^2 + 4) + (-2x^4 - x^3 + 3x^2) + (3x^2 + 2x + 1)$

91. $(3y^2 - 2y + 4) + [(2y^2 - 3y + 2) - (y^2 + 4y + 3)]$

92. $(-t^2 - t - 1) - [(t^2 + 3t - 1) - (-2t^2 + 4)]$

93. Add:
$$\begin{array}{r} 3x^3 - 2x^2 + 4x - 3 \\ -2x^3 + 3x^2 + 3x - 2 \\ 5x^3 - 7x^2 + 7x - 12 \\ \hline \end{array}$$

94. Add:
$$\begin{array}{r} 7a^3 + 3a + 7 \\ -2a^3 + 4a^2 - 13 \\ 3a^3 - 3a^2 + 4a + 5 \\ \hline \end{array}$$

95. Add:
$$\begin{array}{r} -2y^4 - 2y^3 + 4y^2 - 3y + 10 \\ -3y^4 + 7y^3 - y^2 + 14y - 3 \\ -3y^3 - 5y^2 - 5y + 7 \\ -4y^4 + y^3 - 13y^2 + 14y - 2 \\ \hline \end{array}$$

96. Add:
$$\begin{array}{r} 17t^4 + 3t^3 - 2t^2 - 3t + 4 \\ -12t^4 - 2t^3 + 3t^2 - 5t - 17 \\ -2t^4 - 7t^3 + 4t^2 + 12t - 5 \\ 5t^4 + t^3 + 5t^2 - 13t + 12 \\ \hline \end{array}$$

97. Subtract:
$$\begin{array}{r} 3x^2 - 4x + 17 \\ 2x^2 + 4x - 5 \\ \hline \end{array}$$

98. Subtract:
$$\begin{array}{r} -2y^2 - 4y + 3 \\ 3y^2 + 10y - 5 \\ \hline \end{array}$$

99. Subtract:
$$\begin{array}{r} -5y^3 + 4y^2 - 11y + 3 \\ -2y^3 - 14y^2 + 17y - 32 \\ \hline \end{array}$$

100. Subtract:
$$\begin{array}{r} 17x^4 - 3x^2 - 65x - 12 \\ 23x^4 + 14x^2 + 3x - 23 \\ \hline \end{array}$$

In Exercises 101–106, simplify each expression.

101. $2(x^3 + x^2) + 3(2x^3 - x^2)$

102. $3(y^2 + 2y) - 4(y^2 - 4)$

103. $-5(2x^3 + 7x^2 + 4x) - 2(3x^3 - 4x^2 - 4x)$

104. $-3(3a^2 + 4b^3 + 7) + 4(5a^2 - 2b^3 + 3)$

105. $4(3z^2 - 4z + 5) + 6(-2z^2 - 3z + 4) - 2(4z^2 + 3z - 5)$

106. $-3(4x^3 - 2x^2 + 4) - 4(3x^3 + 4x^2 + 3x) + 5(3x - 4)$

107. Find the difference when $3x^2 + 4x - 3$ is subtracted from the sum of $-2x^2 - x + 7$ and $5x^2 + 3x - 1$.

108. Find the difference when $8x^3 + 2x^2 - 1$ is subtracted from the sum of $x^2 + x + 2$ and $2x^3 - x + 9$.

109. Find the sum when $2x^2 - 4x + 3$ minus $8x^2 + 5x - 3$ is added to $-2x^2 + 7x - 4$.

110. Find the sum when $7x^3 - 4x$ minus $x^2 + 2$ is added to $5 + 3x$.

2.4   MULTIPLYING POLYNOMIALS

We first consider multiplying a monomial by a monomial. In the examples, the commutative and associative properties of multiplication enable us to rearrange the factors and the parentheses.

Example 1   **a.** $(3x^2)(6x^3) = 3 \cdot x^2 \cdot 6 \cdot x^3$
$$= (3 \cdot 6)(x^2 \cdot x^3)$$
$$= 18x^5$$

b. $(-8x)(2y)(xy) = -8 \cdot x \cdot 2 \cdot y \cdot x \cdot y$
$$= (-8 \cdot 2) \cdot x \cdot x \cdot y \cdot y$$
$$= -16x^2y^2$$

c. $(2a^3b)(-7b^2c)(-12ac^4) = 2 \cdot a^3 \cdot b \cdot (-7) \cdot b^2 \cdot c \cdot (-12) \cdot a \cdot c^4$
$$= 2(-7)(-12) \cdot a^3 \cdot a \cdot b \cdot b^2 \cdot c \cdot c^4$$
$$= 168a^4b^3c^5 \qquad \blacksquare$$

The results in Example 1 suggest that to multiply monomials we first multiply the numerical factors and then multiply the variable factors.

To find the product of a monomial and a polynomial, we use the distributive property or the extended distributive property.

$$a(b + c + d + \cdots) = ab + ac + ad + \cdots$$

Example 2   **a.** $3x^2(6xy + 3y^2) = 3x^2 \cdot 6xy + 3x^2 \cdot 3y^2$
$$= 18x^3y + 9x^2y^2$$

b. $5x^3y^2(xy^3 - 2x^2y) = 5x^3y^2 \cdot xy^3 - 5x^3y^2 \cdot 2x^2y$
$$= 5x^4y^5 - 10x^5y^3$$

c. $-2ab^2(3bz - 2az + 4z^3) = -2ab^2 \cdot 3bz - (-2ab^2) \cdot 2az + (-2ab^2) \cdot 4z^3$
$$= -6ab^3z + 4a^2b^2z - 8ab^2z^3 \qquad \blacksquare$$

The results in Example 2 suggest that, to multiply a polynomial by a monomial, we multiply each term of the polynomial by the monomial.

To multiply a polynomial by another polynomial, we use the distributive property repeatedly.

Example 3 **a.** $(3x + 2)(4x + 9) = (3x + 2) \cdot 4x + (3x + 2) \cdot 9$
$$= 12x^2 + 8x + 27x + 18$$
$$= 12x^2 + 35x + 18$$

b. $(2a - b)(3a^2 - 4ab + b^2) = (2a - b)3a^2 - (2a - b)4ab + (2a - b)b^2$
$$= 6a^3 - 3a^2b - 8a^2b + 4ab^2 + 2ab^2 - b^3$$
$$= 6a^3 - 11a^2b + 6ab^2 - b^3 \qquad \blacksquare$$

The results of Example 3 suggest that, to multiply one polynomial by another, we multiply each term of one polynomial by each term of the other polynomial. It is often convenient to organize the work vertically, as in the following example.

Example 4 **a.** Multiply:

$$3x + 2$$
$$4x + 9$$

$4x(3x + 2) \longrightarrow 12x^2 + 8x$
$9(3x + 2) \longrightarrow + 27x + 18$
$$12x^2 + 35x + 18$$

b. Multiply:

$$3a^2 - 4ab + b^2$$
$$2a - b$$

$2a(3a^2 - 4ab + b^2) \longrightarrow 6a^3 - 8a^2b + 2ab^2$
$-b(3a^2 - 4ab + b^2) \longrightarrow - 3a^2b + 4ab^2 - b^3$
$$6a^3 - 11a^2b + 6ab^2 - b^3$$

c. Multiply:

$$2x^2 + 3xy + y^3$$
$$x - 2y$$

$x(2x^2 + 3xy + y^3) \longrightarrow 2x^3 + 3x^2y + xy^3$
$-2y(2x^2 + 3xy + y^3) \longrightarrow - 4x^2y - 6xy^2 - 2y^4$
$$2x^3 - x^2y + xy^3 - 6xy^2 - 2y^4 \qquad \blacksquare$$

Multiplying one binomial by another requires that each term of one binomial be multiplied by each term of the other binomial. This fact can be emphasized by drawing arrows to show the indicated products. For example, to multiply the binomials $3x + 2$ and $x + 4$, we can write

First terms   Last terms

$(3x + 2)(x + 4) = 3x \cdot x + 3x \cdot 4 + 2 \cdot x + 2 \cdot 4$
Inner terms $\qquad = 3x^2 + 12x + 2x + 8$
$\qquad\qquad\quad = 3x^2 + 14x + 8$
Outer terms

Note that the product of the first terms is $3x^2$, the product of the outer terms is $12x$, the product of the inner terms is $2x$, and the product of the last terms is 8.

This scheme is often called the **FOIL method of multiplying two binomials**. FOIL is an acronym for **F**irst terms, **O**uter terms, **I**nner terms, and **L**ast terms. Of course, the resulting terms of the product must be combined, if possible.

Example 5   Find the products:

    a. $(x + 3)(x + 3)$,   **b.** $(y - 4)(y - 4)$, and   **c.** $(a + 6)(a - 6)$.

Solution   Multiply each term of one binomial by each term of the other binomial and simplify.

        F   L

a. $(x + 3)(x + 3) = x^2 + 3x + 3x + 3^2$

          I

        O $= x^2 + 6x + 9$

        F   L

b. $(y - 4)(y - 4) = y^2 - 4y - 4y + 4^2$

          I

        O $= y^2 - 8y + 16$

        F   L

c. $(a + 6)(a - 6) = a^2 - 6a + 6a - 6^2$

          I

        O $= a^2 - 36$

The products discussed in Example 5 are called **special products**. Because they occur so often, it is worthwhile to learn their forms. In the exercises, you will be asked to verify the following product formulas:

Special Product Formulas.

$$(x + y)^2 = (x + y)(x + y) = x^2 + 2xy + y^2$$
$$(x - y)^2 = (x - y)(x - y) = x^2 - 2xy + y^2$$
$$(x + y)(x - y) = x^2 - y^2$$

Because $x^2 + 2xy + y^2 = (x + y)^2$ and $x^2 - 2xy + y^2 = (x - y)^2$, the two trinomials are called **perfect square trinomials**.

The expressions $(x + y)^2$ and $(x - y)^2$ have trinomials for their products. It is common for students to forget to write the middle terms in these products. Remember that $(x + y)^2 \neq x^2 + y^2$ and that $(x - y)^2 \neq x^2 - y^2$. The product $(x + y)(x - y)$, however, is the binomial $x^2 - y^2$.

It is easy to multiply binomials by sight if we use the FOIL method. We first find the product of the first terms, then find the products of the outer terms and the inner terms and add them, if possible, and finally find the product of the last terms.

Example 6   Find the products: **a.** $(2x - 3)(3x + 2)$ and   **b.** $(3x + 1)(3x + 4)$ by sight.

Solution   **a.**
$$6x^2 \qquad -6$$
$$(2x - 3)(3x + 2) = 6x^2 - 5x - 6$$
$$-9x$$
$$4x$$

Note that the middle term in the trinomial $6x^2 - 5x - 6$ comes from combining the products $-9x$ and $+4x$:

$$-9x + 4x = -5x$$

b.
$$9x^2 \qquad +4$$
$$(3x + 1)(3x + 4) = 9x^2 + 15x + 4$$
$$3x$$
$$12x$$

Note that the middle term in the trinomial $9x^2 + 15x + 4$ comes from combining the products $+3x$ and $+12x$:

$$+3x + 12x = 15x \qquad \blacksquare$$

At first glance, the expression $3[x^2 - 2(x + 3)]$ does not look like a polynomial. But if we simplify the expression by removing the parentheses and the brackets, the expression takes the form of a polynomial:

$$3[x^2 - 2(x + 3)] = 3[x^2 - 2x - 6]$$
$$= 3x^2 - 6x - 18$$

If an expression has one set of grouping symbols enclosed within another set, it is usually wise to eliminate the inner set first.

Example 7   Find the product of $-2[y^3 + 3(y^2 - 2)]$ and $5[y^2 - 2(y + 1)]$.

Solution   First change each expression to polynomial form:

$$
\begin{array}{c|c}
-2[y^3 + 3(y^2 - 2)] & 5[y^2 - 2(y + 1)] \\
-2[y^3 + 3y^2 - 6] & 5[y^2 - 2y - 2] \\
-2y^3 - 6y^2 + 12 & 5y^2 - 10y - 10
\end{array}
$$

Then, do the multiplication:

$$
\begin{array}{r}
-2y^3 - 6y^2 + 12 \\
5y^2 - 10y - 10 \\
\hline
-10y^5 - 30y^4 \qquad\qquad + 60y^2 \\
+ 20y^4 + 60y^3 \qquad\qquad - 120y \\
+ 20y^3 + 60y^2 \qquad - 120 \\
\hline
-10y^5 - 10y^4 + 80y^3 + 120y^2 - 120y - 120
\end{array}
$$

■

The following two examples show how to use the methods previously discussed to multiply expressions that are not polynomials.

Example 8   Find the product of $x^{-2} + y$ and $x^2 - y^{-2}$.

Solution   Multiply each term of the second expression by each term of the first expression, and simplify:

$$
\begin{aligned}
(x^{-2} + y)(x^2 - y^{-2}) &= x^{-2}x^2 - x^{-2}y^{-2} + yx^2 - yy^{-2} \\
&= x^0 - x^{-2}y^{-2} + yx^2 - y^{1+(-2)} \\
&= 1 - x^{-2}y^{-2} + yx^2 - y^{-1}
\end{aligned}
$$

■

Example 9   Find the product of $x^n + 2x$ and $x^n + 3x^{-n}$.

Solution   Multiply each term of the second expression by each term of the first expression, and simplify:

$$
\begin{aligned}
(x^n + 2x)(x^n + 3x^{-n}) &= x^n x^n + x^n(3x^{-n}) + 2x(x^n) + 2x(3x^{-n}) \\
&= x^{2n} + 3x^n x^{-n} + 2x^{1+n} + 6xx^{-n} \\
&= x^{2n} + 3x^0 + 2x^{n+1} + 6x^{1+(-n)} \\
&= x^{2n} + 3 + 2x^{n+1} + 6x^{1-n}
\end{aligned}
$$

■

■ **EXERCISE 2.4**

In Exercises 1–30, find each product.

1. $(2a^2)(-3ab)$

2. $(-3x^2y)(3xy)$

3. $(-3ab^2c)(5ac^2)$

4. $(-2m^2n)(-4mn^3)$

5. $(4a^2b)(-5a^3b^2)(6a^4)$

6. $(2x^2y^3)(4xy^5)(-5y^6)$

7. $(3x^3y^5)(2xy^2)^2$

8. $(a^3b^2c)^3(ab^2c^3)$

9. $(5x^3y^2)^4\left(\dfrac{1}{5}x^{-2}\right)^2$

10. $(4a^{-2}b^{-1})^2(2a^3b^4)^4$

11. $(-5xx^2)(-3xy)^4$

12. $(-2a^2ab^2)^3(-3ab^2b^2)$

13. $[(-2x^3y)(5x^2y^2)]^2$

14. $[(3x^2y^3)(4xy^5)]^3$

15. $3(x + 2)$

16. $-5(a + b)$

17. $-a(a - b)$

18. $y^2(y - 1)$

19. $3x(x^2 + 3x)$

20. $-2x(3x^2 - 2)$

21. $-2x(3x^2 - 3x + 2)$

22. $3a(4a^2 + 3a - 4)$

23. $5a^2b^3(2a^4b - 5a^0b^3)$

24. $-2a^3b(3a^0b^4 - 2a^2b^3)$

25. $7rst(r^2 + s^2 - t^2)$

26. $3x^2yz(x^2 - 2y + 3z^2)$

27. $-4x^2y^3(3x^2 - 4xy + y^2)$

28. $-2x^2(3x^4 - 2x^2 - 7)$

29. $4m^2n(-3mn)(m + n)$

30. $-3a^2b^3c(2bc^4)(3a + b - c)$

In Exercises 31–64, find each product. If possible, find the product by sight.

31. $(x + 2)(x + 2)$

32. $(x - 3)(x - 3)$

33. $(a - 4)(a - 4)$

34. $(y + 5)(y + 5)$

35. $(a + b)(a + b)$

36. $(a - 2b)(a - 2b)$

37. $(2x - y)(2x - y)$

38. $(3m + 4n)(3m + 4n)$

39. $(x + 2)(x - 2)$

40. $(z + 3)(z - 3)$

41. $(a + b)(a - b)$

42. $(2x + 3y)(2x - 3y)$

43. $(x + 2)(x + 3)$

44. $(y - 3)(y + 4)$

45. $(z - 7)(z - 2)$

46. $(x + 3)(x - 5)$

47. $(2a + 1)(a - 2)$

48. $(3b - 1)(2b - 1)$

49. $(3y - z)(2y - z)$

50. $(2m + n)(3m + n)$

51. $(2x - 3y)(x + 2y)$

52. $(3y + 2z)(y - 3z)$

53. $(3 - 2x)(3 + 4x)$

54. $(2x - 5)(2x + 5)$

55. $(3x + y)(3x - 3y)$

56. $(2x - 1)(3x + 2)$

57. $(4a - 3)(2a + 5)$

58. $(3a + 2)(2a - 7)$

59. $(u - v)^2$

60. $(u + v)^2$

61. $(2x + 1)^2$

62. $(3y - 2)^2$

63. $(3x + 2y)^2$

64. $(3x - 2y)^2$

In Exercises 65–80, find each product.

65. $(3y + 1)(2y^2 + 3y + 2)$

66. $(a + 2)(3a^2 + 4a - 2)$

67. $(2a - b)(3a^2 - 2ab + 2b^2)$

68. $(4x - 3y)(x^2 - 2xy + y^2)$

69. $(a + b + c)(2a - b - 2c)$

70. $(x - 2y - 3z)(3x + 2y + z)$

71. $(x + 2y + 3z)^2$

72. $(3x - 2y - z)^2$

73. $(r + s)^2(r - s)^2$

74. $r(r + s)(r - s)^2$

75. $(2x - 1)[2x^2 - 3(x + 2)]$

76. $(x + 1)^2[x^2 - 2(x + 2)]$

77. $[2x - 3(x^2 - x)]^3$

78. $-[y - 2(y + y^2)]^3$

79. $(a + b)(a - b)(a - 3b)$

80. $(x - y)(x + 2y)(x - 2y)$

In Exercises 81–90, find each product. Write all answers without negative exponents.

81. $x^3(2x^2 + x^{-2})$

82. $x^{-4}(2x^{-3} - 5x^2)$

83. $x^3y^{-6}z^{-2}(3x^{-2}y^2z - x^3y^{-4})$

84. $ab^{-2}c^{-3}(a^{-4}bc^3 + a^{-3}b^4c^3)$

85. $(x^{-1} + y)(x^{-1} - y)$

86. $(x^{-1} - y)(x^{-1} - y)$

87. $(2x^{-3} + y^3)(2x^3 - y^{-3})$

88. $(5x^{-4} - 4y^2)(5x^2 - 4y^{-4})$

89. $(2x^2 - 3y^{-2})(2x^2 + 3y^2)^2$

90. $(-3x^{-4} + 2y^{-1})(x + y^{-2})^2$

In Exercises 91–102, find each indicated product. Consider n to be a whole number.

91. $x^n(x^{2n} - x^n)$

92. $a^{2n}(a^n + a^{2n})$

93. $(x^n + 1)(x^n - 1)$

94. $(x^n - a^n)(x^n + a^n)$

95. $(x^n - y^n)(x^n - y^{-n})$

96. $(x^n + y^n)(x^n + y^{-n})$

97. $(x^{2n} + y^{2n})(x^{2n} - y^{2n})$

98. $(a^{3n} - b^{3n})(a^{3n} + b^{3n})$

99. $(2x^n - y^{2n})(3x^{-n} + y^{-2n})$

100. $(3x^{2n} + 2x^n - 1)^2$

101. $(x^n + y^n - 1)(x^n - y^n + 1)$

102. $(1 - x^n)(x^{-n} - 1)$

In Exercises 103–108, simplify each given expression.

103. $(3x - 4)^2 - (2x + 3)^2$

104. $(3y + 1)^2 + (2y - 4)^2$

105. $3(x - 3y)^2 + 2(3x + y)^2$

106. $2(x - y^2)^2 - 3(y^2 + 2x)^2$

107. $5(2y - z)^2 + 4(y + 2z)^2$

108. $3(x + 2z)^2 - 2(2x - z)^2$

109. Verify that $(x + y)^2 = x^2 + 2xy + y^2$.

110. Verify that $(x - y)^2 = x^2 - 2xy + y^2$.

111. Verify that $(x + y)(x - y) = x^2 - y^2$.

112. Verify that $(x + y + z)^2 = x^2 + y^2 + z^2 + 2xy + 2xz + 2yz$.

In Exercises 113–116, use a calculator to find each product.

113. $(3.21x - 7.85)(2.87x + 4.59)$

114. $(7.44y + 56.7)(-2.1y - 67.3)$

115. $(-17.3y + 4.35)^2$

116. $(-0.31x + 29.3)(-81x - 0.2)$

2.5 DIVIDING POLYNOMIALS

We begin by considering the quotient of two monomials.

Example 1 Simplify the expression $3a^2b^3 \div 2a^3b$.

Solution 1 Rewrite the expression as a fraction, rewrite both the numerator and the denominator, and divide out all common factors:

$$\frac{3a^2b^3}{2a^3b} = \frac{3aabbb}{2aaab}$$

$$= \frac{3\cancel{a}\cancel{a}b\cancel{b}b}{2\cancel{a}\cancel{a}a\cancel{b}}$$

$$= \frac{3b^2}{2a}$$

Solution 2   Rewrite the expression as a fraction and use the rules of exponents:

$$\frac{3a^2b^3}{2a^3b} = \frac{3}{2}a^{-1}b^2$$

$$= \frac{3}{2}\left(\frac{1}{a}\right)\frac{b^2}{1}$$

$$= \frac{3b^2}{2a}$$

∎

The same ideas are used to divide any polynomial by a monomial.

Example 2   Divide $4x^3y^2 + 3xy^5 - 12xy$ by $3x^2y^3$.

Solution   Rewrite the expression as a fraction, and then as the sum of three separate fractions:

$$\frac{4x^3y^2 + 3xy^5 - 12xy}{3x^2y^3} = \frac{4x^3y^2}{3x^2y^3} + \frac{3xy^5}{3x^2y^3} + \frac{-12xy}{3x^2y^3}$$

Pick one of the methods used in the previous example and simplify each of the three fractions on the right-hand side of the equals sign:

$$\frac{4x^3y^2 + 3xy^5 - 12xy}{3x^2y^3} = \frac{4x}{3y} + \frac{y^2}{x} + \frac{-4}{xy^2}$$

$$= \frac{4x}{3y} + \frac{y^2}{x} - \frac{4}{xy^2}$$

∎

In the next example, we consider dividing a polynomial by another polynomial.

Example 3   Divide $x^2 + 7x + 12$ by $x + 4$.

Solution   In Example 2, you divided a polynomial by a monomial. In this example, you must divide a polynomial by a binomial. There is an **algorithm** (a repeating series of steps) to use when the divisor is not a monomial. The algorithm follows closely the method of long division used when dividing numbers. The division of $x^2 + 7x + 12$ by $x + 4$ can be written in the form

$$x + 4 \overline{)\, x^2 + 7x + 12}$$

Here is how the division process works:

Step 1

$$x + 4 \overline{)\, x^2 + 7x + 12}$$ with x above

How many times does x divide x^2? $x^2/x = x$. Place the x above the division symbol.

Step 2

$$x + 4 \overline{\smash{)}\ x^2 + 7x + 12}$$
$$\underline{x^2 + 4x}$$

Multiply each term in the divisor by x. Place the product under $x^2 + 7x$ as indicated, and draw a line.

Step 3

$$x + 4 \overline{\smash{)}\ x^2 + 7x + 12}$$
$$(-)$$
$$\underline{(-)\ x^2 + 4x}$$
$$3x + 12$$

Subtract $x^2 + 4x$ from $x^2 + 7x$ by adding the negative of $x^2 + 4x$ to $x^2 + 7x$. Bring down the next term.

Step 4

$$x\ +3$$
$$x + 4 \overline{\smash{)}\ x^2 + 7x + 12}$$
$$x^2 + 4x$$
$$3x + 12$$

How many times does x divide $3x$? $3x/x = +3$. Place the $+3$ above the division symbol.

Step 5

$$x\ +3$$
$$x + 4 \overline{\smash{)}\ x^2 + 7x + 12}$$
$$\underline{x^2 + 4x}$$
$$3x + 12$$
$$3x + 12$$

Multiply each term in the divisor by 3. Place the product under $3x + 12$ as indicated, and draw a line.

Step 6

$$x\ +3$$
$$x + 4 \overline{\smash{)}\ x^2 + 7x + 12}$$
$$\underline{x^2 + 4x}$$
$$3x + 12$$
$$(-)$$
$$\underline{(-)\ 3x + 12}$$
$$0$$

Subtract $3x + 12$ from $3x + 12$ by adding the negative of $3x + 12$.

The division process terminates when the result of the subtraction is either a constant or a polynomial with degree less than the degree of the divisor. Thus, the quotient is $x + 3$ and the remainder is 0.

It is always a good idea to check the quotient in a division problem. To do this, multiply the divisor (the number that you divided by) by the quotient (the answer). If the result is the dividend (the part under the division symbol), the answer is correct. Because

$$\underbrace{(x + 4)}_{\text{divisor}} \cdot \underbrace{(x + 3)}_{\text{quotient}} = \underbrace{x^2 + 7x + 12}_{\text{dividend}}$$

the answer checks. The quotient is $x + 3$.   ∎

Example 4   Divide $2a^3 + 9a^2 + 5a - 6$ by $2a + 3$.

Solution   *Step 1*

$$
\begin{array}{r}
a^2 \\
2a + 3 \overline{\smash{)}\, 2a^3 + 9a^2 + 5a - 6}
\end{array}
$$

How many times does $2a$ divide $2a^3$? $2a^3/2a = a^2$. Place the a^2 in the quotient.

Step 2

$$
\begin{array}{r}
a^2 \\
2a + 3 \overline{\smash{)}\, 2a^3 + 9a^2 + 5a - 6} \\
\underline{2a^3 + 3a^2}
\end{array}
$$

Multiply each term in the divisor by a^2. Place the product under $2a^3 + 9a^2$, and draw a line.

Step 3

$$
\begin{array}{r}
a^2 \\
2a + 3 \overline{\smash{)}\, 2a^3 + 9a^2 + 5a - 6} \\
(-) \\
(-) \quad \underline{2a^3 + 3a^2} \\
6a^2 + 5a
\end{array}
$$

Subtract $2a^3 + 3a^2$ by adding its negative. Bring down the next term.

Step 4

$$
\begin{array}{r}
a^2 + 3a \\
2a + 3 \overline{\smash{)}\, 2a^3 + 9a^2 + 5a - 6} \\
2a^3 + 3a^2 \\
\underline{} \\
6a^2 + 5a
\end{array}
$$

How many times does $2a$ divide $6a^2$? $6a^2/2a = +3a$. Place the $+3a$ in the quotient.

Step 5

$$
\begin{array}{r}
a^2 + 3a \\
2a + 3 \overline{\smash{)}\, 2a^3 + 9a^2 + 5a - 6} \\
2a^3 + 3a^2 \\
\underline{} \\
6a^2 + 5a \\
\underline{6a^2 + 9a}
\end{array}
$$

Multiply each term in the divisor by $+3a$. Place the product under $6a^2 + 5a$, and draw a line.

Step 6

$$
\begin{array}{r}
a^2 + 3a \\
2a + 3 \overline{\smash{)}\, 2a^3 + 9a^2 + 5a - 6} \\
\underline{2a^3 + 3a^2} \\
6a^2 + 5a \\
(-) \\
(-) \, \underline{6a^2 + 9a} \\
-4a - 6
\end{array}
$$

Subtract $6a^2 + 9a$ by adding its negative. Bring down the next term.

Step 7

$$
\begin{array}{r}
a^2 + 3a \ - 2 \\
2a + 3 \overline{\smash{)}\, 2a^3 + 9a^2 + 5a - 6} \\
\underline{2a^3 + 3a^2} \\
6a^2 + 5a \\
\underline{6a^2 + 9a} \\
-4a - 6
\end{array}
$$

How many times does $2a$ divide $-4a$? $-4a/2a = -2$. Place the -2 in the quotient.

Step 8

$$
\begin{array}{r}
a^2 + 3a \; - 2 \\
2a + 3 \overline{)\; 2a^3 + 9a^2 + 5a - 6} \\
\underline{2a^3 + 3a^2} \\
6a^2 + 5a \\
6a^2 + 9a \\
\underline{} \\
-4a - 6 \\
\underline{-4a - 6}
\end{array}
$$

Multiply each term in the divisor by -2. Place the product under $-4a - 6$, and draw a line.

Step 9

$$
\begin{array}{r}
a^2 + 3a \; - 2 \\
2a + 3 \overline{)\; 2a^3 + 9a^2 + 5a - 6} \\
\underline{2a^3 + 3a^2} \\
6a^2 + 5a \\
6a^2 + 9a \\
\underline{} \\
-4a - 6 \\
(+)(+) \\
\underline{-4a - 6} \\
0
\end{array}
$$

Subtract $-4a - 6$ by adding its negative.

Because the remainder is 0, the quotient is $a^2 + 3a - 2$.
This work can be checked by verifying that

$$
\underbrace{(2a + 3)}_{\text{divisor}} \; \cdot \; \underbrace{(a^2 + 3a - 2)}_{\text{quotient}} \; \underset{=}{=} \; \underbrace{2a^3 + 9a^2 + 5a - 6}_{\text{dividend}}
$$

■

Example 5   Divide $3x^3 + 2x^2 - 3x + 8$ by $x - 2$.

Solution

$$
\begin{array}{r}
3x^2 + 8x \; + 13 \\
x - 2 \overline{)\; 3x^3 + 2x^2 - \; 3x + \; 8} \\
\underline{3x^3 - 6x^2} \\
8x^2 - \; 3x \\
\underline{8x^2 - 16x} \\
13x + \; 8 \\
\underline{13x - 26} \\
34
\end{array}
$$

This division gives a quotient of $3x^2 + 8x + 13$ and a remainder of 34. It is common to form a fraction with the remainder as numerator and the divisor as denominator, and to write the answer as

$$
3x^2 + 8x + 13 + \frac{34}{x - 2}
$$

To check this answer, verify that

$$(x - 2)\left(3x^2 + 8x + 13 + \frac{34}{x - 2}\right) = 3x^3 + 2x^2 - 3x + 8$$ ∎

Example 6   Divide $-9x + 10x^2 + 8x^3 - 9$ by $3 + 2x$.

Solution   The division algorithm works most efficiently when the polynomials in both the dividend and the divisor are written in descending powers of x. Use the commutative property of addition to rearrange the terms. Then, do the division:

$$
\begin{array}{r}
4x^2 - x - 3 \\
2x + 3 \overline{\smash{)}\,8x^3 + 10x^2 - 9x - 9} \\
\underline{8x^3 + 12x^2 } \\
-2x^2 - 9x \\
\underline{-2x^2 - 3x } \\
-6x - 9 \\
\underline{-6x - 9} \\
0
\end{array}
$$

Hence,

$$\frac{-9x + 10x^2 + 8x^3 - 9}{3 + 2x} = 4x^2 - x - 3$$

Check this answer.   ∎

Example 7   Divide $8x^3 + 1$ by $2x + 1$.

Solution   Note that the terms involving x^2 and x are missing in the dividend of $8x^3 + 1$. You must either include the terms $0x^2$ and $0x$ in the dividend or leave spaces for them. After this adjustment, the division is routine.

$$
\begin{array}{r}
4x^2 - 2x + 1 \\
2x + 1 \overline{\smash{)}\,8x^3 + 0x^2 + 0x + 1} \\
\underline{8x^3 + 4x^2 } \\
-4x^2 + 0x \\
\underline{-4x^2 - 2x } \\
+2x + 1 \\
\underline{+2x + 1} \\
0
\end{array}
$$

Hence,

$$\frac{8x^3 + 1}{2x + 1} = 4x^2 - 2x + 1$$

Check this answer.   ∎

Example 8   Divide $-17x^2 + 5x + x^4 + 2$ by $x^2 - 1 + 4x$.

Solution   Rewrite the problem with both the divisor and the dividend in descending powers of x. Leave spaces for the missing terms in the dividend. Then perform the division as follows:

$$
\begin{array}{r}
x^2 - 4x \\
x^2 + 4x - 1 \overline{)\, x^4 - 17x^2 + 5x + 2} \\
\underline{x^4 + 4x^3 - x^2 } \\
-4x^3 - 16x^2 + 5x \\
\underline{-4x^3 - 16x^2 + 4x } \\
x + 2
\end{array}
$$

This division gives a quotient of $x^2 - 4x$ and a remainder of $x + 2$. Hence,

$$\frac{-17x^2 + 5x + x^4 + 2}{x^2 - 1 + 4x} = x^2 - 4x + \frac{x + 2}{x^2 + 4x - 1}$$

Check this answer.   ∎

EXERCISE 2.5

In Exercises 1–18, perform each indicated operation. Express all answers without using negative exponents.

1. $4x^2y^3 \div 8x^5y^2$

2. $25x^4y^7 \div 5xy^9$

3. $\dfrac{33a^{-2}b^2}{44a^2b^{-2}}$

4. $\dfrac{-63a^4b^{-3}}{81a^{-3}b^3}$

5. $\dfrac{45x^{-2}y^{-3}t^0}{-63x^{-1}y^4t^2}$

6. $\dfrac{112a^0b^2c^{-3}}{48a^4b^0c^4}$

7. $\dfrac{-65a^{2n}b^nc^{3n}}{-15a^nb^{-n}c}$

8. $\dfrac{-32x^{-3n}y^{-2n}z}{40x^{-2}y^{-n}z^{n+1}}$

9. $(4x^2 - x^3) \div 6x$

10. $(5y^4 + 45y^3) \div 15y^2$

11. $\dfrac{4x^2y^3 + 2x^3y^2}{6xy}$

12. $\dfrac{9a^3y^2 - 18a^4y^3}{27a^2y^2}$

13. $\dfrac{24x^6y^7 - 12x^5y^{12} + 36xy}{48x^2y^3}$

14. $\dfrac{9x^4y^3 + 18x^2y - 27xy^4}{9x^3y^3}$

15. $\dfrac{3a^{-2}b^3 - 6a^2b^{-3} + 9a^{-2}}{12a^{-1}b}$

16. $\dfrac{4x^3y^{-2} + 8x^{-2}y^2 - 12y^4}{12x^{-1}y^{-1}}$

17. $\dfrac{x^ny^n - 3x^{2n}y^{2n} + 6x^{3n}y^{3n}}{x^ny^n}$

18. $\dfrac{2a^n - 3a^nb^{2n} - 6b^{4n}}{a^nb^{n-1}}$

In Exercises 19–54, use the division algorithm to find each quotient.

19. $\dfrac{x^2 + 5x + 6}{x + 3}$

20. $\dfrac{x^2 - 5x + 6}{x - 3}$

21. $(x^2 + 10x + 21) \div (x + 3)$

22. $(x^2 + 10x + 21) \div (x + 7)$

23. $\dfrac{6x^2 - x - 12}{2x + 3}$

24. $\dfrac{6x^2 - x - 12}{2x - 3}$

25. $\dfrac{3x^3 - 2x^2 + x + 6}{x - 1}$

26. $\dfrac{4a^3 + a^2 - 3a + 7}{a + 1}$

27. $\dfrac{6x^3 + 11x^2 - x - 2}{3x - 2}$

28. $\dfrac{6x^3 + 11x^2 - x + 10}{2x + 3}$

29. $\dfrac{6x^3 - x^2 - 6x - 9}{2x - 3}$

30. $\dfrac{16x^3 + 16x^2 - 9x - 5}{4x + 5}$

31. $(2a + 1 + a^2) \div (a + 1)$

32. $(a - 15 + 6a^2) \div (2a - 3)$

33. $(6y - 4 + 10y^2) \div (5y - 2)$

34. $(-10xy + x^2 + 16y^2) \div (x - 2y)$

35. $\dfrac{-18x + 12 + 6x^2}{x - 1}$

36. $\dfrac{27x + 23x^2 + 6x^3}{2x + 3}$

37. $\dfrac{-9x^2 + 8x + 9x^3 - 4}{3x - 2}$

38. $\dfrac{6x^2 + 8x^3 - 13x + 3}{4x - 3}$

39. $\dfrac{13x + 16x^4 + 3x^2 + 3}{4x + 3}$

40. $\dfrac{3x^2 + 9x^3 + 4x + 4}{3x + 2}$

41. $(a^3 + 1) \div (a - 1)$

42. $(27a^3 - 8b^3) \div (3a - 2b)$

43. $\dfrac{15a^3 - 29a^2 + 16}{3a - 4}$

44. $\dfrac{4x^3 - 12x^2 + 17x - 12}{2x - 3}$

45. $y - 2 \overline{)\, -24y + 24 + 6y^2}$

46. $3 - a \overline{)\, 21a - a^2 - 54}$

47. $2x + y \overline{)\, 32x^5 + y^5}$

48. $3x - y \overline{)\, 81x^4 - y^4}$

49. $x^2 - 2 \overline{)\, x^6 - x^4 + 2x^2 - 8}$

50. $x^2 + 3 \overline{)\, x^6 + 2x^4 - 6x^2 - 9}$

51. $(x^4 + 2x^3 + 4x^2 + 3x + 2) \div (x^2 + x + 2)$

52. $(2x^4 + 3x^3 + 3x^2 - 5x - 3) \div (2x^2 - x - 1)$

53. $x + x^2 + 2 \overline{)\, x^3 + 3x + 5x^2 + 6 + x^4}$

54. $x^3 + 1 + 2x \overline{)\, x^5 + 3x + 2}$

In Exercises 55–56, use a calculator to find each quotient.

55. $x - 2 \overline{)\, 9.8x^2 - 3.2x - 69.3}$

56. $2.5x - 3.7 \overline{)\, -22.25x^2 - 38.9x - 16.65}$

2.6   SYNTHETIC DIVISION

There is a shortcut method, called **synthetic division**, that we can use to divide a polynomial by a binomial of the form $x - r$. To see how this method works, we consider the division of $4x^3 - 5x^2 - 11x + 20$ by $x - 2$.

$$
\begin{array}{r}
4x^2 + 3x - 5 \\
x - 2 \overline{)\, 4x^3 - 5x^2 - 11x + 20} \\
\underline{4x^3 - 8x^2} \\
3x^2 - 11x \\
\underline{3x^2 - 6x} \\
-5x + 20 \\
\underline{-5x + 10} \\
10 \quad \text{(remainder)}
\end{array}
$$

$$
\begin{array}{r}
4 \quad 3 \quad -5 \\
1 - 2 \overline{)\, 4 \quad -5 \quad -11 \quad 20} \\
\underline{4 \quad -8} \\
3 \quad -11 \\
3 \quad -6 \\
-5 \quad 20 \\
-5 \quad 10 \\
10 \quad \text{(remainder)}
\end{array}
$$

On the left is the familiar long-division process, and on the right is the skeleton form of that division. All references to the variable x have been removed. The various powers of x can be remembered without actually writing them because the exponents of the terms in the divisor, dividend, and quotient were written in descending order.

We can further shorten the version on the right. The numbers printed in color need not be written because they are duplicates of the numbers immediately above them. Thus, we can write the division in the following form:

$$
\begin{array}{r}
4 \quad\ \ 3 \quad -5 \\
\hline
1-2 \overline{)\ 4 \quad -5 \quad -11 \quad 20} \\
-8 \\
\hline
3 \\
-6 \\
\hline
-5 \\
10 \\
\hline
10
\end{array}
$$

We can shorten the process still further by compressing the work vertically, and eliminating the 1 (the coefficient of x in the divisor):

$$
\begin{array}{r}
4 \quad\ \ 3 \quad -5 \\
\hline
-2 \overline{)\ 4 \quad -5 \quad -11 \quad 20} \\
-8 \quad -6 \quad 10 \\
\hline
3 \quad -5 \quad 10
\end{array}
$$

There is no reason why the quotient, represented by the numbers 4   3 -5, must appear *above* the long division. If we write the 4 on the bottom line, the bottom line gives the coefficients of the quotient, and it also gives the remainder. The entire top line can be eliminated. The division now appears as follows:

$$
\begin{array}{r}
\underline{-2}\ \big|\ \ 4 \quad -5 \quad -11 \quad 20 \\
-8 \quad -6 \quad 10 \\
\hline
4 \quad\ \ 3 \quad -5 \quad 10
\end{array}
$$

The bottom line was obtained by subtracting the middle line from the top line. If we were to replace the -2 in the divisor by a $+2$, the division process would reverse the signs of every entry in the middle line. Then, the bottom line could be obtained by addition. Thus, we have this final form of the synthetic division.

$$
\begin{array}{r}
\underline{+2}\ \big|\ \ 4 \quad -5 \quad -11 \quad\ \ 20 \\
8 \quad\ \ 6 \quad -10 \\
\hline
4 \quad\ \ 3 \quad -5 \ \big|\ \ 10
\end{array}
$$

The coefficients of the dividend

The coefficients of the quotient, and the remainder to the right of the vertical bar

Thus,

$$
\frac{4x^3 - 5x^2 - 11x + 20}{x - 2} = 4x^2 + 3x - 5 + \frac{10}{x - 2}
$$

Example 1   Use synthetic division to divide $6x^2 + 5x - 2$ by $x - 5$.

Solution   Begin by writing the coefficients of the dividend, and the 5 from the divisor, in the following form:

$$\underline{5 \rvert}\quad 6 \quad 5 \quad -2$$

Then follow these steps:

$$\underline{5 \rvert}\quad 6 \quad 5 \quad -2 \qquad\qquad \text{Begin by bringing down the 6.}$$
$$ 6$$

$$\underline{5 \rvert}\quad 6 \quad 5 \quad -2 \qquad\qquad \text{Multiply 5 and 6, to get 30.}$$
$$ 30$$
$$ 6$$

$$\underline{5 \rvert}\quad 6 \quad 5 \quad -2 \qquad\qquad \text{Add 5 and 30, to get 35.}$$
$$ 30$$
$$ 6 \quad 35$$

$$\underline{5 \rvert}\quad 6 \quad 5 \quad -2 \qquad\qquad \text{Multiply 5 and 35, to get 175.}$$
$$ 30 \quad 175$$
$$ 6 \quad 35$$

$$\underline{5 \rvert}\quad 6 \quad 5 \quad -2 \qquad\qquad \text{Add } -2 \text{ and 175, to get 173.}$$
$$ 30 \quad 175$$
$$ 6 \quad 35 \quad \rvert \quad 173$$

The numbers 6 and 35 represent the quotient: $6x + 35$. The number 173 is the remainder. Thus,

$$\frac{6x^2 + 5x - 2}{x - 5} = 6x + 35 + \frac{173}{x - 5}$$

Check this answer.   ∎

Example 2   Use synthetic division to divide $5x^3 + x^2 - 3$ by $x - 2$.

Solution   Begin by writing

$$\underline{2 \rvert}\quad 5 \quad 1 \quad 0 \quad -3 \qquad \text{Write 0 for the coefficient of } x, \text{ the missing term.}$$

Then complete the division as follows:

$$\underline{2 \rvert}\ 5 \ \ 1 \ \ 0 \ \ -3 \qquad \underline{2 \rvert}\ 5 \ \ 1 \ \ 0 \ \ -3 \qquad \underline{2 \rvert}\ 5 \ \ 1 \ \ 0 \ \ \ \ -3$$
$$ 10 10 \ \ 22 10 \ \ 22 \ \ 44$$
$$ 5 \ \ 11 5 \ \ 11 \ \ 22 5 \ \ 11 \ \ 22 \ \rvert\ \ 41$$

Thus,

$$\frac{5x^3 + x^2 - 3}{x - 2} = 5x^2 + 11x + 22 + \frac{41}{x - 2}$$

Check this answer.   ■

Example 3   Use synthetic division to divide $5x^2 + 6x^3 + 2 - 4x$ by $x + 2$.

Solution   First, rewrite the dividend with the exponents in descending order: $6x^3 + 5x^2 - 4x + 2$. Then rewrite the divisor in $x - r$ form: $x - (-2)$. Using synthetic division, begin by writing

$$-2\big|\ \ 6 \qquad 5 \quad -4 \qquad\quad 2$$

Then complete the division:

$$-2\big|\ \ \begin{array}{ccc} 6 & 5 & -4 & 2 \\ & -12 & 14 & -20 \\ \hline 6 & -7 & 10 & \big|\ -18 \end{array}$$

Thus,

$$\frac{5x^2 + 6x^3 + 2 - 4x}{x + 2} = 6x^2 - 7x + 10 + \frac{-18}{x + 2}$$

Check this answer.   ■

Synthetic division is important in mathematics because of the following theorem, called the **Remainder Theorem**.

Remainder Theorem.   If a polynomial $P(x)$ is divided by $x - r$, then the remainder is $P(r)$.

We will illustrate the Remainder Theorem in Example 4.

Example 4   If $P(x) = 2x^3 - 3x^2 - 2x + 1$, determine   **a.** $P(3)$ and   **b.**   the remainder when $P(x)$ is divided by $x - 3$.

Solution   **a.** $P(3) = 2(3)^3 - 3(3)^2 - 2(3) + 1$

$= 2(27) - 3(9) - 6 + 1$

$= 54 - 27 - 5$

$= 22$

b. Use synthetic division to find the remainder when $P(x) = 2x^3 - 3x^2 - 2x + 1$ is divided by $x - 3$.

$$
\begin{array}{r|rrrr}
3 & 2 & -3 & -2 & 1 \\
  &   & 6 & 9 & 21 \\
\hline
  & 2 & 3 & 7 & \mid\ 22
\end{array}
$$

The remainder is 22.

The results of parts a and b show that, when $P(x)$ is divided by $x - 3$, the remainder is $P(3)$.   ∎

It is often easier to calculate $P(r)$ by using synthetic division than by substituting r for x in $P(x)$. This is especially true if r is a number such as 2.3, 0.06, or 3.698.

■ EXERCISE 2.6

In Exercises 1–14, use synthetic division to perform each division.

1. $(x^2 + x - 2) \div (x - 1)$

2. $(x^2 + x - 6) \div (x - 2)$

3. $x - 4 \overline{)\, x^2 - 7x + 12}$

4. $x - 5 \overline{)\, x^2 - 6x + 5}$

5. $(x^2 + 8 + 6x) \div (x + 4)$

6. $(x^2 - 15 - 2x) \div (x + 3)$

7. $x + 2 \overline{)\, x^2 - 5x + 14}$

8. $x + 6 \overline{)\, x^2 + 13x + 42}$

9. $(3x^3 - 10x^2 + 5x - 6) \div (x - 3)$

10. $(2x^3 - 9x^2 + 10x - 3) \div (x - 3)$

11. $(2x^3 - 5x - 6) \div (x - 2)$

12. $(4x^3 + 5x^2 - 1) \div (x + 2)$

13. $x + 1 \overline{)\, 5x^2 + 6x^3 + 4}$

14. $x - 4 \overline{)\, 4 - 3x^2 + x}$

In Exercises 15–20, use a calculator and synthetic division to perform each division.

15. $x - 0.2 \overline{)\, 7.2x^2 - 2.1x + 0.5}$

16. $x - 0.4 \overline{)\, 8.1x^2 + 3.2x - 5.7}$

17. $x + 1.7 \overline{)\, 2.7x^2 + x - 5.2}$

18. $x + 2.5 \overline{)\, 1.3x^2 - 0.5x - 2.3}$

19. $x + 57 \overline{)\, 9x^3 - 25}$

20. $x - 2.3 \overline{)\, 0.5x^3 + x}$

In Exercises 21–28, let $P(x) = 2x^3 - 4x^2 + 2x - 1$. Evaluate the polynomial by substituting the given value of x into the polynomial and simplifying. Then evaluate the polynomial by using the Remainder Theorem and synthetic division.

21. $P(1)$

22. $P(2)$

23. $P(-2)$

24. $P(-1)$

25. $P(3)$

26. $P(-4)$

27. $P(0)$

28. $P(4)$

In Exercises 29–36, let $Q(x) = x^4 - 3x^3 + 2x^2 + x - 3$. Evaluate the polynomial by substituting the given value of x into the polynomial and simplifying. Then evaluate the polynomial by using the Remainder Theorem and synthetic division.

29. $Q(-1)$

30. $Q(1)$

31. $Q(2)$

32. $Q(-2)$

33. $Q(3)$

34. $Q(0)$

35. $Q(-3)$

36. $Q(-4)$

In Exercises 37–44, use the Remainder Theorem and synthetic division to find P(r).

37. $P(x) = x^3 - 4x^2 + x - 2;\ r = 2$

38. $P(x) = x^3 - 3x^2 + x + 1;\ r = 1$

39. $P(x) = 2x^3 + x + 2;\ r = 3$

40. $P(x) = x^3 + x^2 + 1;\ r = -2$

41. $P(x) = x^4 - 2x^3 + x^2 - 3x + 2;\ r = -2$

42. $P(x) = x^5 + 3x^4 - x^2 + 1;\ r = -1$

43. $P(x) = 3x^5 + 1;\ r = -\frac{1}{2}$

44. $P(x) = 5x^7 - 7x^4 + x^2 + 1;\ r = 2$

45. Calculate 2^6 by using synthetic division to evaluate the polynomial $P(x) = x^6$ at $x = 2$.

46. Calculate $(-3)^5$ by using synthetic division to evaluate the polynomial $P(x) = x^5$ at $x = -3$.

CHAPTER SUMMARY

Key Words

algorithm   (2.5)

base of an exponential
    expression   (2.1)

binomial   (2.3)

coefficient   (2.1)

degree of a polynomial   (2.3)

exponent   (2.1)

FOIL method for multiplying
    binomials   (2.4)

like terms   (2.3)

monomial   (2.3)

polynomial   (2.3)

power of x   (2.1)

similar terms   (2.3)

trinomial   (2.3)

Key Ideas

(2.1)   For any natural n, $x^n = \overbrace{x \cdot x \cdot x \cdot \cdots \cdot x}^{n \text{ factors of } x}$

If m and n are integers and there are no divisions by zero, then

a. $x^m x^n = x^{m+n}$

b. $(x^m)^n = x^{mn}$

c. $(xy)^n = x^n y^n$

d. $\left(\dfrac{y}{x}\right)^n = \dfrac{y^n}{x^n}$

e. $x^0 = 1$

f. $x^{-n} = \dfrac{1}{x^n}$

g. $\dfrac{x^m}{x^n} = x^{m-n}$

h. $\left(\dfrac{y}{x}\right)^{-n} = \left(\dfrac{x}{y}\right)^n$

(2.2)   A number is written in scientific notation if it is expressed as a number between 1 and 10 multiplied by an appropriate power of 10.

(2.3)   A polynomial in x is the sum of one or more terms of the form ax^n, where a is a real number and n is a whole number.

A polynomial in several variables, say x, y, and z, is a sum of one or more terms of the form $ax^m y^n z^p$, where a is a real number and m, n, and p are whole numbers.

The **degree of a polynomial** is the degree of the term with highest degree contained within the polynomial.

If $P(x)$ is a polynomial in x, then $P(r)$ is the value of the polynomial at $x = r$.

To add like terms, add their numerical coefficients and use the same variables with the same exponents.

To add polynomials, add their like terms.

To subtract polynomials, add the negative of the subtrahend to the other polynomial.

(2.4)   To multiply monomials, multiply their numerical factors and multiply their variable factors.

To multiply a polynomial by a monomial, multiply each term of the polynomial by the monomial.

To multiply polynomials, multiply each term of one polynomial by each term of the other polynomial.

If one set of grouping symbols is contained within another set, remove the inner set first.

(2.5)   To find the quotient of two monomials, express the quotient as a fraction and use the rules of exponents to simplify.

(2.6)   Synthetic division can be used to divide polynomials by binomials of the form $x - r$.

If a polynomial $P(x)$ is divided by $x - r$, then the remainder is $P(r)$.

REVIEW EXERCISES

In Review Exercises 1–28, use the rules of exponents to simplify each quantity. Write all answers without using negative exponents.

1. 3^6

2. -2^6

3. $(-4)^3$

4. $-(-5)^4$

5. $(3x^4)(-2x^2)$

6. $(-x^5)(3x^3)$

7. $x^{-4}x^3$

8. $x^{-10}x^{12}$

9. $(3x^2)^3$

10. $(4x^4)^4$

11. $(-2x^2)^5$

12. $-(-3x^3)^5$

13. $(x^2)^{-5}$

14. $(x^{-4})^{-5}$

15. $(3x^{-3})^{-2}$

16. $(2x^{-4})^4$

17. $\dfrac{x^6}{x^4}$

18. $\dfrac{x^{12}}{x^7}$

19. $\dfrac{a^7}{a^{12}}$

20. $\dfrac{a^4}{a^7}$

21. $\dfrac{y^{-3}}{y^4}$

22. $\dfrac{y^5}{y^{-4}}$

23. $\dfrac{x^{-5}}{x^{-4}}$

24. $\dfrac{x^{-6}}{x^{-9}}$

25. $(3x^2y^3)^2$

26. $(-4a^3b^2)^{-4}$

27. $\left(\dfrac{3x^2}{4y^3}\right)^{-3}$

28. $\left(\dfrac{4y^{-2}}{5y^{-3}}\right)^3$

In Review Exercises 29–30, write each numeral in scientific notation.

29. 19,300,000,000

30. 0.0000000273

In Review Exercises 31–32, write each numeral in standard notation.

31. 7.2×10^7

32. 8.3×10^{-9}

In Review Exercises 33–36, find the required value if $P(x) = -x^2 + 4x + 6$.

33. $P(0)$

34. $P(1)$

35. $P(-t)$

36. $P(z)$

37. Give the degree of $P(x) = 3x^5 + 4x^3 + 2$.

38. Give the degree of $9x^2y + 13x^3y^2 + 8x^4y^4$.

In Review Exercises 39–42, simplify each expression.

39. $(3x^2 + 4x + 9) - (2x^2 - 2x + 7) + (4x^2 - 3x - 2)$

40. $(4x^3 + 4x^2 + 7) - (-2x^3 - x - 2) + (-5x^3 - 3x^2)$

41. $(2x^2 - 5x + 9) - (x^2 - 3) - (-3x^2 + 4x - 7)$

42. $(7x^3 - 6x^2 + 4x - 3) - (7x^3 + 6x^2 + 4x - 3)$

In Review Exercises 43–50, find each product.

43. $(8a^2b^2)(-2abc)$

44. $(-3xy^2z)(2xz^3)$

45. $2xy^2(x^3y - 4xy^5)$

46. $a^2b(a^2 + 2ab + b^2)$

47. $(8x - 5)(2x + 3)$

48. $(3x^2 + 2)(2x - 4)$

49. $(5x^2 - 4x + 5)(3x^2 - 2x + 10)$

50. $(3x^2 + x - 2)(x^2 - x + 2)$

In Review Exercises 51–58, perform each division.

51. Divide $(3x^3 - 4x^2 + 3x + 2)$ by $(x + 3)$.

52. Divide $(64x^3 + 125y^3)$ by $(4x + 5y)$.

53. $x - 1 \,)\overline{\, x^5 - 1 \,}$

54. $x + 2 \,)\overline{\, x^5 + 32 \,}$

55. $3x - 2 \,)\overline{\, 6x^3 + 5x^2 - 3x + 8 \,}$

56. $x^2 - 1 \,)\overline{\, x^4 + x^2 - 2 \,}$

57. $x^2 + 2x + 3 \,)\overline{\, x^4 - x^2 - 3 \,}$

58. $3x + x^3 + 1 \,)\overline{\, 3x^3 - x^2 + 5 \,}$

In Review Exercises 59–60, use synthetic division to perform each division.

59. $x - 4 \,)\overline{\, x^3 - 13x - 12 \,}$

60. $x + 1 \,)\overline{\, x^4 + x^2 + 1 \,}$

In Review Exercises 61–62, let $P(x) = 3x^2 - 2x + 3$. Use synthetic division and the Remainder Theorem to find each value.

61. $P(2)$

62. $P(-1)$

CHAPTER TWO TEST

In Problems 1–6, simplify each expression. Write all answers without using negative exponents. Assume that no denominators are zero.

1. x^3x^5

2. $(x^2y^3)^3$

3. $(m^{-4})^2$

4. $\left(\dfrac{a^3}{b^2}\right)^4$

5. $3x^0$

6. $\left(\dfrac{m^2n^3}{m^4n^{-2}}\right)^{-2}$

In Problems 7–8, write each number in scientific notation.

7. 4,700,000

8. 0.00000023

In Problems 9–10, write each number in standard notation.

9. 6.53×10^5

10. 24.5×10^{-3}

11. The moon is approximately 235,000 miles from the earth. Use scientific notation to express this distance in kilometers. (*Hint:* 1 mile $\approx$ 1.6 kilometers.)

12. Give the degree of the polynomial $3x^2y^3 + 4x^3y^7 - 7x^4y^5$.

In Problems 13–14, let $P(x) = -3x^2 + 2x - 1$ and find each value.

13. $P(2)$

14. $P(-1)$

In Problems 15–16, $x = 3$ and $y = -2$. Find the value of each expression.

15. $x^2 - y^2$

16. $\dfrac{4x^2 + y^2}{-xy^2}$

In Problems 17–28, perform the indicated operations.

17. $(2y^2 + 4y + 3) + (3y^2 - 3y - 4)$

18. $(-3u^2 + 2u - 7) - (u^2 + 7)$

19. $3(2a^2 - 4a + 2) - 4(-a^2 - 3a - 4)$

20. Add: $8x^2 + 4x - 9$
$-2x^2 - 6x + 8$

21. $(3x^3y^2z)(-2xy^{-1}z^3)$

22. $-5a^2b(3ab^3 - 2ab^4)$

23. $(z + 4)(z - 4)$

24. $(3x - 2)(4x + 3)$

25. $(2x + 1)(x^2 - x - 3)$

26. $(x^n + y^n)(2x^n - y^n)$

27. $\dfrac{18x^2y^3 - 12x^3y^2 + 9xy}{-3xy^4}$

28. $2x - 1 \overline{)\, 6x^3 + 5x^2 - 2}$

29. Find the remainder in the division

$$\frac{x^3 - 4x^2 + 5x + 3}{x + 1}$$

30. Find the remainder when $4x^3 + 3x^2 + 2x - 1$ is divided by $x - 2$ by completing the synthetic division

$$\underline{2}\,\big|\quad 4 \quad 3 \quad 2 \quad -1$$

CAREERS AND MATHEMATICS

BANKER Practically every bank has a group of officers who make decisions affecting bank operations: the president who directs overall operations; one or more vice presidents who act as general managers or are in charge of bank departments, such as trust or credit; a comptroller or cashier who, as an executive officer, is generally responsible for all bank property; and treasurers and other senior officers as well as junior officers, who supervise sections within departments.

These officers make decisions within a framework of policy set by a board of directors and existing laws and regulations. They must have a broad knowledge of business activities to relate to the operations of their departments since their customers will include a variety of individuals and businesses applying for a loan, seeking investment advice, organizing trusts, setting up pensions, and so on. Besides supervising these financial services, officers advise individuals and businesses and participate in community projects.

Qualifications Bank officer and management positions are filled by management trainees, and by promoting outstanding bank clerks or tellers. A college degree in finance or liberal arts, including accounting, economics, commercial law, political science, and statistics, is necessary. A Master of Business Administration (MBA) is preferred by some banks, although people with backgrounds as diverse as nuclear physics and forestry are hired by some banks to meet the needs of the complex, high-technology industries with which they deal.

Job outlook Through the mid-1990s, employment of bank officers is expected to increase faster than the average for other occupations due to expanding bank services, both domestic and international, and the increasing dependence on computers.

Example application Exponential expressions appear in banking problems that involve compound interest. If some amount of money P is deposited in a bank at a rate of $r\%$ (expressed as a decimal) for t years, compounded k times per year, the formula

$$A = P\left(1 + \frac{r}{k}\right)^{kt}$$

gives the amount A in the account after t years.

If \$1000 is deposited in an account that earns 10% annual interest, compounded twice a year, find the amount in the account after one year.

Solution In this problem, $P = \$1000$, $r = 0.10$, $t = 1$, and $k = 2$. Substitute these values into the preceding formula, and simplify. A calculator will be helpful.

$$A = P\left(1 + \frac{r}{k}\right)^{kt}$$
$$A = 1000\left(1 + \frac{0.10}{2}\right)^{2}$$
$$A = 1000(1 + 0.05)^{2}$$

$$A = 1000(1.05)^2$$
$$A = 1000(1.1025)$$
$$A = 1102.50$$

At the end of one year, there will be $1102.50 in the account.

Exercises

1. If $10,000 is deposited in an account earning 8% annual interest, compounded once a year, how much will be in the account at the end of two years?
2. If $10,000 is deposited in an account earning 8% annual interest, compounded twice a year, how much will be in the account at the end of two years?
3. If $10,000 is deposited in an account earning 8% annual interest, compounded four times a year, how much will be in the account at the end of two years?
4. Refer to Exercises 1–3. How much more is earned by compounding twice a year, instead of once a year? How much more is earned by compounding four times a year, instead of twice a year?

(Answers: **1.** $11,664 **2.** $11,698.59 **3.** $11,716.59
4. $34.59; $18)

3 ▷ Factoring Polynomials

In the previous chapter we discussed how to multiply polynomials by polynomials. We now reverse that procedure and discuss how to split products apart and undo multiplications. The process of finding the individual factors of a known product is called **factoring**.

3.1 THE GREATEST COMMON FACTOR

If a natural number a divides a natural number b (without a remainder), then a is called a **factor** of b. The natural number factors of 6, for example, are 6, 3, 2, and 1 because each of these numbers divides 6. Recall that, if the only natural number factors of a natural number p (where $p > 1$) are 1 and p, then p is called a **prime number**. The set of prime numbers is the set

$$\mathbf{P} = \{2, 3, 5, 7, 11, 13, 17, 19, 23, 29, \dots\}$$

To factor a natural number means to write the number as a product of other natural numbers. If each of the factors in the product is prime, we say that the natural number is written in **prime-factored form**. The statements

$$60 = 6 \cdot 10 = 2 \cdot 3 \cdot 2 \cdot 5 = 2^2 \cdot 3 \cdot 5$$
$$84 = 4 \cdot 21 = 2 \cdot 2 \cdot 3 \cdot 7 = 2^2 \cdot 3 \cdot 7$$

and

$$180 = 10 \cdot 18 = 2 \cdot 5 \cdot 3 \cdot 6 = 2 \cdot 5 \cdot 3 \cdot 3 \cdot 2 = 2^2 \cdot 3^2 \cdot 5$$

show the prime-factored forms of 60, 84, and 180. If a quantity is written in prime-factored form, we say it is in **completely factored form**.

The largest natural number that divides 60, 84, and 180 is called the **greatest common factor** or **greatest common divisor** of these three numbers. Because 60, 84, and 180 all have at least two factors of 2 and one factor of 3, the greatest common factor of these three numbers is $2^2 \cdot 3 = 12$. We note that

$$\frac{60}{12} = 5, \qquad \frac{84}{12} = 7, \qquad \text{and} \qquad \frac{180}{12} = 15$$

There is no natural number greater than 12 that divides 60, 84, and 180.

Likewise, algebraic monomials have greatest common factors. We consider three monomials with their prime factorizations:

$$6a^2b^3c = 3 \cdot 2 \cdot a \cdot a \cdot b \cdot b \cdot b \cdot c$$
$$9a^3b^2c = 3^2 \cdot a \cdot a \cdot a \cdot b \cdot b \cdot c$$
$$18a^4c^3 = 2 \cdot 3^2 \cdot a \cdot a \cdot a \cdot a \cdot c \cdot c \cdot c$$

Because each monomial has at least one factor of 3, two factors of a, and one factor of c in common, their greatest common factor is

$$3^1 \cdot a^2 \cdot c^1 = 3a^2c$$

To find the greatest common factor of several monomials, we follow these steps:

1. Completely factor each monomial.
2. Use each common factor the least number of times it appears in any one monomial.
3. Find the product of the factors found in step 2 to obtain the greatest common factor.

Recall that the distributive property provides a method for multiplying a polynomial by a monomial. For example,

$$2x^3y^3(3x^2 - 4y^3) = 2x^3y^3 \cdot 3x^2 - 2x^3y^3 \cdot 4y^3$$
$$= 6x^5y^3 - 8x^3y^6$$

If the product of a multiplication is $6x^5y^3 - 8x^3y^6$, we can use the distributive property backwards to find the individual factors.

$$6x^5y^3 - 8x^3y^6 = 2x^3y^3 \cdot 3x^2 - 2x^3y^3 \cdot 4y^3$$
$$= 2x^3y^3(3x^2 - 4y^3)$$

Because $2x^3y^3$ is the greatest common factor of the terms of $6x^5y^3 - 8x^3y^6$, this process is called **factoring out the greatest common factor**.

Example 1 Factor $25a^3b + 15ab^3$.

Solution First factor each monomial:

$$25a^3b = 5 \cdot 5 \cdot a \cdot a \cdot a \cdot b$$
$$15ab^3 = 5 \cdot 3 \cdot a \cdot b \cdot b \cdot b$$

Because each term has at least one factor of 5, one factor of a, and one factor of b, and because there are no other common factors, $5ab$ is the greatest common factor of the two terms. Use the distributive property to factor out the $5ab$:

$$25a^3b + 15ab^3 = 5ab \cdot 5a^2 + 5ab \cdot 3b^2$$
$$= 5ab(5a^2 + 3b^2)$$

∎

Example 2   Factor $3xy^2z^3 + 6xz^2 - 9xyz^4$.

Solution   First factor each monomial:

$$3xy^2z^3 = 3 \cdot x \cdot y \cdot y \cdot z \cdot z \cdot z$$
$$6xz^2 = 3 \cdot 2 \cdot x \cdot z \cdot z$$
$$-9xyz^4 = -3 \cdot 3 \cdot x \cdot y \cdot z \cdot z \cdot z \cdot z$$

Because each term has at least one factor of 3, one factor of x, and two factors of z, and because there are no other common factors, $3xz^2$ is the greatest common factor of the three terms. Use the distributive property to factor out the $3xz^2$:

$$3xy^2z^3 + 6xz^2 - 9xyz^4 = 3xz^2 \cdot y^2z + 3xz^2 \cdot 2 - 3xz^2 \cdot 3yz^2$$
$$= 3xz^2(y^2z + 2 - 3yz^2)$$ ∎

Example 3   Factor $x^3y^3z^3 + xyz$.

Solution   Because each term has at least one factor of x, one factor of y, and one factor of z, and because there are no other common factors, the greatest common factor of $x^3y^3z^3$ and xyz is xyz. Thus, the expression $x^3y^3z^3 + xyz$ can be factored as follows:

$$x^3y^3z^3 + xyz = xyz \cdot x^2y^2z^2 + xyz \cdot 1$$
$$= xyz(x^2y^2z^2 + 1)$$

It is important to understand where the "1" comes from. The last term, xyz, of the given binomial has an understood coefficient of 1. When the xyz is factored out, the 1 must be made explicit.   ∎

Example 4   Factor out the negative of the greatest common factor of $-6u^2v^3 + 8u^3v^2$.

Solution   The greatest common factor of the two terms is $2u^2v^2$. Thus, the negative of the greatest common factor is $-2u^2v^2$. To factor out $-2u^2v^2$, proceed as follows:

$$-6u^2v^3 + 8u^3v^2 = -2u^2v^2 \cdot 3v + 2u^2v^2 \cdot 4u$$
$$= -2u^2v^2 \cdot 3v - (-2u^2v^2) \cdot 4u$$
$$= -2u^2v^2(3v - 4u)$$ ∎

If a polynomial cannot be factored, we call the polynomial a **prime polynomial** or an **irreducible polynomial**.

Example 5   Factor $3x^2 + 4y + 7$.

Solution   Factor each monomial:

$$3x^2 = 3 \cdot x \cdot x$$
$$4y = 2 \cdot 2 \cdot y$$
$$7 = 7$$

Because there are no common factors other than 1, the given polynomial cannot be factored. It is an example of a prime polynomial. ∎

Sometimes the common factor in an expression is a polynomial with more than one term. For example, in the expression

$$x(a + b) + y(a + b)$$

the binomial $a + b$ is a factor of both terms. Hence the expression factors as

$$x(a + b) + y(a + b) = (a + b) \cdot x + (a + b) \cdot y$$
$$= (a + b)(x + y)$$

Example 6   Factor $a(x - y + z) - b(x - y + z) + 3(x - y + z)$.

Solution   Determine that $x - y + z$ is the greatest common factor, and use the distributive property to factor it out:

$$a(x - y + z) - b(x - y + z) + 3(x - y + z)$$
$$= (x - y + z) \cdot a - (x - y + z) \cdot b + (x - y + z) \cdot 3$$
$$= (x - y + z)(a - b + 3)$$ ∎

In advanced courses we must sometimes factor out an exponential expression with a variable exponent. For example, to factor x^{2n} from $x^{4n} + x^{3n} + x^{2n}$, we write the trinomial in the form

$$x^{2n} \cdot x^{2n} + x^{2n} \cdot x^n + x^{2n} \cdot 1$$

and factor out the x^{2n}:

$$x^{4n} + x^{3n} + x^{2n} = x^{2n} \cdot x^{2n} + x^{2n} \cdot x^n + x^{2n} \cdot 1$$
$$= x^{2n}(x^{2n} + x^n + 1)$$

Example 7   Factor $a^{-2}b^{-2}$ from $a^{-2}b - a^3 b^{-2}$.

Solution   Write the expression $a^{-2}b - a^3 b^{-2}$ in the form

$$a^{-2}b^{-2} \cdot b^3 - a^{-2}b^{-2} \cdot a^5$$

and factor out the $a^{-2}b^{-2}$:

$$a^{-2}b - a^3 b^{-2} = a^{-2}b^{-2} \cdot b^3 - a^{-2}b^{-2} \cdot a^5$$
$$= a^{-2}b^{-2}(b^3 - a^5)$$ ∎

▬ EXERCISE 3.1 ▬▬▬▬▬▬▬▬▬▬▬▬▬▬▬▬▬▬▬▬▬▬▬▬▬▬▬▬▬▬▬▬▬▬

In Exercises 1–8, find the prime factorization of each number.

1.	6	**2.**	10	**3.**	135	**4.**	98
5.	128	**6.**	357	**7.**	325	**8.**	288

In Exercises 9–16, find the greatest common factor of each set of quantities.

9. 36, 48

10. 45, 75

11. 42, 36, 98

12. 16, 40, 60

13. $4a^2b, 8a^3c$

14. $6x^3y^2z, 9xyz^2$

15. $18x^4y^3z^2, -12xy^2z^3$

16. $6x^2y^3, 24xy^3, 40x^2y^2z^3$

In Exercises 17–46, factor each expression completely. If a polynomial is prime, so indicate.

17. $2x + 8$

18. $3y - 9$

19. $2x^2 - 6x$

20. $3y^3 + 6y^2$

21. $5xy + 10xy^2$

22. $7x^2 + 14x$

23. $15x^2y - 10x^2y^2$

24. $9x^3y^2 - 12x^2y$

25. $12r^2s^3t^4 + 15rt^6$

26. $13ab^2c^3 - 26a^3b^2c$

27. $24r^2s^3 - 12r^3s^2t + 6rst^2$

28. $18x^2y^2z^2 + 12xy^2z^2 - 24x^4y^4z^3$

29. $45x^{10}y^3 - 63x^7y^7 + 81x^{10}y^{10}$

30. $48u^6v^6 - 16u^4v^4 - 3u^6v^3$

31. $25x^3 - 14y^3 + 36x^3y^3$

32. $9m^4n^3p^2 + 18m^2n^3p^4 - 27m^3n^4p$

33. $24a^3b^5 + 32a^5b^3 - 64a^5b^5c^5$

34. $32a^4 + 9b^2 + 5a^4b^2$

35. $4(x + y) + t(x + y)$

36. $5(a - b) - t(a - b)$

37. $(a - b)r - (a - b)s$

38. $(x + y)u + (x + y)v$

39. $3(m + n + p) + x(m + n + p)$

40. $x^2(x - y - z) + y(x - y - z) - z(x - y - z)$

41. $(a + b)x - a(x + b)$

42. $a(x - b) + b(x + a)$

43. $(x + y)(x + y) + z(x + y)$

44. $(a - b)^2 + (a - b)$

45. $(u + v) - (u + v)^2$

46. $a(x - y) - (x - y)(x - y)$

In Exercises 47–56, factor out the negative of the greatest common factor.

47. $-x + y$

48. $-x^2 - x$

49. $-18a^2b - 12ab^2$

50. $-15y^3 + 25y^2$

51. $-63u^3v^6z^9 + 28u^2v^7z^2 - 21u^3v^3z^4$

52. $-56x^4y^3z^2 - 72x^3y^4z^5 + 80xy^2z^3$

53. $-a(x + y) + b(x + y)$

54. $-bx(a - b) - cx(a - b)$

55. $-32x^3(m - n + p) - 40x^2(m - n + p) + 16x^4(m - n + p)$

56. $-45a^2b^3(x + y - z) + 81a^3b^2(x + y - z) - 90a^4b^2(x + y - z)$

57. Factor x^2 from $x^{n+2} + x^{n+3}$.

58. Factor y^3 from $y^{n+3} + y^{n+5}$.

59. Factor y^n from $y^{n+2} - y^{n+3}$.

60. Factor x^n from $x^{n+3} - x^{n+5}$.

61. Factor x^{-2} from $x^4 - 5x^6$.

62. Factor y^{-4} from $7y^4 + y$.

63. Factor y^{-2n} from $y^{2n} + 1 + y^{-2n}$.

64. Factor x^{-3n} from $x^{6n} + x^{3n} + 1$.

In Exercises 65–68, use long division.

65. Show that $x + y$ is a factor of $x^4 - y^4$.

66. Show that $x - y$ is a factor of $x^5 - y^5$.

67. Show that $x + y$ is a factor of $x^7 + y^7$.

68. Show that $x^2 + y^2$ is a factor of $x^8 - y^8$.

*If the greatest common factor of several terms is 1, the terms are called **relatively prime**. In Exercises 69–76, tell whether the terms in each set are relatively prime.*

69. 14, 45

70. 24, 63, 112

71. 60, 28, 36

72. 55, 49, 78

73. $12x^2y, 5ab^3, 35x^2b^3$

74. $18uv, 25rs, 12rsuv$

75. $9(a - b), 16(a + b), 25(a + b + c)$

76. $44(x + y - z), 99(x - y + z), 121(x + y + z)$

3.2   THE DIFFERENCE OF TWO SQUARES; THE SUM AND DIFFERENCE OF TWO CUBES

There are special product formulas that lead to factoring formulas for the difference of two squares and the sum and difference of two cubes. We begin this section by using a special product formula discussed in Section 2.4 to factor the difference of two squares.

The Difference of Two Squares

If we multiply a binomial of the form $x + y$ by a binomial of the form $x - y$, we obtain another binomial:

1. $(x + y)(x - y) = x^2 - y^2$

The binomial $x^2 - y^2$ is called the **difference of two squares** because x^2 represents the square of x, y^2 represents the square of y, and the binomial $x^2 - y^2$ represents the difference of these squares. Because of the symmetric property of equality, Equation 1 can be written in reverse order to give a formula for factoring the difference of two squares.

> **Formula for Factoring the Difference of Two Squares.**
>
> $$x^2 - y^2 = (x + y)(x - y)$$

This formula points out that the difference of the squares of two quantities such as x and y always factors into the sum of these quantities multiplied by the difference of these quantities.

To factor $49x^2 - 16$, for example, we write $49x^2 - 16$ in the form $(7x)^2 - 4^2$ and apply the formula for factoring the difference of two squares:

$$49x^2 - 16 = (7x)^2 - 4^2$$
$$= (7x + 4)(7x - 4)$$

This result can be verified by multiplying $7x + 4$ by $7x - 4$ and obtaining the product $49x^2 - 16$:

$$(7x + 4)(7x - 4) = 49x^2 - 28x + 28x - 16$$
$$= 49x^2 - 16$$

We note that, if $49x^2 - 16$ is divided by $7x + 4$, the quotient is $7x - 4$, and, if $49x^2 - 16$ is divided by $7x - 4$, the quotient is $7x + 4$.

Expressions such as $(7x)^2 + 4^2$ that represent the sum of two squares cannot be factored by using integer coefficients only. Thus, the binomial $49x^2 + 16$ is a prime binomial.

Example 1   Factor $64x^4 - 25y^2$.

Solution   Because $64x^4$ is the square of $8x^2$, and $25y^2$ is the square of $5y$, the binomial represents the difference of two squares. Its two factors are the sum of $8x^2$ and $5y$, and the difference of $8x^2$ and $5y$.

$$64x^4 - 25y^2 = (8x^2)^2 - (5y)^2$$
$$= (8x^2 + 5y)(8x^2 - 5y)$$

Verify by multiplication that $(8x^2 + 5y)(8x^2 - 5y) = 64x^4 - 25y^2$.   ∎

Example 2   Factor $a^4 - 1$.

Solution   Because the binomial represents the difference of the squares of a^2 and 1, it factors into the sum of a^2 and 1, and the difference of a^2 and 1:

$$a^4 - 1 = (a^2)^2 - 1^2$$
$$= (a^2 + 1)(a^2 - 1)$$

The factor $a^2 + 1$ represents the sum of two squares and cannot be factored. However, the factor $a^2 - 1$ represents the difference of two squares and can be factored as $(a + 1)(a - 1)$. Thus,

$$a^4 - 1 = (a^2 + 1)(a^2 - 1)$$
$$= (a^2 + 1)(a + 1)(a - 1)$$ ∎

Example 3   Factor $(x + y)^4 - z^4$.

Solution   This expression represents the difference of two squares and can be factored as follows:

$$(x + y)^4 - z^4 = [(x + y)^2]^2 - (z^2)^2$$
$$= [(x + y)^2 + z^2][(x + y)^2 - z^2]$$

The factor $(x + y)^2 + z^2$ represents the sum of two squares and cannot be factored. However, the factor $(x + y)^2 - z^2$ represents the difference of two squares and can be factored as $(x + y + z)(x + y - z)$. Thus,

$$(x + y)^4 - z^4 = [(x + y)^2 + z^2][(x + y)^2 - z^2]$$
$$= [(x + y)^2 + z^2](x + y + z)(x + y - z)$$ ∎

If it is possible to factor out a common factor before factoring the difference of two squares, we shall always do so. The factoring process is easier if all common factors are factored out first.

Example 4   Factor $2x^4y - 32y$.

Solution   Proceed as follows:

$$2x^4y - 32y = 2y(x^4 - 16) \qquad \text{Factor out } 2y.$$
$$= 2y(x^2 + 4)(x^2 - 4) \qquad \text{Factor } x^4 - 16.$$
$$= 2y(x^2 + 4)(x + 2)(x - 2) \qquad \text{Factor } x^2 - 4.$$ ∎

The Sum and Difference of Two Cubes

Two other special product formulas are

$$(x + y)(x^2 - xy + y^2) = x^3 + y^3$$

and

$$(x - y)(x^2 + xy + y^2) = x^3 - y^3$$

To verify the first formula, we multiply $x^2 - xy + y^2$ by $x + y$:

$$(x + y)(x^2 - xy + y^2) = x \cdot x^2 - x \cdot xy + x \cdot y^2 + y \cdot x^2 - y \cdot xy + y \cdot y^2$$
$$= x^3 - x^2y + xy^2 + x^2y - xy^2 + y^3$$
$$= x^3 + y^3$$

Likewise, the second formula can be verified by multiplication.

If we use the symmetric property of equality to write the previous special product formulas in reverse order, we have formulas for factoring the **sum of two cubes** and the **difference of two cubes.**

Formulas for Factoring the Sum and Difference of Two Cubes.

$$x^3 + y^3 = (x + y)(x^2 - xy + y^2)$$
$$x^3 - y^3 = (x - y)(x^2 + xy + y^2)$$

Note that the first factor in the factorization of $x^3 + y^3$ is $x + y$; the second factor is x^2 *minus* xy plus y^2. The first factor in the factorization of $x^3 - y^3$ is $x - y$; the second factor is x^2 *plus* xy plus y^2.

Example 5   Factor $a^3 + 8$.

Solution   Use the formula $x^3 + y^3 = (x + y)(x^2 - xy + y^2)$ with a in place of x and 2 in place of y:

$$a^3 + 8 = a^3 + 2^3$$
$$= (a + 2)(a^2 - a2 + 2^2)$$
$$= (a + 2)(a^2 - 2a + 4)$$

Verify this result by multiplication.   ∎

Example 6   Factor $27a^3 - 64b^3$.

Solution   Use the formula $x^3 - y^3 = (x - y)(x^2 + xy + y^2)$ with $3a$ in place of x and $4b$ in place of y:

$$27a^3 - 64b^3 = (3a)^3 - (4b)^3$$
$$= (3a - 4b)[(3a)^2 + (3a)(4b) + (4b)^2]$$
$$= (3a - 4b)(9a^2 + 12ab + 16b^2)$$ ∎

Example 7   Factor $a^3 - (c + d)^3$.

Solution $a^3 - (c + d)^3 = [a - (c + d)][a^2 + a(c + d) + (c + d)^2]$
$$= (a - c - d)(a^2 + ac + ad + c^2 + 2cd + d^2) \qquad \blacksquare$$

Example 8   Factor $x^6 - 64$.

Solution   This expression is the difference of two squares and factors into the product of a sum and a difference:

$$x^6 - 64 = (x^3)^2 - 8^2$$
$$= (x^3 + 8)(x^3 - 8)$$

Each of these factors further, however, for one is the sum of two cubes and the other is the difference of two cubes:

$$x^6 - 64 = (x + 2)(x^2 - 2x + 4)(x - 2)(x^2 + 2x + 4) \qquad \blacksquare$$

Example 9   Factor $2a^5 + 128a^2$.

Solution   First factor out the common monomial factor of $2a^2$ to obtain

$$2a^5 + 128a^2 = 2a^2(a^3 + 64)$$

Then factor $a^3 + 64$ as the sum of two cubes to obtain

$$2a^5 + 128a^2 = 2a^2(a + 4)(a^2 - 4a + 16) \qquad \blacksquare$$

Example 10   Factor $16r^{6m} - 54t^{3n}$.

Solution   Proceed as follows:

$$16r^{6m} - 54t^{3n} = 2(8r^{6m} - 27t^{3n}) \qquad \text{Factor out a 2.}$$
$$= 2[(2r^{2m})^3 - (3t^n)^3] \qquad \text{Write } 8r^{6m} \text{ as } (2r^{2m})^3$$
$$\text{and write } 27t^{3n} \text{ as } (3t^n)^3.$$
$$= 2(2r^{2m} - 3t^n)(4r^{4m} + 6r^{2m}t^n + 9t^{2n}) \qquad \text{Factor } (2r^{2m})^3 - (3t^n)^3. \qquad \blacksquare$$

■ **EXERCISE 3.2**

In Exercises 1–20, factor each expression completely. If a polynomial is prime, so indicate.

1. $x^2 - 4$ **2.** $y^2 - 9$ **3.** $9y^2 - 64$ **4.** $16x^4 - 81y^2$

5. $x^2 + 25$ **6.** $144a^2 - b^4$ **7.** $625a^2 - 169b^4$ **8.** $4y^2 + 9z^4$

9. $81a^4 - 49b^2$ **10.** $64r^6 - 121s^2$ **11.** $36x^4y^2 - 49z^4$ **12.** $100a^2b^4c^6 - 225d^8$

13. $(x + y)^2 - z^2$ **14.** $a^2 - (b - c)^2$ **15.** $(a - b)^2 - (c + d)^2$ **16.** $(m + n)^4 - (p - q)^2$

17. $x^4 - y^4$ **18.** $16a^4 - 81b^4$ **19.** $256x^4y^4 - z^8$ **20.** $225a^4 - 196b^8c^{12}$

In Exercises 21–30, factor each expression completely. Factor out all common monomial factors first.

21. $2x^2 - 288$ **22.** $8x^2 - 72$ **23.** $2x^3 - 32x$ **24.** $3x^2 - 243$

25. $5x^3 - 125x$ **26.** $6x^4 - 216x^2$ **27.** $r^2s^2t^2 - t^2x^4y^2$ **28.** $16a^4b^3c^4 - 64a^2bc^6$

29. $2(c - d)x^2 - 18(c - d)$ **30.** $3a^3(x + y) - 27a(x + y)$

In Exercises 31–42, factor each expression completely.

31. $r^3 + s^3$ **32.** $t^3 - v^3$ **33.** $x^3 - 8y^3$ **34.** $27a^3 + b^3$

35. $64a^3 - 125b^6$ **36.** $8x^6 + 125y^3$ **37.** $27x^3y^6 + 216z^9$ **38.** $1000a^6 - 343b^3c^6$

39. $27a^3 + (x + y)^6$ **40.** $64(x - y)^3 - 125z^6$ **41.** $x^6 + y^6$ **42.** $x^9 + y^9$

In Exercises 43–52, factor each expression completely. Factor out all common monomials first.

43. $5x^3 + 625$ **44.** $2x^3 - 128$ **45.** $4x^5 - 256x^2$ **46.** $2x^6 + 54x^3$

47. $128u^2v^3 - 2t^3u^2$ **48.** $56rs^2t^3 + 7rs^2v^6$ **49.** $(a+b)x^3 + 27(a+b)$ **50.** $rs^2(c - d)^3 + 27rs^2$

51. $6(a + b)^3 - 6z^3$ **52.** $18(x - y)^3 + 144(c - d)^3$

In Exercises 53–62, factor each expression completely. Assume that m and n are natural numbers.

53. $x^{2m} - y^{4n}$ **54.** $a^{4m} - b^{8n}$ **55.** $100a^{4m} - 81b^{2n}$ **56.** $25x^{8m} - 36y^{4n}$

57. $x^{3n} - 8$ **58.** $a^{3m} + 64$ **59.** $a^{3m} + b^{3n}$ **60.** $x^{6m} - y^{3n}$

61. $2x^{6m} + 16y^{3m}$ **62.** $24 + 3c^{3n}$

3.3   FACTORING TRINOMIALS

Many trinomials can be factored by using two special product formulas discussed in Section 2.4.

1. $(x + y)(x + y) = x^2 + 2xy + y^2$

2. $(x - y)(x - y) = x^2 - 2xy + y^2$

Recall that $x^2 + 2xy + y^2$ and $x^2 - 2xy + y^2$ are called **perfect square trinomials**. To factor the perfect square trinomial $z^2 + 6z + 9$, for example, we note that the trinomial can be written in the form $z^2 + 2(3)z + 3^2$. If $x = z$ and $y = 3$, this form matches the right-hand side of Formula 1. Thus, $z^2 + 6z + 9$ factors as

$$z^2 + 6z + 9 = z^2 + 2(3)z + 3^2$$
$$= (z + 3)(z + 3)$$

This result can be verified by multiplication:

$$(z + 3)(z + 3) = z^2 + 3z + 3z + 9$$
$$= z^2 + 6z + 9$$

Likewise, the perfect square trinomial $a^2 - 4ab + 4b^2$ can be written in the form

$$a^2 - 2(2b)a + (2b)^2$$

If $x = a$ and $y = 2b$, this form matches the right-hand side of Formula 2. Thus, $a^2 - 4ab + 4b^2$ factors as

$$a^2 - 4ab + 4b^2 = a^2 - 2(2b)a + (2b)^2$$
$$= (a - 2b)(a - 2b)$$

This result also can be verified by multiplication.

Many second-degree trinomials cannot be factored by using special product formulas. We begin our discussion of these **general trinomials** by considering trinomials whose lead coefficient (the coefficient of the squared term) is 1.

Factoring Trinomials with a Lead Coefficient of 1

To develop a strategy for factoring trinomials with a lead coefficient of 1, we recall that multiplying two binomials requires that each term of one binomial be multiplied by each term of the other binomial. For example,

$$(x + 3)(x + 1) = x^2 + x + 3x + 3$$
$$= x^2 + 4x + 3$$

In general, multiplying two binomials of the form $x + a$ and $x + b$ gives the following result:

$$(x + a)(x + b) = x^2 + bx + ax + ab$$
$$= x^2 + (b + a)x + ab \qquad \text{Factor out } x.$$

In the product, the coefficient of x^2 is 1, the coefficient of x is the sum of a and b, and the constant term is the product of a and b.

To factor the trinomial $x^2 + 7x + 12$, for example, we must find two binomials $x + a$ and $x + b$ such that

$$x^2 + 7x + 12 = (x + a)(x + b)$$

where the product of a and b is 12, and the sum of a and b is 7.

$$ab = 12 \qquad \text{and} \qquad a + b = 7$$

To find such numbers a and b, we list the possible factorizations of 12:

The one to choose

$$12(1) \qquad 6(2) \qquad 4(3) \qquad -12(-1) \qquad -6(-2) \qquad -4(-3)$$

Only in the factorization 4(3) do the factors have a sum of 7. Hence, $a = 4$, $b = 3$, and

$$x^2 + 7x + 12 = (x + a)(x + b)$$
$$\mathbf{3.} \quad x^2 + 7x + 12 = (x + 4)(x + 3)$$

This factorization can be verified by multiplying $x + 4$ by $x + 3$ and observing that the product is $x^2 + 7x + 12$:

$$(x + 4)(x + 3) = x^2 + 3x + 4x + 12$$
$$= x^2 + 7x + 12$$

Because of the commutative property of multiplication, the order of the factors in Equation 3 is not important. Equation 3 can also be written as

$$x^2 + 7x + 12 = (x + 3)(x + 4)$$

In general, to factor trinomials with lead coefficients of 1, we follow these steps:

---

1. Write the trinomial in descending powers of one variable.
2. List the factorizations of the third term of the trinomial.
3. Pick the factorization in which the sum of the factors is the coefficient of the middle term.

Example 1   Factor $x^2 - 6x + 8$.

Solution   Because this trinomial is already written in descending powers of x, you can proceed to step 2 and list the possible factorizations of the third term, 8.

The one to choose

$$8(1) \qquad 4(2) \qquad -8(-1) \qquad -4(-2)$$

The factorization in which the sum of the factors is -6 (the coefficient of the middle term of $x^2 - 6x + 8$) is $-4(-2)$. Hence, $a = -4$, $b = -2$, and

$$x^2 - 6x + 8 = (x + a)(x + b)$$
$$= (x - 4)(x - 2)$$

This result can be verified by multiplication:

$$(x - 4)(x - 2) = x^2 - 2x - 4x + 8$$
$$= x^2 - 6x + 8$$

Example 2   Factor $-x + x^2 - 12$.

Solution   Begin by writing the trinomial in descending powers of x:

$$-x + x^2 - 12 = x^2 - x - 12$$

Because the coefficient of the first term is 1, you can proceed to step 2 and list the possible factorizations of the third term:

The one to choose
$$\downarrow$$

$$12(-1) \quad 6(-2) \quad 4(-3) \quad 1(-12) \quad 2(-6) \quad 3(-4)$$

The factorization in which the sum of the factors is -1 (the coefficient of the middle term of $x^2 - x - 12$) is $3(-4)$. Hence, $a = 3$, $b = -4$, and

$$-x + x^2 - 12 = (x + a)(x + b)$$
$$= (x + 3)(x - 4)$$

Verify the result by multiplication. ∎

Example 3 Factor $30x - 4xy - 2xy^2$.

Solution Begin by writing the trinomial in descending powers of y:

$$30x - 4xy - 2xy^2 = -2xy^2 - 4xy + 30x$$

Each term in this trinomial shares a common monomial factor of $-2x$, which can be factored out:

$$30x - 4xy - 2xy^2 = -2x(y^2 + 2y - 15)$$

Because the lead coefficient of $y^2 + 2y - 15$ is 1, you can find the factorization of this trinomial by finding two factors of -15 whose sum is 2:

The one to choose
$$\downarrow$$

$$15(-1) \quad 5(-3) \quad 1(-15) \quad 3(-5)$$

Thus,

$$30x - 4xy - 2xy^2 = -2x(y^2 + 2y - 15)$$
$$= -2x(y + 5)(y - 3)$$

It is important to write the factor $-2x$ in each factorization. Verify this result by multiplication. ∎

Factoring Trinomials with a Lead Coefficient other than 1

There are more combinations of factors to consider when factoring trinomials with a lead coefficient other than 1. To factor $5x^2 + 7x + 2$, for example, we must find two binomials of the form $ax + b$ and $cx + d$ such that

$$5x^2 + 7x + 2 = (ax + b)(cx + d)$$

Because the first term of the trinomial $5x^2 + 7x + 2$ is $5x^2$, the first terms of the binomial factors must be $5x$ and x:

$$5x^2 + 7x + 2 = (5x + b)(x + d)$$

Because the product of the last terms must be 2, and because the sum of the products of the outer and inner terms must be $7x$, we must find two numbers whose product is 2 that will give a middle term of $7x$:

$$5x^2 + 7x + 2 = (5x + b)(x + d)$$
$$O + I = 7x$$

Because both $2(1)$ and $(-2)(-1)$ give a product of 2, there are four possible combinations to consider:

$$(5x + 2)(x + 1) \qquad (5x - 2)(x - 1)$$
$$(5x + 1)(x + 2) \qquad (5x - 1)(x - 2)$$

Of these possibilities, only the first one gives the proper middle term of $7x$. Thus,

$$5x^2 + 7x + 2 = (5x + 2)(x + 1)$$

Verify this result by multiplication:

$$(5x + 2)(x + 1) = 5x^2 + 5x + 2x + 2$$
$$= 5x^2 + 7x + 2$$

Example 4 Factor $3a^2 - 4a - 4$.

Solution Because the first term of the trinomial is $3a^2$, the first terms of the binomial factors must be $3a$ and a:

$$3a^2 - 4a - 4 = (3a + ?)(a + ?)$$

The product of the last terms must be -4, and the sum of the products of the outer terms and the inner terms must be $-4a$:

$$3a^2 - 4a - 4 = (3a + ?)(a + ?)$$
$$O + I = -4a$$

Because $1(-4)$, $-1(4)$, and $-2(2)$ all give a product of -4, there are six possible combinations to consider:

$$(3a + 1)(a - 4) \qquad (3a - 4)(a + 1)$$
$$(3a - 1)(a + 4) \qquad (3a + 4)(a - 1)$$
$$(3a - 2)(a + 2) \qquad (3a + 2)(a - 2)$$

Of these possibilities, only the last gives the required middle term of $-4a$. Thus,

$$3a^2 - 4a - 4 = (3a + 2)(a - 2)$$

Verify this result by multiplication.   ■

It is not easy to give specific rules for factoring general trinomials because some guesswork is often necessary. However, the following hints are helpful:

To factor a general trinomial, follow these steps:
1. Write the trinomial in descending powers of one variable.
2. Factor out any greatest common factor (including -1 if that is necessary to make the coefficient of the first term positive).
3. When the sign of the first term of a trinomial is $+$ and the sign of the third term is $+$, the signs between the terms of each binomial factor are the same as the sign of the middle term of the trinomial. When the sign of the first term is $+$ and the sign of the third term is $-$, one of the signs between the terms of the binomial factors is $+$ and one is $-$.
4. Mentally try various combinations of first terms and last terms until you find one that works. If you exhaust all the possibilities, the trinomial does not factor using only integer coefficients.
5. Check the factorization by multiplication.

Example 5   Factor $24y + 10xy - 6x^2y$.

Solution   Begin by writing the trinomial in descending powers of x and factoring out the common factor of $-2y$:

$$24y + 10xy - 6x^2y = -6x^2y + 10xy + 24y$$
$$= -2y(3x^2 - 5x - 12)$$

Because the sign of the third term of $3x^2 - 5x - 12$ is $-$, the signs between the binomial factors will be opposite. Because the first term is $3x^2$, the first terms of the binomial factors must be $3x$ and x:

$$24y + 10xy - 6x^2y = -2y(\overset{\overbrace{\qquad\quad}^{3x^2}}{3x}\quad)(x\quad)$$

The product of the last terms must be -12, and the sum of the outer terms and the inner terms must be $-5x$:

$$24y + 10xy - 6x^2y = -2y(\underset{O + I = -5x}{\overset{\overbrace{\qquad\qquad\quad}^{-12}}{3x\quad ?)(x\quad ?}})$$

Because $1(-12)$, $2(-6)$, $3(-4)$, $12(-1)$, $6(-2)$, and $4(-3)$ all give a product of -12, there are 12 possible combinations to consider:

$$(3x + 1)(x - 12) \qquad (3x - 12)(x + 1)$$
$$(3x + 2)(x - 6) \qquad (3x - 6)(x + 2)$$
$$(3x + 3)(x - 4) \qquad (3x - 4)(x + 3)$$
$$(3x + 12)(x - 1) \qquad (3x - 1)(x + 12)$$
$$(3x + 6)(x - 2) \qquad (3x - 2)(x + 6)$$

The one to choose $\longrightarrow (3x + 4)(x - 3) \qquad (3x - 3)(x + 4)$

After mentally trying these combinations, you will find that only $(3x + 4)(x - 3)$ gives the proper middle term of $-5x$. Thus,

$$24y + 10xy - 6x^2y = -2y(3x^2 - 5x - 12)$$
$$= -2y(3x + 4)(x - 3)$$

Verify this result by multiplication. ∎

Example 6 Factor $6y + 13x^2y + 6x^4y$.

Solution Write the trinomial in descending powers of x and factor out the common factor of y to obtain

$$6y + 13x^2y + 6x^4y = 6x^4y + 13x^2y + 6y$$
$$= y(6x^4 + 13x^2 + 6)$$

Because the coefficients of the first and last terms of the trinomial $6x^4 + 13x^2 + 6$ are positive, the signs between the terms of each binomial will be $+$. The first term of the trinomial is $6x^4$, so the first terms of the binomial factors must be either $2x^2$ and $3x^2$, or perhaps x^2 and $6x^2$. Because the product of the last terms of the binomial factors must be 6, you must find two numbers whose product is 6 that will lead to a middle term of $13x^2$. After mentally trying some combinations, you will find the one that works.

$$6y + 13x^2y + 6x^4y = y(6x^4 + 13x^2 + 6)$$
$$= y(2x^2 + 3)(3x^2 + 2)$$

Verify this result by multiplication. ∎

Example 7 Factor the trinomial $(x + y)^2 + 7(x + y) + 12$.

Solution Note that the trinomial $(x + y)^2 + 7(x + y) + 12$ can be written as $z^2 + 7z + 12$, where $z = x + y$. The trinomial $z^2 + 7z + 12$ factors as $(z + 4)(z + 3)$. To find the factorization of $(x + y)^2 + 7(x + y) + 12$, substitute $x + y$ for z in the expression $(z + 4)(z + 3)$ to obtain

$$(x + y)^2 + 7(x + y) + 12 = (x + y + 4)(x + y + 3)$$ ∎

Example 8   Factor $x^{2n} + x^n - 2$.

Solution   Because the first term is x^{2n}, the first terms of the binomial factors must be x^n and x^n:

$$x^{2n} + x^n - 2 = (x^n \qquad)(x^n \qquad)$$

Because the third term of the trinomial is -2, the last terms of the binomial factors must have opposite signs, have a product of -2, and lead to a middle term of x^n. The only combination that works is

$$x^{2n} + x^n - 2 = (x^n + 2)(x^n - 1)$$

Verify this result by multiplication.   ∎

The Key Number Method

The **key number method** of factoring trinomials eliminates much of the guesswork associated with the trial-and-error method. We illustrate the key number method by considering a trinomial of the form $ax^2 + bx + c$, whose factorization will be of the form $(mx + n)(px + q)$.

To factor the trinomial $6x^2 + 7x - 3$, where $a = 6$, $b = 7$, and $c = -3$, we follow these steps:

1.   Find the product of a and c. This is the *key number*. In this example,

$$\begin{aligned} ac &= 6(-3) \\ &= -18 \end{aligned}$$

2.   Find two factors of the key number whose sum is b. In this example, $b = 7$ and the key number, -18, factors as

$$18(-1), \quad 9(-2), \quad 6(-3), \quad 3(-6), \quad 2(-9), \quad \text{or} \quad 1(-18)$$

The only pair of factors whose sum is 7 is $9(-2)$:

$$9 + (-2) = 7$$

3.   The numbers 9 and -2 will be the coefficients of the products of the outer and inner terms of the trinomial's factors:

$$(mx \quad n)(px \quad q)$$

To obtain the coefficient 9 of the outer product $9x$, we must use one of the following factorizations for values of m and q:

$$9(1), \quad 3(3), \quad \text{or} \quad 1(9)$$

But if $m = 9$, we cannot obtain a first term of $6x^2$, and if $q = 9$, we cannot obtain a last term of -3. Hence, the factorizations $9(1)$ and $1(9)$ can be eliminated, and the factors leading to the outer product must be $3(3)$:

Substitute 3 for m and 3 for q.

The other numbers are now determined. The number n must be -1 to give -3 as the product of the last terms, and the number p must be 2 to give $6x^2$ as the product of the first terms. Thus, the factorization of $6x^2 + 7x - 3$ is

$$(3x - 1)(2x + 3)$$

Check this result by multiplication.

Example 9   Use the key number method to factor $4x^2 - 16x + 15$.

Solution   In this example, $a = 4$, $b = -16$, and $c = 15$. The key number is $ac = 4(15) = 60$. Two factors of 60 that have a sum of -16 are -10 and -6:

$$(-10)(-6) = 60 \qquad \text{and} \qquad -10 + (-6) = b = -16$$

Thus, if the product of the outer terms is $-10x$, the product of the inner terms is $-6x$. (We could just as well have made the product of the outer terms $-6x$ and the product of the inner terms $-10x$.)

To obtain the coefficient -10 of the outer product, $-10x$, we must use one of the following factorizations for m and q:

$$10(-1), \quad 5(-2), \quad 2(-5), \quad \text{or} \quad 1(-10)$$

However, values of 10 or 5 for m cannot give a first term of $4x^2$, and a value of -10 for q cannot give a last term of 15. Thus, the factors leading to the outer product must be $2(-5)$:

Substitute 2 for m and -5 for q.

The other numbers are now determined. The number n must be -3 to give 15 as the product of the last terms, and the number p must be 2 to give $4x^2$ as the product of the first terms. Thus, the factorization of $4x^2 - 16x + 15$ is

$$(2x - 3)(2x - 5)$$

Check this result by multiplication.   ■

■ EXERCISE 3.3

In Exercises 1–10, use a special product formula to factor each perfect square trinomial.

1. $x^2 + 2x + 1$ **2.** $y^2 - 2y + 1$ **3.** $a^2 - 18a + 81$ **4.** $b^2 + 12b + 36$
5. $4y^2 + 4y + 1$ **6.** $9x^2 + 6x + 1$ **7.** $9b^2 - 12b + 4$ **8.** $4a^2 - 12a + 9$
9. $9z^2 + 24z + 16$ **10.** $16z^2 - 24z + 9$

In Exercises 11–22, factor each trinomial. If a trinomial is prime, so indicate.

11. $x^2 + 5x + 6$ **12.** $y^2 + 7y + 6$ **13.** $x^2 - 4x + 4$ **14.** $c^2 - 6c + 9$
15. $b^2 + 8b + 18$ **16.** $x^2 - 14x + 49$ **17.** $x^2 - x - 30$ **18.** $a^2 + 4a - 45$
19. $a^2 + 5a - 50$ **20.** $b^2 + 9b - 36$ **21.** $y^2 - 4y - 21$ **22.** $x^2 + 4x - 28$

In Exercises 23–34, factor each trinomial. Factor out all common monomials first. If the coefficient of the first term is negative, begin by factoring out -1.

23. $3x^2 + 12x - 63$ **24.** $2y^2 + 4y - 48$
25. $a^2b^2 - 13ab^2 + 22b^2$ **26.** $a^2b^2x^2 - 18a^2b^2x + 81a^2b^2$
27. $b^2x^2 - 12bx^2 + 35x^2$ **28.** $c^3x^2 + 11c^3x - 42c^3$ **29.** $-a^2 + 4a + 32$ **30.** $-x^2 - 2x + 15$
31. $-3x^2 + 15x - 18$ **32.** $-2y^2 - 16y + 40$ **33.** $-4x^2 + 4x + 80$ **34.** $-5a^2 + 40a - 75$

In Exercises 35–72, factor each trinomial. Factor out all common monomials first, including -1 if the first term is negative. If a trinomial is prime, so indicate.

35. $6y^2 + 7y + 2$ **36.** $6x^2 - 11x + 3$ **37.** $8a^2 + 6a - 9$ **38.** $15b^2 + 4b - 4$
39. $6x^2 - 5x - 4$ **40.** $18y^2 - 3y - 10$ **41.** $5x^2 + 4x + 1$ **42.** $6z^2 + 17z + 12$
43. $8x^2 - 10x + 3$ **44.** $4a^2 + 20a + 3$ **45.** $6z^2 + 7z - 20$ **46.** $7x^2 - 23x + 6$
47. $a^2 - 3ab - 4b^2$ **48.** $b^2 + 2bc - 80c^2$ **49.** $2y^2 + yt - 6t^2$ **50.** $3x^2 - 10xy - 8y^2$
51. $3x^3 - 10x^2 + 3x$ **52.** $6y^2 + 7y + 2$ **53.** $-3a^2 + ab + 2b^2$ **54.** $-2x^2 + 3xy + 5y^2$
55. $9t^2 + 3t - 2$ **56.** $3t^3 - 3t^2 + t$ **57.** $9x^2 - 12x + 4$ **58.** $4a^2 + 28a + 49$
59. $-4x^2 - 9 + 12x$ **60.** $6x + 4 + 9x^2$ **61.** $15x^2 + 2 - 13x$ **62.** $-90x^2 + 2 - 8x$
63. $5a^2 + 45b^2 - 30ab$ **64.** $x^2 + 324y^2 - 36xy$ **65.** $8x^2z + 6xyz + 9y^2z$ **66.** $x^3 - 60xy^2 + 7x^2y$
67. $15x^2 + 74x - 5$ **68.** $15x^2 - 7x - 30$ **69.** $21x^4 - 10x^3 - 16x^2$ **70.** $16x^3 - 50x^2 + 36x$
71. $6x^2y^2 - 17xyz + 12z^2$ **72.** $6u^2v^2 - uvz + 12z^2$

In Exercises 73–84, factor each trinomial.

73. $x^4 + 8x^2 + 15$ **74.** $x^4 + 11x^2 + 24$ **75.** $y^4 - 13y^2 + 30$ **76.** $y^4 - 13y^2 + 42$

77. $a^4 - 13a^2 + 36$ **78.** $b^4 - 17b^2 + 16$ **79.** $z^4 - z^2 - 12$ **80.** $c^4 - 8c^2 - 9$

81. $x^6 + 4x^3 + 3$ **82.** $a^6 + a^3 - 2$ **83.** $y^6 - 9y^3 + 8$ **84.** $x^6 + 9x^3 + 8$

In Exercises 85–94, factor each expression.

85. $(x + 1)^2 + 2(x + 1) + 1$ **86.** $(a + b)^2 - 2(a + b) + 1$

87. $(a + b)^2 - 2(a + b) - 24$ **88.** $(x - y)^2 + 3(x - y) - 10$

89. $6(x + y)^2 - 7(x + y) - 20$ **90.** $2(x - z)^2 + 9(x - z) + 4$

91. $5(x^2 - 4x + 4) - 4(x - 2) - 1$ **92.** $6(x^2 + 2x + 1) + 5(x + 1) + 1$

93. $(4x^2 - 8x + 4) + (-5x + 5) + 1$ **94.** $(5x^2 + 30x + 45) + (6x + 18) + 1$

In Exercises 95–102, factor each expression. Assume that n is a natural number.

95. $x^{2n} + 2x^n + 1$ **96.** $x^{4n} - 2x^{2n} + 1$ **97.** $2a^{6n} - 3a^{3n} - 2$ **98.** $b^{2n} - b^n - 6$

99. $x^{4n} + 2x^{2n}y^{2n} + y^{4n}$ **100.** $y^{6n} + 2y^{3n}z + z^2$ **101.** $6x^{2n} + 7x^n - 3$ **102.** $12y^{4n} + 10y^{2n} + 2$

3.4   FACTORING BY GROUPING

Suppose we wish to factor an expression such as

$$ac + ad + bc + bd$$

Although there is no factor common to all four terms, there is a common factor of a in $ac + ad$ and a common factor of b in $bc + bd$. If we factor out these common factors, we obtain

$$ac + ad + bc + bd = a(c + d) + b(c + d)$$

Because each term in the expression $a(c + d) + b(c + d)$ has a common factor of $c + d$, it is not completely factored. To write the expression in completely factored form, we factor out the common factor of $c + d$ and obtain

$$ac + ad + bc + bd = a(c + d) + b(c + d)$$
$$= (c + d)(a + b)$$

The grouping in this type of problem is not always unique. For example, if we write the expression $ac + ad + bc + bd$ in the form

$$ac + bc + ad + bd$$

and factor c from the first two terms and d from the last two terms, we obtain

$$ac + bc + ad + bd = c(a + b) + d(a + b)$$

We can then factor out the common factor of $a + b$ to obtain the same result:

$$ac + bc + ad + bd = (a + b)(c + d)$$

The method used in the previous examples is called **factoring by grouping**.

Example 1   Factor $3ax^2 + 3bx^2 + a + 5bx + 5ax + b$.

Solution   Although there is no factor common to all six terms, $3x^2$ can be factored out of the first two terms, and $5x$ can be factored out of the fourth and fifth terms. Hence,

$$3ax^2 + 3bx^2 + a + 5bx + 5ax + b = 3x^2(a + b) + a + 5x(b + a) + b$$

This result can be written in the form

$$= 3x^2(a + b) + 5x(a + b) + (a + b)$$

The binomial $a + b$ is common to all three terms of the preceding expression and can be factored out. Thus,

$$3ax^2 + 3bx^2 + a + 5bx + 5ax + b = (a + b)(3x^2 + 5x + 1)$$

Because $3x^2 + 5x + 1$ is a prime polynomial, the factorization is complete.   ■

To factor an expression completely, it is often necessary to factor more than once, as the following example illustrates.

Example 2   Factor $3x^3y - 4x^2y^2 - 6x^2y + 8xy^2$.

Solution   Begin by factoring out the common factor of xy:

$$3x^3y - 4x^2y^2 - 6x^2y + 8xy^2 = xy(3x^2 - 4xy - 6x + 8y)$$

It would be incorrect to stop here because the expression is not in completely factored form. You can factor $3x^2 - 4xy - 6x + 8y$ by grouping:

$$3x^3y - 4x^2y^2 - 6x^2y + 8xy^2$$
$$= xy(3x^2 - 4xy - 6x + 8y)$$
$$= xy[x(3x - 4y) - 2(3x - 4y)] \qquad \text{Factor } x \text{ from } 3x^2 - 4xy \text{ and}$$
$$\qquad\qquad\qquad\qquad\qquad\qquad\qquad -2 \text{ from } -6x + 8y.$$
$$= xy(3x - 4y)(x - 2) \qquad\qquad \text{Factor out } 3x - 4y.$$

Because no more factoring can be done, the factorization is complete.   ■

Example 3   Factor $x^3 + 5x^2 + 6x + x^2y + 5xy + 6y$.

Solution   Factor x from the first three terms and y from the last three and proceed as follows:

$$x^3 + 5x^2 + 6x + x^2y + 5xy + 6y$$
$$= x(x^2 + 5x + 6) + y(x^2 + 5x + 6)$$
$$= (x^2 + 5x + 6)(x + y) \qquad \text{Factor out } x^2 + 5x + 6.$$
$$= (x + 3)(x + 2)(x + y) \qquad \text{Factor } x^2 + 5x + 6.$$

    ■

Example 4   Factor $x^4 + 2x^3 + x^2 + x + 1$.

Solution   Factor x^2 from the first three terms and proceed as follows:

$$x^4 + 2x^3 + x^2 + x + 1 = x^2(x^2 + 2x + 1) + (x + 1)$$
$$= x^2(x + 1)(x + 1) + (x + 1) \qquad \text{Factor } x^2 + 2x + 1.$$
$$= (x + 1)[x^2(x + 1) + 1] \qquad \text{Factor out } x + 1.$$
$$= (x + 1)(x^3 + x^2 + 1)$$

◼

Example 5   Factor $x^2 + 12x + 36 - y^2$.

Solution   The first three terms factor as $(x + 6)(x + 6)$. Thus,

$$x^2 + 12x + 36 - y^2 = (x + 6)^2 - y^2$$

The expression $(x + 6)^2 - y^2$ is the difference of two squares and can be factored accordingly:

$$x^2 + 12x + 36 - y^2 = (x + 6)^2 - y^2$$
$$= [(x + 6) + y][(x + 6) - y]$$
$$= (x + 6 + y)(x + 6 - y)$$

◼

The next examples use the technique of adding and subtracting the same quantity from an expression to make it factor as the difference of two squares.

Example 6   Factor $x^4 + x^2 + 1$.

Solution   This trinomial cannot be factored as the product of two binomials because there is no combination that will give a middle term of $1x^2$. However, if the middle term were $2x^2$, the factorization would be easy. You can make the middle term $2x^2$ by adding x^2 to the trinomial. To ensure that adding x^2 does not change the value of the trinomial, you must also subtract x^2:

$$x^4 + x^2 + 1 = x^4 + x^2 + x^2 + 1 - x^2$$
$$= x^4 + 2x^2 + 1 - x^2$$

Because the first three terms of the previous expression factor as $(x^2 + 1)(x^2 + 1)$, you can proceed as follows:

$$x^4 + x^2 + 1 = x^4 + 2x^2 + 1 - x^2$$
$$= (x^2 + 1)^2 - x^2$$
$$= [(x^2 + 1) + x][(x^2 + 1) - x] \quad \text{Factor the difference of two squares.}$$
$$= (x^2 + x + 1)(x^2 - x + 1) \quad \text{Write each factor in descending powers of } x.$$

◼

Example 7   Factor $x^6 - 1$.

Solution   The expression $x^6 - 1$ is the difference of two cubes and factors accordingly:

$$x^6 - 1 = (x^2)^3 - 1^3 = (x^2 - 1)(x^4 + x^2 + 1)$$

The first factor is the difference of two squares and can be factored as follows:

$$x^6 - 1 = (x + 1)(x - 1)(x^4 + x^2 + 1)$$

The last factor is the trinomial of Example 6 and can be factored as

$$(x^2 + x + 1)(x^2 - x + 1)$$

Thus,

$$x^6 - 1 = (x + 1)(x - 1)(x^2 + x + 1)(x^2 - x + 1)$$

Note that this problem could have been factored as the difference of two squares first, and then as the sum and difference of two cubes:

$$x^6 - 1 = (x^3 + 1)(x^3 - 1)$$
$$= (x + 1)(x^2 - x + 1)(x - 1)(x^2 + x + 1) \qquad \blacksquare$$

The method of factoring by grouping can be used to help factor trinomials of the form $ax^2 + bx + c$. For example, to factor the trinomial $6x^2 + 7x - 3$, we proceed as follows:

1. First determine the product ac: $6(-3) = -18$. This number is often called the **key number**.
2. Find two factors of the key number -18 whose sum is $b = 7$:

$$9(-2) = -18 \qquad \text{and} \qquad 9 + (-2) = 7$$

3. Use the factors 9 and -2 as coefficients of the terms to be placed between $6x^2$ and -3:

$$6x^2 + 7x - 3 = 6x^2 + 9x - 2x - 3$$

4. Factor by grouping:

$$6x^2 + 9x - 2x - 3 = 3x(2x + 3) - (2x + 3)$$
$$= (2x + 3)(3x - 1) \qquad \text{Factor out } 2x + 3.$$

We can verify this factorization by multiplication.

▬ EXERCISE 3.4 ▬▬▬▬▬▬▬▬▬▬

In Exercises 1–24, factor each expression by grouping. You may have to rearrange some terms first.

1. $ax + bx + ay + by$
2. $ar - br + as - bs$
3. $x^2 + yx + 2x + 2y$
4. $2c + 2d - cd - d^2$
5. $3c - cd + 3d - c^2$
6. $x^2 + 4y - xy - 4x$
7. $a^2 - 4b + ab - 4a$
8. $7u + v^2 - 7v - uv$
9. $ax + bx - a - b$
10. $x^2y - ax - xy + a$
11. $x^2 + xy + xz + xy + y^2 + zy$
12. $ab - b^2 - bc + ac - bc - c^2$

13. $x^2y + xy^2 + 2xyz + xy^2 + y^3 + 2y^2z$
14. $a^3 - 2a^2b + a^2c - a^2b + 2ab^2 - abc$
15. $2n^4p - 2n^2 - n^3p^2 + np + 2mn^3p - 2mn$
16. $a^2c^3 + ac^2 + a^3c^2 - 2a^2bc^2 - 2bc^2 + c^3$
17. $x^4 - 2x^3 + x^2 + x - 1$
18. $a^4 + 4a^3 + 4a^2 + 2a + 4$
19. $a^2 - b^2 + 2a - 2b$
20. $m^2 + 3m + 3n - n^2$
21. $y + 4 + ay + y + 4a + 4$
22. $bx + 2x - 3b - 6 + x - 3$
23. $x^2 - 1 + 3x^5 + 6x^4 + 3x^3$
24. $3y^2 + 6y + 3y^6 + 12y^5 + 12y^4$

In Exercises 25–34, factor a trinomial and then factor the difference of two squares. You may have to rearrange some terms first.

25. $x^2 + 4x + 4 - y^2$
26. $x^2 - 6x + 9 - 4y^2$
27. $x^2 + 2x + 1 - 9z^2$
28. $x^2 + 10x + 25 - 16z^2$
29. $4a^2 - 4ab + b^2 - c^2$
30. $a^2 - 6ab + 9b^2 - 25c^2$
31. $a^2 - b^2 + 8a + 16$
32. $a^2 + 14a - 25b^2 + 49$
33. $4x^2 - z^2 + 4xy + y^2$
34. $x^2 - 4xy - 4z^2 + 4y^2$

In Exercises 35–42, add and subtract some term and then factor as the difference of two squares.

35. $x^4 + 5x^2 + 9$
36. $x^4 + 7x^2 + 16$
37. $4a^4 + 1 + 3a^2$
38. $a^4 + 6a^2 + 25$
39. $a^4 + 4b^4$
40. $4a^4 + b^4$
41. $x^4 - 7x^2 + 1$
42. $x^4 - 12x^2 + 4$
43. Factor $x^6 - 64$
44. Factor $x^6 - y^6$.

In Exercises 45–50, use factoring by grouping to help factor each trinomial.

45. $a^2 - 17a + 16$
46. $b^2 - 4b - 21$
47. $2u^2 + 5u + 3$
48. $6y^2 + 5y - 6$
49. $20r^2 - 7rs - 6s^2$
50. $20u^2 + 19uv + 3v^2$

3.5 SUMMARY OF FACTORING TECHNIQUES

In this section we discuss ways to approach a randomly chosen factoring problem. For example, suppose we wish to factor the trinomial

$$x^2y^2z^3 + 7xy^2z^3 + 6y^2z^3$$

We begin by attempting to identify the problem type. The first type to look for is **factoring out a common monomial**. Because the trinomial has a common monomial factor of y^2z^3, we factor it out:

$$x^2y^2z^3 + 7xy^2z^3 + 6y^2z^3 = y^2z^3(x^2 + 7x + 6)$$

We then note that $x^2 + 7x + 6$ is a trinomial that can be factored as $(x + 6)(x + 1)$. Thus,

$$x^2y^2z^3 + 7xy^2z^3 + 6y^2z^3 = y^2z^3(x^2 + 7x + 6)$$
$$= y^2z^3(x + 6)(x + 1)$$

To identify the type of a factoring problem, follow these steps:

---

1. Factor out all common monomial factors.
2. If an expression has two terms, check to see if the problem type is
   a. the **difference of two squares**: $x^2 - y^2 = (x + y)(x - y)$
   b. the **sum of two cubes**: $x^3 + y^3 = (x + y)(x^2 - xy + y^2)$
   c. the **difference of two cubes**: $x^3 - y^3 = (x - y)(x^2 + xy + y^2)$
3. If an expression has three terms, check to see if the problem type is a **perfect trinomial square**: $x^2 + 2xy + y^2 = (x + y)(x + y)$
   $$x^2 - 2xy + y^2 = (x - y)(x - y)$$
   If the trinomial is not a perfect trinomial square, attempt to factor the trinomial as a **general trinomial**.
4. If an expression has four or more terms, try to **factor the expression by grouping**.
5. Continue until each individual factor is prime.
6. Check the results by multiplying.

Example 1   Factor $48a^4c^3 - 3b^4c^3$.

Solution   Begin by factoring out the common monomial factor of $3c^3$:

$$48a^4c^3 - 3b^4c^3 = 3c^3(16a^4 - b^4)$$

Because the expression $16a^4 - b^4$ has two terms, check to see if it is the difference of two squares, which it is. As the difference of two squares, it factors as $(4a^2 + b^2)(4a^2 - b^2)$. Thus,

$$48a^4c^3 - 3b^4c^3 = 3c^3(16a^4 - b^4)$$
$$= 3c^3(4a^2 + b^2)(4a^2 - b^2)$$

The binomial $4a^2 + b^2$ is the sum of two squares and cannot be factored. However, the binomial $4a^2 - b^2$ is again the difference of two squares and factors as $(2a + b)(2a - b)$. Thus,

$$48a^4c^3 - 3b^4c^3 = 3c^3(16a^4 - b^4)$$
$$= 3c^3(4a^2 + b^2)(4a^2 - b^2)$$
$$= 3c^3(4a^2 + b^2)(2a + b)(2a - b)$$

Because each of the individual factors is prime, the given expression is in completely factored form.   ∎

Example 2   Factor $x^5y + x^2y^4 - x^3y^3 - y^6$.

Solution   Begin by factoring out the common monomial factor of y:

$$x^5y + x^2y^4 - x^3y^3 - y^6 = y(x^5 + x^2y^3 - x^3y^2 - y^5)$$

Because the expression $x^5 + x^2y^3 - x^3y^2 - y^5$ has four terms, try factoring by grouping to obtain

$$\begin{aligned} x^5y + x^2y^4 - x^3y^3 - y^6 &= y(x^5 + x^2y^3 - x^3y^2 - y^5) \\ &= y[x^2(x^3 + y^3) - y^2(x^3 + y^3)] \\ &= y(x^3 + y^3)(x^2 - y^2) \qquad \text{Factor out } x^3 + y^3. \end{aligned}$$

Finally, factor $x^3 + y^3$ (the sum of two cubes) and $x^2 - y^2$ (the difference of two squares) to obtain

$$x^5y + x^2y^4 - x^3y^3 - y^6 = y(x + y)(x^2 - xy + y^2)(x + y)(x - y)$$

Because each of the individual factors is prime, the given expression is in completely factored form.   ■

■ EXERCISE 3.5

In Exercises 1–44, factor each polynomial completely. If the polynomial is prime, so indicate.

1.	$x^2 + 8x + 16$			**2.**	$20 + 11x - 3x^2$		
3.	$8x^3y^3 - 27$			**4.**	$3x^2y + 6xy^2 - 12xy$		
5.	$xy - ty + xs - ts$	**6.**	$bc + b + cd + d$	**7.**	$25x^2 - 16y^2$	**8.**	$27x^9 - y^3$
9.	$12x^2 + 52x + 35$	**10.**	$12x^2 + 14x - 6$	**11.**	$6x^2 - 14x + 8$	**12.**	$12x^2 - 12$
13.	$56x^2 - 15x + 1$	**14.**	$7x^2 - 57x + 8$	**15.**	$4x^2y^2 + 4xy^2 + y^2$	**16.**	$100z^2 - 81t^2$
17.	$x^3 + (a^2y)^3$	**18.**	$4x^2y^2z^2 - 26x^2y^2z^3$	**19.**	$2x^3 - 54$	**20.**	$4(xy)^3 + 256$
21.	$ae + bf + af + be$			**22.**	$a^2x^2 + b^2y^2 + b^2x^2 + a^2y^2$		
23.	$2(x + y)^2 + (x + y) - 3$			**24.**	$(x - y)^3 + 125$		
25.	$625x^4 - 256y^4$			**26.**	$2(a - b)^2 + 5(a - b) + 3$		
27.	$36x^4 - 36$	**28.**	$6x^2 - 63 - 13x$	**29.**	$2x^6 + 2y^6$	**30.**	$x^4 - x^4y^4$
31.	$a^4 - 13a^2 + 36$			**32.**	$x^4 - 17x^2 + 16$		
33.	$x^2 + 6x + 9 - y^2$			**34.**	$x^2 + 10x + 25 - y^8$		
35.	$4x^2 + 4x + 1 - 4y^2$	**36.**	$9x^2 - 6x + 1 - 25y^2$	**37.**	$z^4 + 7z^2 + 16$	**38.**	$x^4 + 9x^2 + 25$
39.	$x^5 + x^2 - x^3 - 1$			**40.**	$x^5 - x^2 - 4x^3 + 4$		
41.	$x^5 - 9x^3 + 8x^2 - 72$			**42.**	$x^5 - 4x^3 - 8x^2 + 32$		
43.	$2x^5z - 2x^2y^3z - 2x^3y^2z + 2y^5z$			**44.**	$x^2y^3 - 4x^2y - 9y^3 + 36y$		

CHAPTER SUMMARY

Key Words

completely factored form	(3.1)	*irreducible polynomial*	(3.1)
difference of two cubes	(3.2)	*prime-factored form*	(3.1)
difference of two squares	(3.2)	*prime polynomial*	(3.1)
greatest common divisor	(3.1)	*relatively prime terms*	(3.1)
greatest common factor	(3.1)	*sum of two cubes*	(3.2)

Key Ideas

(3.1)   Always factor out all common monomial factors as your first step in a factoring problem.

Use the distributive property to factor out common monomial factors.

(3.2)   The **difference of the squares** of two quantities factors into the product of the sum and difference of those quantities: $x^2 - y^2 = (x + y)(x - y)$.

The **sum of two cubes** factors as $x^3 + y^3 = (x + y)(x^2 - xy + y^2)$.

The **difference of two cubes** factors as $x^3 - y^3 = (x - y)(x^2 + xy + y^2)$.

(3.3)   Use the special product formulas

$$x^2 + 2xy + y^2 = (x + y)(x + y)$$
$$x^2 - 2xy + y^2 = (x - y)(x - y)$$

to factor perfect trinomial squares. If a polynomial is not a perfect trinomial square, use the methods for factoring general trinomials.

(3.4)   If an expression has four or more terms, try to factor the expression by grouping.

If a trinomial cannot be factored directly, consider adding to and subtracting from the trinomial some appropriate quantity to form the difference of two squares.

REVIEW EXERCISES

In Review Exercises 1–50, factor each polynomial completely. Factor out all common monomials first, including -1 if the coefficient of the first term is negative. If a polynomial is prime, so indicate.

1. $4x + 8$ **2.** $3x^2 - 6x$ **3.** $5x^2y^3 - 10xy^2$ **4.** $7a^4b^2 + 49a^3b$

5. $-8x^2y^3z^4 - 12x^4y^3z^2$ **6.** $12a^6b^4c^2 + 15a^2b^4c^6$

7. $27x^3y^3z^3 + 81x^4y^5z^2 - 90x^2y^3z^7$ **8.** $-36a^5b^4c^2 + 60a^7b^5c^3 - 24a^2b^3c^7$

9. $5x^2(x + y)^3 - 15x^3(x + y)^4$ **10.** $-49a^3b^2(a - b)^4 + 63a^2b^4(a - b)^3$

11. $z^2 - 16$ **12.** $y^2 - 121$ **13.** $x^2y^4 - 64z^6$ **14.** $a^2b^2 + c^2$

15. $(x + z)^2 - t^2$ **16.** $(a + b)^4 - c^2$ **17.** $2x^4 - 98$ **18.** $3x^6 - 300x^2$

19. $x^3 + 343$ **20.** $a^3 - 125$ **21.** $8y^3 - 512$ **22.** $4x^3y + 108yz^3$

23. $y^2 + 21y + 20$ **24.** $z^2 - 11z + 30$ **25.** $-x^2 - 3x + 28$ **26.** $y^2 - 5y - 24$

27. $4a^2 - 5a + 1$ **28.** $3b^2 + 2b + 1$ **29.** $7x^2 + x + 2$ **30.** $-15x^2 + 14x + 8$

31. $y^3 + y^2 - 2y$ **32.** $2a^4 + 4a^3 - 6a^2$ **33.** $-3x^2 - 9x - 6$ **34.** $8x^2 - 4x - 24$

35. $15x^2 - 57xy - 12y^2$ **36.** $30x^2 + 65xy + 10y^2$

37. $24x^2 - 23xy - 12y^2$ **38.** $14x^2 + 13xy - 12y^2$ **39.** $x^4 + 13x^2 + 49$ **40.** $x^4 + 17x^2 + 81$

41. $xy + 2y + 4x + 8$ **42.** $ac + bc + 3a + 3b$

43. $x^4 + 4y + 4x^2 + x^2y$ **44.** $a^5 + b^2c + a^2c + a^3b^2$ **45.** $z^2 - 4 + zx - 2x$ **46.** $x^2 + 2x + 1 - p^2$

47. $x^2 + 4x + 4 - 4p^4$ **48.** $y^2 + 3y + 2 + 2x + xy$ **49.** $a^4 + 4a^2 + 16$ **50.** $z^4 + 64$

CHAPTER THREE TEST

1. Find the prime factorization of 228.

2. Which of the following polynomials is prime?

$$2x + 4 \qquad 3x^2 + 2x \qquad 3x^2 + 2$$

In Problems 3–6, factor each expression completely.

3. $3xy^2 + 6x^2y$

4. $12a^3b^2c - 3a^2b^2c^2 + 6abc^3$

5. $(u - v)r + (u - v)s$

6. $-2x^2y(r + s) + 4xy^2(r + s)$

7. Factor y^n from $x^2y^{n+2} + y^n$.

8. Factor b^n from $a^nb^n - ab^{-n}$.

In Problems 9–16, factor each expression completely.

9. $x^2 - 49$ **10.** $2x^2 - 32$ **11.** $4y^4 - 64$ **12.** $b^3 + 125$

13. $b^3 - 27$ **14.** $3u^3 - 24$ **15.** $9z^2 - 16t^4$ **16.** $8x^3 + 27y^3$

In Problems 17–26, factor each trinomial completely. Assume that n is a natural number.

17. $x^2 + 8x + 15$ **18.** $x^2 - 3x - 18$ **19.** $2a^2 - 5a - 12$ **20.** $6b^2 + b - 2$

21. $6u^2 + 9u - 6$ **22.** $20r^2 - 15r - 5$ **23.** $6x^2 + xy - y^2$ **24.** $5x^2 + 3xy - 2y^2$

25. $x^{2n} + 2x^n + 1$ **26.** $2x^{2n} + 3x^n - 2$

In Problems 27–32, factor each expression completely.

27. $ax - xy + ay - y^2$

28. $ax + ay + bx + by - cx - cy$

29. $x^2 + 6x + 9 - y^2$

30. $x^4 + 3x^2 + 4$

31. $x^4 - 20x^2 + 64$

32. $x^4 + 2x^3 - 2x - 1$

CUMULATIVE REVIEW EXERCISES (CHAPTERS 1–3)

In Exercises 1–2, classify each number as a natural number, a whole number, an integer, a rational number, an irrational number, a real number, a positive number, or a negative number. Each number will be in many classifications.

1. $\dfrac{5}{3}$

2. $-\sqrt{11}$

In Exercises 3–4, simplify each expression and classify each result as an even integer, an odd integer, a prime number, or a composite number. Each result will be in more than one category.

3. $\dfrac{27}{9}$

4. $\dfrac{15 + 5}{5}$

In Exercises 5–6, write each expression without using absolute value symbols, and simplify.

5. $-(|5| - |3|)$

6. $\dfrac{|-5| + |-3|}{-|4|}$

7. Write the terminating decimal 0.875 in fractional form.

8. Draw a number line and graph the prime numbers from 50 to 60.

In Exercises 9–12, indicate which property of equality or property of real numbers justifies each statement.

9. If $3 = x$ and $x = y$, then $3 = y$.

10. $3(x + y) = 3x + 3y$

11. $(a + b) + c = c + (a + b)$

12. $(ab)c = a(bc)$

13. Find the additive inverse of -5.

14. Find the multiplicative inverse of $-\frac{2}{3}$.

In Exercises 15–18, perform the indicated operations.

15. $2 + 4 \cdot 5$

16. $\dfrac{8 - 4}{2 - 4}$

17. $20 \div (-10 \div 2)$

18. $\dfrac{6 + 3(6 + 4)}{2(3 - 9)}$

In Exercises 19–22, perform the indicated operations. Simplify each answer, if possible.

19. $-\dfrac{5}{6} \cdot \dfrac{3}{20}$

20. $\dfrac{2}{3}\left(-\dfrac{4}{5} \div \dfrac{28}{35}\right)$

21. $-\left(\dfrac{1}{3} + \dfrac{3}{4}\right)\left(\dfrac{5}{3} + \dfrac{1}{2}\right)$

22. $\dfrac{1}{2} + \left(\dfrac{2}{3} - \dfrac{3}{4}\right)$

In Exercises 23–26, simplify each expression and write all answers without using negative exponents. Assume that no denominators are 0.

23. $(x^2y^3)^4$

24. $\dfrac{c^4 c^8}{(c^5)^2}$

25. $\left(-\dfrac{a^3 b^{-2}}{ab}\right)^{-1}$

26. $\left(\dfrac{-3a^3 b^{-2}}{6a^{-2}b^3}\right)^0$

27. Write the number 0.00000497 using scientific notation.

28. Write the number 9.32×10^8 using standard notation.

29. Classify the polynomial $3 + x + 4x^2$ as a monomial, binomial, or trinomial.

30. Give the degree of the polynomial $3 + x^2y + 17x^3y^4$.

31. If $P(x) = -3x^3 + x - 4$, find $P(-2)$.

32. Evaluate $\dfrac{x^2 - y^2}{2x + y}$ if $x = 2$ and $y = -3$.

In Exercises 33–38, perform the indicated operations and simplify.

33. $(3x^2 - 2x + 7) + (-2x^2 + 2x + 5) + (3x^2 - 4x + 2)$

34. $(-5x^2 + 3x + 4) - (-2x^2 + 3x + 7)$

35. $(3x + 4)(2x - 5)$

36. $(2x^n - 1)(x^n + 2)$

37. $(x^2 + 9x + 20) \div (x + 5)$

38. $(2x^2 + 4x - x^3 + 3) \div (x - 1)$

In Exercises 39–50, factor each expression.

39. $3r^2s^3 - 6rs^4$

40. $5(x - y) - a(x - y)$

41. $xu + yv + xv + yu$

42. $81x^4 - 16y^4$

43. $8x^3 - 27y^6$

44. $6x^2 + 5x - 6$

45. $9x^2 - 30x + 25$

46. $15x^2 - x - 6$

47. $27a^3 + 8b^3$

48. $6x^2 + x - 35$

49. $x^2 + 10x + 25 - y^4$

50. $x^4 + 8x^2 + 36$

CAREERS AND MATHEMATICS

COMPUTER PROGRAMMER

Computers process vast quantities of information rapidly and accurately when they are given programs to follow. Computer programmers write those programs, which logically list the steps the machine must follow to organize data, solve a problem, or do some other task.

Programmers work from descriptions (prepared by systems analysts) of the task that the computer system should perform. The programmer then writes the specific program from the description, by breaking down each step into a series of coded instructions using one of several possible computer languages.

Application programmers are usually business, engineering, or science oriented. System programmers maintain the **software** that controls the computer system. Because of their knowledge of operating systems, system programmers often help application programmers determine the source of problems that may occur with their programs.

Qualifications

Most programmers have taken special courses in computer programming or have degrees in computer or information science, mathematics, engineering, or the physical sciences. Graduate degrees are required for some jobs.

Some employers who use computers for business applications do not require college degrees, but prefer applicants who have had college courses in data processing.

Employers look for people who can think logically, are capable of exacting, analytical work, and who demonstrate ingenuity and imagination when solving problems.

Job outlook

The need for computer programmers will increase as business, government, schools, and scientific organizations develop new applications and require improvements in the software they already use. System programmers will be needed to develop and maintain the complex operating systems required by new, more powerful computer languages and by computer networking. Job prospects are excellent through the mid-1990s for college graduates who have had computer-related courses, and are especially good for those who have a major in computer science or a related field. Graduates of two-year programs in data-processing technologies also have good prospects, primarily in business applications.

Example application

The polynomial

$$3x^4 + 2x^3 + 5x^2 + 7x + 1$$

can be written as

$$3 \cdot x \cdot x \cdot x \cdot x + 2 \cdot x \cdot x \cdot x + 5 \cdot x \cdot x + 7 \cdot x + 1$$

to illustrate that it involves 10 multiplications and 4 additions. Multiplications are more time-consuming on a computer than are additions. Rewrite the polynomial to require fewer multiplications.

Solution   Factor the common factor of x from the first four terms of the given polynomial as follows.

$$3x^4 + 2x^3 + 5x^2 + 7x + 1$$
$$x(3x^3 + 2x^2 + 5x + 7) + 1$$

Now factor the common x from the first three terms of the polynomial appearing within the parentheses.

$$x[x(3x^2 + 2x + 5) + 7] + 1$$

Again, factor an x from the first two terms within the set of parentheses to get

$$x\{x[x(3x + 2) + 5] + 7\} + 1$$

To emphasize the number of multiplications, we rewrite the previous expression as

$$x \cdot \{x \cdot [x \cdot (3 \cdot x + 2) + 5] + 7\} + 1$$

Each colored dot represents one multiplication. There are now only four multiplications, although there are still four additions. If the polynomial were to be evaluated for many different values of x, the savings in computer time afforded by using the revised form of the polynomial would be substantial.

Exercises

1. Evaluate the polynomial of the example at $x = -1$. Do this twice—once by direct substitution into the original polynomial, and once again, by substitution into the revised form.
2. Write $5x^7 + 3x^4 + 9x + 2$ in revised form.
3. Evaluate the polynomial of Exercise 2 at $x = 1$, by using each method.
4. Evaluate the polynomial of Exercise 2 at $x = -2$, by using each method.

(Answers: **1.**   0   **2.** $x \cdot \{x \cdot [x \cdot x \cdot (5 \cdot x \cdot x \cdot x + 3)] + 9\} + 2$
3.   19   **4.** -608)

4 Equations and Inequalities

In this chapter we consider equations and inequalities, two of the most basic ideas in algebra. Equations and inequalities are used in almost every academic and vocational field, especially in science, business, economics, and electronics.

4.1 LINEAR EQUATIONS AND THEIR SOLUTIONS

An **equation** is a statement indicating that two mathematical expressions are equal. The equation $2 + 4 = 6$ is an example of a true equation, and $2 + 4 = 7$ is an example of a false equation. An equation such as

$$7x - 3 = 4$$

is either true or false depending on the value of x, called a **variable**. If $x = 1$, the equation is true because the value 1 satisfies the equation:

$$7(1) - 3 = 4$$
$$7 - 3 = 4$$
$$4 = 4$$

However, the equation is false for all other values of x.

The set of all numbers that are permissible replacements for a variable is called the **domain** of the variable. The numbers in the domain of the variable that satisfy the equation make up the **solution set** of the equation. The elements of the solution set are called **roots** or **solutions** of the equation. Finding the solution set or the roots of an equation is called **solving the equation**.

Example 1 Determine the domain of x and the solution set for the equation $2x + 4 = 10$.

Solution Because any number can be substituted for x, the domain of x is the set of real numbers:

The domain of $x = \{x : x \text{ is a real number.}\}$

The solution set of the equation is $\{3\}$, because 3 is the only real number that satisfies the equation:

$$2x + 4 = 10$$
$$2(3) + 4 \stackrel{?}{=} 10$$
$$6 + 4 \stackrel{?}{=} 10$$
$$10 = 10$$

∎

Example 2   Determine the domain of x and the solution set for the equation $\dfrac{8}{x} = 4$.

Solution   The domain of x is the set of all real numbers except 0. If 0 is substituted for x, the equation is neither true nor false. In fact, the equation is nonsense because division by 0 is undefined.

The domain of $x = \{x : x$ is a nonzero real number.$\}$

The solution set of the equation is $\{2\}$, because 2 is the only number that satisfies the equation:

$$\frac{8}{x} = 4$$

$$\frac{8}{2} \overset{?}{=} 4$$

$$4 = 4 \qquad\qquad\qquad\qquad ∎$$

When every number in the domain of the variable is a solution of the equation, we call the equation an **identity**. If only some of the numbers in the domain are solutions, the equation is called a **conditional equation**. If no numbers in the domain are solutions, the equation is called an **impossible equation** or a **contradiction**. For example, if x is a real number, then

$x + 2 = x + 2$ is an identity because every number in the domain of x is a solution.

$2x + 4 = 10$ is a conditional equation because only the number 3 in the domain of x is a solution.

$x + 2 = x + 1$ is an impossible equation because no number in the domain of x is a solution.

Solving Equations

The method used for finding the solution set of a conditional equation involves replacing the equation with simpler ones having the same solution set. Such equations are called **equivalent equations**.

Definition.   Two equations are called **equivalent equations** if they have the same solution set.

The process of replacing conditional equations with simpler but equivalent ones continues until a trivial equation appears with solutions that are obvious. This method requires the following properties of equality along with the properties of equality listed in Section 1.2.

The Addition and Subtraction Properties of Equality.   If any quantity is added to or subtracted from both sides of an equation, a new equation is formed that is equivalent to the original equation.

In symbols, if a, b, and c are real numbers and $a = b$, then

$$a + c = b + c \qquad \text{and} \qquad a - c = b - c$$

The Multiplication and Division Properties of Equality.   If both sides of an equation are multiplied or divided by the same nonzero constant, a new equation is formed that is equivalent to the original equation.

In symbols, if a, b, and c are real numbers, $a = b$, and $c \neq 0$, then

$$ac = bc \qquad \text{and} \qquad \frac{a}{c} = \frac{b}{c}$$

The easiest equations to solve are the **first-degree** or **linear equations**.

Definition.   A **linear equation in one variable**, say x, is any equation that can be written in the form

$$ax + c = 0 \qquad (a \text{ and } c \text{ are real numbers and } a \neq 0)$$

To solve the linear equation $2x + 8 = 0$, for example, we first add -8 to both sides of the equation and then combine terms:

$$2x + 8 = 0$$
$$2x + 8 + (-8) = 0 + (-8)$$
$$2x = -8$$

We then divide both sides of the equation by 2 and simplify:

$$\frac{2x}{2} = \frac{-8}{2}$$
$$x = -4$$

To show that -4 is a solution, we substitute -4 for x in the original equation and simplify:

$$2x + 8 = 0$$
$$2(-4) + 8 \overset{?}{=} 0$$
$$-8 + 8 \overset{?}{=} 0$$
$$0 = 0$$

Because -4 satisfies the equation, it is a solution. Furthermore, -4 is the only number that satisfies the equation. Thus, the solution set is $\{-4\}$. In general, every linear equation has exactly one solution.

The next examples involve linear equations.

Example 3   Solve the equation $3x + 6 = 24$.

Solution   You must find a number x that satisfies the equation. This goal is achieved by writing a series of simplified equations, all of which are equivalent to the original equation:

$$3x + 6 = 24$$
$$3x + 6 + (-6) = 24 + (-6) \qquad \text{Add } -6 \text{ to both sides.}$$
$$3x = 18 \qquad \text{Combine terms.}$$
$$\frac{3x}{3} = \frac{18}{3} \qquad \text{Divide both sides by 3.}$$
$$x = 6 \qquad \text{Simplify.}$$

To check this result, substitute 6 for x in the original equation and show that it satisfies the equation:

$$3x + 6 = 24$$
$$3(6) + 6 \overset{?}{=} 24$$
$$18 + 6 \overset{?}{=} 24$$
$$24 = 24$$

Because 6 satisfies the equation, it is a solution, and the solution set of the original equation is $\{6\}$.   ■

Example 4   Solve $3(2x - 1) = 2x + 9$ for x.

Solution   Use the distributive property to remove parentheses, and proceed as follows:

$$3(2x - 1) = 2x + 9$$
$$6x - 3 = 2x + 9 \qquad \text{Remove parentheses.}$$
$$6x - 3 + 3 = 2x + 9 + 3 \qquad \text{Add 3 to both sides.}$$
$$6x = 2x + 12 \qquad \text{Combine terms.}$$
$$6x + (-2x) = 2x + (-2x) + 12 \qquad \text{Add } -2x \text{ to both sides.}$$
$$4x = 12 \qquad \text{Combine terms.}$$
$$\frac{4x}{4} = \frac{12}{4} \qquad \text{Divide both sides by 4.}$$
$$x = 3 \qquad \text{Simplify.}$$

To check this result, substitute 3 for x in the original equation and verify that 3 makes the equation a true statement:

$$3(2x - 1) = 2x + 9$$
$$3(2 \cdot 3 - 1) \overset{?}{=} 2 \cdot 3 + 9$$
$$3(5) \overset{?}{=} 6 + 9$$
$$15 = 15$$

Because 3 satisfies the equation, it is a solution, and the solution set of the original equation is $\{3\}$. ∎

To solve a linear equation, it is helpful to follow these steps:

1. If the equation contains fractions, multiply both sides of the equation by a suitable number to eliminate the fractions.
2. Use the distributive property to remove all sets of parentheses and combine like terms.
3. Undo all indicated additions and subtractions to get all variables on one side of the equation and all numbers on the other side. Combine terms, if necessary.
4. Undo all indicated multiplications and divisions to cause the coefficient of the variable to be 1.
5. Check the result by replacing the variable with the possible solution and verifying that the number satisfies the equation.

Example 5 Solve the equation $\dfrac{5}{3}(x - 3) = \dfrac{3}{2}(x - 2) + 2$.

Solution *Step 1*

Because 6 is the smallest number that can be divided by both 2 and 3, multiply both sides of the equation by 6 to eliminate the fractions:

$$\frac{5}{3}(x - 3) = \frac{3}{2}(x - 2) + 2$$
$$6\left[\frac{5}{3}(x - 3)\right] = 6\left[\frac{3}{2}(x - 2) + 2\right]$$
$$10(x - 3) = 9(x - 2) + 12$$

Step 2

Remove parentheses and combine terms:

$$10x - 30 = 9x - 18 + 12$$
$$10x - 30 = 9x - 6$$

Step 3

Undo all additions and subtractions by adding $-9x$ and 30 to both sides to get the variables and numbers on opposite sides of the equals sign. Then, combine terms.

$$10x - 30 + (-9x) + 30 = 9x - 6 + (-9x) + 30$$
$$x = 24$$

Because the coefficient of x in the above equation is 1, *Step 4* is unnecessary.

Step 5

Check the result by substituting **24** for x in the original equation and simplifying:

$$\frac{5}{3}(x - 3) = \frac{3}{2}(x - 2) + 2$$

$$\frac{5}{3}(24 - 3) \overset{?}{=} \frac{3}{2}(24 - 2) + 2$$

$$\frac{5}{3}(21) \overset{?}{=} \frac{3}{2}(22) + 2$$

$$5(7) \overset{?}{=} 33 + 2$$

$$35 = 35$$

Because 24 satisfies the equation, it is a solution, and the solution set of the original equation is $\{24\}$. ∎

Example 6   Solve the equation $\dfrac{x + 2}{5} - 4x = \dfrac{8}{5} - \dfrac{x + 9}{2}$.

Solution

$$\frac{x + 2}{5} - 4x = \frac{8}{5} - \frac{x + 9}{2}$$

$$10\left(\frac{x + 2}{5} - 4x\right) = 10\left(\frac{8}{5} - \frac{x + 9}{2}\right) \qquad \text{Multiply both sides by 10.}$$

$$2(x + 2) - 40x = 2(8) - 5(x + 9) \qquad \text{Remove parentheses and simplify.}$$

$$2x + 4 - 40x = 16 - 5x - 45 \qquad \text{Remove parentheses.}$$

$$-38x + 4 = -5x - 29 \qquad \text{Combine terms.}$$

$$-33x = -33 \qquad \text{Add } 5x \text{ and } -4 \text{ to both sides.}$$

$$\frac{-33x}{-33} = \frac{-33}{-33} \qquad \text{Divide both sides by } -33.$$

$$x = 1 \qquad \text{Simplify.}$$

Check:

$$\frac{x + 2}{5} - 4x = \frac{8}{5} - \frac{x + 9}{2}$$

$$\frac{1 + 2}{5} - 4(1) \overset{?}{=} \frac{8}{5} - \frac{1 + 9}{2} \qquad \text{Substitute 1 for } x \text{ in the original equation.}$$

$$\frac{3}{5} - 4 \overset{?}{=} \frac{8}{5} - 5$$

$$\frac{3}{5} - \frac{20}{5} \overset{?}{=} \frac{8}{5} - \frac{25}{5}$$

$$-\frac{17}{5} = -\frac{17}{5}$$

Because 1 satisfies the equation, it is a solution, and the solution set of the original equation is $\{1\}$. ∎

Example 7   Solve the equation $\dfrac{x - 1}{3} + 4x = \dfrac{3}{2} + \dfrac{13x - 2}{3}$.

Solution

$$\frac{x - 1}{3} + 4x = \frac{3}{2} + \frac{13x - 2}{3}$$

$$6\left(\frac{x - 1}{3} + 4x\right) = 6\left(\frac{3}{2} + \frac{13x - 2}{3}\right) \qquad \text{Multiply both sides by 6.}$$

$$2(x - 1) + 6(4x) = 9 + 2(13x - 2) \qquad \text{Use the distributive property.}$$

$$2x - 2 + 24x = 9 + 26x - 4 \qquad \text{Remove parentheses.}$$

$$26x - 2 = 26x + 5 \qquad \text{Combine terms.}$$

$$-2 = 5 \qquad \text{Add } -26x \text{ to both sides.}$$

Because $-2 \neq 5$, the original equation has no solutions. Its solution set is $\emptyset$.
   If the final equation in this example had been an identity such as $-2 = -2$ or $5 = 5$, the original equation would be an identity also. ∎

Repeating Decimals

In Section 1.1 we claimed that every repeating decimal can be written as a rational number in fractional form. We now show that this statement is true by demonstrating a process that uses linear equations, by which we can convert any repeating decimal to fractional form. For example, to write the decimal $0.2\,54\,54\,\overline{54}$ as a rational number in fractional form, we first note that the decimal has a repeating block of two digits. We then form an equation by setting x equal to the decimal:

1. $x = 0.2\,54\,54\,54 \ldots$

We then form another equation by multiplying both sides of Equation 1 by 10^2, which is 100:

2. $100x = 25.4\,54\,54\,54\ldots$

We then subtract each side of Equation 1 from the corresponding side of Equation 2 to obtain

$$
\begin{array}{r}
100x = 25.4\,54\,54\,54\ldots \\
x = 0.2\,54\,54\,54\ldots \\
\hline
99x = 25.2
\end{array}
$$

Finally, we solve the linear equation $99x = 25.2$ for x and simplify the fraction:

$$x = \frac{25.2}{99} = \frac{252}{990} = \frac{\cancel{18} \cdot 14}{\cancel{18} \cdot 55} = \frac{14}{55}$$

We can use a calculator to verify that the decimal representation of $\frac{14}{55}$ is

$0.2\,54\,54\,\overline{54}$

The key step in the process was multiplying both sides of Equation 1 by 10^2. If there had been n digits in the repeating block of the decimal, we would have multiplied both sides of Equation 1 by 10^n.

Word Problems

The solution of word problems often leads to linear equations. In this section we consider two integer problems and a perimeter problem.

Example 8   If the sum of three consecutive odd integers is 213, what are the integers?

Analysis: Recall that consecutive odd integers are odd integers that differ by 2; numbers such as 11, 13, 15, and 17 are consecutive odd integers. Thus, if x is an odd integer, the algebraic expressions

$x,\quad x+2,\quad \text{and}\quad x+4$

represent three consecutive odd integers. The sum of these integers can be expressed in two ways, as either

$x + (x+2) + (x+4)$ or   213

Solution   Let x represent the first odd integer.
Then $x + 2$ represents the second odd integer.
Then $x + 4$ represents the third odd integer.

The first odd integer	+	the second odd integer	+	the third odd integer	=	the sum of the integers.

$$x \quad + \quad x + 2 \quad + \quad x + 4 \quad = \quad 213$$

$$3x + 6 = 213 \qquad \text{Combine terms.}$$
$$3x = 207 \qquad \text{Add } -6 \text{ to both sides.}$$
$$x = 69 \qquad \text{Divide both sides by 3.}$$
$$x + 2 = 71$$
$$x + 4 = 73$$

The three consecutive odd integers are 69, 71, and 73.

Check: The numbers 69, 71, and 73 are consecutive odd integers that do have a sum of 213.   ∎

Example 9   A man wants to cut a 27-foot rope into three pieces. He wants the longest piece to be 2 feet longer than twice the shortest and the middle-sized piece to be twice as long as the shortest. How long will each piece be?

Analysis: The information is given in terms of the length of the shortest piece of rope. Thus, pick a variable to represent the length of the shortest piece. Then express the lengths of the other pieces in terms of the variable.

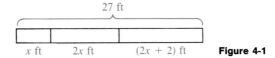

27 ft

x ft $2x$ ft $(2x + 2)$ ft   **Figure 4-1**

Solution   Let x represent the length of the shortest piece of rope.
Then $2x$ represents the length of the middle-sized piece, and $2x + 2$ represents the length of the longest piece.

Sketch the facts described in the problem as in Figure 4-1. From the figure you can see that the sum of the individual lengths must equal the total length of the rope:

The length of the shortest piece	+	the length of the middle-sized piece	+	the length of the longest piece	=	the total length.

$$x \quad + \quad 2x \quad + \quad 2x + 2 \quad = \quad 27$$

$$5x + 2 = 27 \qquad \text{Combine terms.}$$
$$5x = 25 \qquad \text{Add } -2 \text{ to both sides.}$$
$$x = 5 \qquad \text{Divide both sides by 5.}$$

Thus, the shortest piece is 5 feet long, the middle-sized piece is 2(5) or 10 feet long, and the longest piece is 2(5) + 2 or 12 feet long.

Check: Because 5 feet, 10 feet, and 12 feet total 27 feet, because the middle piece is twice as long as the shortest, and because the longest piece is 2 feet longer than twice the shortest, the solution checks.   ∎

Example 10   A rectangle is 6 meters longer than it is wide. If the perimeter of the rectangle is 28 meters, find the dimensions of the rectangle.

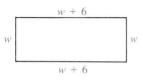

Figure 4-2

Analysis: The perimeter of a rectangle is the distance around it. If w is chosen to represent the width of the rectangle, then its length is $w + 6$. Making a sketch such as Figure 4-2 is helpful. The perimeter can be expressed either as $w + (w + 6) + w + (w + 6)$ or as 28.

Solution   Let w represent the width of the rectangle.
Then $w + 6$ represents the length of the rectangle.
Then you have

One width	+	one length	+	another width	+	another length	=	the perimeter.

$$w \;+\; w + 6 \;+\; w \;+\; w + 6 \;=\; 28$$

$$4w + 12 = 28 \quad \text{Combine terms.}$$
$$4w = 16 \quad \text{Add } -12 \text{ to both sides.}$$
$$w = 4 \quad \text{Divide both sides by 4.}$$
$$w + 6 = 10$$

The dimensions of the rectangle are 4 meters by 10 meters.

Check: If a rectangle has a width of 4 meters and a length of 10 meters, the length is 6 meters greater than the width, and the perimeter is $4 + 10 + 4 + 10 = 28$ meters.   ∎

▬ EXERCISE 4.1 ▬▬▬▬▬▬▬▬▬▬▬▬▬▬▬▬▬▬▬▬▬▬▬▬▬▬

In Exercises 1–8, find the domain of the variable. ***Do not solve the equation.***

1. $3x + 2 = 7$

2. $7x - 2 = 2$

3. $\dfrac{1}{x - 2} = 3$

4. $\dfrac{x + 2}{x - 3} = 3$

5. $\dfrac{3x - 2}{(x + 3)(x - 2)} = 4$

6. $\dfrac{8x + 5}{(x + 12)(x + 1)} = 5$

7. $\dfrac{3x^2 - 15}{x(x + 10)(x - 10)} = 32$

8. $\dfrac{3x}{x + 9} = \dfrac{x}{x^2 + 4}$

In Exercises 9–60, solve each equation and check the result.

9. $x + 6 = 8$ **10.** $y - 7 = 3$ **11.** $10 = z - 4$ **12.** $12 = 7 + t$

13. $2u = 6$ **14.** $3v = 12$ **15.** $\dfrac{x}{4} = 7$ **16.** $\dfrac{x}{6} = 8$

17. $\dfrac{3}{4}x = \dfrac{1}{2}$ **18.** $\dfrac{2}{3}x = \dfrac{5}{4}$ **19.** $2x + 1 = 13$ **20.** $2x - 4 = 16$

21. $\dfrac{3}{4}x - 3 = -9$ **22.** $\dfrac{4}{5}x + 5 = 17$ **23.** $2r - 5 = 1 - r$ **24.** $5s - 13 = s - 1$

25. $\dfrac{1}{2}a - 4 = -1 + 2a$ **26.** $2x + 3 = \dfrac{2}{3}x - 1$ **27.** $3a - 22 = -2a - 7$ **28.** $a + 18 = 6a - 3$

29. $3(x + 1) = 15$ **30.** $-2(x + 5) = 30$ **31.** $5(5 - a) = 37 - 2a$ **32.** $4a + 17 = 7(a + 2)$

33. $4(y + 1) = -2(4 - y)$ **34.** $5x + 4 = x + 24$ **35.** $3(y - 4) - 6 = 0$ **36.** $2x + (2x - 3) = 5$

37. $3(y - 5) + 10 = 2(y + 4)$ **38.** $2(5x + 2) = 3(3x - 2)$

39. $9(x + 2) = -6(4 - x) + 18$ **40.** $3(x + 2) - 2 = -(5 + x) + x$

41. $3(x - 4) + 6 = -2(x + 4) + 5x$ **42.** $8(3a - 5) - 4(2a + 3) = 12$

43. $\dfrac{x}{2} - \dfrac{x}{3} = 4$ **44.** $\dfrac{x}{2} + \dfrac{x}{3} = 10$ **45.** $\dfrac{x}{6} + 1 = \dfrac{x}{3}$ **46.** $\dfrac{3}{2}(y + 4) = \dfrac{20 - y}{2}$

47. $5 - \dfrac{x + 2}{3} = 7 - x$ **48.** $3x - \dfrac{2(x + 3)}{3} = 16 - \dfrac{x + 2}{2}$

49. $\dfrac{4x - 2}{2} = \dfrac{3x + 6}{3}$ **50.** $2(x - 3) = \dfrac{3}{2}(x - 4) + \dfrac{x}{2}$

51. $\dfrac{a + 1}{3} + \dfrac{a - 1}{5} = \dfrac{2}{15}$ **52.** $\dfrac{2z + 3}{3} + \dfrac{3z - 4}{6} = \dfrac{z - 2}{2}$

53. $\dfrac{5a}{2} - 12 = \dfrac{a}{3} + 1$ **54.** $\dfrac{5a}{6} - \dfrac{5}{2} = -\dfrac{1}{2} - \dfrac{a}{6}$

55. $y(y + 2) = (y + 1)^2 - 1$ **56.** $x(x - 3) = (x - 1)^2 - (5 + x)$

57. $0.05x + 0.04(5000 - x) = 220$ (*Hint:* Multiply both sides by 100.)

58. $0.06x + 0.08(20,000 - x) = 1,500$

59. $0.05x + 0.1x + 0.25(100 - x) = 24.5$ **60.** $0.09y + 0.14(10,000 - y) = 1,275$

In Exercises 61–64, write each repeating decimal number as a fraction. Simplify the answer, if necessary.

61. $0.33\overline{3}$ **62.** $0.29\,29\,\overline{29}$ **63.** $-0.34\,89\,89\,\overline{89}$ **64.** $-2.3\,47\,47\,\overline{47}$

In Exercises 65–80, solve each word problem.

65. If the sum of two consecutive integers is 75, what are the integers?

66. If the sum of three consecutive integers is 318, what are the integers?

67. The sum of three integers is 114. If the first integer is 24 larger than the second, and the second integer is 6 larger than the third, what is the second integer?

68. The sum of three consecutive even integers is 384. What are the integers?

69. The sum of the first and third integers of three consecutive odd integers is 58. What is the middle integer?

70. The sum of the second and fourth integers of four consecutive integers is 46. What is the sum of all the integers?

71. Jim wants to saw a 17-foot board into three pieces. The longest piece is to be three times as long as the shortest, and the middle-sized piece is to be 2 feet longer than the shortest. How long should each piece be?

72. A 30-foot steel beam is to be cut into two pieces. The longer piece is to be 2 feet more than three times as long as the shorter. How long will each piece be?

73. A 60-foot rope is cut into four pieces with each successive piece being twice as long as the previous one. How long is the longest piece?

74. A 185-foot cable is cut into three pieces: two pieces of equal length and a third piece that is 5 feet longer than the sum of the equal pieces. How long is the longer piece?

75. The length of a rectangle is twice the width. If the perimeter is 72 centimeters, what are the dimensions of the rectangle?

76. The width of a rectangle is one-third its length. The perimeter of the rectangle is 96 meters. What are the dimensions of the rectangle?

77. The length of a rectangle is 6 feet greater than its width. The perimeter of the rectangle is 72 feet. Find its area. (*Hint:*   The area of a rectangle is the product of its length and its width: $A = lw$.)

78. A man has 624 feet of fencing to enclose a small pasture. Because a river runs along one side, he will have to use fencing only on three sides. What will be the dimensions of the enclosed pasture if the length runs along the river and is to be double the width?

79. A man has 150 feet of fencing to build a rectangular, two-section pen as shown in Illustration 1. If one end is to be square, what are the outside dimensions of the entire pen?

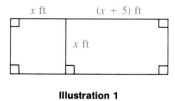

Illustration 1

80. A woman wants to fence in a 20-foot by 30-foot swimming pool and have a walkway of uniform width all the way around. How wide will the walkway be if the woman uses 180 feet of fencing?

4.2 APPLICATIONS

In this section we continue the discussion of word problems. In a broader sense, it is our purpose to discuss problem-solving techniques that can be used in many different situations. The following steps provide a strategy for solving a variety of problems:

1.   Read the problem several times to analyze the facts given. What information is given? What are you asked to find? Often a sketch or diagram will help you visualize the facts of the problem.

2. Pick a variable to indicate a quantity that must be found and write a sentence telling what the variable represents. Express all other quantities mentioned in the problem as expressions involving this single variable.
3. Organize the data and find a way to express a quantity in two different ways.
4. Write an equation showing that the two quantities found in step 3 are equal.
5. Solve the equation.
6. Check the result in the words of the problem.

Although this list is helpful, it does not apply to all situations totally. You may have to modify these steps slightly.

The word problems discussed in this section fall into several categories.

Area Problems

Example 1   The width of a rectangle is 4 centimeters less than its length. If the width is increased by 2 centimeters and the length is increased by 8 centimeters, the area is increased by 84 square centimeters. Find the dimensions of the original rectangle.

Analysis: Recall that the area of a rectangle is given by the formula $A = lw$. If the length of the original rectangle is l, then the width is $l - 4$, and the area is $l(l - 4)$. See Figure 4-3. If the width were increased by 2 centimeters and the length by 8 centimeters, the width and length of the larger rectangle would be $l - 2$ and $l + 8$. Hence, the area of the larger rectangle would be $(l - 2)(l + 8)$. If 84 square centimeters is added to the area of the original rectangle, the result is the same as the area of the larger rectangle.

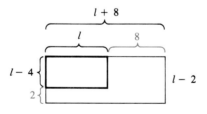

Figure 4-3

Solution   Let l represent the length of the original rectangle.
Then $l - 4$ represents the width of the original rectangle.
Then $l - 2$ represents the width of the larger rectangle.
Then $l + 8$ represents the length of the larger rectangle.

The area of the original rectangle	+	84	=	the area of the larger rectangle.

$$l(l-4) \qquad + \quad 84 \quad = \quad (l-2)(l+8)$$

$$l^2 - 4l + 84 = l^2 + 6l - 16 \qquad \text{Remove parentheses.}$$

$$-4l + 84 = 6l - 16 \qquad \text{Add } -l^2 \text{ to both sides.}$$

$$100 = 10l \qquad \text{Add } 4l + 16 \text{ to both sides.}$$

$$10 = l \qquad \text{Divide both sides by 10.}$$

$$6 = l - 4$$

The dimensions of the original rectangle are 6 centimeters by 10 centimeters.

Check: The area of the original rectangle is $6 \cdot 10 = 60$ square centimeters. The area of the larger rectangle is $8 \cdot 18 = 144$ square centimeters. The larger rectangle is 84 square centimeters greater than the original rectangle. ■

Lever Problems

Example 2 Sarah, David, and Heidi are positioned on a balanced seesaw as in Figure 4-4. If Sarah weighs 35 pounds and Heidi weighs 95 pounds, how much does David weigh?

Analysis: The key to this problem is a fact from physics. If the seesaw is to remain in balance, Sarah's weight times her distance from the fulcrum plus David's weight times his distance from the fulcrum must equal Heidi's weight times her distance from the fulcrum.

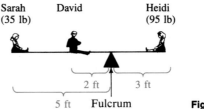

Figure 4-4

Solution Let w represent David's weight in pounds.

Sarah's weight times distance	+	David's weight times distance	=	Heidi's weight times distance.

$$35 \cdot 5 \qquad + \qquad w \cdot 2 \qquad = \qquad 95 \cdot 3$$

$$175 + 2w = 285 \qquad \text{Simplify.}$$

$$2w = 110 \qquad \text{Add } -175 \text{ to both sides.}$$

$$w = 55 \qquad \text{Divide both sides by 2.}$$

David weighs 55 pounds.

Check: $35 \cdot 5 + 55 \cdot 2 = 175 + 110 = 285$
$95 \cdot 3 = 285$ ∎

Coin Problems

Example 3 Judy has a certain number of quarters. She has four times as many dimes. She also has five times as many nickels as dimes. Altogether she has $11.55. How many quarters, dimes, and nickels does she have?

Analysis: The key to this problem is to set the value of the quarters plus the value of the dimes plus the value of the nickels equal to the value $11.55. If q is the number of quarters, $0.25q$ is the value of the quarters. Because there are $4q$ dimes, the value of the dimes is $0.10(4q)$. Because there are $5(4q) = 20q$ nickels, the value of the nickels is $0.05(20q)$. Set the sum of these values equal to $11.55.

Solution Let q represent the number of quarters.
Then $4q$ represents the number of dimes.
Then $5(4q) = 20q$ represents the number of nickels.

The value of the quarters	+	the value of the dimes	+	the value of the nickels	=	the total value.

$$0.25q \quad + \quad 0.10(4q) \quad + \quad 0.05(20q) \quad = \quad 11.55$$

Multiply both sides by 100. $25q + 10(4q) + 5(20q) = 1155$
Simplify. $25q + 40q + 100q = 1155$
Combine terms. $165q = 1155$
Divide both sides by 165. $q = 7$
$4q = 28$
$20q = 140$

Judy has 7 quarters, 28 dimes, and 140 nickels.

Check: The value of the 7 quarters is $1.75, the value of the 28 dimes is $2.80, and the value of the 140 nickels is $7.00. The total value is $11.55.   ∎

Investment Problems

Example 4 Steven has $15,000 to invest for one year. He invests some of it at 9% and the rest at 8%. If his total income from these investments is $1260, how much did he invest at each rate?

Analysis: The key to this problem is to add the interest from the 9% investment to the interest from the 8% investment, and set that sum equal to the total interest earned. For simple interest, the interest earned is computed by the formula $i = prt$, where i is the interest, p is the principal, r is the rate of interest, and t is the length of time (in this case, $t = 1$). Hence, if x dollars are invested at 9%, the interest earned is $0.09x$. If x dollars are invested at 9%, then the remaining money $(15000 - x)$ is invested at 8%. The amount of interest earned on that money is $0.08(15000 - x)$. The sum of these two amounts of money equals $1260.

Solution   Let x be the number of dollars invested at 9%.
Then $15000 - x$ is the number of dollars invested at 8%.

The interest earned at 9%	+	the interest earned at 8%	=	the total interest.

$$0.09x \quad + \quad 0.08(15000 - x) \quad = \quad 1260.00$$

Multiply both sides by 100. $9x + 8(15000 - x) = 126000$

Remove parentheses. $9x + 120000 - 8x = 126000$

Combine terms. $x + 120000 = 126000$

Add -120000 to both sides. $x = 6000$

$$15000 - x = 9000$$

Steven invested $6000 at 9% and $9000 at 8%.

Check: The interest on $6000 is $0.09(\$6000) = \$540$. The interest on $9000 is $0.08(\$9000) = \$720$. The total interest is $1260. ∎

Uniform Motion Problems

Example 5   A car leaves city A traveling toward city B at the rate of 55 miles per hour. At the same time, another car leaves city B traveling toward city A at the rate of 50 miles per hour. How long will it take them to meet if the cities are 157.5 miles apart?

Analysis: Uniform motion problems are based on the formula $d = rt$, where d is distance, r is rate, and t is time. It is often helpful to organize the information of a uniform motion problem in a chart like Figure 4-5. You know that the faster car is traveling at 55 miles per hour and that the slower car is traveling at 50 miles per hour. You also know that they travel for the same amount of time, say t hours. The distance that the faster car travels is $55t$ miles and the

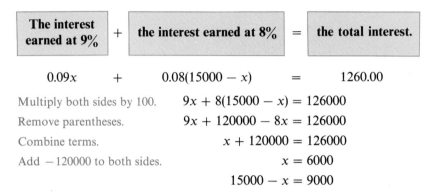

	r	$\cdot$	t	$=$	d
(1) Faster car	55		t		$55t$
(2) Slower car	50		t		$50t$

Figure 4-5

distance that the slower car travels is $50t$ miles. The sum of these distances equals 157.5 miles, the distance between the cities.

Solution Let t represent the time that each car travels.
Then $55t$ represents the distance traveled by the faster car.
Then $50t$ represents the distance traveled by the slower car.

The distance the faster car goes	+	the distance the slower car goes	=	the distance between cities.

$$55t \quad + \quad 50t \quad = \quad 157.5$$

Combine terms. $\qquad\qquad 105t = 157.5$
Divide both sides by 105. $\qquad t = 1.5$

The two cars will meet in 1.5 hours.

Check: Car 1 travels $1.5(55) = 82.5$ miles. Car 2 travels $1.5(50) = 75$ miles. The total distance traveled is 157.5 miles. ∎

Mixture Problems

Example 6 A candy store owner notices that he has 20 pounds of cashews that are getting stale. They are not selling because they cost a high price of $6 per pound. In order to sell the cashews, the store owner decides to mix peanuts with them to lower the price per pound. If peanuts sell for $1.50 per pound, how many pounds of peanuts must be mixed with the cashews to make a mixture that he could sell for $3 per pound?

Analysis: This problem is very similar to the previous uniform motion problem, except that it is based on the formula $V = pn$, where V represents value, p represents price per pound, and n represents the number of pounds. Let x represent the number of pounds of peanuts to be used. Then $20 + x$ represents the total pounds in the mixture. Enter the known information in the chart of Figure 4-6. Note that the value of the cashews plus the value of the peanuts is equal to the value of the mixture.

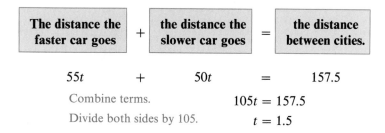

	p ·	n =	V
Cashews	6.00	20	120.00
Peanuts	1.50	x	$1.5x$
Mixture	3.00	$20 + x$	$3(20 + x)$

Figure 4-6

Solution   Let x represent the number of pounds of peanuts to be used. Then $20 + x$ represents the number of pounds in the mixture.

The value of the cashews	+	the value of the peanuts	=	the value of the mixture.

$$120 \qquad + \qquad 1.5x \qquad = \qquad 3(20 + x)$$

Multiply both sides by 10. $1200 + 15x = 30(20 + x)$

Remove parentheses. $1200 + 15x = 600 + 30x$

Add $-15x - 600$ to both sides. $600 = 15x$

Divide both sides by 15. $40 = x$

The store owner should mix 40 pounds of peanuts with the 20 pounds of cashews.

Check: The cashews are valued at $\$6(20) = \$120$. The peanuts are valued at $\$1.50(40) = \$60$. The mixture is valued at $\$3(60) = \$180$. The value of the cashews added to the value of the peanuts does equal the value of the mixture. ∎

Example 7   A car radiator has a capacity of 15 quarts and is filled with a mixture of 80% water and 20% antifreeze. How many quarts of solution must be drained and replaced with pure antifreeze to bring the mixture in the radiator up to a 50% antifreeze solution?

Analysis: The key to this problem is to note that whatever pure antifreeze finally ends up in the radiator could have come from one of two sources—either from what was left in the partially drained radiator, or from the pure antifreeze that was added. If x represents the amount of solution to be drained (and, therefore, the amount to be replaced with pure antifreeze), then the amount of pure antifreeze remaining in the radiator after draining is the amount present originally minus the amount drained. This is 20% of 15 (the amount originally present) minus 20% of x (the amount drained). The amount of pure antifreeze added is x, and the amount needed when you are done is 50% of 15.

Solution   Let x represent the number of quarts to be drained and then replaced.

The amount of pure antifreeze in original solution	−	the amount of pure antifreeze to be drained	+	the amount of pure antifreeze to be added	=	the amount of pure antifreeze in final solution.

$$0.20(15) \qquad - \qquad 0.20(x) \qquad + \qquad 1.00x \qquad = \qquad 0.50(15)$$

Multiply both sides by 10.	$2(15) - 2x + 10x = 5(15)$
Simplify and combine terms.	$30 + 8x = 75$
Add -30 to both sides.	$8x = 45$
Divide both sides by 8.	$x = \dfrac{45}{8}$

You must drain $\frac{45}{8}$, or $5\frac{5}{8}$, quarts.

Check: You begin with $0.20(15) = 3$ quarts of antifreeze. You drain $0.20(\frac{45}{8}) = \frac{9}{8}$ quarts of antifreeze. You add $1.00(\frac{45}{8}) = 5\frac{5}{8}$ quarts of antifreeze. The total is $3 - \frac{9}{8} + \frac{45}{8} = 7\frac{1}{2}$ quarts of antifreeze, which is half the capacity of the radiator.

■

■ EXERCISE 4.2

Solve and check each word problem.

1. A rectangle has a length of 9 centimeters and an area of 54 square centimeters. Find its width.

2. A rectangle is 6 feet longer than it is wide, and its area is 30 square feet more than the square of its width. Find the dimensions of the rectangle.

3. The width of a rectangle is 2 inches shorter than its length. If both the length and the width are increased by 5 inches, the area is increased by 75 square inches. What are the dimensions of the original rectangle?

4. A rectangle is 1 foot longer than it is wide. If the length is increased by 2 feet and the width is decreased by 6 feet, the area is decreased by 50 square feet. What are the dimensions of the original rectangle?

5. A rectangle is 11 meters longer than it is wide. If the length is decreased by 12 meters and the width is increased by 10 meters, the area is decreased by 24 square meters. What is the perimeter of the original rectangle?

6. If the height of a triangle with a base of 8 inches is tripled, its area is increased by 96 square inches. Find the height of the triangle. (*Hint:*   The area is given by the formula $A = \frac{1}{2}bh$, where A is the area, b is the base, and h is the height.)

7. A seesaw is 20 feet long and the fulcrum is in the center. If an 80-pound boy sits at one end, how far will the boy's 160-pound father have to sit from the fulcrum to balance the seesaw?

8. Two girls, one weighing 110 pounds and the other 88 pounds, sit on opposite ends of an 18-foot seesaw. Where must the fulcrum be placed so that the seesaw balances?

9. A lady uses a 10-foot bar to lift a 210-pound stone. If she places another rock 3 feet from the stone to act as the fulcrum, how much force must she exert to move the stone?

10. Two men wish to lift a 2500-pound car. To do so, they use a 12-foot steel bar and a fulcrum placed 3 feet from the car. If one of the two men weighs 200 pounds and pushes on the far end of the bar, and the other man, who weighs 150 pounds, pushes also, can they lift the car?

11. Sally and Sue, each weighing 110 pounds, sit at opposite ends of a 14-foot seesaw, and Sue's brother Jim, who weighs twice as much, sits 3 feet in front of Sue. Where should the fulcrum be placed to balance the seesaw?

12. Jim and Bob sit at opposite ends of an 18-foot seesaw, with the fulcrum at its center. Jim weighs 160 pounds, and Bob weighs 200 pounds. Kimberly sits 4 feet in front of Jim, and the seesaw balances. How much less does Kimberly weigh than Bob?

13. A person has a certain number of nickels, twice as many dimes, and four times as many pennies. How many does the person have of each if their total value is $2.90?

14. Sarah has the same number of quarters and dimes. She also has some nickels. If the value of her 68 coins is $8.15, how many quarters does she have?

15. A collection of 24 coins, only quarters and nickels, has a value of $3.40. How many quarters and nickels are there?

16. A collection of nickels, dimes, and quarters has a value of $11.10. There are three times as many dimes as nickels, and twice as many quarters as dimes. How many are there of each?

17. A handful of dimes and nickels, eight coins in all, would be worth 10¢ less if all the dimes were nickels and all the nickels were dimes. How many dimes and nickels are there?

18. Fred has some dimes and some quarters, ten coins in all. If his dimes were quarters and his quarters were dimes, he would have 30¢ less. How much money does he have?

19. Heidi invested $12,000, some at 8% and some at 9%. How much is invested at each rate if her income from the two investments is $1060 each year?

20. Sam invested $14,000, some at 7% and some at 10%. His annual income from these investments is $1280. How much did he invest at each rate?

21. A teacher wishes to earn $1500 per year in supplemental income from a cash gift of $16,000. She puts $6000 in a credit union that pays 7%. What rate must she earn on the remainder to achieve her goal?

22. Paul split an inheritance between two investments, one paying 7% and the other paying 10%. Paul invested twice as much in the 10% investment as in the 7% investment. His combined annual income from these two investments was $4050. How much did Paul inherit?

23. George has some money to invest. If George could invest an extra $3000, he would be eligible for a 21% special account. Otherwise, he could invest his money at 18% interest. The 21% account would yield twice as much income annually as the 18% investment. If George cannot find the extra money, how much does he have available to invest?

24. A bus driver wishes to earn $3500 per year in supplemental income from an inheritance of $40,000. The driver puts $10,000 in a bank paying 8%. What rate must he earn on the remainder to achieve his goal?

25. For a certain movie, student tickets were $2 each and adult tickets were $4 each. If 200 tickets were sold and the total receipts were $750, how many student and how many adult tickets were sold?

26. At a school play 140 tickets were sold. Adult tickets cost $2.50 each and student tickets cost $1.50 each. How many of each were sold if the receipts were $290?

27. Some college students take a bus to an airport to catch a plane. Their bus traveled at 55 miles per hour for 2 hours. Their plane cruised at 450 miles per hour. How long were they in the air if the total distance traveled was 1460 miles?

28. A cyclist leaves a city riding at the rate of 18 miles per hour. One hour later, a car leaves the same city going at the rate of 44 miles per hour and in the same direction. How long will it take the car to overtake the cyclist?

29. A boat can go 12 miles per hour in still water. If it goes upstream for 3 hours against a current of 4 miles per hour, how long will it take the boat to return?

30. Sarah walked north at the rate of 3 miles per hour and returned at the rate of 4 miles per hour. How many miles did she walk if the round trip took 3.5 hours?

31. Grant traveled a distance of 400 miles in a time of 8 hours. Part of the time the rate of speed was 45 miles per hour, and part of the time the rate of speed was 55 miles per hour. How long did Grant travel at each rate?

32. A motorboat can go 18 miles per hour in still water. If it can go 80 miles downstream in 4 hours and return in 5 hours, what is the speed of the current?

33. The owner of a candy store wishes to make 30 pounds of a mixture of two candies to sell for $1.00 per pound. If one candy sells for $0.95 per pound and the other candy sells for $1.10 per pound, how many pounds of each should he use?

34. A mixture of candy is made to sell for $0.89 per pound. If 32 pounds of a cheaper candy, selling for $0.80 per pound, are used along with 12 pounds of a more expensive candy, what is the price per pound of the better candy?

35. How much water should be added to 20 ounces of a 15% solution of alcohol to dilute it to a 10% solution?

36. How much water must be boiled away to increase the concentration of 300 gallons of a salt solution from 2% to 3%?

37. Cream is approximately 22% butterfat. How many gallons of cream must be mixed with milk testing at 2% butterfat to give 20 gallons of milk containing 4% butterfat?

38. How much acid must be added to 60 grams of a solution that is 65% acid to obtain a new solution that is 75% acid?

4.3   LITERAL EQUATIONS

Equations that contain many variables are called **literal equations**. Often, these equations are **formulas** such as $A = \frac{1}{2}bh$, the formula for finding the area of a triangle with a base of b units and a height of h units. Suppose we wish to use this formula to find the heights of several triangles whose areas and bases are known. It would be very tedious to substitute values of A and b into the formula and then repeatedly solve the formula $A = \frac{1}{2}bh$ for h. It would be better to solve the formula for h first, and then substitute values for A and b, and compute h directly.

To solve a formula for a variable means to isolate that variable on one side of an equation and isolate all other quantities on the other side of the equation. The first example shows how to solve the formula $A = \frac{1}{2}bh$ for h.

Example 1   Solve the formula $A = \dfrac{1}{2}bh$ for h.

Solution
$$A = \frac{1}{2}bh$$

$$2A = bh \qquad \text{Multiply both sides by 2.}$$

$$\frac{2A}{b} = h \qquad \text{Divide both sides by } b.$$

$$h = \frac{2A}{b} \qquad \text{Use the symmetric property of equality.} \qquad ■$$

Example 2   The formula $A = p + prt$ gives the amount of money in a savings account at the end of a specific time. A represents the amount, p the principal, r the rate, and t the time. Solve the formula for t.

Solution

$$A = p + prt$$

$$A - p = prt \qquad \text{Add } -p \text{ to both sides.}$$

$$\frac{A - p}{pr} = t \qquad \text{Divide both sides by } pr.$$

$$t = \frac{A - p}{pr}$$ ∎

Example 3   Solve the formula $A = p + prt$ for p.

Solution   In this example, you must factor the p out of the two terms on the right-hand side of the equation. Then, you can solve for p.

$$A = p + prt$$

$$A = p(1 + rt) \qquad \text{Factor } p \text{ out of } p + prt.$$

$$\frac{A}{1 + rt} = p \qquad \text{Divide both sides by } 1 + rt.$$

$$p = \frac{A}{1 + rt}$$ ∎

Example 4   Solve the formula $\dfrac{1}{r} = \dfrac{1}{r_1} + \dfrac{1}{r_2}$ for r.

Solution   This formula contains three variables: $r, r_1,$ and r_2. The variables r_1 and r_2 contain subscripts of 1 and 2, respectively. The purpose of the subscripts is to identify two variables that represent similar quantities. The given formula is used in electronics to calculate the combined resistance r of two resistors in parallel. The variable r_1 represents the resistance of the first resistor, and the variable r_2 represents the resistance of the second resistor. Solve for r as follows:

$$\frac{1}{r} = \frac{1}{r_1} + \frac{1}{r_2}$$

$$\frac{rr_1r_2}{r} = \frac{rr_1r_2}{r_1} + \frac{rr_1r_2}{r_2} \qquad \text{Multiply both sides by } rr_1r_2.$$

$$r_1r_2 = rr_2 + rr_1 \qquad \text{Simplify each fraction.}$$

$$r_1r_2 = r(r_2 + r_1) \qquad \text{Factor out } r \text{ on the right-hand side.}$$

$$\frac{r_1r_2}{r_2 + r_1} = r \qquad \text{Divide both sides by } r_2 + r_1.$$

$$r = \frac{r_1r_2}{r_2 + r_1}$$ ∎

Example 5   A saleslady in a high-fashion dress shop earns \$200 per week plus a 5% commission on the value of the merchandise she sells. What dollar volume must she sell each week to earn $250, $300, and $350 in three successive weeks?

Solution   The saleslady's weekly earnings e are computed with the formula

$$e = 200 + 0.05v$$

where v represents the value of the merchandise she sells. Begin by solving the formula for v:

$$e = 200 + 0.05v$$

$$e - 200 = 0.05v$$

$$v = \frac{e - 200}{0.05}$$

Then substitute $250, $300, and $350 for e and solve for v:

$$v = \frac{e - 200}{0.05} \qquad v = \frac{e - 200}{0.05} \qquad v = \frac{e - 200}{0.05}$$

$$v = \frac{250 - 200}{0.05} \qquad v = \frac{300 - 200}{0.05} \qquad v = \frac{350 - 200}{0.05}$$

$$= 1000 \qquad\qquad = 2000 \qquad\qquad = 3000$$

She must sell $1000 worth of merchandise the first week, $2000 worth the second week, and $3000 worth the third week.   ∎

■ EXERCISE 4.3

In Exercises 1–40, solve each equation for the variable indicated.

1. $A = lw$ for w

2. $p = 4s$ for s

3. $A = \pi r^2$ for r^2

4. $A = \frac{1}{2}bh$ for b

5. $V = \frac{1}{3}Bh$ for B

6. $V = \pi r^2 h$ for h

7. $I = prt$ for t

8. $I = prt$ for r

9. $p = 2l + 2w$ for w

10. $p = 2l + 2w$ for l

11. $A = \frac{1}{2}h(b_1 + b_2)$ for b_1

12. $A = \frac{1}{2}h(b_1 + b_2)$ for b_2

13. $z = \frac{x - \mu}{\sigma}$ for x

14. $z = \frac{x - \mu}{\sigma}$ for μ

15. $y = mx + b$ for x

16. $y = mx + b$ for m

17. $l = a + (n - 1)d$ for n

18. $l = a + (n - 1)d$ for d

19. $r_1 r_2 = rr_2 + rr_1$ for r_1

20. $r_1 r_2 = rr_2 + rr_1$ for r

21. $\sigma^2 = \frac{\Sigma x^2}{n} - \mu^2$ for Σx^2

22. $\sigma^2 = \frac{\Sigma x^2}{n} - \mu^2$ for n

23. $S = \dfrac{a - lr}{1 - r}$ for r

24. $S = \dfrac{a - lr}{1 - r}$ for l

25. $a = \dfrac{(n - 2)180}{n}$ for n

26. $C = \dfrac{5}{9}(F - 32)$ for F

27. $P = L + \dfrac{s}{f}i$ for s

28. $P = L + \dfrac{s}{f}i$ for f

29. $\dfrac{1}{r} = \dfrac{1}{r_1} + \dfrac{1}{r_2}$ for r_2

30. $S = \dfrac{n(a + l)}{2}$ for n

31. $y - y_1 = m(x - x_1)$ for x

32. $y - y_1 = m(x - x_1)$ for x_1

33. $H = \dfrac{2ab}{a + b}$ for a

34. $H = \dfrac{2ab}{a + b}$ for b

35. $\dfrac{x^2}{a^2} + \dfrac{y^2}{b^2} = 1$ for a^2

36. $\dfrac{x^2}{a^2} - \dfrac{y^2}{b^2} = 1$ for y^2

37. $y = a(x^2 - h) + k$ for h

38. $y = a(x^2 - h) + k$ for x^2

39. $V = \dfrac{1}{3}h(B_1 + B_2 + \sqrt{B_1 B_2})$ for h

40. $V = \pi h^2 \left(r - \dfrac{h}{3} \right)$ for r

41. Solve the formula $F = \frac{9}{5}C + 32$ for C and find the Celsius temperatures that correspond to Fahrenheit temperatures of $32°$, $70°$, and $212°$.

42. A man intends to invest $1000. Solve the formula $A = p + prt$ for t and find the time required to double his money at the rates of 5%, 7%, and 10%.

43. The cost of electricity in a certain city is given by the formula

$$C = 0.07n + 6.50$$

where C is the cost and n is the number of kilowatt hours used. Solve the formula for n and find the number of kilowatt hours used for costs of $49.90, $75.10, and $125.50.

44. A monthly water bill in a certain city is calculated by using the formula

$$n = \frac{5,000C - 17,500}{6}$$

where n is the number of gallons used and C is the monthly cost. Solve the formula for C and compute the bill for quantities used of 500, 1200, and 2500 gallons.

45. While waiting for his car to be repaired, John rents a car for $12 per day plus 10¢ per mile. If John keeps the car for 2 days, how many miles can he drive for a total cost of $30? How many miles can he drive for a total cost of $40?

46. Jill earns $17 per day delivering pizzas. She is paid $5 per day plus 60¢ for each pizza delivered. How many more deliveries must she make each day to increase her earnings to $23 per day?

47. Bob's father will pay him $1750 and give him a used car if Bob agrees to work 5 months. When Bob quit after 3 months, his father paid him $810 and the car. How much was the car worth?

48. The landlord of a duplex apartment collected $8730 rent in one year by renting both units. One apartment rents for $60 more per month than the other, but was vacant for 3 months. What is the monthly rent of the more expensive apartment?

4.4   SOLVING EQUATIONS BY FACTORING

An equation such as $3x^2 + 4x - 7 = 0$ or $-5y^2 + 3y + 8 = 0$ is called a **quadratic** or **second-degree equation**.

Definition.   A **quadratic equation** is an equation that can be written in the form $ax^2 + bx + c = 0$, where a, b, and c are real numbers and $a \neq 0$.

Many quadratic equations can be solved by factoring. To do so, we use the **zero-factor theorem**.

The Zero-Factor Theorem.   If a and b are real numbers, and

$$\text{if } ab = 0, \quad \text{then} \quad a = 0 \text{ or } b = 0.$$

This theorem points out that, if the product of two or more numbers is 0, then at least one of them must be 0.

For example, to use the zero-factor theorem to solve the equation

1. $x^2 + 5x + 6 = 0$

we factor the left-hand side to obtain

$$(x + 3)(x + 2) = 0$$

Because the product of the factors $x + 3$ and $x + 2$ is 0, then (by the zero-factor theorem) at least one of the factors is 0. Thus, we can set each factor equal to 0 and solve each resulting equation for x:

$$x + 3 = 0 \qquad \text{or} \qquad x + 2 = 0$$
$$x = -3 \qquad\qquad\qquad x = -2$$

To check the work we substitute -3 and -2 for x in Equation 1 and verify that each number satisfies it.

$$x^2 + 5x + 6 = 0 \qquad\qquad x^2 + 5x + 6 = 0$$
$$(-3)^2 + 5(-3) + 6 \overset{?}{=} 0 \qquad (-2)^2 + 5(-2) + 6 \overset{?}{=} 0$$
$$9 - 15 + 6 \overset{?}{=} 0 \qquad\qquad 4 - 10 + 6 \overset{?}{=} 0$$
$$0 = 0 \qquad\qquad\qquad\qquad 0 = 0$$

Because both -3 and -2 satisfy the equation, both numbers are solutions.

If either b or c is 0 in an equation of the form $ax^2 + bx + c = 0$, the equation is called an **incomplete quadratic equation**.

Example 1   Solve the equation $3x^2 + 6x = 0$.

Solution   This equation is an incomplete quadratic equation with c (the constant term) equal to 0. To solve this equation, factor the left-hand side, set each factor equal to 0, and solve each resulting equation for x.

$$3x^2 + 6x = 0$$
$$3x(x + 2) = 0$$
$$3x = 0 \quad \text{or} \quad x + 2 = 0$$
$$x = 0 \quad \quad \quad x = -2$$

Verify that both solutions check.   ∎

Example 2   Solve the equation $x^2 - 16 = 0$.

Solution   This equation is an incomplete quadratic equation with b (the coefficient of x) equal to 0. To solve this equation, factor the difference of two squares on the left-hand side, set each factor equal to 0, and solve each resulting equation.

$$x^2 - 16 = 0$$
$$(x + 4)(x - 4) = 0$$
$$x + 4 = 0 \quad \text{or} \quad x - 4 = 0$$
$$x = -4 \quad \quad \quad x = 4$$

Verify that both solutions check.   ∎

Many equations that do not appear to be quadratic can be put into quadratic form and then solved by factoring.

Example 3   Solve the equation $x = \dfrac{6}{5} - \dfrac{6}{5}x^2$.

Solution

$$x = \frac{6}{5} - \frac{6}{5}x^2$$

$5x = 6 - 6x^2$	Multiply both sides by 5.
$6x^2 + 5x - 6 = 0$	Add $6x^2 - 6$ to both sides.
$(3x - 2)(2x + 3) = 0$	Factor $6x^2 + 5x - 6$.
$3x - 2 = 0 \quad \text{or} \quad 2x + 3 = 0$	Set each factor equal to zero.
$3x = 2 \quad\quad\quad\quad 2x = -3$	
$x = \dfrac{2}{3} \quad\quad\quad\quad x = -\dfrac{3}{2}$	

Check: $x = \dfrac{6}{5} - \dfrac{6}{5}x^2$ $\bigg|$ $x = \dfrac{6}{5} - \dfrac{6}{5}x^2$

$$\dfrac{2}{3} \overset{?}{=} \dfrac{6}{5} - \dfrac{6}{5}\left(\dfrac{2}{3}\right)^2 \qquad -\dfrac{3}{2} \overset{?}{=} \dfrac{6}{5} - \dfrac{6}{5}\left(-\dfrac{3}{2}\right)^2$$

$$\dfrac{2}{3} \overset{?}{=} \dfrac{6}{5} - \dfrac{6}{5}\left(\dfrac{4}{9}\right) \qquad -\dfrac{3}{2} \overset{?}{=} \dfrac{6}{5} - \dfrac{6}{5}\left(\dfrac{9}{4}\right)$$

$$\dfrac{2}{3} \overset{?}{=} \dfrac{6}{5} - \dfrac{8}{15} \qquad -\dfrac{3}{2} \overset{?}{=} \dfrac{6}{5} - \dfrac{27}{10}$$

$$\dfrac{2}{3} \overset{?}{=} \dfrac{18}{15} - \dfrac{8}{15} \qquad -\dfrac{3}{2} \overset{?}{=} \dfrac{12}{10} - \dfrac{27}{10}$$

$$\dfrac{2}{3} \overset{?}{=} \dfrac{10}{15} \qquad -\dfrac{3}{2} \overset{?}{=} -\dfrac{15}{10}$$

$$\dfrac{2}{3} = \dfrac{2}{3} \qquad -\dfrac{3}{2} = -\dfrac{3}{2}$$

Both solutions check.   ■

Often the factoring method can be used to solve equations that contain polynomials with degree greater than two.

Example 4   Solve the equation $6x^3 - x^2 - 2x = 0$.

Solution   Because x is a common factor, factor an x from the third-degree polynomial on the left-hand side and proceed as follows:

$$6x^3 - x^2 - 2x = 0$$
$$x(6x^2 - x - 2) = 0$$
$$x(3x - 2)(2x + 1) = 0 \qquad \text{Factor } 6x^2 - x - 2.$$
$$x = 0 \quad \text{or} \quad 3x - 2 = 0 \quad \text{or} \quad 2x + 1 = 0 \qquad \text{Set each factor equal to 0.}$$
$$x = \dfrac{2}{3} \qquad\qquad x = -\dfrac{1}{2}$$

Verify that all three solutions check.   ■

Example 5   Solve the equation $x^4 - 5x^2 + 4 = 0$.

Solution   Factor the trinomial on the left-hand side and proceed as follows:

$$x^4 - 5x^2 + 4 = 0$$
$$(x^2 - 1)(x^2 - 4) = 0$$
$$(x + 1)(x - 1)(x + 2)(x - 2) = 0 \qquad \text{Factor } x^2 - 1 \text{ and } x^2 - 4.$$

$$x + 1 = 0 \quad \text{or} \quad x - 1 = 0 \quad \text{or} \quad x + 2 = 0 \quad \text{or} \quad x - 2 = 0$$
$$x = -1 \qquad\qquad x = 1 \qquad\qquad x = -2 \qquad\qquad x = 2$$

Set each factor equal to 0.

Verify that all four solutions check.   ■

Example 6   The height of a triangle is 3 times its base. The area of the triangle is 96 square meters. Find its base and height.

Solution   Let x be the positive number that represents the length of the base of the triangle. Then $3x$ represents the height of the triangle. Substitute x for b, $3x$ for h, and **96** for A in the formula for the area of a triangle, and solve for x.

$$A = \frac{1}{2} bh$$

$$96 = \frac{1}{2} x(3x)$$

$$192 = 3x^2 \qquad\qquad \text{Multiply both sides by 2.}$$
$$64 = x^2 \qquad\qquad \text{Divide both sides by 3.}$$
$$0 = x^2 - 64 \qquad\qquad \text{Add } -64 \text{ to both sides.}$$
$$0 = (x + 8)(x - 8) \qquad\qquad \text{Factor } x^2 - 64.$$
$$x + 8 = 0 \quad \text{or} \quad x - 8 = 0 \qquad \text{Set each factor equal to 0.}$$
$$x = -8 \qquad\qquad x = 8$$

A triangle cannot have a base of -8 meters. Thus, the base of the triangle is 8 meters, and the height is 3(8) meters, or 24 meters.

Check: The area of a triangle with a base of 8 meters and a height of 24 meters is 96 square meters:

$$A = \frac{1}{2} bh$$

$$= \frac{1}{2} (8)(24)$$

$$= 4(24)$$

$$= 96$$

The solution checks.   ■

■ **EXERCISE 4.4** ▬▬▬▬▬▬▬▬▬▬▬▬▬▬▬▬▬▬▬▬▬▬▬▬▬▬▬▬

In Exercises 1–24, solve each equation by factoring.

1. $4x^2 + 8x = 0$ **2.** $x^2 - 9 = 0$ **3.** $y^2 - 16 = 0$ **4.** $5y^2 - 10y = 0$

5. $x^2 + x = 0$ **6.** $x^2 - 3x = 0$ **7.** $5y^2 - 25y = 0$ **8.** $y^2 - 36 = 0$

9. $z^2 + 8z + 15 = 0$ **10.** $w^2 + 7w + 12 = 0$ **11.** $y^2 - 7y + 6 = 0$ **12.** $n^2 - 5n + 6 = 0$

13. $y^2 - 7y + 12 = 0$ **14.** $x^2 - 3x + 2 = 0$ **15.** $x^2 + 6x + 8 = 0$ **16.** $x^2 + 9x + 20 = 0$

17. $3m^2 + 10m + 3 = 0$ **18.** $2r^2 + 5r + 3 = 0$ **19.** $2y^2 - 5y + 2 = 0$ **20.** $2x^2 - 3x + 1 = 0$

21. $2x^2 - x - 1 = 0$ **22.** $2x^2 - 3x - 5 = 0$ **23.** $3s^2 - 5s - 2 = 0$ **24.** $8t^2 + 10t - 3 = 0$

In Exercises 25–36, write each equation in quadratic form and solve it by factoring.

25. $x(x - 6) + 9 = 0$ **26.** $x^2 + 8(x + 2) = 0$ **27.** $8a^2 = 3 - 10a$ **28.** $5z^2 = 6 - 13z$

29. $b(6b - 7) = 10$ **30.** $2y(4y + 3) = 9$ **31.** $\dfrac{3a^2}{2} = \dfrac{1}{2} - a$ **32.** $x^2 = \dfrac{1}{2}(x + 1)$

33. $x^2 + 1 = \dfrac{5}{2}x$ **34.** $\dfrac{3}{5}(x^2 - 4) = -\dfrac{9}{5}x$ **35.** $x\left(3x + \dfrac{22}{5}\right) = 1$ **36.** $x\left(\dfrac{x}{11} - \dfrac{1}{7}\right) = \dfrac{6}{77}$

In Exercises 37–48, solve each equation by factoring.

37. $x^3 + x^2 = 0$ **38.** $2x^4 + 8x^3 = 0$ **39.** $y^3 - 49y = 0$ **40.** $2z^3 - 200z = 0$

41. $x^3 - 4x^2 - 21x = 0$ **42.** $x^3 + 8x^2 - 9x = 0$ **43.** $z^4 - 13z^2 + 36 = 0$ **44.** $y^4 - 10y^2 + 9 = 0$

45. $3a(a^2 + 5a) = -18a$ **46.** $7t^3 = 2t\left(t + \dfrac{5}{2}\right)$ **47.** $\dfrac{x^2(6x + 37)}{35} = x$ **48.** $x^2 = -\dfrac{4x^3(3x + 5)}{3}$

In Exercises 49–54, solve each equation by factoring by grouping.

49. $x^3 + 3x^2 - x - 3 = 0$ **50.** $x^3 - x^2 - 4x + 4 = 0$

51. $2r^3 + 3r^2 - 18r - 27 = 0$ **52.** $3s^3 - 2s^2 - 3s + 2 = 0$

53. $3y^3 + y^2 = 4(3y + 1)$ **54.** $w^3 + 16 = w(w + 16)$

In Exercises 55–66, solve each word problem.

55. The product of two consecutive even integers is 288. Find the integers.

56. The product of two consecutive odd integers is 143. Find the integers.

57. The sum of the squares of two consecutive positive integers is 85. Find the integers.

58. The sum of the squares of three consecutive positive integers is 77. Find the integers.

59. A rectangle with an area of 96 square meters is 4 meters longer than it is wide. Find its perimeter.

60. One side of a rectangle is three times longer than another. If its area is 147 square centimeters, what are its dimensions?

61. The area of a square is numerically equal to its perimeter. How long is one side of the square?

62. A rectangle is 2 inches longer than it is wide. Numerically, its area exceeds its perimeter by 11. What is the perimeter of the rectangle?

63. An artist intends to paint a 60-square-foot mural on a wall that is 18 feet long and 11 feet high. What will be the dimensions of the mural if the artist leaves a border of uniform width around it?

64. A woman plans to use one-fourth of her 48-foot by 100-foot rectangular backyard as a garden. What will be the perimeter of the garden if the length is to be 40 feet greater than the width?

65. A rectangular room is twice as long as it is wide. It is divided into two rectangular parts by a partition that is 12 feet from one end of the room. If the other part of the room contains 560 square feet, find the dimensions of the room.

66. If the length of one side of a square is increased by 4 inches, the area of the square becomes nine times greater. What was the perimeter of the original square?

(continued)

67. Find a quadratic equation whose roots are 3 and 5.

68. Find a quadratic equation whose roots are -2 and 6.

69. Find a quadratic equation whose roots are   0 and -5.

70. Find a quadratic equation whose roots are $\frac{1}{2}$ and $\frac{1}{3}$.

4.5   ABSOLUTE VALUE EQUATIONS

In this section we will discuss how to solve equations that involve absolute values. We begin by reviewing the definition of the absolute value of x.

Definition.   If x is a positive number or 0, then $|x| = x$.
If x is a negative number, then $|x| = -x$.

The definition of the absolute value of any real number describes a method for associating a nonnegative real number with a real number. If a number x is positive or 0, then the number is its own absolute value. On the other hand, if a number x is negative, then the negative of x (which is positive) is the absolute value of x. In either case, $|x|$ is nonnegative (positive or 0):

$$|x| \geq 0 \qquad \textbf{for all real numbers } x$$

Example 1   Find   **a.** $|9|$,   **b.** $|-5|$,   **c.** $|0|$, and   **d.** $|2 - \pi|$.

Solution   **a.**   Because 9 is positive, 9 is its own absolute value:

$$|9| = 9$$

b.   Because -5 is negative, the negative of -5 is the absolute value:

$$|-5| = -(-5) = 5$$

c.   Part one of the definition states that

$$|0| = 0$$

d.   Because π is approximately 3.14, it follows that $2 - \pi$ is negative. Thus,

$$|2 - \pi| = -(2 - \pi) = \pi - 2 \qquad \blacksquare$$

The placement of a $-$ sign in an expression containing an absolute value symbol is important. For example, $|-19| = 19$, but $-|19| = -(19) = -19$.

Example 2   Find   **a.** $-|-10|$,   **b.** $-|13|$, and   **c.** $-(-|-3|)$.

Solution   **a.** $-|-10| = -(10) = -10$

b. $-|13| = -(13) = -13$

c. $-(-|-3|) = -(-(3)) = 3 \qquad \blacksquare$

Consider the equation $|x| = 5$. If x is positive or 0, then $|x| = x$ and $x = 5$. On the other hand, if x is negative, then $|x| = -x$ and $-x = 5$ or $x = -5$. Hence, we have $x = 5$ or $x = -5$.

We generalize this idea in the following theorem.

Theorem.   If k is a positive constant, then
$$|x| = k \quad \text{is equivalent to} \quad x = k \text{ or } x = -k$$

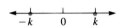

Figure 4-7

The absolute value of x can be interpreted as the distance on a number line that a point is from the origin. The solutions of the equation $|x| = k$ are represented by the two points that lie exactly k units from the origin. See Figure 4-7.

The expression $|x - 3|$ in the equation $|x - 3| = 7$ can be interpreted as a distance also. The equation $|x - 3| = 7$ indicates that a point on a number line with a coordinate of $x - 3$ is 7 units from the origin. Thus, $x - 3$ can be either 7 or -7. This gives two equations, and each can be solved separately:

$$x - 3 = 7 \qquad \text{or} \qquad x - 3 = -7$$
$$x = 10 \qquad\qquad\qquad x = -4$$

The solutions to the equation $|x - 3| = 7$ are $x = 10$ and $x = -4$. Note that, if either 10 or -4 is substituted for x in the expression $|x - 3|$, the result is 7. Hence, both values check.

Example 3   Solve the equation $|3x - 2| = 5$.

Solution   Write $|3x - 2| = 5$ as

$$3x - 2 = 5 \qquad \text{or} \qquad 3x - 2 = -5$$

and solve each possibility for x:

$$3x - 2 = 5 \qquad \text{or} \qquad 3x - 2 = -5$$
$$3x = 7 \qquad\qquad\qquad 3x = -3$$
$$x = \frac{7}{3} \qquad\qquad\qquad x = -1$$

Verify that both solutions check.   ∎

Example 4   Solve the equation $\left|7x + \dfrac{1}{2}\right| = -4$.

Solution   Because the absolute value of any number cannot be negative, the expression $|7x + \frac{1}{2}|$ cannot equal -4. This equation is an impossible equation and has no solutions.   ∎

Example 5   Solve the equation $|5x + 3| = |3x + 25|$.

Solution   This equation is true if $5x + 3$ and $3x + 25$ are equal or if they are negatives of each other. Write an equation for each condition and solve each one for x.

$$
\begin{array}{c|c}
5x + 3 = 3x + 25 & 5x + 3 = -(3x + 25) \\
2x = 22 & 5x + 3 = -3x - 25 \\
x = 11 & 8x = -28 \\
& x = -\dfrac{28}{8} \\
& x = -\dfrac{7}{2}
\end{array}
$$

or

Verify that both solutions check.   ∎

■ EXERCISE 4.5

In Exercises 1–12, find the value of each expression.

1. $|8|$

2. $|-18|$

3. $|-12|$

4. $|15|$

5. $-|2|$

6. $-|-20|$

7. $-|-30|$

8. $-|5|^2$

9. $-(-|50|)$

10. $-(-|-60|)$

11. $|\pi - 4|$

12. $|2\pi - 4|$

In Exercises 13–24, select the smaller of the two numbers.

13. $|2|, \quad |5|$

14. $|-6|, \quad |2|$

15. $|5|, \quad |-8|$

16. $|6|, \quad |3|$

17. $|-2|, \quad |10|$

18. $|-6|, \quad -|6|$

19. $|-3|, \quad -|-4|$

20. $|-3|, \quad |-2|$

21. $-|-5|, \quad -|-7|$

22. $-|-8|, \quad -|20|$

23. $-x, \quad |x + 1|$ (x is a negative integer)

24. $y, \quad |y - 1|$ (y is a positive integer)

In Exercises 25–50, solve each equation, if possible, and check the answers.

25. $|x| = 8$

26. $|x| = 9$

27. $|x - 3| = 6$

28. $|x + 4| = 8$

29. $|2x - 3| = 5$

30. $|4x - 4| = 20$

31. $|3x + 2| = 16$

32. $|5x - 3| = 22$

33. $\left|\dfrac{7}{2}x + 3\right| = -5$

34. $|2x + 10| = 0$

35. $\left|\dfrac{x}{2} - 1\right| = 3$

36. $\left|\dfrac{4x - 64}{4}\right| = 32$

37. $|3x + 24| = 0$

38. $|x - 21| = -8$

39. $\left|\dfrac{3x + 48}{3}\right| = 12$

40. $\left|\dfrac{x}{2} + 2\right| = 4$

41. $|2x + 1| = |3x + 3|$

42. $|5x - 7| = |4x + 1|$

43. $|3x - 1| = |x + 5|$

44. $|3x + 1| = |x - 5|$

45. $\left|\dfrac{x}{2} + 2\right| = \left|\dfrac{x}{2} - 2\right|$

46. $|7x + 12| = |x - 6|$

47. $\left|x + \dfrac{1}{3}\right| = |x - 3|$

48. $\left|x - \dfrac{1}{4}\right| = |x + 4|$

49. $|3x + 7| = -|8x - 2|$

50. $-|17x + 13| = |3x - 14|$

51. Construct several examples to illustrate that $|a \cdot b| = |a| \cdot |b|$.

52. Construct several examples to illustrate that $\left|\dfrac{a}{b}\right| = \dfrac{|a|}{|b|}$.

53. Construct several examples to illustrate that $|a + b|$ does not always equal $|a| + |b|$.

54. Construct several examples to illustrate that $|a - b|$ does not always equal $|a| - |b|$.

4.6   LINEAR INEQUALITIES

So far, we have discussed statements indicating that quantities are equal. We now turn our attention to statements indicating that two quantities are not equal. Such statements are called **inequalities**. Recall the following symbols and their meanings.

$a < b$ means   "*a* is less than *b*."

$a > b$ means   "*a* is greater than *b*."

$a \le b$ means   "*a* is less than or equal to *b*."

$a \ge b$ means   "*a* is greater than or equal to *b*."

Because 2 is less than 3, we write $2 < 3$. To indicate that a number x is greater than 4 or equal to 4, we write $x \ge 4$. By definition, the symbol $a < b$ means that "*a* is less than *b*," but it also means that "*b* is greater than *a*." Note that, if $a < b$, then a lies to the left of b on a number line.

Statements such as $x^2 + 1 > 0$ that are true for all values of x are called **absolute inequalities**. Statements such as $3x + 2 < 8$ that are true for some x, but not all x, are called **conditional inequalities**. We begin the discussion on solving conditional inequalities by listing the fundamental properties of inequalities.

The Trichotomy Property for Real Numbers.   For two real numbers a and b, exactly one of the following three statements is true:

$a < b,$ $a = b,$ or $a > b$

The trichotomy property indicates that one and only one of three statements is true about any two real numbers. Either the first number is less than the second, the first number is equal to the second, or the first number is greater than the second.

If a, b, and c are real numbers with $a < b$ and $b < c$, then $a < c$.

> If a, b, and c are real numbers with $a > b$ and $b > c$, then $a > c$.

The previous two statements cite the **transitive property** for $<$ and $>$. The relationships $\leq$ and $\geq$ are also transitive.

> **Property 1 of Inequalities.**   Any real number can be added to (or subtracted from) both sides of an inequality to produce another inequality with the *same* order (direction).

Consider the true inequality $3 < 12$. If any number x is added to both sides of this inequality, another true inequality results, $3 + x < 12 + x$, and the "is less than" symbol remains an "is less than" symbol. Adding x to both sides does not change the order (direction) of the inequality.

Likewise, subtracting x from both sides of an inequality does not change the order (direction) of the inequality. Thus, we have $3 - x < 12 - x$.

> **Property 2 of Inequalities.**   If both sides of an inequality are multiplied (or divided) by a positive number, another inequality results that has the *same* order (direction) as the original inequality.

Consider the true inequality $-4 < 6$. If both sides of this inequality are multiplied by any positive number such as $+7$, another true inequality results that has the *same* order:

$$-4 < 6$$
$$7(-4) < 7(6) \qquad \text{Multiply both sides by 7.}$$
$$-28 < 42 \qquad \text{Simplify.}$$

Likewise, if both sides of $-4 < 6$ are divided by any positive number such as $+2$, another true inequality results that has the *same* order:

$$-4 < 6$$
$$\frac{-4}{2} < \frac{6}{2} \qquad \text{Divide both sides by 2.}$$
$$-2 < 3 \qquad \text{Simplify.}$$

> **Property 3 of Inequalities.**   If both sides of an inequality are multiplied (or divided) by a negative number, another inequality results, but with the *opposite* order (direction) of the original inequality.

Consider the true inequality $-4 < 6$. If both sides of this inequality are multiplied by any negative number such as -7, another inequality results that has the *opposite* order.

$$-4 < 6$$

$$-7(-4) > -7(6) \qquad \text{Multiply both sides by } -7 \text{ and}$$
$$\text{reverse the order of the inequality.}$$

$$28 > -42 \qquad \text{Simplify.}$$

Likewise, if both sides of $-4 < 6$ are divided by a negative number such as -2, another true inequality results that has the *opposite* order.

$$-4 < 6$$

$$\frac{-4}{-2} > \frac{6}{-2} \qquad \text{Divide both sides by } -2 \text{ and reverse the order of the inequality.}$$

$$2 > -3 \qquad \text{Simplify.}$$

If an inequality can be expressed in either the form $ax + c < 0$ or $ax + c > 0$ with $a \neq 0$, then it is called a **linear inequality**. Solve linear inequalities as if they were linear equations, but with one exception. If you multiply or divide both sides by a *negative* number, remember to reverse the order of the inequality.

Example 1   Solve the linear inequality $3(2x - 9) < 9$.

Solution   Solve the inequality as if it were an equation:

$$3(2x - 9) < 9$$
$$6x - 27 < 9 \qquad \text{Remove parentheses.}$$
$$6x < 36 \qquad \text{Add 27 to both sides.}$$
$$x < 6 \qquad \text{Divide both sides by 6.}$$

It is common to show the solution set as a graph on a number line. All values of x that are less than 6 can be shown by the arrow in Figure 4-8. The open circle at 6 indicates that 6 is not included in the solution set.

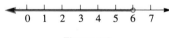

Figure 4-8   ■

Example 2   Solve the linear inequality $-4(3x + 2) \leq 16$.

Solution   Solve the inequality as if it were an equation:

$$-4(3x + 2) \leq 16$$
$$-12x - 8 \leq 16 \qquad \text{Remove parentheses.}$$
$$-12x \leq 24 \qquad \text{Add 8 to both sides.}$$
$$x \geq -2 \qquad \text{Divide both sides by } -12 \text{ and reverse the } \leq \text{ symbol.}$$

Note that the order of the inequality is reversed when both sides are divided by -12. The graph of the solution set is shown in Figure 4-9. The closed circle at -2 indicates that -2 is included in the solution set.

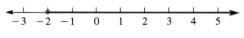

Figure 4-9

Example 3   Solve the inequality $\frac{2}{3}(x + 2) > \frac{4}{5}(x - 3)$.

Solution

$$\frac{2}{3}(x + 2) > \frac{4}{5}(x - 3)$$

$$15 \cdot \frac{2}{3}(x + 2) > 15 \cdot \frac{4}{5}(x - 3) \qquad \text{Multiply both sides by 15.}$$

$$10(x + 2) > 12(x - 3) \qquad \text{Simplify.}$$

$$10x + 20 > 12x - 36 \qquad \text{Remove parentheses.}$$

$$-2x + 20 > -36 \qquad \text{Add } -12x \text{ to both sides.}$$

$$-2x > -56 \qquad \text{Add } -20 \text{ to both sides.}$$

$$x < 28 \qquad \text{Divide both sides by } -2 \text{ and reverse the } > \text{ symbol.}$$

The graph of the solution set is shown in Figure 4-10.

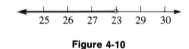

Figure 4-10

Double Inequalities

There are times when it is convenient to say that x is between two numbers— for example, that x is between -3 and 8. This can be expressed by the *double inequality*

$$-3 < x < 8$$

This double inequality is understood to contain two distinct linear inequalities:

$$-3 < x \quad and \quad x < 8$$

These two inequalities mean that "x is greater than -3 *and* x is less than 8." The word *and* indicates that the two inequalities must hold true simultaneously. This result is generalized in the following statement.

> The double inequality $c < x < d$ is equivalent to $c < x$ *and* $x < d$.

Example 4   Solve the inequality $-3 \leq 2x + 5 < 7$.

Solution   This inequality can be solved by isolating x between the inequality symbols in the double inequality:

$$-3 \leq 2x + 5 < 7$$
$$-8 \leq 2x < 2 \qquad \text{Add } -5 \text{ to all three sides.}$$
$$-4 \leq x < 1 \qquad \text{Divide all three sides by 2.}$$

The graph of the solution set is shown in Figure 4-11.

Figure 4-11

Example 5   Solve the inequality $x + 3 < 2x - 1 < 4x - 3$.

Solution   Because it is impossible to isolate x between the "is less than" symbols, it is necessary to convert this double inequality into two linear inequalities, and solve each one separately.

$$x + 3 < 2x - 1 \qquad \text{and} \qquad 2x - 1 < 4x - 3$$
$$4 < x \qquad\qquad\qquad\qquad 2 < 2x$$
$$\qquad\qquad\qquad\qquad\qquad\qquad 1 < x$$

Only those values of x that are greater than 4 and also greater than 1 are in the solution set. Because all numbers that are greater than 4 are also greater than 1, the solutions are all numbers x, where $x > 4$. The graph of the solution set is shown in Figure 4-12.

Figure 4-12

To indicate that x is *not* between two numbers such as -3 and 8 requires the following two separate inequalities:

$$x \leq -3 \quad or \quad x \geq 8$$

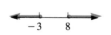

Figure 4-13

This statement indicates that x must be less than or equal to -3 *or* x must be greater than or equal to 8. It is incorrect to string the inequalities together as $8 \leq x \leq -3$ because that would imply that $8 \leq -3$, which is impossible. Consequently, these two inequalities *cannot* be written as a double inequality. The graph of the statement $x \leq -3$ or $x \geq 8$ appears in Figure 4-13.

Note that the word *or* in the statement $x \leq -3$ or $x \geq 8$ indicates that only one of the inequalities needs to be true.

■ EXERCISE 4.6 ■

In Exercises 1–42, solve each inequality and graph its solution set.

1. $x + 4 < 5$ **2.** $x - 5 > 2$ **3.** $-3x - 1 \le 5$ **4.** $-2x + 6 \ge 16$

5. $5x - 3 > 7$ **6.** $7x - 9 < 5$ **7.** $8 - 9y \ge -y$ **8.** $4 - 3x \le x$

9. $-3(a + 2) > 2(a + 1)$ **10.** $-4(y - 1) < y + 8$

11. $3(z - 2) \le 2(z + 7)$ **12.** $5(3 + z) > -3(z + 3)$

13. $-11(2 - b) < 4(2b + 2)$ **14.** $-9(h - 3) + 2h \le 8(4 - h)$

15. $\dfrac{1}{2}y + 2 \ge \dfrac{1}{3}y - 4$ **16.** $\dfrac{1}{4}x - \dfrac{1}{3} \le x + 2$ **17.** $\dfrac{3}{4}x - 4 < \dfrac{4}{5}x + 1$ **18.** $\dfrac{7}{8}(a - 3) < \dfrac{3}{4}(a + 3)$

19. $\dfrac{2}{3}x + \dfrac{3}{2}(x - 5) \le x$ **20.** $\dfrac{5}{9}(x + 3) - \dfrac{4}{3}(x - 3) \ge x - 1$

21. $-2 < -b + 3 < 5$ **22.** $4 < -t - 2 < 9$ **23.** $15 > 2x - 7 > 9$ **24.** $25 > 3x - 2 > 7$

25. $-6 < -3(x - 4) \le 24$ **26.** $-4 \le -2(x + 8) < 8$

27. $0 \ge \dfrac{1}{2}x - 4 > 6$ **28.** $-6 \le \dfrac{1}{3}a + 1 < 0$ **29.** $0 \le \dfrac{4 - x}{3} \le 2$ **30.** $-2 \le \dfrac{5 - 3x}{2} \le 2$

31. $x + 3 < 3x - 1 < 2x + 2$ **32.** $x - 1 \le 2x + 4 \le 3x - 1$

33. $4x \ge -x + 5 \ge 3x - 4$ **34.** $\dfrac{1}{2}x > -\dfrac{1}{3}x > x + 2$

35. $5(x + 1) \le 4(x + 3) < 3(x - 1)$ **36.** $-5(2 + x) < 4x + 1 < 3x$

37. $3x + 2 < 8$ or $2x - 3 > 11$ **38.** $3x + 4 < -2$ or $3x + 4 > 10$

39. $-4(x + 2) \ge 12$ or $3x + 8 < 11$ **40.** $5(x - 2) \ge 0$ and $-3x < 9$

41. $x < -3$ and $x > 3$ **42.** $x < 3$ or $x > -3$

43. Construct examples to illustrate that $|x| + |y| \ge |x + y|$.

44. Under what conditions is $|x| + |y| = |x + y|$?

45. Under what conditions is $|x| + |y| > |x + y|$? **46.** Are the relations "$\le$" and "$\ge$" reflexive?

47. Are the relations "$<$" and "$>$" reflexive? **48.** Are the relations "$<$" and "$>$" symmetric?

49. Are the relations "$\le$" and "$\ge$" symmetric? **50.** If $x < 3$, must it be true that $x^2 < 9$?

51. The $27 wholesale cost of a clock radio added to the profit must be no more than $42 or the radio will not sell. What are the possible profits?

52. A train can travel 80 miles per hour for no more than 3 hours. How far might the train travel in that time?

53. One side of a regular pentagon (a figure with five equal sides) must be at least 37 inches but cannot exceed 52 inches. What is the range of values for the perimeter?

54. Jim can afford to spend no more than $330 on a record player and some records. If the record player costs $175 and records are $8.50 each, what is the greatest number of records he can buy?

55. A student has exam scores of 70, 77, and 85. What score on a fourth exam would give him an average of 80 or better?

56. The length of a rectangle is 3 feet greater than its width. If the perimeter must be at least 34 feet but not more than 54 feet, what is the possible range of values of the width?

4.7 INEQUALITIES CONTAINING ABSOLUTE VALUES

Consider the inequality $|x| < 5$. If $x \geq 0$, then $|x| = x$ and $|x| < 5$ is equivalent to $x < 5$. If $x < 0$, then $|x| = -x$ and $-x < 5$, which is equivalent to $x > -5$. Thus, if x is positive or 0, then x must be less than 5, and if x is negative, then x must be greater than -5. This implies that x must be between -5 and 5. The two inequalities $x < 5$ and $x > -5$ are equivalent to the double inequality $-5 < x < 5$. We generalize the results of this example in the following theorem.

Theorem.   If k is a positive constant, then

$$|x| < k \quad \text{is equivalent to} \quad -k < x < k$$

The solutions to the inequality $|x| < k$ are the coordinates of the points on a number line that are less than k units from the origin. See Figure 4-14.

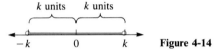

Figure 4-14

Example 1   Solve the inequality $|2x - 3| < 9$.

Solution   Rewrite the inequality as a double inequality and solve for x:

$$|2x - 3| < 9 \quad \text{is equivalent to} \quad -9 < 2x - 3 < 9$$

$$-9 < 2x - 3 < 9$$
$$-6 < 2x < 12 \qquad \text{Add 3 to all sides.}$$
$$-3 < x < 6 \qquad \text{Divide all sides by 2.}$$

Figure 4-15

Any number between -3 and 6, not including either -3 or 6, is in the solution set for this inequality. The graph appears in Figure 4-15. ■

The previous theorem is also true if the "$<$" symbol is replaced by the "$\leq$" symbol. Thus,

$$|x| \leq k \quad \text{is equivalent to} \quad -k \leq x \leq k$$

Example 2   Solve the inequality $|3x + 2| \leq 5$.

Solution   Rewrite the expression as a double inequality and solve for x:

$$|3x + 2| \leq 5 \quad \text{is equivalent to} \quad -5 \leq 3x + 2 \leq 5$$

$$-5 \leq 3x + 2 \leq 5$$
$$-7 \leq 3x \leq 3 \qquad \text{Add } -2 \text{ to all sides.}$$
$$-\frac{7}{3} \leq x \leq 1 \qquad \text{Divide all sides by 3.}$$

Figure 4-16

Figure 4-16 is the graph of the solution set. ■

We have considered the relationships $|x| = k$ and $|x| < k$. We now consider the inequality $|x| > k$, where k is a positive constant. If $x \geq 0$, then $|x| = x$ and $|x| > k$ is equivalent to $x > k$. If $x < 0$, then $|x| = -x$ and $-x > k$, which is equivalent to $x < -k$. Thus, if x is positive or 0, then x is greater than k; if x is negative, then x is less than $-k$. This implies that we can write the solutions to the inequality $|x| > k$ as

$$x < -k \quad \text{or} \quad x > k$$

The solutions to the inequality $|x| > k$ are the coordinates of the points on the number line that are greater than k units from the origin. See Figure 4-17.

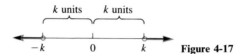

Figure 4-17

The solutions of the inequality $|x| > k$ are $x < -k$ or $x > k$. The *or* indicates an either/or situation. It is necessary for x to satisfy only one of the two conditions to be in the solution set.

Theorem.   If k is a nonnegative constant, then

$$|x| > k \quad \text{is equivalent to} \quad x < -k \text{ or } x > k$$

Example 3   Solve the inequality $|5x - 10| > 20$.

Solution   Write the inequality as two separate inequalities and solve each one for x:

$$|5x - 10| > 20 \quad \text{is equivalent to} \quad 5x - 10 < -20 \text{ or } 5x - 10 > 20$$

$$5x - 10 < -20 \qquad \text{or} \qquad 5x - 10 > 20$$
$$5x < -10 \qquad\qquad\qquad 5x > 30 \qquad \text{Add 10 to both sides.}$$
$$x < -2 \qquad\qquad\qquad x > 6 \qquad \text{Divide both sides by 5.}$$

Figure 4-18

Thus, x is either less than -2 or greater than 6: $x < -2$ or $x > 6$. Figure 4-18 is the graph of the solution set. ∎

The previous theorem is also true if the ">" and "<" symbols are replaced by "$\geq$" and "$\leq$" symbols, respectively. Thus,

Figure 4-19

$$|x| \geq k \quad \text{is equivalent to} \quad x \leq -k \text{ or } x \geq k$$

Example 4   Solve the inequality $|3x - 5| \geq -2$.

Solution   Because the absolute value of any number is nonnegative and because any nonnegative number is larger than -2, the inequality is true for all x. Figure 4-19 is the graph of the solution set for this absolute inequality. ∎

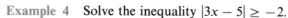

Example 5   Solve the inequality $\left|\dfrac{x-3}{5}\right| \geq 6$.

Solution   Rewrite the inequality as two separate inequalities and solve each one for x:

$$\left|\frac{x-3}{5}\right| \geq 6 \quad \text{is equivalent to} \quad \frac{x-3}{5} \leq -6 \quad \text{or} \quad \frac{x-3}{5} \geq 6$$

$$\begin{array}{lll}
\dfrac{x-3}{5} \leq -6 & \text{or} \quad \dfrac{x-3}{5} \geq 6 & \\[2mm]
x - 3 \leq -30 & x - 3 \geq 30 & \text{Multiply both sides by 5.} \\[1mm]
x \leq -27 & x \geq 33 & \text{Add 3 to both sides.}
\end{array}$$

Figure 4-20

Figure 4-20 is the graph of the solution set.   ∎

Example 6   Solve the inequality $\left|\dfrac{2}{3}x - 2\right| - 3 > 6$.

Solution

$$\left|\frac{2}{3}x - 2\right| - 3 > 6$$

$$\left|\frac{2}{3}x - 2\right| > 9 \qquad\qquad \text{Add 3 to both sides.}$$

$$\begin{array}{lll}
\dfrac{2}{3}x - 2 < -9 \quad \text{or} \quad \dfrac{2}{3}x - 2 > 9 & & \begin{array}{l}\text{Rewrite as two separate} \\ \text{inequalities.}\end{array} \\[3mm]
\dfrac{2}{3}x < -7 \qquad\qquad \dfrac{2}{3}x > 11 & & \text{Add 2 to both sides.} \\[3mm]
2x < -21 \qquad\qquad 2x > 33 & & \text{Multiply both sides by 3.} \\[3mm]
x < -\dfrac{21}{2} \qquad\qquad x > \dfrac{33}{2} & & \text{Divide both sides by 2.}
\end{array}$$

Figure 4-21

Figure 4-21 is the graph of the solution set.   ∎

▬ EXERCISE 4.7 ▬▬▬▬▬▬▬▬▬▬▬▬▬▬▬▬▬▬▬▬▬

In Exercises 1–40, solve each inequality and graph its solution set.

1. $\|2x\| < 8$	**2.** $\|3x\| < 27$	**3.** $\|x+9\| \leq 12$	**4.** $\|x-8\| \leq 12$
5. $\|3x+2\| < -3$	**6.** $\|3x-2\| < 10$	**7.** $\|4x-1\| \leq 7$	**8.** $\|5x-12\| < -5$
9. $\|5x\| > 5$	**10.** $\|7x\| > 7$	**11.** $\|x-12\| \geq 24$	**12.** $\|x+5\| \geq 7$
13. $\|3x+2\| > 14$	**14.** $\|2x-5\| > 25$	**15.** $\|4x+3\| > -5$	**16.** $\|4x+3\| > 0$
17. $-\|2x-3\| < -7$	**18.** $-\|3x+1\| < -8$	**19.** $\|8x-3\| > 0$	**20.** $\|7x+2\| > -8$
21. $\left\|\dfrac{x-2}{3}\right\| \leq 4$	**22.** $\left\|\dfrac{x-2}{3}\right\| > 4$	**23.** $\|3x+1\| + 2 < 6$	**24.** $\|3x-2\| + 2 \geq 0$

25. $3|2x + 5| \geq 9$ **26.** $-2|3x - 4| < 16$ **27.** $|5x - 1| + 4 \leq 0$ **28.** $-|5x - 1| + 2 < 0$

29. $\left|\frac{1}{3}x + 7\right| + 5 > 6$ **30.** $\left|\frac{1}{2}x - 3\right| - 4 < 2$ **31.** $\left|\frac{1}{5}x - 5\right| + 4 > 4$ **32.** $\left|\frac{1}{6}x + 6\right| + 2 < 2$

33. $\left|\frac{3}{5}x + \frac{7}{3}\right| < 2$ **34.** $\left|\frac{7}{3}x - \frac{3}{5}\right| \geq 1$ **35.** $\left|3\left(\frac{x + 4}{4}\right)\right| > 0$ **36.** $3\left|\frac{1}{3}(x - 2)\right| + 2 \leq 3$

37. $\left|\frac{1}{7}x + 1\right| \leq 0$ **38.** $|2x + 1| + 2 \leq 2$ **39.** $\left|\frac{x - 5}{10}\right| \leq 0$ **40.** $\left|\frac{3}{5}x - 2\right| + 3 \leq 3$

CHAPTER SUMMARY

Key Words

absolute inequalities   (4.6)
conditional equation   (4.1)
conditional inequalities   (4.6)
contradiction   (4.1)
domain   (4.1)
equation   (4.1)
equivalent equations   (4.1)
first-degree equation   (4.1)
formula   (4.3)
identity   (4.1)
impossible equation   (4.1)

incomplete quadratic equation   (4.4)
linear equation   (4.1)
linear inequality   (4.6)
literal equation   (4.3)
quadratic equation   (4.4)
roots of an equation   (4.1)
second-degree equation   (4.4)
solution of an equation   (4.1)
solution set   (4.1)
variable   (4.1)

Key Ideas

(4.1)   An **equation** is a statement indicating that two quantities are equal.

An **identity** is an equation that is true for all values of its variable.

A **conditional equation** is true for some but not all values of its variable.

An **impossible equation** is true for no values of its variable.

The **domain** of a variable is the set of numbers that, when substituted for the variable, give statements that are defined and can be judged as true or false.

The numbers in the domain of a variable that make an equation true are called **solutions** or **roots** of the equation.

If a and b are real numbers and $a = b$, then

$$a + c = b + c \qquad a - c = b - c \qquad ac = bc \qquad \frac{a}{c} = \frac{b}{c} \quad (c \neq 0)$$

Every repeating decimal can be written as a rational number in fractional form.

(4.2)   To solve a word problem, try to express a quantity in two ways and, thereby, form an equation. Then solve the equation.

(4.3)   **Literal equations** are equations that contain several variables. Many literal equations are formulas.

Use the methods for solving equations to solve literal equations for an indicated variable.

(4.4)   To solve a **quadratic equation** by factoring, write the equation in quadratic form, factor the polynomial, and use the zero-factor theorem.

(4.5) $\begin{cases} \text{If } x \geq 0, \text{ then } |x| = x. \\ \text{If } x < 0, \text{ then } |x| = -x. \end{cases}$

If k is a positive constant,

$$|x| = k \quad \text{is equivalent to} \quad x = k \text{ or } x = -k.$$

(4.6)   The **trichotomy property** states that, if a and b are two real numbers, then $a < b$, $a = b$, or $a > b$.

The relationships $<$, $>$, $\leq$, and $\geq$ obey the transitive property.

Solve a **linear inequality** in the same way as a linear equation. However, remember to change the order of the inequality if both sides are multiplied or divided by a negative number.

Solve a **double inequality** by isolating the variable between the inequality symbols, or by solving each inequality separately.

(4.7)   If k is a positive constant, then

$$|x| < k \quad \text{is equivalent to} \quad -k < x < k$$
$$|x| > k \quad \text{is equivalent to} \quad x < -k \text{ or } x > k$$

REVIEW EXERCISES

In Review Exercises 1–8, solve and check each equation.

1. $5x + 12 = 37$

2. $-3x - 7 = 20$

3. $4(y - 1) = 28$

4. $3(x + 7) = 42$

5. $13(x - 9) - 2 = 7x - 5$

6. $\dfrac{8(x - 5)}{3} = 2(x - 4)$

7. $\dfrac{3y}{4} - 13 = -\dfrac{y}{3}$

8. $\dfrac{2y}{5} + 5 = \dfrac{14y}{10}$

In Review Exercises 9–12, solve and check each word problem.

9.   The sum of three consecutive even integers is 270. What are the integers?

10. A rectangle is 4 meters longer than it is wide. If the perimeter of the rectangle is 28 meters, find its area.

11. Sally has $25,000 to invest. She invests some money at 10% interest and the rest at 9%. If her total annual income from these two investments is $2430, how much did she invest at each rate?

12. How much water must be added to 20 liters of a 12% alcohol solution to dilute it to an 8% solution?

In Review Exercises 13–16, solve for the variable indicated.

13. $V = \dfrac{4}{3}\pi r^3$ for r^3

14. $V = \dfrac{1}{3}\pi r^2 h$ for h

15. $V = \dfrac{1}{6}H(S_0 + 4S_1 + S_2)$ for S_0

16. $V = \pi h^2\left(r - \dfrac{h}{3}\right)$ for r

In Review Exercises 17–20, solve each equation by factoring. Check each solution.

17. $12x^2 + 4x - 5 = 0$

18. $7y^2 - 37y + 10 = 0$

19. $t^2(15t - 2) = 8t$

20. $3u^3 = u(19u + 14)$

In Review Exercises 21–24, solve and check each equation.

21. $|3x + 1| = 10$

22. $\left|\dfrac{3}{2}x - 4\right| = 9$

23. $|3x + 2| = |2x - 3|$

24. $|5x - 4| = |4x - 5|$

In Review Exercises 25–28, solve each inequality and graph its solution set.

25. $\dfrac{1}{3}y - 2 \geq \dfrac{1}{2}y + 2$

26. $\dfrac{7}{4}(x + 3) < \dfrac{3}{8}(x - 3)$

27. $3 < 3x + 4 < 10$

28. $4x > 3x + 2 > x - 3$

In Review Exercises 29–32, solve each inequality and graph its solution set.

29. $|2x + 7| < 3$

30. $|3x - 8| \geq 4$

31. $\left|\dfrac{3}{2}x - 14\right| \geq 0$

32. $\left|\dfrac{2}{3}x + 14\right| < 0$

CHAPTER FOUR TEST

1. Find the domain of x in the following equation. **Do not solve the equation.**

$$\frac{x - 2}{(x + 2)(x - 3)} = 7$$

In Problems 2–3, solve each equation.

2. $9(x + 4) + 4 = 4(x - 5)$

3. $\dfrac{y - 1}{5} + 2 = \dfrac{2y - 3}{3}$

4. A 20-foot pipe is to be cut into three pieces. One piece is to be twice as long as another, and the third piece is to be six times as long as the shortest. What will be the length of the longest piece?

5. A rectangle with a perimeter of 26 centimeters is 5 centimeters longer than it is wide. What is its area?

6. Bob invests part of $10,000 at 9% annual interest and the rest at 8%. His annual income from these investments is $860. How much does he invest at 8%?

7. Two students drive separate cars from one city to another. One student drives at 50 mph and the other at 55 mph. How far apart are the cities if the slower driver needs an extra 36 minutes to complete the trip?

8. How many liters of water are needed to dilute 20 liters of a 5% salt solution to a 1% solution?

9. Solve $P = L + \dfrac{s}{f} i$ for i.

10. Solve $r_1 r_2 = r_2 r + r_1 r$ for r.

In Problems 11–12, solve each equation.

11. $x^2 - 5x - 6 = 0$

12. $t^2(1 + t) - 2t = 10t$

13. The area of a square is numerically three times as great as its perimeter. Find its perimeter.

14. The sum of the squares of two consecutive integers is 61. Find the product of the integers.

In Problems 15–16, write each expression without using absolute value symbols.

15. $|5 - 8|$

16. $|4\pi - 4|$

In Problems 17–22, solve each equation or inequality. Graph each solution set on the number line.

17. $|2x + 3| = 11$

18. $|3x + 4| = |x + 12|$

19. $-2(2x + 3) \geq 14$

20. $-2 < \dfrac{x - 4}{3} < 4$

21. $|x + 3| \leq 4$

22. $|2x - 4| > 2$

STATISTICIAN

Statisticians devise, carry out, and interpret the numerical results of surveys and experiments, and then apply their knowledge to subject areas such as economics, human behavior, natural science, or engineering.

Often they are able to obtain accurate information about a group of people or things by surveying a small portion, called a **sample**, rather than the whole group. This technique requires statisticians to decide where and how to get the data, how to determine the type and size of the sample group, and how to develop the survey questionnaire or reporting form.

Qualifications

A bachelor's degree in statistics or mathematics is the minimum educational requirement for many beginning jobs in statistics. For other entry level statistical jobs, however, a bachelor's degree with a major in an applied field such as economics or natural science and a minor in statistics is preferable.

Job outlook

Employment opportunities for persons who combine training in statistics with knowledge of a field of application such as manufacturing, engineering, scientific research, or business are expected to be favorable through the mid-1990s.

Example application

A researcher wishes to estimate the mean (average) property tax paid by homeowners living in a certain town. To do so, she decides to select a **random sample** of resident homeowners, and compute the mean tax paid by homeowners in that sample. How large must the sample be so the researcher can be 95% sure that the sample mean will be within $35 of the population mean? Assume the standard deviation (σ) of all tax bills is $110.

Solution

From elementary statistics the researcher has the formula

$$1.96\,\frac{\sigma}{\sqrt{N}} < E$$

where σ is the standard deviation, E is the maximum acceptable error, and N is the sample size. She substitutes 110 for σ and 35 for E in the previous formula, and solves for N.

$$1.96\,\frac{110}{\sqrt{N}} < 35$$

$$1.96(110) < 35\sqrt{N} \qquad \text{Multiply both sides by } \sqrt{N}.$$

$$\frac{1.96(110)}{35} < \sqrt{N} \qquad \text{Divide both sides by 35.}$$

$$6.16 < \sqrt{N} \qquad \text{Simplify.}$$

$$37.9 < N \qquad \text{Square both sides.}$$

To be 95% sure that the sample mean will be within $35 of the population mean, the researcher must have a random sample of at least 38 resident homeowners.

Exercises

1. In the example, what must the sample size be to keep the error less than $20?
2. What must the sample size be in the example to keep the error less than $10?
3. If the sample size in the example were 12, what would the maximum error be?
4. If the sample size in the example were 48, what would the maximum error be?

(Answers: **1.** 117 **2.** 465 **3.** $62.24 **4.** $31.12)

5 Rational Expressions

Expressions such as $\frac{3}{5}$ and $-\frac{7}{4}$ that indicate the quotient of two numbers are called **arithmetic fractions**. Algebraic fractions such as

$$\frac{x^2 + 2}{x - 3} \quad \text{and} \quad \frac{a^3 + 2a^2 + 7}{2a^2 - 5a + 4}$$

that indicate the quotient of two polynomials are called **rational expressions**. Because division by 0 is undefined, the value of a polynomial occurring in the denominator of a rational expression can never be 0. For example, x cannot be 7 in the fraction

$$\frac{3x}{x - 7}$$

because this value would cause the denominator of the fraction to be 0. Likewise, the number -2 cannot be substituted for m in the fraction

$$\frac{5m + n}{8m + 16}$$

because the denominator would be 0: $8(-2) + 16 = 0$.

The factoring skills learned in Chapter 3 will be of extreme value as we develop the properties of rational expressions.

5.1 SIMPLIFYING RATIONAL EXPRESSIONS

Rational expressions—algebraic fractions with polynomial numerators and polynomial denominators—behave exactly like arithmetic fractions. They can be simplified, multiplied, divided, added, and subtracted using the same rules that govern arithmetic fractions.

To simplify a fraction means to reduce it to lowest terms. To do so, we use Property 3 of fractions, first introduced in Section 1.4. This property enables us to divide out all factors that are common to both the numerator and the denominator of a fraction.

> **Property 3 of Fractions.**   If $b \neq 0$ and $k \neq 0$, then
>
> $$\frac{a}{b} = \frac{ak}{bk}$$

Example 1 Simplify the following rational expressions by reducing them to lowest terms:

$$\textbf{a.}\quad \frac{10k}{25k^2} \quad \text{and} \quad \textbf{b.}\quad \frac{-8y^3z^5}{6y^4z^3}$$

Solution Factor each numerator and denominator, and use Property 3 of fractions to divide out all common factors:

$$\textbf{a.}\quad \frac{10k}{25k^2} = \frac{5 \cdot 2 \cdot k}{5 \cdot 5 \cdot k \cdot k}$$

$$= \frac{\cancel{5} \cdot 2 \cdot \cancel{k}}{\cancel{5} \cdot 5 \cdot \cancel{k} \cdot k}$$

$$= \frac{2}{5k}$$

$$\textbf{b.}\quad \frac{-8y^3z^5}{6y^4z^3} = \frac{-2 \cdot 4 \cdot y \cdot y \cdot y \cdot z \cdot z \cdot z \cdot z \cdot z}{2 \cdot 3 \cdot y \cdot y \cdot y \cdot y \cdot z \cdot z \cdot z}$$

$$= \frac{-\cancel{2} \cdot 4 \cdot \cancel{y} \cdot \cancel{y} \cdot \cancel{y} \cdot \cancel{z} \cdot \cancel{z} \cdot \cancel{z} \cdot z \cdot z}{\cancel{2} \cdot 3 \cdot \cancel{y} \cdot \cancel{y} \cdot \cancel{y} \cdot y \cdot \cancel{z} \cdot \cancel{z} \cdot \cancel{z}}$$

$$= -\frac{4z^2}{3y} \qquad \blacksquare$$

Example 2 Simplify the rational expression $\dfrac{x^2 - 16}{x + 4}$.

Solution Factor the numerator and divide out all common factors:

$$\frac{x^2 - 16}{x + 4} = \frac{\cancel{(x + 4)}(x - 4)}{1\cancel{(x + 4)}} = \frac{x - 4}{1} = x - 4 \qquad \blacksquare$$

Example 3 Simplify the fraction $\dfrac{2x^2 + 11x + 12}{3x^2 + 11x - 4}$.

Solution Factor both the numerator and the denominator and divide out all common factors:

$$\frac{2x^2 + 11x + 12}{3x^2 + 11x - 4} = \frac{(2x + 3)\cancel{(x + 4)}}{(3x - 1)\cancel{(x + 4)}} = \frac{2x + 3}{3x - 1}$$

CAUTION! Do not divide out the x's in this result. The x in the numerator is a factor of the first *term* only and not a factor of the entire numerator. Likewise, the x in the denominator is not a factor of the entire denominator. ■

To simplify the fraction $\dfrac{b - a}{a - b}$, we can factor -1 from the numerator and divide out any factors common to both the numerator and the denominator:

$$\frac{b - a}{a - b} = \frac{-(-b + a)}{a - b} = \frac{-\overset{1}{\cancel{(a - b)}}}{\underset{1}{\cancel{a - b}}} = \frac{-1}{1} = -1$$

In general, we have the following theorem:

Theorem.   The quotient of any nonzero quantity and its negative is -1.

Example 4   Simplify the fraction $\dfrac{3x^2 - 10xy - 8y^2}{4y^2 - xy}$.

Solution   Factor both the numerator and the denominator and apply the previous theorem:

$$\frac{3x^2 - 10xy - 8y^2}{4y^2 - xy} = \frac{(3x + 2y)\overset{-1}{\cancel{(x - 4y)}}}{y\underset{1}{\cancel{(4y - x)}}}$$

Because $x - 4y$ and $4y - x$ are negatives, their quotient is -1.

$$= \frac{-(3x + 2y)}{y}$$

$$= \frac{-3x - 2y}{y}$$ ■

Many fractions we shall encounter are already in simplified form. For example, to attempt to simplify the fraction

$$\frac{x^2 + xa + 2x + 2a}{x^2 + x - 6}$$

we factor both the numerator and the denominator and divide out any common factors that exist:

$$\frac{x^2 + xa + 2x + 2a}{x^2 + x - 6} = \frac{x(x + a) + 2(x + a)}{(x - 2)(x + 3)} = \frac{(x + a)(x + 2)}{(x - 2)(x + 3)}$$

Because there are no factors common to both the numerator and the denominator, this fraction cannot be simplified. It is already in lowest terms.

Example 5   Simplify the fraction $\dfrac{(x^2 + 2x)(x^2 + 2x - 3)}{(x^2 + x - 2)(x^2 + 3x)}$.

Solution   Factor the polynomials in both the numerator and the denominator and divide out all common factors:

$$\frac{(x^2 + 2x)(x^2 + 2x - 3)}{(x^2 + x - 2)(x^2 + 3x)} = \frac{\cancel{x}(\cancel{x + 2})(\cancel{x + 3})(\cancel{x - 1})}{(\cancel{x + 2})(\cancel{x - 1})\cancel{x}(\cancel{x + 3})}$$

$$= 1 \qquad \blacksquare$$

Remember that only *factors* that are common to the *entire* numerator and the *entire* denominator can be divided out. *Terms* that are common to both the numerator and denominator *cannot* be divided out. For example, consider this correct simplification:

$$\frac{3 + 7}{3} = \frac{10}{3}$$

It would be incorrect to divide out the common *term* of 3 in the above simplification. Note that doing so gives an incorrect answer:

$$\cancel{\frac{3 + 7}{3} = \frac{\overset{1}{\cancel{3}} + 7}{\underset{1}{\cancel{3}}} = \frac{1 + 7}{1}} = 8$$

The 3's in the fraction

$$\frac{5 + 3(2)}{3(4)}$$

cannot be divided out either. The 3 in the numerator is a factor of the second term only. To be divided out, it must be a factor of the entire numerator. It is not correct to divide out the y in the fraction

$$\frac{x^2 y + 6x}{y}$$

either. The y is not a factor of the entire numerator.

EXERCISE 5.1

In Exercises 1–58, simplify each fraction if possible. If a fraction is already in lowest terms, so indicate.

1. $\dfrac{12}{18}$ **2.** $\dfrac{25}{55}$ **3.** $-\dfrac{112}{36}$ **4.** $-\dfrac{49}{21}$

5. $\dfrac{288}{312}$ **6.** $\dfrac{144}{72}$ **7.** $-\dfrac{244}{74}$ **8.** $-\dfrac{512}{236}$

9. $\dfrac{12x^3}{3x}$

10. $-\dfrac{15a^2}{25a^3}$

11. $\dfrac{-24x^3y^4}{18x^4y^3}$

12. $\dfrac{15a^5b^4}{21b^3c^2}$

13. $\dfrac{(3x^3)^2}{9x^4}$

14. $\dfrac{8(x^2y^3)^3}{2(xy^2)^2}$

15. $-\dfrac{11x(x-y)}{22(x-y)}$

16. $\dfrac{x(x-2)^2}{(x-2)^3}$

17. $\dfrac{9y^2(y-z)}{21y(y-z)^2}$

18. $\dfrac{(a-b)(b-c)(c-d)}{(c-d)(b-c)(a-b)}$

19. $\dfrac{x+y}{x^2-y^2}$

20. $\dfrac{x-y}{x^2-y^2}$

21. $\dfrac{5x-10}{x^2-4x+4}$

22. $\dfrac{y-xy}{xy-x}$

23. $\dfrac{12-3x^2}{x^2-x-2}$

24. $\dfrac{x^2+2x-15}{x^2-25}$

25. $\dfrac{3x+6y}{x+2y}$

26. $\dfrac{x^2+y^2}{x+y}$

27. $\dfrac{x^3+8}{x^2-2x+4}$

28. $\dfrac{x^2+3x+9}{x^3-27}$

29. $\dfrac{x^2+2x+1}{x^2+4x+3}$

30. $\dfrac{6x^2+x-2}{8x^2+2x-3}$

31. $\dfrac{3m-6n}{3n-6m}$

32. $\dfrac{ax+by+ay+bx}{a^2-b^2}$

33. $\dfrac{4x^2+24x+32}{16x^2+8x-48}$

34. $\dfrac{a^2-4}{a^3-8}$

35. $\dfrac{3x^2-3y^2}{x^2+2y+2x+yx}$

36. $\dfrac{x^2+x-30}{x^2-x-20}$

37. $\dfrac{4x^2+8x+3}{6+x-2x^2}$

38. $\dfrac{6x^2+13x+6}{6-5x-6x^2}$

39. $\dfrac{a^3+27}{4a^2-36}$

40. $\dfrac{a-b}{b^2-a^2}$

41. $\dfrac{2x^2-3x-9}{2x^2+3x-9}$

42. $\dfrac{6x^2-7x-5}{2x^2+5x+2}$

43. $\dfrac{(m+n)^3}{m^2+2mn+n^2}$

44. $\dfrac{x^3-27}{3x^2-8x-3}$

45. $\dfrac{x^4-y^4}{(x^2+2xy+y^2)(x^2+y^2)}$

46. $\dfrac{-4x-4+3x^2}{4x^2-2-7x}$

47. $\dfrac{4a^2-9b^2}{2a^2-ab-6b^2}$

48. $\dfrac{x^2+2xy}{x+2y+x^2-4y^2}$

49. $\dfrac{x-y}{x^3-y^3-x+y}$

50. $\dfrac{2x^2+2x-12}{x^3+3x^2-4x-12}$

51. $\dfrac{x^6-y^6}{x^4+x^2y^2+y^4}$

52. $\dfrac{6xy-4x-9y+6}{6y^2-13y+6}$

53. $\dfrac{x^4+15x^2+64}{x^2+8-x}$

54. $\dfrac{x^2+9-x}{x^4+17x^2+81}$

55. $\dfrac{(x^2-1)(x+1)}{(x^2-2x+1)^2}$

56. $\dfrac{(x^2+2x+1)(x^2-2x+1)}{(x^2-1)^2}$

57. $\dfrac{(2x^2+3xy+y^2)(3a+b)}{(x+y)(2xy+2bx+y^2+by)}$

58. $\dfrac{(x-1)(6ax+9x+4a+6)}{(3x+2)(2ax-2a+3x-3)}$

5.2 MULTIPLYING AND DIVIDING RATIONAL EXPRESSIONS

Recall that Property 4 of fractions, first introduced in Section 1.4, asserted that fractions are multiplied and divided according to the following rules:

> **Property 4 of Fractions.** If no denominators are 0, then
>
> $$\frac{a}{b}\cdot\frac{c}{d}=\frac{a\cdot c}{b\cdot d}=\frac{ac}{bd} \quad \text{and} \quad \frac{a}{b}\div\frac{c}{d}=\frac{a}{b}\cdot\frac{d}{c}=\frac{ad}{bc}$$

These rules were assumed to be true in Section 1.4. We now explain why they are reasonable. We begin with the multiplication rule.

Recall that the area of a rectangle is the product of its length and width. If the square in Figure 5-1 is divided into 5 equal parts vertically and 4 equal parts horizontally, the square is divided into 20 equal pieces. Each of the 20 pieces represents $\frac{1}{20}$ of the total area of the square. The shaded rectangle has a width of $\frac{3}{4}$ and a length of $\frac{4}{5}$. The area of the shaded rectangle is 12 of these 20 equal pieces. Hence, we have

$$A = l \cdot w$$
$$\frac{12}{20} = \frac{4}{5} \cdot \frac{3}{4}$$

This example justifies the procedure for multiplying fractions: multiply the numerators and multiply the denominators.

Figure 5-1

Example 1   Find the product of $\dfrac{x^2 - 6x + 9}{x}$ and $\dfrac{x^2}{x - 3}$.

Solution   Factor $x^2 - 6x + 9$ and x^2, find the product of the fractions by multiplying their numerators and multiplying their denominators. Then simplify the resulting fraction by dividing out any factors that are common to both the numerator and denominator.

$$\frac{x^2 - 6x + 9}{x} \cdot \frac{x^2}{x - 3} = \frac{(x - 3)(x - 3)}{x} \cdot \frac{xx}{x - 3}$$
$$= \frac{(x - 3)(x - 3)xx}{x(x - 3)}$$
$$= x(x - 3) \qquad \blacksquare$$

Example 2   Find the product of $\dfrac{x^2 - x - 6}{x^2 - 4}$ and $\dfrac{x^2 + x - 6}{x^2 - 9}$.

Solution   Factor each polynomial, find the product of the fractions, and simplify the resulting fraction.

$$\frac{x^2 - x - 6}{x^2 - 4} \cdot \frac{x^2 + x - 6}{x^2 - 9} = \frac{(x - 3)(x + 2)}{(x + 2)(x - 2)} \cdot \frac{(x + 3)(x - 2)}{(x + 3)(x - 3)}$$
$$= \frac{(x - 3)(x + 2)(x + 3)(x - 2)}{(x + 2)(x - 2)(x + 3)(x - 3)}$$
$$= \frac{(x - 3)(x + 2)(x + 3)(x - 2)}{(x + 2)(x - 2)(x + 3)(x - 3)}$$
$$= 1 \qquad \blacksquare$$

Note that in the previous examples the common factors could have been divided out before the multiplication was performed.

Example 3   Multiply: $\dfrac{6x^2 + 5x - 4}{2x^2 + 5x + 3} \cdot \dfrac{8x^2 + 6x - 9}{12x^2 + 7x - 12}$.

Solution   Factor first in order to divide out all factors common to both the numerator and the denominator in either fraction. Then multiply.

$$\frac{6x^2 + 5x - 4}{2x^2 + 5x + 3} \cdot \frac{8x^2 + 6x - 9}{12x^2 + 7x - 12} = \frac{(3x + 4)(2x - 1)}{(2x + 3)(x + 1)} \cdot \frac{(4x - 3)(2x + 3)}{(3x + 4)(4x - 3)}$$

$$= \frac{(3x + 4)(2x - 1)}{(2x + 3)(x + 1)} \cdot \frac{(4x - 3)(2x + 3)}{(3x + 4)(4x - 3)}$$

$$= \frac{2x - 1}{x + 1} \qquad \blacksquare$$

Example 4   Simplify: $(x^2 - 2x) \cdot \dfrac{x}{x^2 - 5x + 6}$.

Solution   Write the first factor as a fraction with a denominator of 1. Then proceed as in the previous examples.

$$(x^2 - 2x) \cdot \frac{x}{x^2 - 5x + 6} = \frac{x^2 - 2x}{1} \cdot \frac{x}{x^2 - 5x + 6}$$

$$= \frac{x(x - 2)}{1} \cdot \frac{x}{(x - 2)(x - 3)}$$

$$= \frac{x^2}{x - 3} \qquad \blacksquare$$

Recall that certain pairs of numbers such as 3 and $\frac{1}{3}$, $\frac{2}{5}$ and $\frac{5}{2}$, and $\frac{-7}{8}$ and $\frac{-8}{7}$ have products that are equal to 1. Such pairs of numbers are called **reciprocals** of each other.

> **Definition.**   If the product of two numbers is 1, those numbers are called **multiplicative inverses** or **reciprocals** of each other.

Dividing by a number is the same as multiplying by its **multiplicative inverse** or **reciprocal**. For example, $6 \div 3$ and $6 \cdot \frac{1}{3}$ both give 2 for the answer. To divide by 3 is to multiply by $\frac{1}{3}$. Similarly, $56 \div \frac{8}{7}$ and $56 \cdot \frac{7}{8}$ give the same result. To divide by $\frac{8}{7}$ is to multiply by $\frac{7}{8}$. The rule for division of fractions assets that, if $b \neq 0$, $c \neq 0$, and $d \neq 0$, then

$$\frac{a}{b} \div \frac{c}{d} \quad \text{is equivalent to} \quad \frac{a}{b} \cdot \frac{d}{c}$$

This is true because c/d and d/c are reciprocals of each other. To divide by c/d is to multiply by d/c. The previous discussion justifies the procedure for dividing fractions: invert the divisor and multiply.

Example 5 Divide: $\dfrac{x^3 + 8}{x + 1} \div \dfrac{x^2 - 2x + 4}{2x^2 - 2}$.

Solution Using the rule for division of fractions, multiply by the reciprocal of the divisor. Then simplify.

$$\frac{x^3 + 8}{x + 1} \div \frac{x^2 - 2x + 4}{2x^2 - 2} = \frac{x^3 + 8}{x + 1} \cdot \frac{2x^2 - 2}{x^2 - 2x + 4}$$

$$= \frac{(x + 2)(x^2 - 2x + 4)}{(x + 1)} \cdot \frac{2(x + 1)(x - 1)}{(x^2 - 2x + 4)}$$

$$= 2(x + 2)(x - 1) \qquad\blacksquare$$

Example 6 Divide: $\dfrac{x^2 - 4}{x - 1} \div (x - 2)$.

Solution Write the divisor as a fraction with a denominator of 1. Then invert the divisor, multiply, factor, and divide out any common factors.

$$\frac{x^2 - 4}{x - 1} \div (x - 2) = \frac{x^2 - 4}{x - 1} \div \frac{x - 2}{1}$$

$$= \frac{x^2 - 4}{x - 1} \cdot \frac{1}{x - 2}$$

$$= \frac{(x + 2)(x - 2) \cdot 1}{(x - 1)(x - 2)}$$

$$= \frac{x + 2}{x - 1} \qquad\blacksquare$$

Example 7 Simplify: $\dfrac{x^2 + 2x - 3}{6x^2 + 5x + 1} \div \dfrac{2x^2 - 2}{2x^2 - 5x - 3} \cdot \dfrac{6x^2 + 4x - 2}{x^2 - 2x - 3}$.

Solution Factor everything you can, and change the division to a multiplication. It is understood that, in the absence of grouping symbols, multiplications and divisions are performed from left to right. Therefore, only the middle fraction is to be inverted. Finally, divide out any common factors and multiply.

$$\frac{x^2 + 2x - 3}{6x^2 + 5x + 1} \div \frac{2x^2 - 2}{2x^2 - 5x - 3} \cdot \frac{6x^2 + 4x - 2}{x^2 - 2x - 3}$$

$$= \frac{(x + 3)(x - 1)}{(3x + 1)(2x + 1)} \cdot \frac{(2x + 1)(x - 3)}{2(x + 1)(x - 1)} \cdot \frac{2(3x - 1)(x + 1)}{(x - 3)(x + 1)}$$

$$= \frac{(x + 3)(3x - 1)}{(3x + 1)(x + 1)} \qquad\blacksquare$$

EXERCISE 5.2

In Exercises 1–46, perform the indicated operations and simplify.

1. $\dfrac{3}{4} \cdot \dfrac{5}{3} \cdot \dfrac{8}{7}$

2. $-\dfrac{5}{6} \cdot \dfrac{3}{7} \cdot \dfrac{14}{25}$

3. $-\dfrac{6}{11} \div \dfrac{36}{55}$

4. $\dfrac{17}{12} \div \dfrac{34}{3}$

5. $\dfrac{x^2 y^2}{cd} \cdot \dfrac{c^{-2} d^2}{x}$

6. $\dfrac{a^{-2} b^2}{x^{-1} y} \cdot \dfrac{a^4 b^4}{x^2 y^3}$

7. $\dfrac{-x^2 y^{-2}}{x^{-1} y^{-3}} \div \dfrac{x^{-3} y^2}{x^4 y^{-1}}$

8. $\dfrac{(a^3)^2}{b^{-1}} \div \dfrac{(a^3)^{-2}}{b^{-1}}$

9. $\dfrac{x^2 + 2x + 1}{x} \cdot \dfrac{x^2 - x}{x^2 - 1}$

10. $\dfrac{a + 6}{a^2 - 16} \cdot \dfrac{3a - 12}{3a + 18}$

11. $\dfrac{2x^2 - x - 3}{x^2 - 1} \cdot \dfrac{x^2 + x - 2}{2x^2 + x - 6}$

12. $\dfrac{9x^2 + 3x - 20}{3x^2 - 7x + 4} \cdot \dfrac{3x^2 - 5x + 2}{9x^2 + 18x + 5}$

13. $\dfrac{x^2 - 16}{x^2 - 25} \div \dfrac{x + 4}{x - 5}$

14. $\dfrac{a^2 - 9}{a^2 - 49} \div \dfrac{a + 3}{a + 7}$

15. $\dfrac{a^2 + 2a - 35}{12x} \div \dfrac{x}{a^2 + 4a - 21}$

16. $\dfrac{x^2 - 4}{2b - bx} \div \dfrac{x^2 + 4x + 4}{2b + bx}$

17. $(x + 1) \cdot \dfrac{1}{x^2 + 2x + 1}$

18. $\dfrac{x^2 - 4}{x} \div (x + 2)$

19. $(x^2 - x - 2) \cdot \dfrac{x^2 + 3x + 2}{x^2 - 4}$

20. $(2x^2 - 9x - 5) \cdot \dfrac{x}{2x^2 + x}$

21. $(2x^2 - 15x + 25) \div \dfrac{(2x - 5)(x + 1)}{x + 1}$

22. $(x^2 - 6x + 9) \div \dfrac{x^2 - 9}{x + 3}$

23. $\dfrac{x^3 + y^3}{x^3 - y^3} \div \dfrac{x^2 - xy + y^2}{x^2 + xy + y^2}$

24. $\dfrac{x^2 - 6x + 9}{x^2 - 4} \div \dfrac{x^2 - 9}{x^2 - 8x + 12}$

25. $\dfrac{m^2 - n^2}{2x^2 + 3x - 2} \cdot \dfrac{2x^2 + 5x - 3}{n^2 - m^2}$

26. $\dfrac{x^2 - y^2}{2x^2 + 2xy + x + y} \cdot \dfrac{2x^2 - 5x - 3}{yx - 3y - x^2 + 3x}$

27. $\dfrac{ax + ay + bx + by}{x^3 - 27} \cdot \dfrac{x^2 + 3x + 9}{xc + xd + yc + yd}$

28. $\dfrac{x^2 + 3x + yx + 3y}{x^2 - 9} \cdot \dfrac{x - 3}{x + 3}$

29. $\dfrac{3x^2 y^2}{6x^3 y} \cdot \dfrac{-4x^7 y^{-2}}{18x^{-2} y} \div \dfrac{36x}{18y^{-2}}$

30. $\dfrac{9ab^3}{7xy} \cdot \dfrac{14xy^2}{27z^3} \div \dfrac{18a^2 b^2 x}{3z^2}$

31. $(4x + 12) \cdot \dfrac{x^2}{2x - 6} \div \dfrac{2}{x - 3}$

32. $(4x^2 - 9) \div \dfrac{2x^2 + 5x + 3}{x + 2} \div (2x - 3)$

33. $\dfrac{2x^2 - 2x - 4}{x^2 + 2x - 8} \cdot \dfrac{3x^2 + 15x}{x + 1} \div \dfrac{4x^2 - 100}{x^2 - x - 20}$

34. $\dfrac{6a^2 - 7a - 3}{a^2 - 1} \div \dfrac{4a^2 - 12a + 9}{a^2 - 1} \cdot \dfrac{2a^2 - a - 3}{3a^2 - 2a - 1}$

35. $\dfrac{2t^2 + 5t + 2}{t^2 - 4t + 16} \div \dfrac{t + 2}{t^3 + 64} \div \dfrac{2t^3 + 9t^2 + 4t}{t + 1}$

36. $\dfrac{a^6 - b^6}{a^4 - a^3 b} \cdot \dfrac{a^3}{a^4 + a^2 b^2 + b^4} \div \dfrac{1}{a}$

37. $\dfrac{x^4 + 7x^2 + 16}{x^6} \cdot \dfrac{x^3}{x^2 + 4 - x}$

38. $\dfrac{x^2 + 6 - 3x}{y^2 - 1} \cdot \dfrac{y^2 + 2y + 1}{x^4 + 3x^2 + 36}$

39. $(x^2 - x - 6) \div (x - 3) \div (x - 2)$

40. $(x^2 - x - 6) \div [(x - 3) \div (x - 2)]$

41. $\dfrac{3x^2 - 2x}{3x + 2} \div (3x - 2) \div \dfrac{3x}{3x - 3}$

42. $(2x^2 - 3x - 2) \div \dfrac{2x^2 - x - 1}{x - 2} \div (x - 1)$

43. $\dfrac{2x^2 + 5x - 3}{x^2 + 2x - 3} \div \left(\dfrac{x^2 + 2x - 35}{x^2 - 6x + 5} \div \dfrac{x^2 - 9x + 14}{2x^2 - 5x + 2} \right)$

44. $\dfrac{x^2 - 4}{x^2 - x - 6} \div \left(\dfrac{x^2 - x - 2}{x^2 - 8x + 15} \cdot \dfrac{x^2 - 3x - 10}{x^2 + 3x + 2} \right)$

45. $\dfrac{x^2 - x - 12}{x^2 + x - 2} \div \dfrac{x^2 - 6x + 8}{x^2 - 3x - 10} \cdot \dfrac{x^2 - 3x + 2}{x^2 - 2x - 15}$

46. $\dfrac{4x^2 - 10x + 6}{x^4 - 3x^3} \div \dfrac{2x - 3}{2x^3} \cdot \dfrac{x - 3}{2x - 2}$

5.3 ADDING AND SUBTRACTING RATIONAL EXPRESSIONS

Property 5 of fractions, first discussed in Section 1.4, asserted that fractions are added and subtracted according to the following rules:

> **Property 5 of Fractions.**   If there are no divisions by 0, then
>
> $$\frac{a}{b} + \frac{c}{b} = \frac{a + c}{b} \quad \text{and} \quad \frac{a}{b} - \frac{c}{b} = \frac{a - c}{b}$$

When we add or subtract two fractions with like denominators, these rules can be used directly: Simply add or subtract the numerators and use the same denominator.

To make these rules reasonable, think of problems such as

$$\frac{2}{7} + \frac{3}{7} = \frac{5}{7} \quad \text{and} \quad \frac{5}{9} - \frac{4}{9} = \frac{1}{9}$$

as problems in the form

2 sevenths + 3 sevenths = 5 sevenths

and

5 ninths − 4 ninths = 1 ninth

Example 1 Simplify: **a.** $\dfrac{17}{22} + \dfrac{13}{22}$,   **b.** $\dfrac{3}{2x} + \dfrac{7}{2x}$, and   **c.** $\dfrac{4x}{x + 2} - \dfrac{7x}{x + 2}$.

Solution **a.** $\dfrac{17}{22} + \dfrac{13}{22} = \dfrac{17 + 13}{22} = \dfrac{30}{22} = \dfrac{15 \cdot 2}{11 \cdot 2} = \dfrac{15}{11}$

b. $\dfrac{3}{2x} + \dfrac{7}{2x} = \dfrac{3 + 7}{2x} = \dfrac{10}{2x} = \dfrac{5}{x}$

c. $\dfrac{4x}{x + 2} - \dfrac{7x}{x + 2} = \dfrac{4x - 7x}{x + 2} = \dfrac{-3x}{x + 2}$ ∎

If we add or subtract fractions with unlike denominators, we must convert them to fractions with the same denominator. To do so, we use Property 3 of fractions to insert whatever factors are required to form a common denominator.

Example 2   Simplify: **a.** $\dfrac{3}{x} + \dfrac{4}{y}$ and   **b.** $\dfrac{4x}{x+2} - \dfrac{7x}{x-2}$.

Solution   **a.** $\dfrac{3}{x} + \dfrac{4}{y} = \dfrac{3y}{xy} + \dfrac{x4}{xy} = \dfrac{3y+4x}{xy}$

b. $\dfrac{4x}{x+2} - \dfrac{7x}{x-2} = \dfrac{4x(x-2)}{(x+2)(x-2)} - \dfrac{(x+2)7x}{(x+2)(x-2)}$

$$= \dfrac{(4x^2 - 8x) - (7x^2 + 14x)}{(x+2)(x-2)}$$

$$= \dfrac{4x^2 - 8x - 7x^2 - 14x}{(x+2)(x-2)}$$

$$= \dfrac{-3x^2 - 22x}{(x+2)(x-2)}$$

Note that the minus sign between the fractions in part b influences *both* terms of the binomial $7x^2 + 14x$. ∎

To add fractions, a common denominator must be found. It is sensible to use the smallest and the least complicated denominator possible. That simplest denominator is called the **least** (or lowest) **common denominator** or, more simply, the **LCD**. We now consider how it can be found.

Suppose the unlike denominators of three fractions are 12, 20, and 35. First, we find the unique prime factorization of each number.

$$12 = 4 \cdot 3 = 2^2 \cdot 3$$
$$20 = 4 \cdot 5 = 2^2 \cdot 5$$
$$35 = 5 \cdot 7$$

Because the least common denominator is the smallest number that can be divided by 12, 20, and 35, it must contain factors of 2^2, 3, 5, and 7. Hence, the least common denominator is

$$\text{LCD} = 2^2 \cdot 3 \cdot 5 \cdot 7$$
$$= 420$$

That is, 420 is the smallest number that can be divided evenly by 12, 20, and 35.

When finding a least common denominator, always factor each denominator first and then create the LCD by using each factor the greatest number of times that it appears in any one denominator. The product of these factors is the LCD.

Example 3   The polynomial denominators of three fractions are $x^2 + 7x + 6$, $x^2 - 36$, and $x^2 + 12x + 36$. Find the LCD.

Solution   Factor each polynomial:

$$x^2 + 7x + 6 = (x + 6)(x + 1)$$
$$x^2 - 36 = (x + 6)(x - 6)$$
$$x^2 + 12x + 36 = (x + 6)(x + 6) = (x + 6)^2$$

Use each factor the greatest number of times that it appears in any one denominator to find the LCD:

$$\text{LCD} = (x + 6)^2(x + 1)(x - 6)$$

There is no need to do any multiplication. Leave the LCD in factored form.

■

Example 4   Add the fractions $\dfrac{x}{x^2 - 2x + 1}$ and $\dfrac{3}{x^2 - 1}$.

Solution   Factor each denominator and find the LCD:

$$x^2 - 2x + 1 = (x - 1)(x - 1) = (x - 1)^2$$
$$x^2 - 1 = (x + 1)(x - 1)$$

The LCD is $(x - 1)^2(x + 1)$. Write each fraction with its denominator in factored form, convert all fractions to fractions with a denominator of $(x - 1)^2(x + 1)$, add the fractions, and simplify.

$$\frac{x}{x^2 - 2x + 1} + \frac{3}{x^2 - 1}$$

$$= \frac{x}{(x - 1)(x - 1)} + \frac{3}{(x + 1)(x - 1)}$$

$$= \frac{x(x + 1)}{(x - 1)(x - 1)(x + 1)} + \frac{3(x - 1)}{(x + 1)(x - 1)(x - 1)}$$

$$= \frac{x^2 + x + 3x - 3}{(x - 1)(x - 1)(x + 1)}$$

$$= \frac{x^2 + 4x - 3}{(x - 1)(x - 1)(x + 1)}$$

Always simplify the final result if possible. In this case, it cannot be simplified.

■

Example 5   Simplify: $\dfrac{2x}{x^2 - 4} - \dfrac{1}{x^2 - 3x + 2} + \dfrac{x + 1}{x^2 + x - 2}$.

Solution   Factor the denominators and determine that the LCD is $(x + 2)(x - 2)(x - 1)$. Write all of the fractions in a form bearing the least common denominator, remove the resulting parentheses in each numerator, perform the indicated subtraction and addition, and simplify.

$$\frac{2x}{x^2 - 4} - \frac{1}{x^2 - 3x + 2} + \frac{x + 1}{x^2 + x - 2}$$

$$= \frac{2x}{(x - 2)(x + 2)} - \frac{1}{(x - 2)(x - 1)} + \frac{(x + 1)}{(x - 1)(x + 2)}$$

$$= \frac{2x(x - 1)}{(x - 2)(x + 2)(x - 1)} - \frac{1(x + 2)}{(x - 2)(x - 1)(x + 2)} + \frac{(x + 1)(x - 2)}{(x - 1)(x + 2)(x - 2)}$$

$$= \frac{2x(x - 1) - 1(x + 2) + (x + 1)(x - 2)}{(x + 2)(x - 2)(x - 1)}$$

$$= \frac{2x^2 - 2x - x - 2 + x^2 - x - 2}{(x + 2)(x - 2)(x - 1)}$$

$$= \frac{3x^2 - 4x - 4}{(x + 2)(x - 2)(x - 1)}$$

Factor the numerator on the chance that the fraction will simplify:

$$= \frac{(3x + 2)(x - 2)}{(x + 2)(x - 2)(x - 1)}$$

In this example, the fraction does simplify. Hence,

$$\frac{2x}{x^2 - 4} - \frac{1}{x^2 - 3x + 2} + \frac{x + 1}{x^2 + x - 2} = \frac{3x + 2}{(x + 2)(x - 1)}$$ ∎

Example 6   Simplify: $3 + \dfrac{7}{x - 2}$.

Solution   Because 3 can be written as $\frac{3}{1}$, you must find the sum of two fractions. The LCD is $x - 2$, and the first term (the constant 3) must be written as a fraction with that denominator. Then, add the fractions.

$$3 + \frac{7}{x - 2} = \frac{3}{1} + \frac{7}{x - 2}$$

$$= \frac{3(x - 2)}{1(x - 2)} + \frac{7}{x - 2}$$

$$= \frac{3x - 6 + 7}{x - 2}$$

$$= \frac{3x + 1}{x - 2}$$ ∎

Example 7   Simplify: $\dfrac{3x}{x - 1} - \dfrac{2x^2 + 3x - 2}{(x + 1)(x - 1)}$.

Solution   Write each fraction in a form bearing the LCD of $(x + 1)(x - 1)$, remove the
resulting parentheses in the first numerator, perform the indicated subtraction,
and simplify.

$$
\frac{3x}{x - 1} - \frac{2x^2 + 3x - 2}{(x + 1)(x - 1)} = \frac{(x + 1)3x}{(x + 1)(x - 1)} - \frac{2x^2 + 3x - 2}{(x + 1)(x - 1)}
$$

$$
= \frac{3x^2 + 3x}{(x + 1)(x - 1)} - \frac{2x^2 + 3x - 2}{(x + 1)(x - 1)}
$$

$$
= \frac{3x^2 + 3x - (2x^2 + 3x - 2)}{(x + 1)(x - 1)}
$$

$$
= \frac{3x^2 + 3x - 2x^2 - 3x + 2}{(x + 1)(x - 1)}
$$

$$
= \frac{x^2 + 2}{(x + 1)(x - 1)}
$$

Note that the minus sign between the fractions influences all three terms of the
trinomial $2x^2 + 3x - 2$. Whenever you subtract one fraction from another, remember to subtract each term of the numerator of the second fraction.   ■

Example 8   Simplify: $\left(\dfrac{x^2}{x - 2} + \dfrac{4}{2 - x}\right)^2$

Solution   Perform the addition of the fractions within the parentheses. To write them as
fractions with a common denominator, multiply both the numerator and the
denominator of the second fraction by -1. Combine these fractions, simplify,
and square the result.

$$
\left(\frac{x^2}{x - 2} + \frac{4}{2 - x}\right)^2 = \left[\frac{x^2}{x - 2} + \frac{(-1)4}{-1(2 - x)}\right]^2
$$

$$
= \left[\frac{x^2}{x - 2} + \frac{-4}{x - 2}\right]^2
$$

$$
= \left[\frac{x^2 - 4}{x - 2}\right]^2
$$

$$
= \left[\frac{(x + 2)(x - 2)}{x - 2}\right]^2
$$

$$
= (x + 2)^2
$$

$$
= x^2 + 4x + 4 \qquad ■
$$

■ EXERCISE 5.3

In Exercises 1–16, perform the indicated operations and simplify, if necessary.

1. $\dfrac{3}{4} + \dfrac{7}{4}$ **2.** $\dfrac{5}{11} + \dfrac{2}{11}$ **3.** $\dfrac{10}{33} - \dfrac{21}{33}$ **4.** $\dfrac{8}{15} - \dfrac{2}{15}$

5. $\dfrac{3}{4y} + \dfrac{8}{4y}$

6. $\dfrac{5}{3z^2} - \dfrac{6}{3z^2}$

7. $\dfrac{3}{a+b} - \dfrac{a}{a+b}$

8. $\dfrac{x}{x+4} + \dfrac{5}{x+4}$

9. $\dfrac{3x}{2x+2} + \dfrac{x+8}{2x+2}$

10. $\dfrac{4y}{y-4} - \dfrac{16}{y-4}$

11. $\dfrac{3x}{x-3} - \dfrac{9}{x-3}$

12. $\dfrac{9x}{x-y} - \dfrac{9y}{x-y}$

13. $\dfrac{5x}{x+1} + \dfrac{3}{x+1} - \dfrac{2x}{x+1}$

14. $\dfrac{4}{a+4} - \dfrac{2a}{a+4} + \dfrac{3a}{a+4}$

15. $\dfrac{3(x^2+x)}{x^2-5x+6} + \dfrac{-3(x^2-x)}{x^2-5x+6}$

16. $\dfrac{2x+4}{x^2+13x+12} - \dfrac{x+3}{x^2+13x+12}$

In Exercises 17–24, the denominators of several fractions are given. Find the LCD.

17. 8, 12, 18

18. 10, 15, 28

19. $x^2 + 3x$, $x^2 - 9$

20. $3y^2 - 6y$, $3y(y-4)$

21. $x^3 + 27$, $x^2 + 6x + 9$

22. $x^3 - 8$, $x^2 - 4x + 4$

23. $2x^2 + 5x + 3$, $4x^2 + 12x + 9$, $x^2 + 2x + 1$

24. $2x^2 + 5x + 3$, $4x^2 + 12x + 9$, $4x + 6$

In Exercises 25–80, perform the indicated operations and simplify, if possible.

25. $\dfrac{1}{2} + \dfrac{1}{3}$

26. $\dfrac{5}{6} + \dfrac{2}{7}$

27. $\dfrac{7}{15} - \dfrac{17}{25}$

28. $\dfrac{8}{9} - \dfrac{5}{12}$

29. $\dfrac{a}{2} + \dfrac{2a}{5}$

30. $\dfrac{b}{6} + \dfrac{3a}{4}$

31. $\dfrac{3a}{2} - \dfrac{4b}{7}$

32. $\dfrac{2m}{3} - \dfrac{4n}{5}$

33. $\dfrac{3}{4x} + \dfrac{2}{3x}$

34. $\dfrac{2}{5a} + \dfrac{3}{2b}$

35. $\dfrac{3a}{2b} - \dfrac{2b}{3a}$

36. $\dfrac{5m}{2n} - \dfrac{3n}{4m}$

37. $\dfrac{a+b}{3} + \dfrac{a-b}{7}$

38. $\dfrac{x-y}{2} + \dfrac{x+y}{3}$

39. $\dfrac{3}{x+2} + \dfrac{5}{x-4}$

40. $\dfrac{2}{a+4} - \dfrac{6}{a+3}$

41. $\dfrac{x+2}{x+5} - \dfrac{x-3}{x+7}$

42. $\dfrac{7}{x+3} + \dfrac{4x}{x+6}$

43. $x + \dfrac{1}{x}$

44. $2 - \dfrac{1}{x+1}$

45. $\dfrac{x}{x^2+5x+6} + \dfrac{x}{x^2-4}$

46. $\dfrac{x}{3x^2-2x-1} + \dfrac{4}{3x^2+10x+3}$

47. $\dfrac{4}{x^2-2x-3} - \dfrac{x}{3x^2-7x-6}$

48. $\dfrac{2a}{a^2-2a-8} + \dfrac{3}{a^2-5a+4}$

49. $\dfrac{8}{x^2-9} + \dfrac{2}{x-3} - \dfrac{6}{x}$

50. $\dfrac{x}{x^2-4} - \dfrac{x}{x+2} + \dfrac{2}{x}$

51. $2x + 3 + \dfrac{1}{x+1}$

52. $x + 1 + \dfrac{1}{x-1}$

53. $1 + x - \dfrac{x}{x-5}$

54. $2 - x + \dfrac{3}{x-9}$

55. $\dfrac{3x}{x-1} - 2x - x^2$

56. $\dfrac{23}{x-1} + 4x - 5x^2$

57. $\dfrac{y+4}{y^2+7y+12} - \dfrac{y-4}{y+3} + \dfrac{47}{y+4}$

58. $\dfrac{x+3}{2x^2-5x+2} - \dfrac{3x-1}{x^2-x-2}$

59. $\dfrac{3}{x+1} - \dfrac{2}{x-1} + \dfrac{x+3}{x^2-1}$

60. $\dfrac{2}{x-2} + \dfrac{3}{x+2} - \dfrac{x-1}{x^2-4}$

61. $\dfrac{x-2}{x^2-3x}+\dfrac{2x-1}{x^2+3x}-\dfrac{2}{x^2-9}$

62. $\dfrac{2}{x-1}-\dfrac{2x}{x^2-1}-\dfrac{x}{x^2+2x+1}$

63. $\dfrac{5}{x^2-25}-\dfrac{3}{2x^2-9x-5}+1$

64. $\dfrac{3x}{2x-1}+\dfrac{x+1}{3x+2}+\dfrac{2x}{6x^3+x^2-2x}$

65. $\dfrac{3x}{x-3}+\dfrac{4}{x-2}-\dfrac{5x}{x^3-5x^2+6x}$

66. $\dfrac{2x-1}{x^2+x-6}-\dfrac{3x-5}{x^2-2x-15}+\dfrac{2x-3}{x^2-7x+10}$

67. $2+\dfrac{4a}{a^2-1}-\dfrac{2}{a+1}$

68. $\dfrac{a}{a-1}-\dfrac{a+1}{2a-2}+a$

69. $\dfrac{x+5}{2x^2-2}+\dfrac{x}{2x+2}-\dfrac{3}{x-1}$

70. $\dfrac{a}{2-a}+\dfrac{3}{a-2}-\dfrac{3a-2}{a^2-4}$

71. $\dfrac{a}{a-b}+\dfrac{b}{a+b}+\dfrac{a^2+b^2}{b^2-a^2}$

72. $\dfrac{7n^2}{m-n}+\dfrac{3m}{n-m}-\dfrac{3m^2-n}{m^2-2mn+n^2}$

73. $\dfrac{3b}{2a-b}+\dfrac{2a-1}{b-2a}-\dfrac{3a^2+b}{b^2-4ab+4a^2}$

74. $\dfrac{m+1}{m^2+2m+1}+\dfrac{m-1}{m^2-2m+1}+\dfrac{2}{m^2-1}$ (*Hint:* Think about this before finding the LCD.)

75. $\left(\dfrac{1}{x-1}+1\right)\left(\dfrac{x-1}{3}\right)$

76. $\left(\dfrac{1}{x-1}+\dfrac{1}{x+1}\right)\left(\dfrac{x+1}{2x+4}\right)$

77. $\left(\dfrac{x}{3}-\dfrac{3}{x}\right)\left(\dfrac{1}{x-3}-\dfrac{1}{x+3}\right)$

78. $\left(\dfrac{4}{y}-\dfrac{y}{4}\right)\left(\dfrac{y}{y+4}+\dfrac{y}{y-4}\right)$

79. $\left(1-\dfrac{3}{x}\right)^2$

80. $\left(\dfrac{3}{a-2}+\dfrac{a}{a+2}\right)^2$

81. Show that $\dfrac{a}{b}+\dfrac{c}{d}=\dfrac{ad+bc}{bd}$.

82. Show that $\dfrac{a}{b}-\dfrac{c}{d}=\dfrac{ad-bc}{bd}$.

5.4 COMPLEX FRACTIONS

A **complex fraction** is a rational expression that has a fraction in its numerator or in its denominator or both. Examples of complex fractions are

$$\frac{\frac{3}{5}}{\frac{6}{7}}\qquad \frac{\frac{x+2}{3}}{x-4}\qquad\text{and}\qquad \frac{\frac{3x^2-2}{2x}}{3x-\frac{2}{y}}$$

There are two methods used to simplify complex fractions. In one method we write the complex fraction in an equivalent form:

$$\frac{\frac{a}{b}}{\frac{c}{d}}=\frac{a}{b}\div\frac{c}{d}$$

and then use the division rule for fractions. In the other method we multiply both the numerator and the denominator of the complex fraction

$$\frac{\dfrac{a}{b}}{\dfrac{c}{d}}$$

by bd, the LCD of $\dfrac{a}{b}$ and $\dfrac{c}{d}$. The next three examples illustrate each method.

Example 1   Simplify the complex fraction $\dfrac{\dfrac{3a}{b}}{\dfrac{6ac}{b^2}}$.

Solution 1   Write the complex fraction in an equivalent form:

$$\frac{\dfrac{3a}{b}}{\dfrac{6ac}{b^2}} = \frac{3a}{b} \div \frac{6ac}{b^2}$$

Then the indicated division can be accomplished as follows:

$$\frac{\dfrac{3a}{b}}{\dfrac{6ac}{b^2}} = \frac{3a}{b} \div \frac{6ac}{b^2}$$

$$= \frac{3a}{b} \cdot \frac{b^2}{6ac}$$

$$= \frac{b}{2c}$$

Solution 2   Multiply both the numerator and the denominator of the complex fraction by b^2, the LCD of $3a/b$ and $6ac/b^2$, and simplify:

$$\frac{\dfrac{3a}{b}}{\dfrac{6ac}{b^2}} = \frac{\dfrac{3a}{b} \cdot b^2}{\dfrac{6ac}{b^2} \cdot b^2}$$

$$= \frac{3ab}{6ac}$$

$$= \frac{b}{2c}$$ ∎

Example 2   Simplify the complex fraction $\dfrac{\dfrac{1}{x}+\dfrac{1}{y}}{\dfrac{1}{x}-\dfrac{1}{y}}$.

Solution 1   Add the fractions in the numerator and in the denominator. Then simplify the fraction as in Solution 1 of Example 1.

$$\frac{\dfrac{1}{x}+\dfrac{1}{y}}{\dfrac{1}{x}-\dfrac{1}{y}}=\frac{\dfrac{1y}{xy}+\dfrac{x1}{xy}}{\dfrac{1y}{xy}-\dfrac{x1}{xy}}$$

$$=\frac{\dfrac{y+x}{xy}}{\dfrac{y-x}{xy}}$$

$$=\frac{y+x}{xy}\div\frac{y-x}{xy}$$

$$=\frac{y+x}{xy}\cdot\frac{xy}{y-x}$$

$$=\frac{y+x}{y-x}$$

Solution 2   Multiply both the numerator and the denominator by xy and simplify. Note that the product xy is the least common denominator of all the fractions that appear in the problem.

$$\frac{\dfrac{1}{x}+\dfrac{1}{y}}{\dfrac{1}{x}-\dfrac{1}{y}}=\frac{\left(\dfrac{1}{x}+\dfrac{1}{y}\right)xy}{\left(\dfrac{1}{x}-\dfrac{1}{y}\right)xy}$$

$$=\frac{\dfrac{xy}{x}+\dfrac{xy}{y}}{\dfrac{xy}{x}-\dfrac{xy}{y}}$$

$$=\frac{y+x}{y-x}\qquad\blacksquare$$

Example 3   Simplify $\dfrac{x^{-1}+y^{-1}}{x^{-2}-y^{-2}}$. If you recall that $a^{-n}=\dfrac{1}{a^{n}}$, then you will recognize that this problem is a disguised complex fraction.

Solution 1   Add the fractions in the numerator and the denominator. Then simplify as in Solution 1 of Example 1.

$$\frac{x^{-1} + y^{-1}}{x^{-2} - y^{-2}} = \frac{\dfrac{1}{x} + \dfrac{1}{y}}{\dfrac{1}{x^2} - \dfrac{1}{y^2}}$$

$$= \frac{\dfrac{y}{xy} + \dfrac{x}{xy}}{\dfrac{y^2}{x^2 y^2} - \dfrac{x^2}{x^2 y^2}}$$

$$= \frac{\dfrac{y + x}{xy}}{\dfrac{y^2 - x^2}{x^2 y^2}}$$

$$= \frac{y + x}{xy} \div \frac{y^2 - x^2}{x^2 y^2}$$

$$= \frac{\cancel{y + x}}{\cancel{xy}} \cdot \frac{\cancel{x}x\cancel{y}y}{(y - x)\cancel{(y + x)}}$$

$$= \frac{xy}{y - x}$$

Solution 2   Multiply both numerator and denominator by $x^2 y^2$, the LCD of all the fractions that appear in the problem, and simplify:

$$\frac{x^{-1} + y^{-1}}{x^{-2} - y^{-2}} = \frac{\dfrac{1}{x} + \dfrac{1}{y}}{\dfrac{1}{x^2} - \dfrac{1}{y^2}}$$

$$= \frac{\left(\dfrac{1}{x} + \dfrac{1}{y}\right) x^2 y^2}{\left(\dfrac{1}{x^2} - \dfrac{1}{y^2}\right) x^2 y^2}$$

$$= \frac{xy^2 + yx^2}{y^2 - x^2}$$

$$= \frac{xy\cancel{(y + x)}}{\cancel{(y + x)}(y - x)}$$

$$= \frac{xy}{y - x} \qquad\qquad \blacksquare$$

Note that $x^{-1} + y^{-1}$ means $1/x + 1/y$, and that $(x + y)^{-1}$ means $1/(x + y)$. Hence, $x^{-1} + y^{-1} \neq (x + y)^{-1}$.

Example 4   Simplify the fraction $\dfrac{\dfrac{2x}{1 - \dfrac{1}{x}} + 3}{3 - \dfrac{2}{x}}$.

Solution   Begin by multiplying both the numerator and denominator of the fraction

$$\frac{2x}{1 - \dfrac{1}{x}}$$

by x. This will eliminate the complex fraction in the numerator of the original fraction.

$$\frac{\dfrac{2x}{1 - \dfrac{1}{x}} + 3}{3 - \dfrac{2}{x}} = \frac{\dfrac{x2x}{x\left(1 - \dfrac{1}{x}\right)} + 3}{3 - \dfrac{2}{x}}$$

$$= \frac{\dfrac{2x^2}{x - 1} + 3}{3 - \dfrac{2}{x}}$$

Then multiply both the numerator and denominator by $x(x - 1)$, the LCD of $2x^2/(x - 1)$, 3, and $2/x$, and simplify:

$$\frac{\dfrac{2x}{1 - \dfrac{1}{x}} + 3}{3 - \dfrac{2}{x}} = \frac{\left(\dfrac{2x^2}{x - 1} + 3\right)x(x - 1)}{\left(3 - \dfrac{2}{x}\right)x(x - 1)}$$

$$= \frac{2x^3 + 3x(x - 1)}{3x(x - 1) - 2(x - 1)}$$

$$= \frac{2x^3 + 3x^2 - 3x}{3x^2 - 5x + 2}$$

This result does not simplify.   ∎

EXERCISE 5.4

Simplify each complex fraction.

1. $\dfrac{\dfrac{1}{2}}{\dfrac{3}{4}}$

2. $-\dfrac{\dfrac{3}{4}}{\dfrac{1}{2}}$

3. $\dfrac{-\dfrac{2}{3}}{\dfrac{6}{9}}$

4. $\dfrac{\dfrac{11}{18}}{\dfrac{22}{27}}$

5. $\dfrac{\dfrac{1}{2}+\dfrac{1}{3}}{\dfrac{1}{4}}$

6. $\dfrac{\dfrac{1}{4}-\dfrac{1}{5}}{\dfrac{1}{3}}$

7. $\dfrac{\dfrac{1}{6}-\dfrac{2}{7}}{\dfrac{1}{7}}$

8. $\dfrac{\dfrac{2}{3}+\dfrac{4}{5}}{\dfrac{1}{3}}$

9. $\dfrac{\dfrac{4x}{y}}{\dfrac{6xz}{y^2}}$

10. $\dfrac{\dfrac{5t^4}{9x}}{\dfrac{2t}{18x}}$

11. $\dfrac{5ab^2}{\dfrac{ab}{25}}$

12. $\dfrac{\dfrac{6a^2b}{4t}}{3a^2b^2}$

13. $\dfrac{\dfrac{x-y}{xy}}{\dfrac{y-x}{x}}$

14. $\dfrac{\dfrac{x^2+5x+6}{3xy}}{\dfrac{x^2-9}{6xy}}$

15. $\dfrac{\dfrac{1}{x}-\dfrac{1}{y}}{xy}$

16. $\dfrac{xy}{\dfrac{1}{x}-\dfrac{1}{y}}$

17. $\dfrac{\dfrac{1}{a}+\dfrac{1}{b}}{\dfrac{1}{a}}$

18. $\dfrac{\dfrac{1}{b}}{\dfrac{1}{a}-\dfrac{1}{b}}$

19. $\dfrac{1+\dfrac{x}{y}}{1-\dfrac{x}{y}}$

20. $\dfrac{\dfrac{x}{y}+1}{1-\dfrac{x}{y}}$

21. $\dfrac{x+1-\dfrac{6}{x}}{\dfrac{1}{x}}$

22. $\dfrac{x-1-\dfrac{2}{x}}{\dfrac{x}{3}}$

23. $\dfrac{5xy}{1+\dfrac{1}{xy}}$

24. $\dfrac{3a}{a+\dfrac{1}{a}}$

25. $\dfrac{a-4+\dfrac{1}{a}}{-\dfrac{1}{a}-a+4}$

26. $\dfrac{a+1+\dfrac{1}{a^2}}{\dfrac{1}{a^2}+a-1}$

27. $\dfrac{1+\dfrac{6}{x}+\dfrac{8}{x^2}}{1+\dfrac{1}{x}-\dfrac{12}{x^2}}$

28. $\dfrac{1-\dfrac{x^2}{x}-\dfrac{2}{x}}{\dfrac{6}{x^2}+\dfrac{1}{x}-1}$

29. $\dfrac{\dfrac{1}{a+1}+1}{\dfrac{3}{a-1}+1}$

30. $\dfrac{2+\dfrac{3}{x+1}}{\dfrac{1}{x}+x+x^2}$

31. $\dfrac{x^{-1}+y^{-1}}{x}$

32. $\dfrac{x^{-1}-y^{-1}}{y}$

33. $\dfrac{y}{x^{-1}-y^{-1}}$

34. $\dfrac{x}{x^{-1}+y^{-1}}$

35. $\dfrac{x^{-1}+y^{-1}}{x^{-1}-y^{-1}}$

36. $\dfrac{(x+y)^{-1}}{x^{-1}+y^{-1}}$

37. $\dfrac{x+y}{x^{-1}+y^{-1}}$

38. $\dfrac{x-y}{x^{-1}-y^{-1}}$

39. $\dfrac{x-y^{-2}}{y-x^{-2}}$

40. $\dfrac{x^{-2}-y^{-2}}{x^{-1}-y^{-1}}$

41. $\dfrac{1+\dfrac{a}{b}}{1-\dfrac{a}{1-\dfrac{a}{b}}}$

42. $\dfrac{1+\dfrac{2}{1-\dfrac{a}{b}}}{1-\dfrac{a}{b}}$

43. $\dfrac{x-\dfrac{1}{x}}{1+\dfrac{1}{\dfrac{1}{x}}}$

44. $\dfrac{\dfrac{a^2+3a+4}{ab}}{2+\dfrac{3+a}{\dfrac{2}{a}}}$

45. $\dfrac{b}{b + \dfrac{2}{2 + \dfrac{1}{2}}}$ **46.** $\dfrac{2y}{y - \dfrac{y}{3 - \dfrac{1}{2}}}$ **47.** $a + \dfrac{a}{1 + \dfrac{a}{a + 1}}$ **48.** $b + \dfrac{b}{1 - \dfrac{b + 1}{b}}$

49. $\dfrac{x - \dfrac{1}{1 - \dfrac{x}{2}}}{\dfrac{3}{x + \dfrac{2}{3}} - x}$ **50.** $\dfrac{\dfrac{2x}{x - \dfrac{1}{x}} - \dfrac{1}{x}}{2x + \dfrac{2x}{1 - \dfrac{1}{x}}}$

5.5   EQUATIONS CONTAINING RATIONAL EXPRESSIONS

Equations often contain fractions that have numerators and denominators that are polynomials. Such equations are called **rational equations**. To solve rational equations, it is usually best to begin by clearing the equation of fractions. To do so, we multiply both sides of the equation by the least common denominator of the fractions that appear in the equation. For example, to solve the equation

$$\frac{3}{5} + \frac{7}{x + 2} = 2$$

we first multiply both sides by $5(x + 2)$ and then simplify to obtain

$$5(x + 2)\left(\frac{3}{5} + \frac{7}{x + 2}\right) = 5(x + 2)2$$
$$3(x + 2) + 5(7) = 10(x + 2)$$
$$3x + 6 + 35 = 10x + 20$$
$$3x + 41 = 10x + 20$$
$$-7x = -21$$
$$x = 3$$

To verify that 3 satisfies the equation, we substitute 3 for x in the original equation and simplify:

$$\frac{3}{5} + \frac{7}{x + 2} = 2$$
$$\frac{3}{5} + \frac{7}{3 + 2} \overset{?}{=} 2$$
$$\frac{3}{5} + \frac{7}{5} \overset{?}{=} 2$$
$$2 = 2$$

Because 3 satisfies the equation, it is a solution, and the solution set is {3}.

182   CHAPTER FIVE   RATIONAL EXPRESSIONS

Example 1   Solve the equation $\dfrac{-x^2+10}{x^2-1}+\dfrac{3x}{x-1}=\dfrac{2x}{x+1}$.

Solution   Clear the fractions by multiplying both sides of the given equation by the LCD of the three fractions. Then proceed as follows:

$$\frac{-x^2+10}{x^2-1}+\frac{3x}{x-1}=\frac{2x}{x+1}$$

$$\frac{-x^2+10}{(x+1)(x-1)}+\frac{3x}{x-1}=\frac{2x}{x+1} \qquad \text{Factor } x^2-1.$$

$$\frac{(x+1)(x-1)(-x^2+10)}{(x+1)(x-1)}+\frac{3x(x+1)(x-1)}{x-1}=\frac{2x(x+1)(x-1)}{x+1} \qquad \begin{array}{l}\text{Multiply both sides}\\ \text{by }(x+1)(x-1).\end{array}$$

$$-x^2+10+3x(x+1)=2x(x-1) \qquad \text{Simplify.}$$

$$-x^2+10+3x^2+3x=2x^2-2x \qquad \text{Remove parentheses.}$$

$$2x^2+10+3x=2x^2-2x \qquad \text{Combine terms.}$$

$$10+3x=-2x \qquad \text{Add }-2x^2\text{ to both sides.}$$

$$10+5x=0 \qquad \text{Add }2x\text{ to both sides.}$$

$$5x=-10 \qquad \text{Add }-10\text{ to both sides.}$$

$$x=-2 \qquad \text{Divide both sides by 5.}$$

Verify that -2 is a solution to the original equation.　■

In the previous examples we multiplied both sides of an equation by a quantity that contained a variable, and we obtained the correct solution to the given equation. Sometimes, however, multiplying both sides of an equation by a quantity that contains a variable leads to false solutions called **extraneous solutions**. This happens when we inadvertently multiply both sides of an equation by 0 and obtain a solution that leads to a 0 in the denominator of a fraction. We must be careful to exclude from the solution set of an equation any value that makes the denominator of a fraction 0 and, thus, is not in the domain of the variable. The next example illustrates a rational equation that has an extraneous solution.

Example 2   Solve the equation $\dfrac{2(x+1)}{x-3}=\dfrac{x+5}{x-3}$.

Solution   Clear the equation of fractions by multiplying both sides by $x-3$. However, if $x=3$, the quantity $x-3$ is equal to 0, and a 0 will appear in the denominator of each fraction. Thus, if you obtain an apparent solution of 3, you must discard it.

$$\frac{2(x+1)}{x-3}=\frac{x+5}{x-3}$$

$$(x - 3)\frac{2(x + 1)}{x - 3} = (x - 3)\frac{x + 5}{x - 3} \qquad \text{Multiply both sides by } x - 3.$$

$$2(x + 1) = x + 5 \qquad\qquad \text{Simplify.}$$

$$2x + 2 = x + 5 \qquad\qquad \text{Remove parentheses.}$$

$$x + 2 = 5 \qquad\qquad\qquad \text{Add } -x \text{ to both sides.}$$

$$x = 3 \qquad\qquad\qquad\quad \text{Add } -2 \text{ to both sides.}$$

Because a solution of 3 leads to a 0 in the denominator of a fraction, it is an extraneous root and must be discarded. This equation has no solutions. Its solution set is $\varnothing$, the empty set. ∎

Example 3 Solve the equation $\dfrac{x + 1}{5} - 2 = -\dfrac{4}{x}$.

Solution Clear the equation of fractions by multiplying both sides by $5x$ and proceed as follows:

$$\frac{x + 1}{5} - 2 = -\frac{4}{x}$$

$$5x\left(\frac{x + 1}{5} - 2\right) = 5x\left(-\frac{4}{x}\right) \qquad \text{Multiply both sides by } 5x.$$

$$x(x + 1) - 10x = -20 \qquad \text{Remove parentheses and simplify.}$$

$$x^2 + x - 10x = -20 \qquad \text{Remove parentheses.}$$

$$x^2 - 9x + 20 = 0 \qquad \text{Combine terms and add 20 to both sides.}$$

$$(x - 5)(x - 4) = 0 \qquad \text{Factor } x^2 - 9x + 20.$$

$$x - 5 = 0 \quad \text{or} \quad x - 4 = 0 \qquad \text{Set each factor equal to 0.}$$

$$x = 5 \qquad\qquad\quad x = 4$$

Because each apparent solution is in the domain of x, both 4 and 5 are solutions of the original equation. The solution set is $\{4, 5\}$. Verify this result by showing that both 4 and 5 satisfy the original equation. ∎

Many applications, such as shared-work problems, lead to rational equations.

Example 4 A drain can empty a swimming pool in 3 days, and a second drain can empty the pool in 2 days. How many days will it take to empty the pool if both drains are used?

Analysis: The key is to note what each drain can do in 1 day. Then, if you add what the first drain can do in 1 day to what the second drain can do in 1 day, the sum is what they can do together in 1 day. Because the first drain can empty the pool in 3 days, it can do $\frac{1}{3}$ of the job in 1 day. Because the second drain can empty the pool in 2 days, it can do $\frac{1}{2}$ of the job in 1 day. If it takes x days for both drains to empty the pool, together they can do $1/x$ of the job in 1 day.

Solution   Let x represent the number of days that it takes both drains to empty the pool. Then form the equation

What drain 1 can do in one day		what drain 2 can do in one day		what they can do together in one day.
$\dfrac{1}{3}$	$+$	$\dfrac{1}{2}$	$=$	$\dfrac{1}{x}$

$$6x\left(\frac{1}{3}+\frac{1}{2}\right) = 6x\left(\frac{1}{x}\right)$$ Multiply both sides by $6x$.

$$2x + 3x = 6$$ Remove parentheses and simplify.

$$5x = 6$$ Combine terms.

$$x = \frac{6}{5}$$ Divide both sides by 5.

It will take both drains $1\frac{1}{5}$ days to drain the pool.

Check: In $\frac{6}{5}$ days the first drain does $\frac{1}{3}(\frac{6}{5})$ of the total job and the second drain does $\frac{1}{2}(\frac{6}{5})$ of the total job. The sum of their efforts, $\frac{2}{5}+\frac{3}{5}$, is equal to one complete job.   ■

Example 5   A man drives 200 miles to a convention. Because of road construction his average speed on the return trip is 10 miles per hour less than his average speed going to the convention. If the return trip took 1 hour longer, how fast did he drive in each direction?

Analysis: Because the distance d traveled is given by the formula

$$d = rt$$

where r is the average rate of speed and t is the time, the time is given by the formula

$$t = \frac{d}{r}$$

Organize the information given in the problem in a chart such as Figure 5-2.

	d	$=$	r	$\cdot$	t
Going	200		r		$\dfrac{200}{r}$
Returning	200		$r - 10$		$\dfrac{200}{r - 10}$

Figure 5-2

Solution   Let r represent the average rate of speed going to the meeting.
Then $r - 10$ represents the average rate of speed on the return trip.
   Because the return trip took 1 hour longer, you can form the following
equation:

The time it took to travel to the convention		the time it took to return.
	$+ 1 =$	

$$\frac{200}{r} \qquad + 1 = \qquad \frac{200}{r - 10}$$

Solve the equation as follows:

$$r(r - 10)\left(\frac{200}{r} + 1\right) = r(r - 10)\frac{200}{r - 10} \qquad \text{Multiply both sides by } r(r - 10).$$

$$200(r - 10) + r(r - 10) = 200r \qquad \text{Remove parentheses and simplify.}$$

$$200r - 2000 + r^2 - 10r = 200r \qquad \text{Remove parentheses.}$$

$$r^2 - 10r - 2000 = 0 \qquad \text{Add } -200r \text{ to both sides.}$$

$$(r - 50)(r + 40) = 0 \qquad \text{Factor } r^2 - 10r - 2000.$$

$$r - 50 = 0 \qquad \text{or} \qquad r + 40 = 0 \qquad \text{Set each factor equal to 0.}$$

$$r = 50 \qquad \qquad r = -40$$

Because a speed cannot be negative, exclude the possible solution of -40. The
man averaged 50 miles per hour going to the convention, thus he averaged
$50 - 10$ or 40 miles per hour returning.

Check: At 50 miles per hour, the 200-mile trip took 4 hours. At 40 miles per
hour, the return trip took 5 hours, which is 1 hour longer.   ■

■ EXERCISE 5.5

In Exercises 1–30, solve each equation. If a solution is extraneous, so indicate.

1. $\dfrac{1}{4} + \dfrac{9}{x} = 1$

2. $\dfrac{1}{3} - \dfrac{10}{x} = -3$

3. $\dfrac{34}{x} - \dfrac{3}{2} = -\dfrac{13}{20}$

4. $\dfrac{1}{2} + \dfrac{7}{x} = 2 + \dfrac{1}{x}$

5. $\dfrac{3}{y} + \dfrac{7}{2y} = 13$

6. $\dfrac{2}{x} + \dfrac{1}{2} = \dfrac{7}{2x}$

7. $\dfrac{x + 1}{x} - \dfrac{x - 1}{x} = 0$

8. $\dfrac{2}{x} + \dfrac{1}{2} = \dfrac{9}{4x} - \dfrac{1}{2x}$

9. $\dfrac{7}{5x} - \dfrac{1}{2} = \dfrac{5}{6x} + \dfrac{1}{3}$

10. $\dfrac{x - 3}{x - 1} - \dfrac{2x - 4}{x - 1} = 0$

11. $\dfrac{3 - 5y}{2 + y} = \dfrac{3 + 5y}{2 - y}$

12. $\dfrac{x}{x - 2} = 1 + \dfrac{1}{x - 3}$

13. $\dfrac{a + 2}{a + 1} = \dfrac{a - 4}{a - 3}$

14. $\dfrac{z + 2}{z + 8} - \dfrac{z - 3}{z - 2} = 0$

15. $\dfrac{x + 2}{x + 3} - 1 = \dfrac{1}{3 - 2x - x^2}$

16. $\dfrac{x - 3}{x - 2} - \dfrac{1}{x} = \dfrac{x - 3}{x}$

17. $\dfrac{x}{x+2} = 1 - \dfrac{3x+2}{x^2+4x+4}$

18. $\dfrac{3+2a}{a^2+6+5a} + \dfrac{2-5a}{a^2-4} = \dfrac{2-3a}{a^2-6+a}$

19. $\dfrac{2}{x-2} + \dfrac{1}{x+1} = \dfrac{1}{x^2-x-2}$

20. $\dfrac{5}{y-1} + \dfrac{3}{y-3} = \dfrac{8}{y-2}$

21. $\dfrac{a-1}{a+3} - \dfrac{1-2a}{3-a} = \dfrac{2-a}{a-3}$

22. $\dfrac{5}{2z^2+z-3} - \dfrac{2}{2z+3} = \dfrac{z+1}{z-1} - 1$

23. $\dfrac{5}{x+4} + \dfrac{1}{x+4} = x - 1$

24. $\dfrac{2}{x-1} + \dfrac{x-2}{3} = \dfrac{4}{x-1}$

25. $\dfrac{3}{x+1} - \dfrac{x-2}{2} = \dfrac{x-2}{x+1}$

26. $\dfrac{x-4}{x-3} + \dfrac{x-2}{x-3} = x - 3$

27. $\dfrac{2}{x-3} + \dfrac{3}{4} = \dfrac{17}{2x}$

28. $\dfrac{30}{y-2} + \dfrac{24}{y-5} = 13$

29. $\dfrac{x+4}{x+7} - \dfrac{x}{x+3} = \dfrac{3}{8}$

30. $\dfrac{5}{x+4} - \dfrac{1}{3} = \dfrac{x-1}{x}$

In Exercises 31–44, solve each word problem.

31. If Laura can paint a house in 5 days and her brother Scott can paint the house in 3 days, how long will it take them to paint the house if they work together?

32. If Kristy can mow a lawn in 4 hours and her younger brother Steven can mow the same lawn in 3 hours, how long will it take if they work together?

33. A pipe can drain a pool in 9 hours. If a second pipe is also used, the pool can be drained in 3 hours. How long does it take the second pipe to drain the pool?

34. One pipe can fill a pond in 3 weeks, and a second pipe can fill the pond in 5 weeks. However, evaporation and seepage can empty the pond in 10 weeks. If both pipes are used, how long will it take to fill the pond?

35. Sally can clean the house in 6 hours, and her father can clean the house in 4 hours. Sally's younger brother, Dennis, can completely mess up the house in 8 hours. If Sally and her father clean and Dennis plays, how long will it take to clean the house?

36. Sam makes hamburgers at a fast food restaurant. He can make 600 burgers in 3 hours. If his manager helps him, together they can make 600 burgers in 2 hours. How long will it take the manager alone to make 400 burgers?

37. A boy can drive a motorboat 45 miles down a river in the same amount of time that he can drive 27 miles upstream. What is the speed of the current in the river if the speed of the boat in still water is 12 miles per hour?

38. If a plane that can fly 340 miles per hour in still air can travel 200 miles downwind in the same amount of time that it can travel 140 miles upwind, what is the velocity of the wind?

39. A train travels 120 miles and returns the same distance in a total time of 5 hours. If the speed of the train averaged 20 miles per hour slower on the return trip, how fast did the train travel in each direction?

40. A woman who can row 3 miles per hour in still water rows 10 miles downstream and returns upstream in a total of 12 hours. What is the speed of the current of the river?

41. If three times a certain integer is added to four times its reciprocal, the result is 8. Find the integer.

42. If three times a number is subtracted from four times its reciprocal, the result is 11. Find the number. (There are two possibilities.)

43. A repairman purchased several washing-machine motors for a total of $224. When the unit cost decreased by $4, he was able to buy one extra motor for the same total price. How many motors did he buy originally?

44. An appliance store manager bought several microwave ovens for a total of $8100 and sold them all at a profit of $90 each. With these total receipts she was able to buy 10 more microwaves than before. What was her per-unit cost on the second order?

CHAPTER SUMMARY

Key Words

complex fractions   (5.4)
extraneous solutions   (5.5)
least common denominator   (5.3)
multiplicative inverses   (5.2)

rational equations   (5.5)
rational expressions   (5.1)
reciprocals   (5.2)

Key Ideas

(5.1) Division by 0 is undefined.

$$\frac{a}{b} = \frac{ak}{bk}, \quad \text{provided that } b \neq 0 \text{ and } k \neq 0.$$

To simplify a fraction, factor the numerator and denominator, and remove all factors common to both the numerator and denominator.

(5.2) $\dfrac{a}{b} \cdot \dfrac{c}{d} = \dfrac{ac}{bd}, \quad$ provided that $b \neq 0$ and $d \neq 0.$

$\dfrac{a}{b} \div \dfrac{c}{d} = \dfrac{a}{b} \cdot \dfrac{d}{c}, \quad$ provided that $b \neq 0$, $d \neq 0$, and $c \neq 0.$

If the product of two numbers is 1, the numbers are called **reciprocals** or **multiplicative inverses** of each other.

If a number is to be divided by a fraction, multiply the number by the reciprocal of the fraction.

(5.3) $\dfrac{a}{b} + \dfrac{c}{b} = \dfrac{a+c}{b}, \quad$ provided that $b \neq 0.$

$\dfrac{a}{b} - \dfrac{c}{b} = \dfrac{a-c}{b}, \quad$ provided that $b \neq 0.$

Fractions must be written with common denominators before they can be combined by addition or subtraction.

To find the least common denominator (LCD) of two fractions, factor each denominator and use each factor the greatest number of times that it appears in any one denominator. The product of these factors is the LCD of the fractions.

(5.4) If a fraction has a fractional numerator or denominator, it is called a **complex fraction**.

(5.5)   Multiplying both sides of an equation by a quantity that contains a variable can lead to **extraneous solutions**.

REVIEW EXERCISES

In Review Exercises 1–6, simplify each fraction.

1. $\dfrac{248x^2y}{576xy^2}$

2. $\dfrac{212m^3n}{588m^2n^3}$

3. $\dfrac{x^2 - 49}{x^2 + 14x + 49}$

4. $\dfrac{x^2 + 6x + 36}{x^3 - 216}$

5. $\dfrac{x^2 - 2x + 4}{2x^3 + 16}$

6. $\dfrac{2m - 2n}{n - m}$

In Review Exercises 7–20, perform the indicated operations and simplify.

7. $\dfrac{x^2 + 4x + 4}{x^2 - x - 6} \cdot \dfrac{x^2 - 9}{x^2 + 5x + 6}$

8. $\dfrac{x^3 - 64}{x^2 + 4x + 16} \div \dfrac{x^2 - 16}{x + 4}$

9. $\dfrac{5y}{x - y} - \dfrac{3}{x - y}$

10. $\dfrac{3x - 1}{x^2 + 2} + \dfrac{3(x - 2)}{x^2 + 2}$

11. $\dfrac{3}{x + 2} + \dfrac{2}{x + 3}$

12. $\dfrac{4x}{x - 4} - \dfrac{3}{x + 3}$

13. $\dfrac{x^2 + 3x + 2}{x^2 - x - 6} \cdot \dfrac{3x^2 - 3x}{x^2 - 3x - 4} \div \dfrac{x^2 + 3x + 2}{x^2 - 2x - 8}$

14. $\dfrac{x^2 - x - 6}{x^2 - 3x - 10} \div \dfrac{x^2 - x}{x^2 - 5x} \cdot \dfrac{x^2 - 4x + 3}{x^2 - 6x + 9}$

15. $\dfrac{2x}{x + 1} + \dfrac{3x}{x + 2} + \dfrac{4x}{x^2 + 3x + 2}$

16. $\dfrac{5x}{x - 3} + \dfrac{5}{x^2 - 5x + 6} + \dfrac{x + 3}{x - 2}$

17. $\dfrac{3(x + 2)}{x^2 - 1} - \dfrac{2}{x + 1} + \dfrac{4(x + 3)}{x^2 - 2x + 1}$

18. $\dfrac{x}{x^2 + 4x + 4} + \dfrac{2x}{x^2 - 4} - \dfrac{x^2 - 4}{x - 2}$

19. $\dfrac{x + 2}{x^2 - 9} - \dfrac{x^2 + 6x + 9}{x^2 + 5x + 6} + \dfrac{3x}{x - 3}$

20. $\dfrac{-2(3 + x)}{x^2 + 6x + 9} + \dfrac{3(x + 2)}{x^2 - 6x + 9} - \dfrac{1}{x^2 - 9}$

In Review Exercise 21–30, simplify each complex fraction.

21. $\dfrac{\dfrac{3}{x} - \dfrac{2}{y}}{xy}$

22. $\dfrac{\dfrac{1}{x} + \dfrac{2}{y}}{\dfrac{2}{x} - \dfrac{1}{y}}$

23. $\dfrac{2x + 3 + \dfrac{1}{x}}{x + 2 + \dfrac{1}{x}}$

24. $\dfrac{6x + 13 + \dfrac{6}{x}}{6x + 5 - \dfrac{6}{x}}$

25. $\dfrac{1 + \dfrac{3}{x}}{x + 3}$

26. $\dfrac{1 - \dfrac{1}{x} - \dfrac{2}{x^2}}{1 + \dfrac{4}{x} + \dfrac{3}{x^2}}$

27. $\dfrac{(x - y)^{-2}}{x^{-2} - y^{-2}}$

28. $\dfrac{x^{-1} + 1}{x + 1}$

29. $\dfrac{x^{-1} + y}{x - y^{-1}}$

30. $\dfrac{x^{-2} + 1}{x^2 + 1}$

In Review Exercises 31–34, solve each equation, if possible.

31. $\dfrac{4}{x} - \dfrac{1}{10} = \dfrac{7}{2x}$

32. $\dfrac{2}{x + 5} - \dfrac{1}{6} = \dfrac{1}{x + 4}$

33. $\dfrac{2(x - 5)}{x - 2} = \dfrac{6x + 12}{4 - x^2}$

34. $\dfrac{7}{x + 9} - \dfrac{x + 2}{2} = \dfrac{x + 4}{x + 9}$

In Review Exercises 35–36, solve each word problem.

35. Traffic reduced Jim's usual speed by 10 miles per hour, which lengthened his 200-mile trip by 1 hour. What is his usual speed?

36. On a 600-mile trip, a pilot can save 30 minutes by increasing her usual speed by 40 miles per hour. What is her usual speed?

CHAPTER FIVE TEST

In Problems 1–4, simplify each fraction.

1. $\dfrac{-12x^2y^3z^2}{18x^3y^4z^2}$

2. $\dfrac{2x+4}{x^2-4}$

3. $\dfrac{3y-6z}{2z-y}$

4. $\dfrac{2x^2+7x+3}{4x+12}$

In Problems 5–14, perform the indicated operations and simplify, if necessary. **Write all answers without using negative exponents.**

5. $\dfrac{x^2y^{-2}}{x^3z^2}\cdot\dfrac{x^2z^4}{y^2z}$

6. $\dfrac{(x+1)(x+2)}{10}\cdot\dfrac{5}{x+2}$

7. $\dfrac{u^2+5u+6}{u^2-4}\cdot\dfrac{u^2-5u+6}{u^2-9}$

8. $\dfrac{x^3+y^3}{4}\div\dfrac{x^2-xy+y^2}{2x+2y}$

9. $\dfrac{xu+2u+3x+6}{u^2-9}\cdot\dfrac{2u-6}{x^2+3x+2}$

10. $\dfrac{a^2+7a+12}{a+3}\div\dfrac{16-a^2}{a-4}$

11. $\dfrac{3t}{t+3}+\dfrac{9}{t+3}$

12. $\dfrac{3w}{w-5}-\dfrac{w+10}{w-5}$

13. $\dfrac{2}{r}+\dfrac{r}{s}$

14. $\dfrac{x+2}{x+1}-\dfrac{x+1}{x+2}$

In Problems 15–16, simplify each complex fraction.

15. $\dfrac{\dfrac{2u^2w^3}{v^2}}{\dfrac{4uw^4}{uv}}$

16. $\dfrac{\dfrac{x}{y}+\dfrac{1}{2}}{\dfrac{x}{2}-\dfrac{1}{y}}$

In Problems 17–18, solve each equation.

17. $\dfrac{2}{x-1}+\dfrac{5}{x+2}=\dfrac{11}{x+2}$

18. $\dfrac{u-2}{u-3}+3=u-\dfrac{u-4}{u-3}$

In Problems 19–20, solve each word problem.

19. A boat sails a distance of 440 nautical miles. If the boat had averaged 11 nautical miles more each day, the trip would have required 2 fewer days. How long did the trip take?

20. A student can earn $300 interest annually by investing in a bank certificate of deposit at a certain interest rate. If she were to receive an annual interest rate that is 4% higher, she could receive the same annual interest by investing $2000 less. How much would she invest at each rate?

CAREERS AND MATHEMATICS

MECHANICAL ENGINEER

Mechanical engineers design and develop power-producing machines such as internal combustion engines, steam and gas turbines, and jet and rocket engines, and power-using machines such as refrigeration and air-conditioning equipment, machine tools, printing presses, and steel rolling mills. Many mechanical engineers do research, test, and design work; others work in maintenance, technical sales, and production operations. Some are administrators or managers; others teach in colleges and universities or work as consultants.

Qualifications

A bachelor's degree in engineering is generally required for beginning engineering jobs. College graduates with a degree in a natural science or mathematics also may qualify for some jobs. Experienced technicians with some engineering education are occasionally able to advance to some engineering jobs.

Job outlook

Employment of mechanical engineers is expected to increase through the mid-1990s due to the growing demand for complex industrial machinery and processes, and to the need for new energy systems and solutions to environmental pollution problems.

Example application

Mechanical power is often transmitted from one location to another by means of a shaft—the drive shaft in an automobile is a good example. It is one concern of the mechanical engineer that the shaft be strong enough so that it will not twist, bend, or snap. One critical value in the required calculations is the shaft's **stiffness**, denoted here by the letter k. The stiffness of a shaft is measured in units of inch pounds per radian.

If a single shaft consists of two sections (see Illustration 1), then the shaft's overall stiffness k is given by the formula

$$k = \frac{1}{\dfrac{1}{k_1} + \dfrac{1}{k_2}}$$

where k_1 and k_2 are the individual stiffnesses of Sections 1 and 2, respectively.

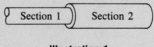

Illustration 1

What is the stiffness of a shaft whose two sections have individual stiffnesses of 3,500,000 in. lb/rad and 4,200,000 in. lb/rad?

Solution

Substitute 3,500,000 for k_1 and 4,200,000 for k_2 into the formula, and evaluate k.

$$k = \cfrac{1}{\cfrac{1}{k_1} + \cfrac{1}{k_2}}$$

$$= \cfrac{1}{\cfrac{1}{3,500,000} + \cfrac{1}{4,200,000}}$$ Substitute values for k_1 and k_2.

$$= \cfrac{1}{\cfrac{1 \cdot 6}{3,500,000 \cdot 6} + \cfrac{1 \cdot 5}{4,200,000 \cdot 5}}$$ Find a common denominator.

$$= \cfrac{1}{\cfrac{6 + 5}{21,000,000}}$$ Add the fractions in the denominator.

$$= \cfrac{1}{\cfrac{11}{21,000,000}}$$ Simplify.

$$= \cfrac{21,000,000}{11}$$ Divide.

$$\approx 1,900,000$$

The effective stiffness of the entire shaft is approximately 1,900,000 in. lb/rad.

Exercises

1. What is the stiffness of a shaft whose two sections have individual stiffnesses of 2,600,000 in. lb/rad and 3,700,000 in. lb/rad?
2. What is the stiffness of a shaft whose two sections have the same stiffness, 4,000,000 in. lb/rad?
3. Show that the formula of this example may also be written as

$$k = \frac{k_1 k_2}{k_1 + k_2}$$

4. The **compliance** of a shaft is the reciprocal of the stiffness. Show that the compliance of the entire shaft is the sum of the compliances of each of its sections.

(Answers: **1.** approximately 1,530,000 in. lb/rad
2. 2,000,000 in. lb/rad)

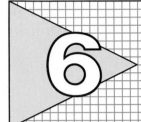

6 Rational Exponents and Radicals

In previous chapters we squared and cubed numbers and raised numbers to other integral powers. We now discuss how to raise numbers to nonintegral powers and how to find roots of numbers.

6.1 RATIONAL EXPONENTS

The rules of integral exponents discussed in Chapter 2 can be extended to include rational (fractional) exponents if we define the exponential expression $a^{1/n}$ in the following way:

Definition. If n is a natural number and $a \geq 0$, then $a^{1/n}$ is the nonnegative number whose nth power is a. In symbols,

$$(a^{1/n})^n = a$$

If n is a natural number and $a \geq 0$, the nonnegative number $a^{1/n}$ is called the **principal nth root** of a. The expression $a^{1/2}$ is called the **square root** of a, and the expression $a^{1/3}$ is called the **cube root** of a.

Example 1
a. $36^{1/2} = 6$ because $6^2 = 36$.

b. $9^{1/2} = 3$ because $3^2 = 9$.

c. $8^{1/3} = 2$ because $2^3 = 8$.

d. $625^{1/4} = 5$ because $5^4 = 625$. Read $625^{1/4}$ as "the fourth root of 625."

e. $0^{1/7} = 0$ because $0^7 = 0$.

f. $-32^{1/5} = -(32^{1/5}) = -(2) = -2$.

g. $(-25)^{1/2}$ is not a real number because the square of no real number is -25. ∎

In the expression $a^{1/n}$, if n is an odd natural number, we can remove the restriction that $a \geq 0$.

> **Definition.** If n is an odd natural number and a is any real number, then $a^{1/n}$ is the real number whose nth power is a. In symbols,
>
> $$(a^{1/n})^n = a$$

Example 2   **a.** $(-8)^{1/3} = -2$ because $(-2)^3 = -8$.

b. $(-32)^{1/5} = -2$ because $(-2)^5 = -32$.

c. $64^{1/3} = 4$ because $4^3 = 64$.

d. $3125^{1/5} = 5$ because $5^5 = 3125$.   ∎

The following chart summarizes the definitions concerning the exponential expression $a^{1/n}$:

> If n is a natural number and a is a real number, then
> If $a > 0$, then $a^{1/n}$ is the positive number such that $(a^{1/n})^n = a$.
> If $a = 0$, then $a^{1/n} = 0$.
> If $a < 0$ $\begin{cases} \text{and } n \text{ is odd, then } a^{1/n} \text{ is the real number such that } (a^{1/n})^n = a. \\ \text{and } n \text{ is even, then } a^{1/n} \text{ is not a real number.} \end{cases}$

The definition of $a^{1/n}$ can be extended to include rational exponents whose numerators are not 1. For example, $4^{3/2}$ can be written as either

$$(4^{1/2})^3 \qquad \text{or} \qquad (4^3)^{1/2}$$

because of a power rule of exponents. In general, we have the following rule:

> If m and n are positive integers and the fraction $\dfrac{m}{n}$ has been simplified to lowest terms, then
>
> $$a^{m/n} = (a^{1/n})^m = (a^m)^{1/n}$$

Because of the previous statement, we can look at the expression $a^{m/n}$ in two different ways:

1. $a^{m/n}$ means the mth power of the nth root of a.
2. $a^{m/n}$ means the nth root of the mth power of a.

For example $16^{3/4}$ and $(-27)^{2/3}$ can be interpreted as

$$16^{3/4} = (16^{1/4})^3 = (2)^3 = 8 \qquad (-27)^{2/3} = [(-27)^{1/3}]^2 = (-3)^2 = 9$$

or as

$$16^{3/4} = (16^3)^{1/4} = (4096)^{1/4} = 8 \qquad (-27)^{2/3} = [(-27)^2]^{1/3} = 729^{1/3} = 9$$

Either way, the answer is the same. As the previous examples suggest, however, it is usually easier to take the root of the base first in order to avoid large numbers.

To be consistent with the definition of negative exponents, we define $a^{-m/n}$ as follows:

Definition. If m and n are positive integers, the fraction $\dfrac{m}{n}$ has been simplified to lowest terms, and $a \neq 0$, then

$$a^{-m/n} = \frac{1}{a^{m/n}} \qquad \text{and} \qquad \frac{1}{a^{-m/n}} = a^{m/n}$$

Example 3   **a.** $64^{-2/3} = \dfrac{1}{64^{2/3}} = \dfrac{1}{(64^{1/3})^2} = \dfrac{1}{4^2} = \dfrac{1}{16}$

b. $16^{-3/2} = \dfrac{1}{16^{3/2}} = \dfrac{1}{(16^{1/2})^3} = \dfrac{1}{4^3} = \dfrac{1}{64}$

c. $(-32)^{-2/5} = \dfrac{1}{(-32)^{2/5}} = \dfrac{1}{[(-32)^{1/5}]^2} = \dfrac{1}{(-2)^2} = \dfrac{1}{4}$

d. $(-16)^{-3/4}$ is not a real number because $(-16)^{1/4}$ is not a real number.   ■

The familiar rules of integral exponents also hold for rational exponents. The following example illustrates the use of each rule for expressions containing rational exponents.

Example 4   Assume that all variables represent positive numbers, and write all answers without using negative exponents.

a. $7^{2/5} \cdot 7^{1/5} = 7^{3/5}$ $a^m a^n = a^{m+n}$

b. $(4^{1/3})^5 = 4^{(1/3)5} = 4^{5/3}$ $(a^m)^n = a^{mn}$

c. $(5 \cdot 4)^{4/3} = 5^{4/3} \cdot 4^{4/3}$ $(ab)^n = a^n b^n$

d. $\dfrac{8^{3/7}}{8^{2/7}} = 8^{3/7 - 2/7} = 8^{1/7}$ $\dfrac{a^m}{a^n} = a^{m-n}$

e. $\left(\dfrac{2}{3}\right)^{2/5} = \dfrac{2^{2/5}}{3^{2/5}}$ $\left(\dfrac{a}{b}\right)^n = \dfrac{a^n}{b^n}$

f. $5^{-2/3} = \dfrac{1}{5^{2/3}}$ $a^{-n} = \dfrac{1}{a^n}$

g. $\dfrac{1}{4^{-2/3}} = 4^{2/3}$ $\dfrac{1}{a^{-n}} = a^n$

h. $(2.5)^0 = 1$ $a^0 = 1$ ∎

The next examples illustrate how to work with exponential expressions with variables in their bases.

Example 5   Assume all variables represent positive numbers. Write all answers without using negative exponents.

a. $(25x)^{1/2} = 25^{1/2}x^{1/2}$
$$= 5x^{1/2}$$

b. $\dfrac{(a^{2/3}b^{1/2})^6}{(y^2)^2} = \dfrac{a^{12/3}b^{6/2}}{y^4}$
$$= \dfrac{a^4 b^3}{y^4}$$

c. $\dfrac{a^{5/3}a^{1/4}}{a^{1/6}} = a^{5/3 + 1/4 - 1/6}$
$$= a^{20/12 + 3/12 - 2/12}$$
$$= a^{21/12}$$
$$= a^{7/4}$$

d. $\dfrac{(12ab^2)^{1/2}(12ab)^{1/2}}{ab^{3/2}} = \dfrac{12^{1/2}a^{1/2}b12^{1/2}a^{1/2}b^{1/2}}{ab^{3/2}}$
$$= \dfrac{12^{1/2}12^{1/2}a^{1/2}a^{1/2}bb^{1/2}}{ab^{3/2}}$$
$$= \dfrac{12^{1/2 + 1/2}a^{1/2 + 1/2}b^{1 + 1/2}}{ab^{3/2}}$$
$$= \dfrac{12ab^{3/2}}{ab^{3/2}}$$
$$= 12$$

e. $\dfrac{a^{x/2}a^{x/3}}{a^{x/4}} = a^{x/2 + x/3 - x/4}$
$$= a^{6x/12 + 4x/12 - 3x/12}$$
$$= a^{7x/12}$$ ∎

Example 6   Assume all variables represent positive numbers. Perform the indicated operations and write all answers without using negative exponents.

a. $a^{4/5}(a^{1/5} + a^{5/4}) = a^{4/5}a^{1/5} + a^{4/5}a^{5/4}$ Use the distributive property.
$$= a^{4/5 + 1/5} + a^{4/5 + 5/4}$$ Use the rule $a^m a^n = a^{m+n}$.
$$= a^{5/5} + a^{16/20 + 25/20}$$ Add exponents.
$$= a + a^{41/20}$$ Simplify.

b. $x^{1/2}(x^{-1/2} + x^{1/2}) = x^{1/2}x^{-1/2} + x^{1/2}x^{1/2}$ Use the distributive property.

$\qquad\qquad\qquad\qquad = x^{1/2-1/2} + x^{1/2+1/2}$ Use the rule $a^m a^n = a^{m+n}$

$\qquad\qquad\qquad\qquad = x^0 + x^1$ Simplify.

$\qquad\qquad\qquad\qquad = 1 + x$

c. $(x^{2/3} + 1)(x^{2/3} - 1) = x^{4/3} - x^{2/3} + x^{2/3} - 1$ Use the FOIL method.

$\qquad\qquad\qquad\qquad = x^{4/3} - 1$ Simplify.   ■

■ EXERCISE 6.1

*In Exercises 1–44, simplify each expression, if possible. **Write all answers without using negative exponents.***

1. $16^{1/2}$ **2.** $25^{1/2}$ **3.** $27^{1/3}$ **4.** $64^{1/3}$

5. $81^{1/4}$ **6.** $625^{1/4}$ **7.** $32^{1/5}$ **8.** $0^{1/5}$

9. $\left(\dfrac{1}{4}\right)^{1/2}$ **10.** $\left(\dfrac{1}{16}\right)^{1/2}$ **11.** $\left(\dfrac{1}{8}\right)^{1/3}$ **12.** $\left(\dfrac{1}{16}\right)^{1/4}$

13. $-16^{1/4}$ **14.** $-125^{1/3}$ **15.** $(-27)^{1/3}$ **16.** $(-125)^{1/3}$

17. $(-64)^{1/3}$ **18.** $(-216)^{1/3}$ **19.** $0^{1/3}$ **20.** $25^{3/2}$

21. $36^{3/2}$ **22.** $27^{2/3}$ **23.** $81^{3/4}$ **24.** $100^{3/2}$

25. $144^{3/2}$ **26.** $1000^{2/3}$ **27.** $\left(\dfrac{1}{8}\right)^{2/3}$ **28.** $\left(\dfrac{4}{9}\right)^{3/2}$

29. $\left(\dfrac{8}{27}\right)^{2/3}$ **30.** $\left(\dfrac{27}{64}\right)^{2/3}$ **31.** $4^{-3/2}$ **32.** $25^{-5/2}$

33. $16^{-3/2}$ **34.** $81^{-3/2}$ **35.** $(-27)^{-2/3}$ **36.** $(-8)^{-2/3}$

37. $(-32)^{-2/5}$ **38.** $(-16)^{-5/2}$ **39.** $\left(\dfrac{1}{4}\right)^{-3/2}$ **40.** $\left(\dfrac{4}{25}\right)^{-3/2}$

41. $\left(\dfrac{27}{8}\right)^{-4/3}$ **42.** $\left(\dfrac{25}{49}\right)^{-3/2}$ **43.** $\left(-\dfrac{8}{27}\right)^{-1/3}$ **44.** $\left(\dfrac{16}{81}\right)^{-3/4}$

*In Exercises 45–56, perform the indicated operations. **Write all answers as a single expression without using negative exponents.***

45. $5^{3/7}5^{2/7}$ **46.** $4^{2/5}4^{2/5}$ **47.** $(4^{1/5})^3$ **48.** $(3^{1/3})^5$

49. $\dfrac{9^{4/5}}{9^{3/5}}$ **50.** $\dfrac{7^{2/3}}{7^{1/2}}$ **51.** $\dfrac{7^{1/2}}{7^0}$ **52.** $5^{1/3}5^{-5/3}$

53. $6^{-2/3}6^{-4/3}$ **54.** $\dfrac{3^{4/3}3^{1/3}}{3^{2/3}}$ **55.** $\dfrac{2^{5/6}2^{1/3}}{2^{1/2}}$ **56.** $\dfrac{5^{1/3}5^{1/2}}{5^{1/3}}$

*In Exercises 57–76, simplify each expression. Assume that all variables represent positive numbers. **Write all answers without using negative exponents.***

57. $(8x)^{1/3}$ **58.** $(32x^3)^{1/5}$ **59.** $(27x^2y)^{2/3}$ **60.** $(81x^2y)^{3/4}$

61. $(25x^2y)^{3/2}$ **62.** $(125x^2y)^{4/3}$ **63.** $(4xy^3)^{-1/2}$ **64.** $(8x^2y)^{-1/3}$

65. $\dfrac{(4x^3y)^{1/2}}{(9xy)^{1/2}}$

66. $\dfrac{(27x^3y)^{1/3}}{(8xy^2)^{2/3}}$

67. $\dfrac{(8xy^2)^{1/2}(2xy)^{1/2}}{xy^{1/2}}$

68. $\dfrac{(4x^2y)^{1/3}(2xy^2)^{1/3}}{(xy^2)^{2/3}}$

69. $x^{a/3}x^{a/2}$

70. $x^{a/4}x^{a/3}$

71. $\dfrac{x^{a/4}}{x^{a/3}}$

72. $\dfrac{x^{a/2}}{x^{a/3}}$

73. $\dfrac{x^{a/5}x^{2a/5}}{x^{4a/5}}$

74. $\dfrac{x^{1/b}x^{2/b}}{x^{4/b}}$

75. $\left(\dfrac{x^{2/a}x^{3/a}}{x^{1/b}}\right)^{ab}$

76. $\left(\dfrac{x^b y^c}{z^b}\right)^{c/b}$

In Exercises 77–98, perform the indicated multiplications to remove parentheses. **Write all answers without using negative exponents.**

77. $y^{1/3}(y^{2/3} + y^{5/3})$

78. $y^{2/5}(y^{-2/5} + y^{3/5})$

79. $a^{2/3}(a^{4/3} + a^{-2/3})$

80. $a^{-2/3}(a^{5/3} + a^{2/3})$

81. $x^{3/5}(x^{7/5} - x^{2/5} + 1)$

82. $x^{4/3}(x^{2/3} + 3x^{5/3} - 4)$

83. $z^{1/3}y^{1/2}(z^{-1/3} + 1 - y^{-1/2})$

84. $z^{-2/3}x^{1/3}(z^{5/3}x^{2/3} - z^{-1/3}x^{-1/3})$

85. $(x^{1/2} + 2)(x^{1/2} - 2)$

86. $(x^{1/2} + y^{1/2})(x^{-1/2} + y^{-1/2})$

87. $(x^{2/3} - x)(x^{2/3} + x)$

88. $(x^{1/2} + y^{1/2})(x^{1/2} - y^{1/2})$

89. $(x^{-1/2} + y^{1/2})(x^{1/2} + y^{-1/2})$

90. $(x^{1/3} + x^2)(x^{1/3} - x^2)$

91. $(x^{2/3} + y^{2/3})^2$

92. $(a^{1/2} - b^{2/3})^2$

93. $(a^{3/2} - b^{3/2})^2$

94. $(x^{-1/2} - x^{1/2})^2$

95. $(x^{1/3} + 1)(x^{2/3} - x^{1/3} + 1)$

96. $(x^{1/3} - 2)(x^{2/3} + 2x^{1/3} + 4)$

97. $[(3 - x^{1/3})(9 + x^{2/3} + 3x^{1/3})]^2$

98. $[x^{-1/3}(x^{-2/3} + x^{1/3})]^2$

6.2   RADICAL EXPRESSIONS

In the previous section we defined the symbol $a^{1/n}$ to be the nth root of a. We now introduce a different symbol for the nth root of a.

Definition.   If n is a natural number greater than 1 and if $a^{1/n}$ is a real number, then

$$a^{1/n} = \sqrt[n]{a}$$

The symbol $\sqrt[n]{a}$ is called a **radical expression** (or radical). The symbol $\sqrt{}$ is called a **radical sign**, the number a is called the **radicand**, and n is called the **index** (or the **order**) of the radical. If the order of a radical is 2, the expression is called a **square root** and we do not write the index. Thus,

$$\sqrt{a} \qquad \text{means} \qquad \sqrt[2]{a}$$

If the index of a radical is 3, the radical is called a **cube root**.

We now restate the definition for the nth root of a number using radical notation:

Definition. If n is a natural number greater than 1 and $a \geq 0$, then $\sqrt[n]{a}$ is the nonnegative number whose nth power is a. In symbols,

$$(\sqrt[n]{a})^n = a$$

The nonnegative number $\sqrt[n]{a}$ is called the **principal nth root** of a.

If 2 is substituted for n in the equation $(\sqrt[n]{a})^n = a$, we have

$$(\sqrt[2]{a})^2 = (\sqrt{a})^2 = \sqrt{a}\,\sqrt{a} = a$$

Thus, if a can be factored into two equal factors, then either of those factors is a square root of a.

In the expression $\sqrt[n]{a}$, if n is an odd natural number greater than 1, we can remove the restriction that $a \geq 0$.

Definition. If n is an odd natural number greater than 1 and if a is any real number, then $\sqrt[n]{a}$ is the real number whose nth power is a. In symbols,

$$(\sqrt[n]{a})^n = a$$

Example 1

a. $\sqrt{36} = 6$ because $6^2 = 36$.

Read $\sqrt{36}$ as "the square root of 36."

b. $\sqrt[4]{81} = 3$ because $3^4 = 81$.

Read $\sqrt[4]{81}$ as "the fourth root of 81."

c. $\sqrt[3]{8} = 2$ because $2^3 = 8$.

Read $\sqrt[3]{8}$ as "the cube root of 8."

d. $\sqrt[3]{-8} = -2$ because $(-2)^3 = -8$.

e. $-\sqrt{16} = -(\sqrt{16}) = -(4) = -4$.

f. $\sqrt[7]{0} = 0$ because $0^7 = 0$.

Read $\sqrt[7]{0}$ as "the seventh root of 0." ■

The following chart summarizes the definitions concerning the radical expression $\sqrt[n]{a}$:

If n is a natural number greater than 1 and a is a real number, then
If $a > 0$, then $\sqrt[n]{a}$ is the positive number such that $(\sqrt[n]{a})^n = a$.
If $a = 0$, then $\sqrt[n]{a} = 0$.
If $a < 0$ $\begin{cases} \text{and } n \text{ is odd, then } \sqrt[n]{a} \text{ is the real number such that } (\sqrt[n]{a})^n = a. \\ \text{and } n \text{ is even, then } \sqrt[n]{a} \text{ is not a real number.} \end{cases}$

In the previous section we saw that $x^{m/n} = (x^m)^{1/n} = (x^{1/n})^m$. The same fact stated in radical notation is

$$x^{m/n} = \sqrt[n]{x^m} = (\sqrt[n]{x})^m \qquad \text{provided } \sqrt[n]{x} \text{ is a real number.}$$

Thus, the nth root of x^m is the same as the mth power of the nth root of x.

Example 2 **a.** $\sqrt[3]{8^2} = \sqrt[3]{64} = 4 \qquad \text{or} \qquad \sqrt[3]{8^2} = (\sqrt[3]{8})^2 = 2^2 = 4$

  b. $\sqrt{9^3} = \sqrt{729} = 27 \qquad \text{or} \qquad \sqrt{9^3} = (\sqrt{9})^3 = 3^3 = 27$ ■

All positive numbers have two square roots. One of them is positive and one of them is negative. For example, a square root of 100 is any number whose square is 100. There are two such numbers: 10 and -10. *The symbol $\sqrt{100}$ represents the positive square root of* 100. Thus,

$$\sqrt{100} = 10$$

The negative square root of 100 is represented by the symbol $-\sqrt{100}$. Thus,

$$-\sqrt{100} = -10$$

In general, the number a^2 has two square roots. They are a and $-a$. One of these numbers is positive and the other is negative. If a might be negative, we cannot write $\sqrt{a^2} = a$. We must use absolute value symbols to guarantee that $\sqrt{a^2}$ is positive. Thus, if a is unrestricted,

$$\sqrt{a^2} = |a|$$

For example, if $a = -3$, we have

$$\sqrt{(-3)^2} = |-3| = 3.$$

A similar argument holds when the index n is any even natural number. The symbol $\sqrt[4]{16}$, for example, means the *positive* fourth root of 16. Thus,

$$\sqrt[4]{16} = 2 \qquad \text{Because } 2^4 = 16$$

The symbol $-\sqrt[4]{16}$ represents the negative fourth root of 16:

$$-\sqrt[4]{16} = -2$$

Example 3 **a.** $\sqrt[4]{256} = |4| = 4$ **b.** $\sqrt[6]{64} = |2| = 2$

  c. $-\sqrt[4]{2^4} = -|2| = -2$ **d.** $\sqrt[4]{(-2)^4} = |-2| = 2$

  e. $-\sqrt[4]{(-3)^4} = -|-3| = -3$ **f.** $-\sqrt[6]{729} = -|3| = -3$ ■

When the index n of a radical is an odd natural number, there is only one real nth root of a and we do not use absolute value symbols. For example,

$$\sqrt[3]{8} = 2 \qquad \text{and} \qquad \sqrt[3]{-8} = -2$$

In general, we have the following rules:

If n is an even natural number, then
$$\sqrt[n]{a^n} = |a|$$
If n is an odd natural number, then
$$\sqrt[n]{a^n} = a$$

Example 4 **a.** $\sqrt{x^2} = |x|$ because $|x|^2 = x^2$ and $|x| \geq 0$.

b. $\sqrt[3]{x^3} = x$ because $x^3 = x^3$.

c. $\sqrt[4]{a^4} = |a|$ because $|a|^4 = a^4$ and $|a| \geq 0$.

d. $\sqrt[5]{-32x^5} = -2x$ because $(-2x)^5 = -32x^5$.

e. $\sqrt{x^4} = x^2$ because $(x^2)^2 = x^4$. No absolute value signs are needed because x^2 cannot be negative. ■

Many properties of exponents have counterparts in radical notation. For example, because $a^{1/n}b^{1/n} = (ab)^{1/n}$, we have

$$\sqrt[n]{a}\,\sqrt[n]{b} = \sqrt[n]{ab}$$

Thus, as long as all expressions represent real numbers, the product of the nth roots of two numbers is equal to the nth root of their product. More formally, we have the following:

Property 1 of Radicals.   If $\sqrt[n]{a}$ and $\sqrt[n]{b}$ are real numbers, then
$$\sqrt[n]{ab} = \sqrt[n]{a}\,\sqrt[n]{b}$$

Note that Property 1 of radicals involves the nth root of the product of two numbers. However, there is no such property for sums. For example,

$$\sqrt{9 + 4} \neq \sqrt{9} + \sqrt{4}$$

because

$$\sqrt{9 + 4} = \sqrt{13} \qquad \text{but} \qquad \sqrt{9} + \sqrt{4} = 3 + 2 = 5$$

Numbers such as 1, 4, 9, 16, 25, and 36 that are squares of integers are called **perfect squares**. Numbers such as 1, 8, 27, and 64 that are cubes of integers are called **perfect cubes**. In like fashion, there are perfect fourth powers, perfect fifth powers, and so on. We can use this concept of perfect powers and Property 1 of radicals to simplify many radical expressions.

Example 5   Simplify   **a.** $\sqrt{12}$,   **b.** $\sqrt{98}$, and   **c.** $\sqrt[3]{54}$.

Solution   **a.**   Factor 12 so that one factor is the largest perfect square that divides 12. In this case, the largest perfect square that divides 12 is 4. Rewrite 12 as $4 \cdot 3$, apply Property 1 of radicals, and simplify:

$$\sqrt{12} = \sqrt{4 \cdot 3} = \sqrt{4}\sqrt{3} = 2\sqrt{3}$$

b.   The largest perfect square factor of 98 is 49. Thus,

$$\sqrt{98} = \sqrt{49 \cdot 2} = \sqrt{49}\sqrt{2} = 7\sqrt{2}$$

c.   The largest perfect cube factor of 54 is 27. Thus,

$$\sqrt[3]{54} = \sqrt[3]{27 \cdot 2} = \sqrt[3]{27}\sqrt[3]{2} = 3\sqrt[3]{2}$$ ∎

Another property of radicals involves the quotient of two radicals. Because $\dfrac{a^{1/n}}{b^{1/n}} = \left(\dfrac{a}{b}\right)^{1/n}$, it follows that

$$\frac{\sqrt[n]{a}}{\sqrt[n]{b}} = \sqrt[n]{\frac{a}{b}} \qquad (b \neq 0)$$

Thus, as long as all expressions represent real numbers, the quotient of the *n*th roots of two numbers is equal to the *n*th root of their quotient. More formally, we have the following:

Property 2 of Radicals.   If $\sqrt[n]{a}$ and $\sqrt[n]{b}$ are real numbers and $b \neq 0$, then

$$\sqrt[n]{\frac{a}{b}} = \frac{\sqrt[n]{a}}{\sqrt[n]{b}}$$

Example 6   Simplify   **a.** $\dfrac{\sqrt{72}}{\sqrt{2}}$ and   **b.** $\sqrt[3]{\dfrac{9}{27}}$.

Solution   **a.**   Write the quotient of the square roots as the square root of a quotient and simplify.

$$\frac{\sqrt{72}}{\sqrt{2}} = \sqrt{\frac{72}{2}} \qquad \text{Use Property 2 of radicals.}$$
$$= \sqrt{36}$$
$$= 6$$

b.   Write the cube root of the quotient as a quotient of two cube roots.

$$\sqrt[3]{\frac{9}{27}} = \frac{\sqrt[3]{9}}{\sqrt[3]{27}} \qquad \text{Use Property 2 of radicals.}$$

$$= \frac{\sqrt[3]{9}}{3}$$

∎

We will use Properties 1 and 2 of radicals to simplify many radical expressions. A radical expression is said to be in simplest form only if each of the following statements is true:

> A radical expression is in simplest form only if
> **1.**   No radical appears in the denominator of a fraction.
> **2.**   No fraction appears in a radicand.
> **3.**   No prime factor of the radicand occurs more times than the index of the radical.
> **4.**   There is no common factor (other than 1) between the index of the radical and all the exponents of the factors of the radicand.

Example 7   Simplify each of the following expressions. Assume that all variables represent positive numbers. Thus, no absolute value signs will be needed.

a. $\sqrt{128a^5}$,   **b.** $\sqrt[3]{24x^5}$,   **c.** $\dfrac{\sqrt{45xy^2}}{\sqrt{5x}}$, and   **d.** $\dfrac{\sqrt[3]{-432x^5}}{\sqrt[3]{8x}}$

Solution   **a.**   Rewrite $128a^5$ as $64a^4 \cdot 2a$, where $64a^4$ is the largest perfect square that divides $128a^5$. Then apply Property 1 of radicals, and simplify.

$$\sqrt{128a^5} = \sqrt{64a^4 \cdot 2a} = \sqrt{64a^4}\sqrt{2a} = 8a^2\sqrt{2a}$$

b.   Rewrite $24x^5$ as $8x^3 \cdot 3x^2$, where $8x^3$ is the largest perfect cube that divides $24x^5$. Then apply Property 1 of radicals, and simplify.

$$\sqrt[3]{24x^5} = \sqrt[3]{8x^3 \cdot 3x^2} = \sqrt[3]{8x^3}\sqrt[3]{3x^2} = 2x\sqrt[3]{3x^2}$$

c.   Use Property 2 of radicals to rewrite the quotient of the square roots as the square root of a quotient. Then simplify.

$$\frac{\sqrt{45xy^2}}{\sqrt{5x}} = \sqrt{\frac{45xy^2}{5x}} = \sqrt{9y^2} = 3y$$

d.   Use Property 2 of radicals to rewrite the quotient of the cube roots as the cube root of a quotient. Then simplify.

$$\frac{\sqrt[3]{-432x^5}}{\sqrt[3]{8x}} = \sqrt[3]{\frac{-432x^5}{8x}} = \sqrt[3]{-54x^4} = \sqrt[3]{-27x^3 \cdot 2x}$$

$$= \sqrt[3]{-27x^3}\sqrt[3]{2x} = -3x\sqrt[3]{2x}$$

∎

To simplify a radical with a complicated radicand, we can use the prime factorization of the radicand to determine its perfect square factors. For example, to simplify $\sqrt{3168x^5y^7}$, we first find the prime factorization of $3168x^5y^7$:

$$3168x^5y^7 = 2^5 \cdot 3^2 \cdot 11 \cdot x^5 \cdot y^7$$

Thus, we have

$$\begin{aligned}
\sqrt{3168x^5y^7} &= \sqrt{2^4 \cdot 3^2 \cdot x^4 \cdot y^6 \cdot 2 \cdot 11 \cdot x \cdot y} \\
&= \sqrt{2^4 \cdot 3^2 \cdot x^4 \cdot y^6} \sqrt{2 \cdot 11 \cdot x \cdot y} \\
&= 2^2 \cdot 3x^2y^3 \sqrt{22xy} \\
&= 12x^2y^3 \sqrt{22xy}
\end{aligned}$$

Rationalizing the Denominator

The radical expression $\sqrt{70}/\sqrt{3}$ is not in simplest form because it contains a radical in its denominator. To simplify this expression we must use a process called **rationalizing the denominator**. To eliminate the radical in the denominator, we multiply both the numerator and the denominator of the fraction by $\sqrt{3}$ to obtain

$$\frac{\sqrt{70}}{\sqrt{3}} = \frac{\sqrt{70} \cdot \sqrt{3}}{\sqrt{3} \cdot \sqrt{3}} = \frac{\sqrt{210}}{3}$$

Because there is no radical in the denominator and because $\sqrt{210}$ cannot be simplified, the radical expression $\sqrt{210}/3$ is in simplest form.

Example 8   Use the process of rationalizing the denominator to simplify $\sqrt{\dfrac{20}{7}}$.

Solution   First write the square root of the quotient as the quotient of two square roots:

$$\sqrt{\frac{20}{7}} = \frac{\sqrt{20}}{\sqrt{7}}$$

Then rationalize the denominator by multiplying both the numerator and the denominator by $\sqrt{7}$:

$$\frac{\sqrt{20}}{\sqrt{7}} = \frac{\sqrt{20} \cdot \sqrt{7}}{\sqrt{7} \cdot \sqrt{7}} = \frac{\sqrt{140}}{7}$$

However, the radical in the numerator of $\sqrt{140}/7$ can be simplified, and the fraction can be written as follows:

$$\frac{\sqrt{140}}{7} = \frac{\sqrt{4 \cdot 35}}{7} = \frac{\sqrt{4} \cdot \sqrt{35}}{7} = \frac{2\sqrt{35}}{7} \qquad \blacksquare$$

Example 9   Rationalize the denominator of the fraction $\dfrac{1}{\sqrt[3]{2}}$.

Solution   To rationalize the denominator, multiply both the numerator and the denominator by a number that will result in a perfect cube under the radical sign. Because $2 \cdot 4 = 8$ is a perfect cube, $\sqrt[3]{4}$ is such a number. Proceed as follows:

$$\frac{1}{\sqrt[3]{2}} = \frac{1 \cdot \sqrt[3]{4}}{\sqrt[3]{2} \cdot \sqrt[3]{4}} = \frac{\sqrt[3]{4}}{\sqrt[3]{8}} = \frac{\sqrt[3]{4}}{2}$$ ■

Example 10   Rationalize the denominator of the fraction $\dfrac{\sqrt[3]{5}}{\sqrt[3]{18}}$.

Solution   To rationalize the denominator, multiply both the numerator and the denominator by a number that will result in a perfect cube under the radical sign in the denominator. The first six perfect cubes are

$$1^3 = 1 \quad 2^3 = 8 \quad 3^3 = 27 \quad 4^3 = 64 \quad 5^3 = 125 \quad \text{and} \quad 6^3 = 216$$

and 216 is the smallest perfect cube that is divisible by 18: $216 \div 18 = 12$. Thus, multiplying both the numerator and the denominator by $\sqrt[3]{12}$ will give a perfect cube under the radical sign in the denominator:

$$\frac{\sqrt[3]{5}}{\sqrt[3]{18}} = \frac{\sqrt[3]{5} \cdot \sqrt[3]{12}}{\sqrt[3]{18} \cdot \sqrt[3]{12}} = \frac{\sqrt[3]{60}}{\sqrt[3]{216}} = \frac{\sqrt[3]{60}}{6}$$ ■

Example 11   Rationalize the denominator of the fraction $\dfrac{\sqrt{5xy^2}}{\sqrt{xy^3}}$. Assume that x and y are positive numbers.

Solution 1 $\dfrac{\sqrt{5xy^2}}{\sqrt{xy^3}} = \sqrt{\dfrac{5xy^2}{xy^3}}$

$\qquad\qquad = \sqrt{\dfrac{5}{y}}$

$\qquad\qquad = \dfrac{\sqrt{5}}{\sqrt{y}}$

$\qquad\qquad = \dfrac{\sqrt{5} \cdot \sqrt{y}}{\sqrt{y} \cdot \sqrt{y}}$

$\qquad\qquad = \dfrac{\sqrt{5y}}{y}$

Solution 2 $\dfrac{\sqrt{5xy^2}}{\sqrt{xy^3}} = \sqrt{\dfrac{5xy^2}{xy^3}}$

$\qquad\qquad = \sqrt{\dfrac{5}{y}}$

$\qquad\qquad = \sqrt{\dfrac{5 \cdot y}{y \cdot y}}$

$\qquad\qquad = \dfrac{\sqrt{5y}}{\sqrt{y^2}}$

$\qquad\qquad = \dfrac{\sqrt{5y}}{y}$ ■

Example 12   Simplify $\sqrt{x^2 + 6x + 9}$. Assume that x is positive.

Solution   Note that the radical can be written in the form $\sqrt{(x+3)^2}$. Because $(x+3)(x+3) = (x+3)^2$ and $x+3$ is positive, it follows that $\sqrt{(x+3)^2} = x+3$. Thus,

$$\sqrt{x^2 + 6x + 9} = \sqrt{(x+3)^2}$$
$$= x + 3$$

EXERCISE 6.2

In Exercises 1–24, simplify each radical.

1. $\sqrt{121}$
2. $\sqrt{144}$
3. $-\sqrt{64}$
4. $-\sqrt{1}$

5. $\sqrt[3]{1}$
6. $\sqrt[3]{-8}$
7. $\sqrt[3]{-125}$
8. $\sqrt[3]{512}$

9. $\sqrt[4]{81}$
10. $\sqrt[6]{64}$
11. $-\sqrt[5]{243}$
12. $-\sqrt[4]{625}$

13. $\sqrt[5]{-32}$
14. $\sqrt[3]{-512}$
15. $\sqrt{9^3}$
16. $\sqrt[3]{8^4}$

17. $\sqrt[4]{16^3}$
18. $\sqrt[3]{(-27)^2}$
19. $\sqrt[5]{-32^2}$
20. $\sqrt[4]{\left(\frac{1}{16}\right)^3}$

21. $\sqrt[3]{-\left(\frac{1}{8}\right)^2}$
22. $\sqrt[3]{\left(\frac{8}{27}\right)^2}$
23. $\sqrt[5]{243^{-2}}$
24. $\sqrt[5]{32^{-3}}$

In Exercises 25–36, simplify each radical. Use absolute value symbols where necessary.

25. $\sqrt{4x^2}$
26. $\sqrt{9a^2}$
27. $\sqrt[3]{8a^3}$
28. $\sqrt[3]{27x^6}$

29. $\sqrt[4]{x^4}$
30. $\sqrt[4]{x^8}$
31. $\sqrt[4]{x^{12}}$
32. $\sqrt{x^8}$

33. $\sqrt[5]{-x^5}$
34. $\sqrt[3]{-x^6}$
35. $\sqrt[3]{-27a^6}$
36. $\sqrt[5]{-32x^5}$

In Exercises 37–88, simplify each radical expression. Assume that all variables represent positive numbers. No absolute value symbols will be necessary.

37. $\sqrt{5}\sqrt{5}$
38. $\sqrt{11}\sqrt{11}$
39. $\sqrt[3]{2}\sqrt[3]{2}\sqrt[3]{2}$
40. $\sqrt[3]{4}\sqrt[3]{4}\sqrt[3]{4}$

41. $\sqrt{t}\sqrt{t}$
42. $-\sqrt{z}\sqrt{z}$
43. $\sqrt{20}$
44. $\sqrt{8}$

45. $\sqrt{24}$
46. $\sqrt{50}$
47. $\sqrt{200}$
48. $\sqrt{250}$

49. $-\sqrt{50x^2}$
50. $-\sqrt{75a^2}$
51. $\sqrt{32b}$
52. $-\sqrt{80c}$

53. $-\sqrt{112a^3}$
54. $\sqrt{147a^5}$
55. $\sqrt{175a^2b^3}$
56. $-\sqrt{128a^3b^5}$

57. $-\sqrt{300xy}$
58. $\sqrt{200x^2y}$
59. $\sqrt[3]{81}$
60. $\sqrt[3]{-72}$

61. $\sqrt[3]{-80}$
62. $\sqrt[3]{270}$
63. $\sqrt[3]{-54x^6}$
64. $-\sqrt[3]{-81a^3}$

65. $-\sqrt[3]{16x^{12}y^3}$
66. $\sqrt[3]{40a^3b^6}$
67. $\sqrt[4]{32x^{12}y^4}$
68. $-\sqrt[4]{243x^{20}y^8}$

69. $-\sqrt[5]{64x^{10}y^5}$
70. $\sqrt[5]{486a^{25}b^{20}}$
71. $-\frac{\sqrt{500}}{\sqrt{5}}$
72. $\frac{\sqrt{128}}{\sqrt{2}}$

73. $\frac{\sqrt{98}}{\sqrt{2}}$
74. $-\frac{\sqrt{75}}{\sqrt{3}}$
75. $\frac{\sqrt{180ab^4}}{\sqrt{5ab^2}}$
76. $\frac{\sqrt{112xy^2}}{\sqrt{7xy}}$

77. $\frac{\sqrt{128x^7}}{\sqrt{8x^2}}$
78. $\frac{\sqrt{245a^7}}{\sqrt{5a}}$
79. $\frac{\sqrt[3]{48}}{\sqrt[3]{6}}$
80. $\frac{\sqrt[3]{64}}{\sqrt[3]{8}}$

81. $\dfrac{\sqrt[3]{189a^4}}{\sqrt[3]{7a}}$ **82.** $\dfrac{\sqrt[3]{243x^2}}{\sqrt[3]{9x}}$ **83.** $\sqrt{\dfrac{7}{9}}$ **84.** $\sqrt{\dfrac{3}{4}}$

85. $\sqrt{\dfrac{3x}{48x^3}}$ **86.** $\sqrt{\dfrac{2a^3}{128a^5}}$ **87.** $\sqrt{\dfrac{10abc^2}{98a^3b^5}}$ **88.** $\sqrt{\dfrac{14tu^2}{128t^3}}$

In Exercises 89–94, you may have to simplify the fraction partially to make the radicand in the denominator a perfect square or a perfect cube.

89. $\sqrt{\dfrac{22x^3}{242x}}$ **90.** $\sqrt{\dfrac{100x^4}{288}}$ **91.** $\sqrt[3]{\dfrac{4a^3}{27a^6}}$ **92.** $\sqrt[3]{\dfrac{3x^9}{8x^{12}}}$

93. $\sqrt[3]{\dfrac{18a^9}{24}}$ **94.** $\sqrt[3]{\dfrac{20b^{12}}{54}}$

In Exercises 95–114, rationalize each denominator. Assume that all variables represent positive numbers.

95. $\dfrac{1}{\sqrt{7}}$ **96.** $\dfrac{2}{\sqrt{6}}$ **97.** $\dfrac{\sqrt{2}}{\sqrt{3}}$ **98.** $\dfrac{\sqrt{3}}{\sqrt{2}}$

99. $\dfrac{\sqrt{5}}{\sqrt{8}}$ **100.** $\dfrac{\sqrt{3}}{\sqrt{50}}$ **101.** $\dfrac{\sqrt{8}}{\sqrt{2}}$ **102.** $\dfrac{\sqrt{27}}{\sqrt{3}}$

103. $\dfrac{1}{\sqrt[3]{2}}$ **104.** $\dfrac{2}{\sqrt[3]{6}}$ **105.** $\dfrac{3}{\sqrt[3]{9}}$ **106.** $\dfrac{2}{\sqrt[3]{a}}$

107. $\dfrac{\sqrt[3]{2}}{\sqrt[3]{3}}$ **108.** $\dfrac{\sqrt[3]{9}}{\sqrt[3]{54}}$ **109.** $\dfrac{\sqrt{8x^2y}}{\sqrt{xy}}$ **110.** $\dfrac{\sqrt{9xy}}{\sqrt{3x^2y}}$

111. $\dfrac{\sqrt{10xy^2}}{\sqrt{2xy^3}}$ **112.** $\dfrac{\sqrt{5ab^2c}}{\sqrt{10abc}}$ **113.** $\dfrac{\sqrt[3]{4a^2}}{\sqrt[3]{2ab}}$ **114.** $\dfrac{\sqrt[3]{9x}}{\sqrt[3]{3xy}}$

In Exercises 115–124, simplify each expression. Assume that all variables represent positive numbers.

115. $\sqrt[4]{x^5y^3z^8}$ **116.** $\sqrt[4]{x^7y^5z^{13}}$ **117.** $\sqrt[3]{-16a^5b^3c^2}$ **118.** $\sqrt[3]{-54a^{11}b^{13}c^{16}}$

119. $\sqrt[5]{-64a^6b^6c^{11}}$ **120.** $\sqrt[5]{486x^7y^8z^9}$ **121.** $\sqrt{x^2+2x+1}$ **122.** $\sqrt{x^2+4x+4}$

123. $\sqrt{x^2+8x+16}$ **124.** $\sqrt{x^4+6x^2+9}$

6.3 ADDING AND SUBTRACTING RADICAL EXPRESSIONS

Two radical expressions with the same index and the same radicand are called **like** or **similar radicals**. Because of the distributive property, we can combine them. For example, to simplify the expression $3\sqrt{2}+2\sqrt{2}$, we use the distributive property to factor out $\sqrt{2}$ and simplify:

$$3\sqrt{2}+2\sqrt{2}=(3+2)\sqrt{2}$$
$$=5\sqrt{2}$$

If two radical expressions have the same index, but different radicands, they often can be adjusted so they have the same radicand. Then they can be combined. For example, to simplify the expression $\sqrt{27} - \sqrt{12}$, we simplify both radicals and then combine the like radicals.

$$\begin{aligned}\sqrt{27} - \sqrt{12} &= \sqrt{9 \cdot 3} - \sqrt{4 \cdot 3} \\ &= \sqrt{9}\sqrt{3} - \sqrt{4}\sqrt{3} \\ &= 3\sqrt{3} - 2\sqrt{3} \\ &= (3 - 2)\sqrt{3} \\ &= \sqrt{3}\end{aligned}$$

As the previous examples suggest, we can use the following rule to add or subtract radicals:

> To add or subtract radicals, simplify each radical and combine all like radicals. To combine like radicals, add their coefficients and keep the same radical.

Example 1   Simplify $2\sqrt{12} - 3\sqrt{48} + 3\sqrt{3}$.

Solution   Simplify each radical separately, factor out the common factor of $\sqrt{3}$, and simplify:

$$\begin{aligned}2\sqrt{12} - 3\sqrt{48} + 3\sqrt{3} &= 2\sqrt{4 \cdot 3} - 3\sqrt{16 \cdot 3} + 3\sqrt{3} \\ &= 2\sqrt{4}\sqrt{3} - 3\sqrt{16}\sqrt{3} + 3\sqrt{3} \\ &= 2(2)\sqrt{3} - 3(4)\sqrt{3} + 3\sqrt{3} \\ &= (4 - 12 + 3)\sqrt{3} \\ &= -5\sqrt{3} \end{aligned}$$ ∎

Example 2   Simplify $\sqrt[3]{16} - \sqrt[3]{54} + \sqrt[3]{24}$.

Solution   Simplify each radical separately and proceed as follows:

$$\begin{aligned}\sqrt[3]{16} - \sqrt[3]{54} + \sqrt[3]{24} &= \sqrt[3]{8 \cdot 2} - \sqrt[3]{27 \cdot 2} + \sqrt[3]{8 \cdot 3} \\ &= \sqrt[3]{8}\sqrt[3]{2} - \sqrt[3]{27}\sqrt[3]{2} + \sqrt[3]{8}\sqrt[3]{3} \\ &= 2\sqrt[3]{2} - 3\sqrt[3]{2} + 2\sqrt[3]{3} \\ &= (2 - 3)\sqrt[3]{2} + 2\sqrt[3]{3} \\ &= -\sqrt[3]{2} + 2\sqrt[3]{3}\end{aligned}$$

Note that you cannot combine $-\sqrt[3]{2}$ and $2\sqrt[3]{3}$ because the radicals have different radicands.   ∎

Example 3   Simplify $\sqrt[3]{16x^4} + \sqrt[3]{54x^4} - \sqrt[3]{-128x^4}$. Assume that $x > 0$.

Solution   Simplify each radical expression separately, factor out the common factor of $\sqrt[3]{2x}$, and simplify:

$$\sqrt[3]{16x^4} + \sqrt[3]{54x^4} - \sqrt[3]{-128x^4} = \sqrt[3]{8x^3}\sqrt[3]{2x} + \sqrt[3]{27x^3}\sqrt[3]{2x} - \sqrt[3]{-64x^3}\sqrt[3]{2x}$$
$$= 2x\sqrt[3]{2x} + 3x\sqrt[3]{2x} + 4x\sqrt[3]{2x}$$
$$= (2x + 3x + 4x)\sqrt[3]{2x}$$
$$= 9x\sqrt[3]{2x}\qquad\blacksquare$$

Example 4   Simplify $3\sqrt[3]{24} - 8\sqrt[3]{81} - 2\sqrt[3]{\dfrac{3}{27}}$.

Solution   Simplify each radical expression separately, and combine terms:

$$3\sqrt[3]{24} - 8\sqrt[3]{81} - 2\sqrt[3]{\frac{3}{27}} = 3\sqrt[3]{8}\sqrt[3]{3} - 8\sqrt[3]{27}\sqrt[3]{3} - 2\frac{\sqrt[3]{3}}{\sqrt[3]{27}}$$
$$= 3(2)\sqrt[3]{3} - 8(3)\sqrt[3]{3} - 2\frac{\sqrt[3]{3}}{3}$$
$$= 6\sqrt[3]{3} - 24\sqrt[3]{3} - \frac{2}{3}\sqrt[3]{3}$$
$$= \frac{18}{3}\sqrt[3]{3} - \frac{72}{3}\sqrt[3]{3} - \frac{2}{3}\sqrt[3]{3}$$
$$= \left(\frac{18}{3} - \frac{72}{3} - \frac{2}{3}\right)\sqrt[3]{3}$$
$$= -\frac{56}{3}\sqrt[3]{3}\qquad\blacksquare$$

■ EXERCISE 6.3

In Exercises 1–52, simplify and combine like radicals, if possible. Assume that all variables represent positive numbers.

1. $4\sqrt{2} + 6\sqrt{2}$
2. $6\sqrt{5} + 3\sqrt{5}$
3. $8\sqrt[5]{7} - 7\sqrt[5]{7}$
4. $10\sqrt[6]{12} - \sqrt[6]{12}$
5. $8\sqrt{x} + 6\sqrt{x}$
6. $10\sqrt{xy} - 2\sqrt{xy}$
7. $\sqrt{3} + \sqrt{27}$
8. $\sqrt{8} + \sqrt{32}$
9. $\sqrt{2} - \sqrt{8}$
10. $\sqrt{20} - \sqrt{125}$
11. $\sqrt{98} - \sqrt{50}$
12. $\sqrt{72} - \sqrt{200}$
13. $3\sqrt{24} + \sqrt{54}$
14. $\sqrt{18} + 2\sqrt{50}$
15. $\sqrt[3]{24} + \sqrt[3]{3}$
16. $\sqrt[3]{16} + \sqrt[3]{128}$
17. $\sqrt[3]{32} - \sqrt[3]{108}$
18. $\sqrt[3]{80} - \sqrt[3]{10000}$
19. $2\sqrt[3]{125} - 5\sqrt[3]{64}$
20. $3\sqrt[3]{27} + 12\sqrt[3]{216}$
21. $3\sqrt[3]{-54} + 8\sqrt[3]{-128}$
22. $5\sqrt[3]{-81} - 7\sqrt[3]{-375}$
23. $14\sqrt[4]{32} - 15\sqrt[4]{162}$
24. $23\sqrt[4]{768} + \sqrt[4]{48}$
25. $\sqrt{98} - \sqrt{50} - \sqrt{72}$
26. $\sqrt{20} + \sqrt{125} - \sqrt{80}$
27. $\sqrt{18} + \sqrt{300} - \sqrt{243}$
28. $\sqrt{80} - \sqrt{128} + \sqrt{288}$
29. $2\sqrt[3]{16} - \sqrt[3]{54} - 3\sqrt[3]{128}$
30. $\sqrt[4]{48} - \sqrt[4]{243} - \sqrt[4]{768}$
31. $\sqrt{25y^2z} - \sqrt{16y^2z}$
32. $\sqrt{x^2y} + \sqrt{9y}$
33. $\sqrt{x^4y^3} - \sqrt{x^4y^5} - \sqrt{x^4y^7}$
34. $\sqrt{ay^3} + \sqrt{ay^5} - \sqrt{ay^7}$

35. $\sqrt[5]{x^6 y^2} + \sqrt[5]{32x^6 y^2} + \sqrt[5]{x^{11} y^2}$

36. $\sqrt[3]{x^4 y^3} + \sqrt[3]{x^7 y^6} - \sqrt[3]{x^{10} y^9}$

37. $\sqrt{x^2 + 4x + 4} + \sqrt{x^2 + 2x + 1}$

38. $\sqrt{4x^2 + 12x + 9} + \sqrt{9x^2 + 6x + 1}$

39. $\sqrt{3x^2 + 6x + 3} + \sqrt{3x^2}$

40. $\sqrt{5x^2 - 10x + 5} - \sqrt{5x^2 + 20x + 20}$

41. $\sqrt{8} + \dfrac{4}{\sqrt{2}} - \sqrt{32}$

42. $\sqrt{12} + \dfrac{6}{\sqrt{3}} - 3\sqrt{243}$

43. $\sqrt{16x} - \dfrac{x}{\sqrt{x}}$

44. $\sqrt{25a^3} + \dfrac{\sqrt{a^4}}{a}$

45. $\sqrt{18} - \sqrt{12} + \dfrac{\sqrt{12}}{6} - \dfrac{\sqrt{8}}{4}$

46. $\sqrt{80} + \sqrt{28} + \dfrac{\sqrt{20}}{5} + \dfrac{\sqrt{28}}{7}$

47. $\dfrac{\sqrt{48a}}{3a} - \dfrac{\sqrt{108a^3}}{9a^2} + \dfrac{\sqrt{12a}}{6a}$

48. $\dfrac{\sqrt{200y}}{4y} + \dfrac{\sqrt{98y}}{2y} - \dfrac{\sqrt{288y^3}}{6y^2}$

49. $\dfrac{\sqrt{108x^3}}{4x^2} + \dfrac{\sqrt{75x}}{x} - \dfrac{\sqrt{147x}}{2x}$

50. $\dfrac{\sqrt{50a^5}}{10a^4} + \dfrac{5\sqrt{2a^3}}{2a^3} - \dfrac{\sqrt{200a}}{10a^2}$

51. $\dfrac{3\sqrt{28x^5 y}}{4x} + \dfrac{2x\sqrt{7xy^3}}{3y} - \dfrac{\sqrt{112x^5 y^3}}{3xy}$

52. $\dfrac{y\sqrt{18x^5}}{2x^2} + \dfrac{7\sqrt{2xy^4}}{4y} - \dfrac{\sqrt{18xy^2}}{12}$

6.4   MULTIPLYING AND DIVIDING RADICAL EXPRESSIONS

If two radical expressions have the same index, they can be multiplied and divided. We begin by showing how to multiply a monomial by a monomial.

Example 1   Multiply $3\sqrt{6}$ by $2\sqrt{3}$.

Solution   Make use of the commutative and associative properties of multiplication to multiply the coefficients and the radicals separately. Remember to simplify any radicals in the product, if possible.

$$
\begin{aligned}
3\sqrt{6} \cdot 2\sqrt{3} &= 3(2)\sqrt{6}\sqrt{3} \\
&= 6\sqrt{18} \\
&= 6\sqrt{9}\sqrt{2} \\
&= 6(3)\sqrt{2} \\
&= 18\sqrt{2}
\end{aligned}
$$ ∎

To multiply a polynomial by a monomial, we use the distributive property to remove parentheses and then simplify each term.

Example 2   Simplify $3\sqrt{3}(4\sqrt{8} - 5\sqrt{27})$.

Solution
$$
\begin{aligned}
3\sqrt{3}(4\sqrt{8} - 5\sqrt{27}) &= 3\sqrt{3} \cdot 4\sqrt{8} - 3\sqrt{3} \cdot 5\sqrt{27} \\
&= 12\sqrt{24} - 15\sqrt{81} \\
&= 12\sqrt{4}\sqrt{6} - 15(9) \\
&= 12(2)\sqrt{6} - 135 \\
&= 24\sqrt{6} - 135
\end{aligned}
$$ ∎

To multiply a binomial by a binomial, we use the FOIL method.

Example 3   Simplify $(\sqrt{7} + \sqrt{2})(\sqrt{7} - 3\sqrt{2})$.

Solution
$$
\begin{aligned}
(\sqrt{7} + \sqrt{2})(\sqrt{7} - 3\sqrt{2}) &= (\sqrt{7})^2 - 3\sqrt{7}\sqrt{2} + \sqrt{2}\sqrt{7} - 3\sqrt{2}\sqrt{2} \\
&= 7 - 3\sqrt{14} + \sqrt{14} - 3(2) \\
&= 7 - 2\sqrt{14} - 6 \\
&= 1 - 2\sqrt{14}
\end{aligned}
$$
∎

Example 4   Multiply $\sqrt{3}x - \sqrt{5}$ by $\sqrt{2}x + \sqrt{10}$.

Solution   Multiply the two binomials using the FOIL method. Then, simplify each of the four terms.

$$
\begin{aligned}
(\sqrt{3}x - \sqrt{5})(\sqrt{2}x + \sqrt{10}) &= \sqrt{3}\sqrt{2}x^2 + \sqrt{3}\sqrt{10}x - \sqrt{5}\sqrt{2}x - \sqrt{5}\sqrt{10} \\
&= \sqrt{6}x^2 + \sqrt{30}x - \sqrt{10}x - \sqrt{50} \\
&= \sqrt{6}x^2 + \sqrt{30}x - \sqrt{10}x - \sqrt{25}\sqrt{2} \\
&= \sqrt{6}x^2 + \sqrt{30}x - \sqrt{10}x - 5\sqrt{2}
\end{aligned}
$$
∎

To divide radical expressions, we rationalize the denominators of all fractions containing radicals and then simplify. In Section 6.2 we considered rationalizing denominators that were monomials. We now consider rationalizing denominators that are binomials. Suppose we wish to rationalize the denominator of the fraction

$$
\frac{1}{\sqrt{2} + 1}
$$

It is not sufficient to multiply the numerator and the denominator by $\sqrt{2}$. (Try it and discover why.) Instead we shall multiply both the numerator and the denominator by $\sqrt{2} - 1$. Note that this binomial is the same binomial as the denominator, but with the opposite sign between its terms.

$$
\begin{aligned}
\frac{1}{\sqrt{2} + 1} &= \frac{1(\sqrt{2} - 1)}{(\sqrt{2} + 1)(\sqrt{2} - 1)} \\
&= \frac{\sqrt{2} - 1}{(\sqrt{2})^2 - 1} \qquad (\sqrt{2} + 1)(\sqrt{2} - 1) = (\sqrt{2})^2 - 1 \\
&= \frac{\sqrt{2} - 1}{2 - 1} \\
&= \frac{\sqrt{2} - 1}{1} \\
&= \sqrt{2} - 1
\end{aligned}
$$

In the previous discussion, we began by multiplying both the numerator and the denominator of the given fraction by $\sqrt{2} - 1$. This binomial is the same as the denominator of the given fraction $\sqrt{2} + 1$, except for the signs between their terms. Such binomials are called **conjugates** of each other.

Definition.   **Conjugate binomials** are binomials that are the same except for the sign between their terms. The conjugate of $a + b$ is $a - b$, and the conjugate of $a - b$ is $a + b$.

Example 5   Rationalize the denominator of the fraction $\dfrac{\sqrt{x} - \sqrt{3}}{\sqrt{x} - \sqrt{2}}$. Assume that x is a positive number.

Solution   Multiply both the numerator and denominator of the fraction by $\sqrt{x} + \sqrt{2}$, which is the conjugate of $\sqrt{x} - \sqrt{2}$. Then simplify.

$$\frac{\sqrt{x} - \sqrt{3}}{\sqrt{x} - \sqrt{2}} = \frac{(\sqrt{x} - \sqrt{3})(\sqrt{x} + \sqrt{2})}{(\sqrt{x} - \sqrt{2})(\sqrt{x} + \sqrt{2})}$$

$$= \frac{x + \sqrt{2x} - \sqrt{3x} - \sqrt{6}}{x - 2}$$ ∎

Example 6   Rationalize the denominator of the fraction $\dfrac{9x^2}{\sqrt{3x}(\sqrt{5} - \sqrt{2})}$. Assume that x represents a positive number.

Solution   Multiply numerator and denominator by $\sqrt{3x}(\sqrt{5} + \sqrt{2})$, and simplify.

$$\frac{9x^2}{\sqrt{3x}(\sqrt{5} - \sqrt{2})} = \frac{9x^2 \cdot \sqrt{3x}(\sqrt{5} + \sqrt{2})}{\sqrt{3x}(\sqrt{5} - \sqrt{2}) \cdot \sqrt{3x}(\sqrt{5} + \sqrt{2})}$$

$$= \frac{9x^2\sqrt{3x}(\sqrt{5} + \sqrt{2})}{\sqrt{3x}\sqrt{3x}(\sqrt{5} - \sqrt{2})(\sqrt{5} + \sqrt{2})}$$

$$= \frac{9x^2\sqrt{3x}(\sqrt{5} + \sqrt{2})}{3x(5 - 2)}$$

$$= \frac{9x^2\sqrt{3x}(\sqrt{5} + \sqrt{2})}{9x}$$

$$= x\sqrt{3x}(\sqrt{5} + \sqrt{2})$$ ∎

■ EXERCISE 6.4

In Exercises 1–24, perform each indicated multiplication and simplify. Assume that all variables represent positive numbers.

1. $\sqrt{2}\sqrt{8}$ **2.** $\sqrt{3}\sqrt{27}$ **3.** $\sqrt{5}\sqrt{10}$ **4.** $\sqrt{7}\sqrt{35}$

5. $\sqrt{3}\sqrt{6}$ **6.** $\sqrt{11}\sqrt{33}$ **7.** $\sqrt[3]{5}\sqrt[3]{25}$ **8.** $\sqrt[3]{7}\sqrt[3]{49}$

9. $\sqrt[3]{9}\sqrt[3]{3}$ **10.** $\sqrt[3]{16}\sqrt[3]{4}$ **11.** $\sqrt[3]{2}\sqrt[3]{12}$ **12.** $\sqrt[3]{3}\sqrt[3]{18}$

13. $\sqrt{ab^3}\sqrt{ab}$ **14.** $\sqrt{8x}\sqrt{2x^3y}$ **15.** $\sqrt{5ab}\sqrt{5a}$ **16.** $\sqrt{15rs^2}\sqrt{10r}$

17. $\sqrt[3]{5r^2s}\sqrt[3]{2r}$ **18.** $\sqrt[3]{3xy^2z^3}\sqrt[3]{9x^3z}$ **19.** $\sqrt[3]{a^5bc^2}\sqrt[3]{16ab^5}$ **20.** $\sqrt[3]{3x^4y}\sqrt[3]{18x}$

21. $\sqrt{x(x+3)}\sqrt{x^3(x+3)}$ **22.** $\sqrt{y^2(x+y)}\sqrt{(x+y)^3}$

23. $\sqrt[3]{6x^2(y+z)^2}\sqrt[3]{18x(y+z)}$ **24.** $\sqrt[3]{9x^2y(z+1)^2}\sqrt[3]{6xy^2(z+1)}$

In Exercises 25–44, perform each indicated multiplication and simplify. Assume that all variables represent positive numbers.

25. $3\sqrt{5}(4-\sqrt{5})$ **26.** $2\sqrt{7}(3\sqrt{7}-1)$ **27.** $3\sqrt{2}(4\sqrt{3}+2\sqrt{7})$ **28.** $-\sqrt{3}(\sqrt{7}-\sqrt{5})$

29. $-2\sqrt{5}(4\sqrt{2}-3\sqrt{3})$ **30.** $3\sqrt{7}(2\sqrt{7}+3\sqrt{3})$

31. $(\sqrt{2}+1)(\sqrt{2}-3)$ **32.** $(2\sqrt{3}+1)(\sqrt{3}-1)$

33. $(4\sqrt{3}+3)(2\sqrt{3}-5)$ **34.** $(7\sqrt{2}+2)(3\sqrt{2}-5)$

35. $(\sqrt{5}+\sqrt{3})(\sqrt{5}+\sqrt{3})$ **36.** $(\sqrt{2}-\sqrt{3})(\sqrt{3}-\sqrt{2})$

37. $(\sqrt{3}-\sqrt{2})(\sqrt{3}+\sqrt{2})$ **38.** $(\sqrt{3}+\sqrt{2})(\sqrt{3}+\sqrt{2})$

39. $(2\sqrt{3}-\sqrt{5})(\sqrt{3}+3\sqrt{5})$ **40.** $(5\sqrt{2}-\sqrt{3})(2\sqrt{2}+2\sqrt{3})$

41. $(3\sqrt{2}-2)^2$ **42.** $(2\sqrt{3}+5)^2$

43. $-2\sqrt{3}(\sqrt{7}+\sqrt{3})^2$ **44.** $\sqrt{2}(2\sqrt{5}+3\sqrt{3})^2$

In Exercises 45–72, perform each division by rationalizing the denominator and simplifying. Assume that all variables represent positive numbers.

45. $\dfrac{1}{\sqrt{2}-1}$ **46.** $\dfrac{2}{\sqrt{3}-1}$ **47.** $\dfrac{-6}{\sqrt{5}+4}$ **48.** $\dfrac{-10}{\sqrt{5}-1}$

49. $\dfrac{2}{\sqrt{3}+1}$ **50.** $\dfrac{2}{\sqrt{5}+1}$ **51.** $\dfrac{25}{\sqrt{6}+1}$ **52.** $\dfrac{50}{\sqrt{7}+1}$

53. $\dfrac{\sqrt{2}}{\sqrt{5}+3}$ **54.** $\dfrac{\sqrt{3}}{\sqrt{3}-2}$ **55.** $\dfrac{\sqrt{7}}{2-\sqrt{5}}$ **56.** $\dfrac{\sqrt{11}}{3+\sqrt{7}}$

57. $\dfrac{2}{\sqrt{7}-\sqrt{5}}$ **58.** $\dfrac{5}{\sqrt{7}-\sqrt{2}}$ **59.** $\dfrac{20}{\sqrt{3}+1}$ **60.** $\dfrac{36}{\sqrt{5}+2}$

61. $\dfrac{\sqrt{3}+1}{\sqrt{3}-1}$ **62.** $\dfrac{\sqrt{2}-1}{\sqrt{2}+1}$ **63.** $\dfrac{\sqrt{7}-2}{\sqrt{2}+\sqrt{7}}$ **64.** $\dfrac{\sqrt{5}+2}{\sqrt{3}-\sqrt{2}}$

65. $\dfrac{2}{\sqrt{x}+1}$ **66.** $\dfrac{3}{\sqrt{x}-2}$ **67.** $\dfrac{x}{\sqrt{x}-4}$ **68.** $\dfrac{2x}{\sqrt{x}+1}$

69. $\dfrac{2z-1}{\sqrt{2z}-1}$ **70.** $\dfrac{3t-1}{\sqrt{3t}+1}$ **71.** $\dfrac{\sqrt{x}-\sqrt{y}}{\sqrt{x}+\sqrt{y}}$ **72.** $\dfrac{\sqrt{x}+\sqrt{y}}{\sqrt{x}-\sqrt{y}}$

6.5 RADICALS WITH DIFFERENT ORDERS

We have discussed how to add, subtract, multiply, and divide radicals with the same order (radicals with the same index). Fractional exponents enable us to simplify certain expressions involving radicals with different orders. For example, to simplify the radical expression $\sqrt[4]{4}$, we note that $4 = 2^2$, write the expression in rational exponent form, and proceed as follows:

$$\sqrt[4]{4} = \sqrt[4]{2^2} = (2^2)^{1/4} = 2^{2(1/4)} = 2^{1/2} = \sqrt{2}$$

Example 1 Assume that x and y represent positive numbers and simplify: **a.** $\sqrt[6]{8}$, **b.** $\sqrt[6]{25}$,   **c.** $\sqrt[12]{x^4}$,   **d.** $\sqrt[20]{x^6 y^2}$, and   **e.** $\sqrt[9]{8x^6}$.

Solution **a.** $\sqrt[6]{8} = \sqrt[6]{2^3} = 2^{3/6} = 2^{1/2} = \sqrt{2}$

b. $\sqrt[6]{25} = \sqrt[6]{5^2} = 5^{2/6} = 5^{1/3} = \sqrt[3]{5}$

c. $\sqrt[12]{x^4} = x^{4/12} = x^{1/3} = \sqrt[3]{x}$

d. $\sqrt[20]{x^6 y^2} = (x^6 y^2)^{1/20} = x^{6/20} y^{2/20} = x^{3/10} y^{1/10} = (x^3 y)^{1/10} = \sqrt[10]{x^3 y}$

e. $\sqrt[9]{8x^6} = \sqrt[9]{2^3 x^6} = (2^3 x^6)^{1/9} = 2^{3/9} x^{6/9} = 2^{1/3} x^{2/3} = (2x^2)^{1/3} = \sqrt[3]{2x^2}$ ∎

By using rational exponents we can add, subtract, multiply, and divide many radicals with different orders. For example, to add the radicals $\sqrt[4]{25}$ and $\sqrt{20}$, we proceed as follows:

$$
\begin{aligned}
\sqrt[4]{25} + \sqrt{20} &= \sqrt[4]{5^2} + \sqrt{4 \cdot 5} \\
&= (5^2)^{1/4} + \sqrt{4}\sqrt{5} \\
&= 5^{2/4} + 2\sqrt{5} \\
&= 5^{1/2} + 2\sqrt{5} \\
&= \sqrt{5} + 2\sqrt{5} \\
&= 3\sqrt{5}
\end{aligned}
$$

To multiply $\sqrt{2}$ by $\sqrt[3]{5}$, we first write each number as a sixth root. To do so, we write each radical as an expression with a rational exponent, change each exponent to a fraction with a denominator of 6, and change back to radical notation:

$$
\begin{aligned}
\sqrt{2} &= 2^{1/2} = 2^{3/6} = (2^3)^{1/6} = \sqrt[6]{2^3} = \sqrt[6]{8} \\
\sqrt[3]{5} &= 5^{1/3} = 5^{2/6} = (5^2)^{1/6} = \sqrt[6]{5^2} = \sqrt[6]{25}
\end{aligned}
$$

We then multiply the sixth roots:

$$\sqrt{2}\sqrt[3]{5} = \sqrt[6]{8}\sqrt[6]{25} = \sqrt[6]{8 \cdot 25} = \sqrt[6]{200}$$

Example 2   Write $\dfrac{\sqrt{3}}{\sqrt[3]{4}}$ as an expression containing a single radical.

Solution   First, eliminate the radical in the denominator by multiplying numerator and denominator by $\sqrt[3]{2}$:

$$\frac{\sqrt{3}}{\sqrt[3]{4}} = \frac{\sqrt{3}\,\sqrt[3]{2}}{\sqrt[3]{4}\,\sqrt[3]{2}}$$

$$= \frac{\sqrt{3}\,\sqrt[3]{2}}{\sqrt[3]{8}}$$

$$= \frac{\sqrt{3}\,\sqrt[3]{2}}{2}$$

Because the numerator is the product of radicals of different orders, rewrite the radicals in forms with fractional exponents having a common denominator, and proceed as follows:

$$\frac{\sqrt{3}}{\sqrt[3]{4}} = \frac{3^{1/2}2^{1/3}}{2}$$

$$= \frac{3^{3/6}2^{2/6}}{2}$$

$$= \frac{\sqrt[6]{3^3}\,\sqrt[6]{2^2}}{2}$$

$$= \frac{\sqrt[6]{3^3 \cdot 2^2}}{2}$$

$$= \frac{\sqrt[6]{27 \cdot 4}}{2}$$

$$= \frac{\sqrt[6]{108}}{2}$$

∎

Example 3   Assume that all variables represent positive numbers, and write $\sqrt{8x^3y}\,\sqrt[3]{9xy^4}$ as an expression containing a single radical.

Solution   First, remove all perfect square factors from the radicand within the square root, and all perfect cube roots from the other radicand:

$$\sqrt{8x^3y}\,\sqrt[3]{9xy^4} = \sqrt{4 \cdot 2 \cdot x^2 \cdot xy}\,\sqrt[3]{9xy \cdot y^3}$$

$$= 2x\sqrt{2xy} \cdot y\sqrt[3]{9xy}$$

$$= 2xy\sqrt{2xy}\,\sqrt[3]{9xy}$$

Write each radical in fractional exponent form, and adjust the fractional exponents to have common denominators. Then, convert back to radical form and find the product of the sixth roots.

$$\sqrt{8x^3y}\,\sqrt[3]{9xy^4} = 2xy(2xy)^{1/2}(9xy)^{1/3}$$
$$= 2xy(2xy)^{3/6}(9xy)^{2/6}$$
$$= 2xy\,\sqrt[6]{(2xy)^3}\,\sqrt[6]{(9xy)^2}$$
$$= 2xy\,\sqrt[6]{(2xy)^3(9xy)^2}$$
$$= 2xy\,\sqrt[6]{8x^3y^3\cdot 81x^2y^2}$$
$$= 2xy\,\sqrt[6]{648x^5y^5} \qquad\blacksquare$$

Example 4   Write $\sqrt[3]{3x}\,\sqrt[5]{2x^2}$ as an expression containing a single radical.

Solution   Convert each radical to a form with fractional exponents having the same denominator, convert back to radical form, and find the product of the fifteenth roots:

$$\sqrt[3]{3x}\,\sqrt[5]{2x^2} = (3x)^{1/3}(2x^2)^{1/5}$$
$$= (3x)^{5/15}(2x^2)^{3/15}$$
$$= \sqrt[15]{(3x)^5}\,\sqrt[15]{(2x^2)^3}$$
$$= \sqrt[15]{243x^5}\,\sqrt[15]{8x^6}$$
$$= \sqrt[15]{243\cdot 8x^5x^6}$$
$$= \sqrt[15]{1944x^{11}} \qquad\blacksquare$$

■ EXERCISE 6.5

In Exercises 1–16, assume that all variables represent positive numbers and simplify each radical.

1. $\sqrt[4]{9}$ **2.** $\sqrt[4]{64}$ **3.** $\sqrt[18]{x^9}$ **4.** $\sqrt[12]{x^6}$

5. $\sqrt[4]{x^8}$ **6.** $\sqrt[5]{y^{10}}$ **7.** $\sqrt[6]{y^{18}}$ **8.** $\sqrt[8]{x^{32}}$

9. $\sqrt[4]{16x^{12}}$ **10.** $\sqrt[6]{27x^3}$ **11.** $\sqrt[8]{81x^4}$ **12.** $\sqrt[10]{243x^{10}}$

13. $\sqrt[9]{-8x^9}$ **14.** $\sqrt[9]{-27x^{18}}$ **15.** $\sqrt[3]{-64a^{12}}$ **16.** $\sqrt[3]{-125a^3b^6}$

In Exercises 17–26, simplify all radicals, if possible, and then combine like radicals, if possible. Assume that all variables represent positive numbers.

17. $\sqrt{12} + \sqrt[4]{9}$ **18.** $\sqrt{8} - \sqrt[4]{4}$ **19.** $\sqrt[4]{64} - \sqrt{32}$ **20.** $\sqrt[4]{144} + \sqrt{12}$

21. $\sqrt[4]{x^6} - \sqrt[6]{x^9}$ **22.** $\sqrt[3]{a^4} + \sqrt[6]{a^8}$ **23.** $\sqrt[6]{x^6y^2} + 2x\sqrt[3]{y}$ **24.** $\sqrt[3]{a^2b^3} + \sqrt[9]{a^6b^9}$

25. $5\sqrt[5]{x^6y^{11}z^{16}} + 12\sqrt[10]{x^{12}y^{22}z^{32}}$ **26.** $13\sqrt[6]{a^8b^6c^{12}} - 4\sqrt[3]{a^4b^3c^6}$

In Exercises 27–42, write each radical as another radical of the order indicated. Assume that all variables represent positive numbers.

27. $\sqrt{3}$;   sixth order   **28.** $\sqrt{2}$;   sixth order

29. $\sqrt{5}$;   eighth order   **30.** $\sqrt{7}$;   eighth order

31. $\sqrt[3]{15}$;   sixth order   **32.** $\sqrt[3]{9}$;   sixth order

33. $\sqrt[3]{4}$; ninth order

34. $\sqrt[3]{7}$; ninth order

35. $\sqrt[6]{3^3}$; second order

36. $\sqrt[8]{7^4}$; fourth order

37. $\sqrt[9]{5^3}$; third order

38. $\sqrt[9]{7^6}$; third order

39. $\sqrt{2xy}$; sixth order

40. $\sqrt{3x^3y}$; sixth order

41. $\sqrt[3]{4x^2y}$; sixth order

42. $\sqrt[3]{9xy^2}$; sixth order

In Exercises 43–66, perform each indicated operation. Express each answer as an expression containing a single radical. Assume that all variables represent positive numbers.

43. $\sqrt{5}\sqrt[3]{7}$

44. $\sqrt[3]{9}\sqrt{2}$

45. $\sqrt[3]{3}\sqrt{5}$

46. $\sqrt[3]{9}\sqrt{5}$

47. $\sqrt[3]{2}\sqrt{2}$

48. $\sqrt{5}\sqrt[3]{5}$

49. $\sqrt[3]{2}\sqrt[5]{8}$

50. $\sqrt[5]{3}\sqrt[3]{2}$

51. $\sqrt{5x}\sqrt[3]{xy}$

52. $\sqrt[3]{4x}\sqrt{xy}$

53. $\sqrt[3]{25x}\sqrt[5]{3y^2}$

54. $\sqrt[3]{3x}\sqrt[5]{2x^2}$

55. $\sqrt[7]{xy^3}\sqrt[5]{x^3y^4}$

56. $\sqrt[9]{x^2y}\sqrt[3]{xy^2}$

57. $\dfrac{\sqrt[3]{3}}{\sqrt{2}}$

58. $\dfrac{\sqrt{2}}{\sqrt[3]{3}}$

59. $\dfrac{\sqrt[5]{3}}{\sqrt[3]{4}}$

60. $\dfrac{\sqrt[5]{4}}{\sqrt[3]{4}}$

61. $\dfrac{\sqrt[5]{3}}{\sqrt[3]{3}}$

62. $\dfrac{\sqrt{3}}{\sqrt[5]{3}}$

63. $\dfrac{\sqrt[7]{4x}}{\sqrt{2x}}$

64. $\dfrac{\sqrt[9]{xy}}{\sqrt{xy}}$

65. $\dfrac{\sqrt[5]{3x}}{\sqrt[3]{3}}$

66. $\dfrac{\sqrt[7]{3x}}{\sqrt[3]{3}}$

6.6   RADICAL EQUATIONS

In this section we will consider equations that contain radicals. To solve such equations we will rely on the following theorem, which is often called the **power rule**.

Theorem.   If a, b, and n are real numbers and $a = b$, then

$$a^n = b^n$$

If we raise both sides of an equation to the same power, the resulting equation may or may not be equivalent to the original equation. For example, if we square both sides of the equation

1. $x = 3$ With a solution set of $\{3\}$

we obtain the equation

2. $x^2 = 9$ With a solution set of $\{3, -3\}$

Equations 1 and 2 are not equivalent because they have different solution sets, and the solution -3 of Equation 2 does not satisfy Equation 1. Because raising both sides of an equation to the same power can introduce extraneous roots that don't check in the original equation, we must check each suspected solution of the original equation.

Example 1   Solve the equation $\sqrt{x + 3} = 4$.

Solution   To eliminate the radical, apply the power rule and square both sides of the equation. Then proceed as follows:

$$\sqrt{x + 3} = 4$$
$$(\sqrt{x + 3})^2 = 4^2$$
$$x + 3 = 16$$
$$x = 13 \qquad \text{Add } -3 \text{ to both sides.}$$

The apparent solution is 13. However, you must check it to verify that it satisfies the original equation.

Check:

$$\sqrt{x + 3} = 4$$
$$\sqrt{13 + 3} \overset{?}{=} 4$$
$$\sqrt{16} \overset{?}{=} 4$$
$$4 = 4$$

Because 13 satisfies the original equation, it is a solution.   ∎

Example 2   Solve the equation $\sqrt{3x + 1} = x - 1$.

Solution   To eliminate the radical, square both sides of the equation. Then proceed as follows:

$$\sqrt{3x + 1} = x - 1$$
$$(\sqrt{3x + 1})^2 = (x - 1)^2$$
$$3x + 1 = x^2 - 2x + 1$$
$$0 = x^2 - 5x \qquad \text{Add } -3x \text{ and } -1 \text{ to both sides.}$$
$$0 = x(x - 5) \qquad \text{Factor } x^2 - 5x.$$
$$x = 0 \quad \text{or} \quad x - 5 = 0 \qquad \text{Set each factor equal to 0.}$$
$$x = 5$$

The apparent solutions are 0 and 5. You must check each one to determine if either is extraneous.

Check:

$$\sqrt{3x + 1} = x - 1 \qquad\qquad \sqrt{3x + 1} = x - 1$$
$$\sqrt{3(0) + 1} \overset{?}{=} 0 - 1 \qquad\qquad \sqrt{3(5) + 1} \overset{?}{=} 5 - 1$$
$$\sqrt{1} \overset{?}{=} -1 \qquad\qquad\qquad \sqrt{16} \overset{?}{=} 4$$
$$1 \neq -1 \qquad\qquad\qquad\quad 4 = 4$$

The apparent solution 0 does not check; it is *extraneous* and must be discarded. The only solution of the original equation is 5.   ∎

Example 3   Solve the equation $\sqrt[3]{x^3 + 7} = x + 1$.

Solution   To eliminate the radical, cube both sides of the equation. Then proceed as follows:

$$\sqrt[3]{x^3 + 7} = x + 1$$
$$(\sqrt[3]{x^3 + 7})^3 = (x + 1)^3$$
$$x^3 + 7 = x^3 + 3x^2 + 3x + 1$$

$0 = 3x^2 + 3x - 6$	Add $-x^3$ and -7 to both sides.
$0 = x^2 + x - 2$	Divide both sides by 3.
$0 = (x + 2)(x - 1)$	Factor $x^2 + x - 2$.
$x + 2 = 0$ or $x - 1 = 0$	Set each factor equal to 0.
$x = -2 \quad \mid \quad x = 1$	

The apparent solutions are -2 and 1. Check each one to determine if either is extraneous.

Check:

$$\sqrt[3]{x^3 + 7} = x + 1 \qquad\qquad \sqrt[3]{x^3 + 7} = x + 1$$
$$\sqrt[3]{(-2)^3 + 7} \overset{?}{=} -2 + 1 \qquad \sqrt[3]{1^3 + 7} \overset{?}{=} 1 + 1$$
$$\sqrt[3]{-8 + 7} \overset{?}{=} -1 \qquad\qquad \sqrt[3]{8} \overset{?}{=} 2$$
$$\sqrt[3]{-1} \overset{?}{=} -1 \qquad\qquad\qquad 2 = 2$$
$$-1 = -1$$

Both solutions satisfy the original equation. The solution set is $\{-2, 1\}$. ∎

When more than one radical appears in an equation, it is often necessary to apply the power rule more than once.

Example 4   Solve the equation $\sqrt{x} + \sqrt{x + 2} = 2$.

Solution   To remove the radicals, both sides of the equation must be squared. This is easier to do if one radical is on each side of the equation. So, add $-\sqrt{x}$ to both sides and proceed as follows:

$$\sqrt{x} + \sqrt{x + 2} = 2$$
$$\sqrt{x + 2} = 2 - \sqrt{x}$$

$(\sqrt{x + 2})^2 = (2 - \sqrt{x})^2$	Square both sides.
$x + 2 = 4 - 4\sqrt{x} + x$	
$2 = 4 - 4\sqrt{x}$	Add $-x$ to both sides.
$-2 = -4\sqrt{x}$	Add -4 to both sides.
$\dfrac{1}{2} = \sqrt{x}$	Divide both sides by -4.
$\dfrac{1}{4} = x$	Square both sides.

Check:

$$\sqrt{x} + \sqrt{x+2} = 2$$

$$\sqrt{\frac{1}{4}} + \sqrt{\frac{1}{4}+2} \stackrel{?}{=} 2$$

$$\frac{1}{2} + \sqrt{\frac{9}{4}} \stackrel{?}{=} 2$$

$$\frac{1}{2} + \frac{3}{2} \stackrel{?}{=} 2$$

$$2 = 2$$

The solution does check. The solution set is $\{\frac{1}{4}\}$.

EXERCISE 6.6

In Exercises 1–50, solve and check each equation. If an equation has no solutions, so indicate.

1. $\sqrt{5x-6} = 2$
2. $\sqrt{7x-10} = 12$
3. $\sqrt{6x+1} + 2 = 7$
4. $\sqrt{6x+13} - 2 = 5$

5. $\sqrt{4(4x+1)} = \sqrt{x+4}$
6. $\sqrt{3(x+4)} = \sqrt{5x-12}$

7. $\sqrt[3]{7n-1} = 3$
8. $\sqrt[3]{12m+4} = 4$
9. $\sqrt[4]{10p+1} = \sqrt[4]{11p-7}$
10. $\sqrt[4]{10y+2} = 2\sqrt[4]{2}$

11. $x = \dfrac{\sqrt{12x-5}}{2}$
12. $x = \dfrac{\sqrt{16x-12}}{2}$

13. $r + 2 = \sqrt{5r+34}$
14. $s + 3 = \sqrt{-4s+20}$

15. $\sqrt{-5x+24} = 6 - x$
16. $\sqrt{-x+2} = x - 2$

17. $\sqrt{y+2} = 4 - y$
18. $\sqrt{22y+86} = y + 9$
19. $\sqrt{x}\sqrt{x+16} = 15$
20. $\sqrt{x}\sqrt{x+6} = 4$

21. $\sqrt[3]{x^3-7} = x - 1$
22. $\sqrt[3]{x^3+56} - 2 = x$

23. $\sqrt[4]{x^4+4x^2-4} = x$
24. $\sqrt[4]{8x-8} + 2 = 0$

25. $\sqrt[4]{12t+4} + 2 = 0$
26. $u = \sqrt[4]{u^4-6u^2+24}$
27. $\sqrt{2y+1} = 1 - 2\sqrt{y}$
28. $\sqrt{u+3} = \sqrt{u-3}$

29. $\sqrt{y+7} + 3 = \sqrt{y+4}$
30. $1 + \sqrt{z} = \sqrt{z+3}$

31. $\sqrt{v} + \sqrt{3} = \sqrt{v+3}$
32. $\sqrt{x} + 2 = \sqrt{x+4}$

33. $2 + \sqrt{u} = \sqrt{2u+7}$
34. $5r + 4 = \sqrt{5r+20} + 4r$

35. $\sqrt{6t+1} - \sqrt{9t} = -1$
36. $\sqrt{4s+1} - \sqrt{6s} = -1$

37. $\sqrt{2x+5} + \sqrt{x+2} = 5$
38. $\sqrt{2x+5} + \sqrt{2x+1} + 4 = 0$

39. $\sqrt{z-1} + \sqrt{z+2} = 3$
40. $\sqrt{16v+1} + \sqrt{8v+1} = 12$

41. $\sqrt{x-5} - \sqrt{x+3} = 4$
42. $\sqrt{x+8} - \sqrt{x-4} = -2$

43. $\sqrt[4]{x^2+2x+1} = \sqrt{x+2}$
44. $\sqrt{x+3} = \sqrt[4]{x^2+5x+6}$

45. $\sqrt{x+1} + \sqrt{3x} = \sqrt{5x+1}$
46. $\sqrt{3x} - \sqrt{x+1} = \sqrt{x-2}$

47. $\sqrt{\sqrt{a} + \sqrt{a+8}} = 2$

48. $\sqrt{\sqrt{2y} - \sqrt{y-1}} = 1$

49. $\dfrac{6}{\sqrt{x+5}} = \sqrt{x}$

50. $\dfrac{\sqrt{2x}}{\sqrt{x+2}} = \sqrt{x-1}$

CHAPTER SUMMARY

Key Words

conjugate binomials (6.4)
cube root (6.1)
index of a radical (6.2)
like radicals (6.3)
order of a radical (6.5)
power rule (6.6)

principal nth root (6.1)
radical expression (6.2)
radicand (6.2)
rationalizing the denominator (6.2)
similar radicals (6.3)
square root (6.1)

Key Ideas

(6.1) If n is a natural number and a is a real number, then

If $a > 0$, then $a^{1/n}$ is the positive number such that $(a^{1/n})^n = a$.

If $a = 0$, then $a^{1/n} = 0$.

If $a < 0$ $\begin{cases} \text{and } n \text{ is odd, then } a^{1/n} \text{ is the real number such that } (a^{1/n})^n = a. \\ \text{and } n \text{ is even, then } a^{1/n} \text{ is not a real number.} \end{cases}$

$a^{m/n} = (a^{1/n})^m = (a^m)^{1/n}$

$a^{-m/n} = \dfrac{1}{a^{m/n}}$ and $\dfrac{1}{a^{-m/n}} = a^{m/n}$ (provided $a \neq 0$)

(6.2) If n is a natural number greater than 1 and a is a real number, then

If $a > 0$, then $\sqrt[n]{a}$ is the positive number such that $(\sqrt[n]{a})^n = a$.

If $a = 0$, then $\sqrt[n]{a} = 0$.

If $a < 0$ $\begin{cases} \text{and } n \text{ is odd, then } \sqrt[n]{a} \text{ is the real number such that } (\sqrt[n]{a})^n = a. \\ \text{and } n \text{ is even, then } \sqrt[n]{a} \text{ is not a real number.} \end{cases}$

If n is an even natural number, then $\sqrt[n]{a^n} = |a|$.

If n is an odd natural number, then $\sqrt[n]{a^n} = a$.

$\sqrt[n]{ab} = \sqrt[n]{a}\,\sqrt[n]{b}$ and, if $b \neq 0$, $\sqrt[n]{\dfrac{a}{b}} = \dfrac{\sqrt[n]{a}}{\sqrt[n]{b}}$

Radicals can be removed from the denominator of a fraction by rationalizing the denominator.

(6.3) Like radicals can be combined by addition and subtraction:

$$3\sqrt{2} + 5\sqrt{2} = 8\sqrt{2}$$

Radicals that are not similar can often be converted to radicals that are similar and then combined:

$$\sqrt{2} + \sqrt{8} = \sqrt{2} + \sqrt{4}\sqrt{2} = \sqrt{2} + 2\sqrt{2} = 3\sqrt{2}$$

(6.4) If two radicals have the same index, they can be multiplied:

$$\sqrt{3x}\sqrt{6x} = \sqrt{18x^2} = 3x\sqrt{2} \qquad \text{provided that } x \geq 0$$

To rationalize the binomial denominator of a fraction, multiply both the numerator and the denominator by the conjugate of the binomial in the denominator.

(6.5) Radicals of different orders can be added, subtracted, multiplied, and divided when they are converted into radicals with common order.

(6.6) If $a = b$, then $a^n = b^n$.

Raising both sides of an equation to the same power can lead to extraneous solutions. Be sure to check all suspected solutions.

REVIEW EXERCISES

In Review Exercises 1–16, perform each operation and simplify the result, if possible. Assume that all variables represent positive numbers.

1. $25^{1/2}$
2. $-36^{1/2}$
3. $9^{3/2}$
4. $16^{3/2}$
5. $(-8)^{1/3}$
6. $-8^{2/3}$
7. $8^{-2/3}$
8. $8^{-1/3}$
9. $-49^{5/2}$
10. $\dfrac{1}{25^{7/2}}$
11. $\left(\dfrac{1}{4}\right)^{-3/2}$
12. $\left(\dfrac{4}{9}\right)^{-3/2}$
13. $(27x^3y)^{1/3}$
14. $(81x^4y^2)^{1/4}$
15. $(25x^6y^4)^{3/2}$
16. $(27u^2v^3)^{-2/3}$

*In Review Exercises 17–20, perform the indicated multiplications to remove parentheses. **Write all answers without using negative exponents.***

17. $u^{1/2}(u^{1/2} - u^{-1/2})$
18. $v^{2/3}(v^{1/3} + v^{4/3})$
19. $(x^{1/2} + y^{1/2})^2$
20. $(a^{2/3} + b^{2/3})(a^{2/3} - b^{2/3})$

In Review Exercises 21–28, simplify each radical.

21. $\sqrt{49}$
22. $-\sqrt{121}$
23. $-\sqrt{36}$
24. $\sqrt{225}$
25. $\sqrt[3]{-27}$
26. $-\sqrt[3]{216}$
27. $\sqrt[4]{625}$
28. $\sqrt[5]{-32}$

In Review Exercises 29–36, simplify each radical. Use absolute value symbols where necessary.

29. $\sqrt{240}$
30. $\sqrt[3]{54}$
31. $\sqrt[4]{32}$
32. $\sqrt[5]{96}$
33. $\sqrt{8x^3}$
34. $\sqrt{18x^4y^3}$
35. $\sqrt[3]{16x^5y^4z^3}$
36. $\sqrt[3]{250x^7y^3z}$

In Review Exercises 37–42, simplify each radical expression. Assume that all variables represent positive numbers.

37. $\sqrt{5}\sqrt{5}$

38. $\sqrt{6}\sqrt{216}$

39. $\sqrt{9x}\sqrt{x}$

40. $\sqrt[3]{3}\sqrt[3]{9}$

41. $-\sqrt[3]{2x^2}\sqrt[3]{4x}$

42. $-\sqrt[4]{256x^5y^{11}}\sqrt[4]{625x^8y^2}$

In Review Exercises 43–46, rationalize each denominator.

43. $\dfrac{1}{\sqrt{3}}$

44. $\dfrac{\sqrt{3}}{\sqrt{5}}$

45. $\dfrac{x}{\sqrt{xy}}$

46. $\dfrac{\sqrt[3]{uv}}{\sqrt[3]{u^5v^7}}$

In Review Exercises 47–48, simplify each radical expression. Assume that all variables represent positive numbers.

47. $\sqrt{x^2 + 6x + 9}$

48. $\sqrt[3]{(x + 1)(x^2 + 2x + 1)}$

In Review Exercises 49–56, simplify and combine like radicals. Assume that all variables represent positive numbers.

49. $\sqrt{2} + \sqrt{8}$

50. $\sqrt{20} - \sqrt{5}$

51. $2\sqrt[3]{3} - \sqrt[3]{24}$

52. $\sqrt[4]{32} + 2\sqrt[4]{162}$

53. $2\sqrt{8} + 2\sqrt{200} + \sqrt{50}$

54. $3\sqrt{27} - 2\sqrt{3} + 5\sqrt{75}$

55. $\sqrt[3]{54} - 3\sqrt[3]{16} + 4\sqrt[3]{128}$

56. $2\sqrt[4]{32} + 4\sqrt[4]{162} - 5\sqrt[4]{512}$

In Review Exercises 57–64, simplify each radical expression. Assume that all variables represent positive numbers.

57. $\sqrt{2}\sqrt{8}$

58. $\sqrt{2}(\sqrt{2} + 3)$

59. $\sqrt{5}(\sqrt{2} - 1)$

60. $\sqrt{7}(\sqrt{3} + \sqrt{2})$

61. $(\sqrt{2} + 1)(\sqrt{2} - 1)$

62. $(\sqrt{3} + \sqrt{2})(\sqrt{3} + \sqrt{2})$

63. $(\sqrt{x} + \sqrt{y})(\sqrt{x} - \sqrt{y})$

64. $(2\sqrt{u} + 3)(3\sqrt{u} - 4)$

In Review Exercises 65–68, rationalize each denominator and simplify, if possible.

65. $\dfrac{2}{\sqrt{2} - 1}$

66. $\dfrac{\sqrt{2}}{\sqrt{3} - 1}$

67. $\dfrac{2x - 32}{\sqrt{x} + 4}$

68. $\dfrac{\sqrt{a} + 1}{\sqrt{a} - 1}$

In Review Exercises 69–76, write each radical as an expression without radicals or as another radical with a smaller index. Assume that all variables represent positive numbers.

69. $\sqrt[4]{25}$

70. $\sqrt[4]{81}$

71. $\sqrt[6]{8}$

72. $\sqrt[6]{27}$

73. $\sqrt[5]{x^{10}}$

74. $\sqrt[6]{64x^{12}}$

75. $\sqrt[4]{x^{10}y^8}$

76. $\sqrt[12]{u^6v^{12}}$

In Review Exercises 77–84, perform each indicated operation. Express each answer as an expression without radicals or as a single radical in simplest form.

77. $\sqrt{5}\sqrt[3]{2}$

78. $\sqrt{2}\sqrt[3]{5}$

79. $\sqrt[3]{3}\sqrt[4]{4}$

80. $\sqrt[3]{3}\sqrt[4]{2}$

81. $\dfrac{\sqrt{5}}{\sqrt[3]{2}}$

82. $\dfrac{\sqrt{2}}{\sqrt[3]{3}}$

83. $\dfrac{\sqrt[5]{5}}{\sqrt{3}}$

84. $\dfrac{\sqrt[5]{3}}{\sqrt[3]{4}}$

In Review Exercises 85–90, solve and check each radical equation.

85. $\sqrt{y + 5} = \sqrt{2y - 17}$

86. $u = \sqrt{25u - 144}$

87. $r = \sqrt{12r - 27}$

88. $\sqrt{z + 1} + \sqrt{z} = 2$

89. $\sqrt{2x + 5} - \sqrt{2x} = 1$

90. $\sqrt[3]{x^3 + 8} = x + 2$

CHAPTER SIX TEST

*In Problems 1–6, simplify each expression. **Write all answers without using negative exponents.** Assume that all variables represent positive numbers.*

1. $16^{1/4}$

2. $27^{2/3}$

3. $36^{-3/2}$

4. $\left(-\dfrac{8}{27}\right)^{-2/3}$

5. $\dfrac{2^{5/3}2^{1/6}}{2^{1/2}}$

6. $\dfrac{(8x^3y)^{1/2}(8xy^5)^{1/2}}{(x^3y^6)^{1/3}}$

In Problems 7–10, simplify each expression. Assume that all variables represent positive numbers.

7. $\sqrt{48}$

8. $\sqrt{250x^3y^5}$

9. $\dfrac{\sqrt[3]{24x^{15}y^4}}{\sqrt[3]{y}}$

10. $\sqrt{\dfrac{3a^5}{48a^7}}$

In Problems 11–14, rationalize each denominator.

11. $\dfrac{1}{\sqrt{5}}$

12. $\dfrac{\sqrt{3}}{\sqrt{8}}$

13. $\dfrac{6}{\sqrt[3]{9}}$

14. $\dfrac{\sqrt{18ab}}{\sqrt{6a^2b}}$

In Problems 15–18, simplify and combine like radicals. Assume that all variables represent positive numbers.

15. $3\sqrt{3} + \sqrt{12} - \sqrt{27}$

16. $2\sqrt[3]{40} - \sqrt[3]{5000} + 4\sqrt[3]{625}$

17. $2\sqrt{48y^5} - 3y\sqrt{12y^3}$

18. $\sqrt[4]{768z^5} + z\sqrt[4]{48z}$

In Problems 19–20, perform each indicated operation and simplify, if possible. Assume that all variables represent positive numbers.

19. $-2\sqrt{xy}\,(3\sqrt{x} + \sqrt{xy^3})$

20. $(3\sqrt{2} + \sqrt{3})(2\sqrt{2} - 3\sqrt{3})$

In Problems 21–22, rationalize each denominator and simplify, if possible. Assume that all variables represent positive numbers.

21. $\dfrac{-\sqrt{2}}{\sqrt{5}+3}$

22. $\dfrac{3r-1}{\sqrt{3r}-1}$

In Problems 23–24, write each radical as a radical with a smaller index.

23. $\sqrt[4]{64}$

24. $\sqrt[8]{625y^4}$

In Problems 25–28, write each expression as a single radical.

25. $\sqrt[4]{36} + \sqrt{54}$

26. $\sqrt[3]{xy^3} - \sqrt[9]{x^3y^9}$

27. $\sqrt[3]{5}\sqrt{7}$

28. $\dfrac{\sqrt[4]{3}}{\sqrt{3}}$

In Problems 29–30, solve and check each equation.

29. $\sqrt[3]{6n+4} - 4 = 0$

30. $1 - \sqrt{u} = \sqrt{u-3}$

▬▬▬▬▬ CUMULATIVE REVIEW EXERCISES (CHAPTERS 4–6) ▬▬▬▬▬

In Exercises 1–4, solve each equation and check the result.

1. $2x - 5 = 11$

2. $\dfrac{2}{3}x - 2 = x + 7$

3. $4(y - 3) + 4 = -3(y + 5)$

4. $2x - \dfrac{3(x - 2)}{2} = 7 - \dfrac{x - 3}{3}$

5. The sum of three consecutive even integers is 90. What are the integers?

6. The length of a rectangle is three times its width. The perimeter of the rectangle is 112 centimeters. What are its dimensions?

7. Solve for a in the formula $S = \dfrac{n(a + l)}{2}$.

In Exercises 8–9, solve each equation by factoring.

8. $x^3 - 4x = 0$

9. $6x^2 + 7 = -23x$

In Exercises 10–11, solve and check each equation.

10. $|4x - 3| = 9$

11. $|2x - 1| = |3x + 4|$

In Exercises 12–16, solve each inequality.

12. $-3(x - 4) \geq x - 32$

13. $-8 < -3x + 1 < 10$

14. $x - 2 \leq 3x + 1 \leq 5x - 4$

15. $|3x - 2| \leq 4$

16. $|2x + 3| > 5$

17. Simplify: $\dfrac{2x^2y + xy - 6y}{3x^2y + 5xy - 2y}$.

18. Perform the indicated multiplication and division, and simplify:

$$\frac{x^2 - 4}{x^2 + 9x + 20} \div \frac{x^2 + 5x + 6}{x^2 + 4x - 5} \cdot \frac{x^2 + 3x - 4}{(x - 1)^2}$$

19. Perform the indicated addition and subtraction, and simplify:

$$\frac{2}{x + y} + \frac{3}{x - y} - \frac{x - 3y}{x^2 - y^2}$$

20. Simplify the complex fraction $\dfrac{\dfrac{a}{b} + b}{a - \dfrac{b}{a}}$.

In Exercises 21–22, solve and check each equation.

21. $\dfrac{5x - 3}{x + 2} = \dfrac{5x + 3}{x - 2}$

22. $\dfrac{3}{x - 2} + \dfrac{x^2}{(x + 3)(x - 2)} = \dfrac{x + 4}{x + 3}$

In Exercises 23–36, simplify each expression. Write all answers without using negative exponents. Assume all variables represent positive numbers.

23. $64^{2/3}$

24. $8^{-1/3}$

25. $\dfrac{x^{5/3}x^{1/2}}{x^{3/4}}$

26. $(x^{2/3} - x^{1/3})(x^{2/3} + x^{1/3})$

27. $\sqrt[3]{-27x^3}$

28. $\sqrt{48t^3}$

29. $\sqrt[3]{\dfrac{128x^4}{2x}}$

30. $\sqrt{x^2 + 8x + 16}$

31. $\sqrt{50} - \sqrt{8} + \sqrt{32}$

32. $-3\sqrt[4]{32} - 2\sqrt[4]{162} + 5\sqrt[4]{48}$

33. $3\sqrt{2}(2\sqrt{3} - 4\sqrt{12})$

34. $\dfrac{5}{\sqrt[3]{x}}$

35. $\dfrac{\sqrt{x} + 2}{\sqrt{x} - 1}$

36. $\sqrt[6]{x^3y^3}$

In Exercises 37–38, solve and check each equation.

37. $5\sqrt{x + 2} = x + 8$

38. $\sqrt{x} + \sqrt{x + 2} = 2$

PHOTOGRAPHER Photographers use camera and film to portray people, objects, places and events. Some specialize in scientific, medical, or engineering photography, and provide illustrations and documentation for publications and research reports. Others specialize in portrait, fashion, or industrial photography and provide the pictures for catalogs and other publications. Photojournalists capture newsworthy events, people, and places, and their work is seen in newspapers and magazines, as well as on television.

Qualifications Employers want applicants with a broad technical understanding of photography as well as imagination, creativity, and a good sense of timing. Some knowledge of mathematics, physics, and chemistry is helpful for understanding lenses, films, lighting, and development processes.

Job outlook Job opportunities will be excellent through the mid-1990s. Business and industry will need photographers to provide visual aids for meetings and reports, sales campaigns, and public-relations work. Law enforcement agencies and scientific and medical research organizations will also require photographers with appropriate technical skills.

Example application In each photographic lens there is an adjustable circular opening called the **aperture**, which controls the amount of light that passes through the lens. Various lenses—wide angle, close-up, and telephoto—are distinguished by their **focal length**. The diameter, d, of the aperture and the focal length, f, of the lens determine the **f–number** of the lens by the formula

$$f\text{–number} = \frac{f}{d}$$

Thus, a lens with a focal length of 12 centimeters and an aperture of 6 centimeters has an f–number of 12/6, or 2. It would be an $f/2$ lens.

Find the aperture of this lens if the *area* of its aperture is cut in half, so as to admit only half as much light.

Solution Determine the area of a circle with diameter d.

$A = \pi r^2$ The formula for the area of a circle.

$A = \pi \left(\dfrac{d}{2}\right)^2$ Substitute $\dfrac{d}{2}$ for r.

$A = \pi \left(\dfrac{6}{2}\right)^2$ Substitute 6 for d.

$A = 9\pi$

To find the diameter of a circle with area equal to one-half of 9π, substitute $\dfrac{9\pi}{2}$ for A in the formula for the area of a circle and solve for d.

$$A = \pi r^2 \qquad \text{The formula for the area of a circle.}$$

$$\frac{9\pi}{2} = \pi \left(\frac{d}{2}\right)^2 \qquad \text{Substitute } \frac{9\pi}{2} \text{ for } A \text{ and } \frac{d}{2} \text{ for } r.$$

$$\frac{9\pi}{2} = \frac{\pi d^2}{4}$$

$$18 = d^2$$

$$3\sqrt{2} = d$$

If the diameter of the aperture were reduced from 6 centimeters to $3\sqrt{2}$ centimeters, the area (and the light admitted) would be cut in half. The f–number of the lens is now found by letting $f = 12$ (as before) and $d = 3\sqrt{2}$ in the formula for an f–number.

$$f\text{--number} = \frac{f}{d}$$

$$= \frac{12}{3\sqrt{2}}$$

$$= 2\sqrt{2}$$

$$\approx 2.8$$

An $f/2.8$ lens admits one-half the light admitted by an $f/2$ lens.

If the light were cut in half again, the f–number would be $2.8\sqrt{2}$, or 4. The next f–number, representing another halving of the light admitted, would be $4\sqrt{2}$, or 5.6. These numbers,

$$f/2, \quad f/2.8, \quad f/4, \quad f5.6, \quad f/8, \quad f/11, \quad f/16$$

are called **f–stops**. They are well known to all professional photographers.

Exercises

1. What would be the f–number of a lens with focal length of 20 cm and aperture of 5 cm?
2. If the focal length of the lens of Exercise 1 were doubled and the aperature held constant, what would be the f–number?
3. What diameter would give a lens with focal length 55 mm an f–number of $f/3.5$?
4. The **speed** of a lens is the square of the reciprocal of the f–number. How many times faster is an $f/2$ lens than an $f/4.5$ lens?

(Answers: **1.** $f/4$ **2.** $f/8$ **3.**   15.7 mm   **4.**   5 times faster)

7 Graphs, Equations of Lines, and Functions

Mathematical expressions are often used to indicate relationships between several quantities. The formula $d = rt$, for example, indicates how distance traveled is related to the rate of speed and time. Because distance depends on rate and time, we say that distance is a function of rate and time. This concept of *function* is one of the fundamental ideas that runs throughout all of mathematics. To introduce this topic, we begin by discussing the rectangular coordinate system.

7.1 THE RECTANGULAR COORDINATE SYSTEM

René Descartes (1596–1650) is credited with the idea of associating ordered pairs of real numbers with points in a geometric plane. His idea is based on two perpendicular number lines, one horizontal and one vertical, that divide the plane into four quadrants numbered as in Figure 7-1. The horizontal number line is called the **x-axis** and the vertical number line is called the **y-axis**. The point where the axes intersect, called the **origin**, is the zero point on each number line. The positive direction on the x-axis is to the right, the positive direction on the y-axis is upward, and the unit distance on each axis is the same. This geometric configuration is called a **rectangular coordinate system**, or a **Cartesian coordinate system**, in honor of its inventor.

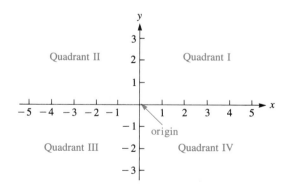

Figure 7-1

To **plot** the point associated with the pair of real numbers (2, 3), we start at the origin and count 2 units to the right, and then 3 units up. See Figure 7-2. The point *P*, which lies in the first quadrant, is called the **graph** of the pair (2, 3). The pair (2, 3) gives the **coordinates** of point *P*. To plot point *Q* with coordinates (−4, 6), we start at the origin and count 4 units to the left, and then 6 units up. Point *Q* lies in the second quadrant. Point *R* with coordinates (6, −4) lies in the fourth quadrant.

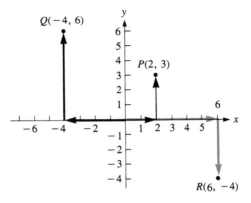

Figure 7-2

Note that the pairs (−4, 6) and (6, −4) represent different points. Because order is important when graphing pairs of real numbers, such pairs are called **ordered pairs**. The first coordinate, *a*, in the ordered pair (*a*, *b*) is called the ***x*-coordinate** or the **abscissa**. The second coordinate, *b*, is called the ***y*-coordinate** or the **ordinate**.

Example 1   Graph the set {(−1, −1), (0, 0), (1, 1), (2, 2)}.

Solution   Draw an *x*-axis and a *y*-axis that are perpendicular, and plot each ordered pair. See Figure 7-3. The four points that are determined form the graph of the given set.

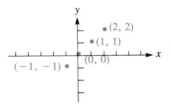

Figure 7-3

It is possible to draw a "picture" or a graph of an equation in two variables. The **graph of an equation** in the two variables *x* and *y* is the set of all points on a rectangular coordinate system with coordinates (x, y) that satisfy the equation.

Example 2   Graph the equation $3x + 2y = 12$.

Solution   Pick some arbitrary values for either x or y, substitute those values in the equation, and solve for the other variable. For example, if $x = 2$, you can find y as follows:

$$3x + 2y = 12$$
$$3(2) + 2y = 12 \qquad \text{Substitute 2 for } x.$$
$$6 + 2y = 12 \qquad \text{Simplify.}$$
$$2y = 6 \qquad \text{Add } -6 \text{ to both sides.}$$
$$y = 3 \qquad \text{Divide both sides by 2.}$$

Thus, one ordered pair that satisfies the equation is $(2, 3)$.
If $y = 6$, you have

$$3x + 2y = 12$$
$$3x + 2(6) = 12 \qquad \text{Substitute 6 for } y.$$
$$3x + 12 = 12 \qquad \text{Simplify.}$$
$$3x = 0 \qquad \text{Add } -12 \text{ to both sides.}$$
$$x = 0 \qquad \text{Divide both sides by 3.}$$

Thus, another ordered pair that satisfies the equation is $(0, 6)$.

The ordered pairs $(2, 3)$, and $(0, 6)$, and others that satisfy the equation $3x + 2y = 12$ are shown in the table of values in Figure 7-4. Plot each of these ordered pairs on a rectangular coordinate system, as in the figure. Note that the resulting points appear to lie on a straight line. It can be shown that all points representing ordered pairs that satisfy the equation $3x + 2y = 12$ do, in fact, lie on a straight line. Draw the line that joins the five points. This line is the graph of the equation.

$3x + 2y = 12$

x	y
2	3
0	6
4	0
6	-3
-2	9

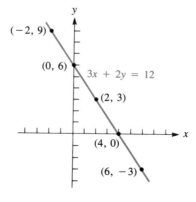

Figure 7-4

Because the line in Example 2 intersects the y-axis at the point $(0, 6)$, the number 6 is called the **y-intercept** of the line. Likewise, 4 is called the **x-intercept**.

Example 3   Use the x- and y-intercepts to graph the equation $2x + 5y = 10$.

Solution   Begin by finding the x- and y-intercepts. To find the y-intercept, substitute **0** for x and solve for y:

$$2x + 5y = 10$$
$$2(0) + 5y = 10 \qquad \text{Substitute 0 for } x.$$
$$5y = 10 \qquad \text{Simplify.}$$
$$y = 2 \qquad \text{Divide both sides by 5.}$$

Because the y-intercept is 2, the line intersects the y-axis at the point $(0, 2)$. To find the x-intercept, substitute **0** for y and solve for x:

$$2x + 5y = 10$$
$$2x + 5(0) = 10 \qquad \text{Substitute 0 for } y.$$
$$2x = 10 \qquad \text{Simplify.}$$
$$x = 5 \qquad \text{Divide both sides by 2.}$$

Because the x-intercept is 5, the line intersects the x-axis at the point $(5, 0)$.

Although these two points, if calculated correctly, are sufficient to draw the line, it is a good idea to find and plot a third ordered pair to act as a check. To find the coordinates of a third point, substitute any convenient number, such as **1**, for y and solve for x:

$$2x + 5y = 10$$
$$2x + 5(1) = 10 \qquad \text{Substitute 1 for } y.$$
$$2x + 5 = 10 \qquad \text{Simplify.}$$
$$2x = 5 \qquad \text{Add } -5 \text{ to both sides.}$$
$$x = \frac{5}{2} \qquad \text{Divide both sides by 2.}$$

Thus, the line passes through the point $(\frac{5}{2}, 1)$.

A table of values and the graph of the equation $2x + 5y = 10$ are shown in Figure 7-5.

$2x + 5y = 10$

x	y
0	2
5	0
$\dfrac{5}{2}$	1

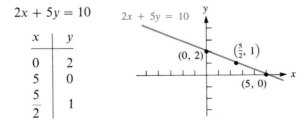

Figure 7-5

Example 4   Graph the equation $y = 3x + 4$.

Solution   Find the x- and y-intercepts: If $x = 0$, then

$$y = 3x + 4$$
$$y = 3(0) + 4$$
$$y = 4 \qquad \text{The } y\text{-intercept is 4.}$$

If $y = 0$, then

$$y = 3x + 4$$
$$0 = 3x + 4$$
$$-4 = 3x$$
$$-\frac{4}{3} = x \qquad \text{The } x\text{-intercept is } -\tfrac{4}{3}.$$

To find the coordinates of a third point, substitute **1** for x, and simplify.

$$y = 3x + 4$$
$$y = 3(1) + 4$$
$$y = 7$$

A table of values and the graph of the equation $y = 3x + 4$ are shown in Figure 7-6.

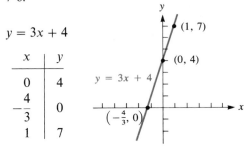

$y = 3x + 4$

x	y
0	4
$-\dfrac{4}{3}$	0
1	7

Figure 7-6

There is a formula, called the **distance formula**, that enables us to find the distance between two points that are graphed on a rectangular coordinate system. Before deriving this formula, we need to review the **Pythagorean theorem** from geometry. This theorem states that, in any right triangle, the square of the hypotenuse (the side opposite the 90° angle) is equal to the sum of the squares of the other two sides. For example, suppose the right triangle shown in Figure 7-7 has sides of 3, 4, and x units. Because of the Pythagorean theorem and the fact that lengths must be positive, we can determine the value of x:

$$3^2 + 4^2 = x^2$$
$$9 + 16 = x^2$$
$$25 = x^2$$
$$5 = x$$

Thus, $x = 5$ units.

Figure 7-7

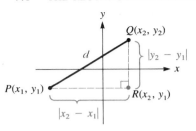

Figure 7-8

We can now derive the distance formula. To find the distance d between points $P(x_1, y_1)$ and $Q(x_2, y_2)$ shown in Figure 7-8, we construct the right triangle PRQ. The distance between P and R is $|x_2 - x_1|$ and the distance between R and Q is $|y_2 - y_1|$. We apply the Pythagorean theorem to the right triangle PRQ to get

$$[d(PQ)]^2 = |x_2 - x_1|^2 + |y_2 - y_1|^2$$
$$= (x_2 - x_1)^2 + (y_2 - y_1)^2 \qquad \text{Because } |x_2 - x_1|^2 = (x_2 - x_1)^2 \text{ and } |y_2 - y_1|^2 = (y_2 - y_1)^2.$$

or

$$d(PQ) = \sqrt{(x_2 - x_1)^2 + (y_2 - y_1)^2}$$

Thus, we have the distance formula. Because it is one of the most important formulas in mathematics, take the time to memorize it.

The Distance Formula.   The distance between two points $P(x_1, y_1)$ and $Q(x_2, y_2)$ is given by the formula

$$d(PQ) = \sqrt{(x_2 - x_1)^2 + (y_2 - y_1)^2}$$

Example 5   Find the distance between points $P(-2, 3)$ and $Q(4, -5)$.

Solution   Use the distance formula. Substitute -2 for x_1, 3 for y_1, 4 for x_2, and -5 for y_2, and simplify:

$$d(PQ) = \sqrt{(x_2 - x_1)^2 + (y_2 - y_1)^2}$$
$$= \sqrt{[4 - (-2)]^2 + (-5 - 3)^2}$$
$$= \sqrt{(4 + 2)^2 + (-5 - 3)^2}$$
$$= \sqrt{6^2 + (-8)^2}$$
$$= \sqrt{36 + 64}$$
$$= \sqrt{100}$$
$$= 10$$

The distance between points P and Q is 10 units.   ∎

Once we have the distance formula, it is easy to prove the **midpoint formula**.

The Midpoint Formula.   The midpoint of the line segment joining points $P(x_1, y_1)$ and $Q(x_2, y_2)$ is the point M with coordinates of

$$\left(\frac{x_1 + x_2}{2}, \frac{y_1 + y_2}{2} \right)$$

The midpoint formula indicates that, to find the midpoint of the line segment PQ, we simply average the x-coordinates of P and Q, and average the y-coordinates of P and Q.

To verify this formula, we assume that point M has coordinates of

$$\left(\frac{x_1 + x_2}{2}, \frac{y_1 + y_2}{2} \right)$$

and show that points P, Q, and M lie on the same line, and that $d(PM) = d(MQ)$ (the distance from P to M is equal to the distance from M to Q). See Figure 7-9.

$$d(PM) = \sqrt{\left(\frac{x_1 + x_2}{2} - x_1 \right)^2 + \left(\frac{y_1 + y_2}{2} - y_1 \right)^2}$$

$$= \sqrt{\left(\frac{x_2 - x_1}{2} \right)^2 + \left(\frac{y_2 - y_1}{2} \right)^2} \qquad \text{Find common denominators and perform the subtractions.}$$

$$= \sqrt{\frac{(x_2 - x_1)^2}{4} + \frac{(y_2 - y_1)^2}{4}}$$

$$= \frac{1}{2} \sqrt{(x_2 - x_1)^2 + (y_2 - y_1)^2} \qquad \text{Because } \sqrt{\frac{1}{4}} = \frac{1}{2}.$$

$$d(MQ) = \sqrt{\left(x_2 - \frac{x_1 + x_2}{2} \right)^2 + \left(y_2 - \frac{y_1 + y_2}{2} \right)^2}$$

$$= \sqrt{\left(\frac{x_2 - x_1}{2} \right)^2 + \left(\frac{y_2 - y_1}{2} \right)^2}$$

$$= \sqrt{\frac{(x_2 - x_1)^2}{4} + \frac{(y_2 - y_1)^2}{4}}$$

$$= \frac{1}{2} \sqrt{(x_2 - x_1)^2 + (y_2 - y_1)^2}$$

$$d(PQ) = \sqrt{(x_2 - x_1)^2 + (y_2 - y_1)^2}$$

Because $d(PM) + d(MQ) = d(PQ)$, points P, Q, and M must lie on the same straight line. Because $d(PM) = d(MQ)$, M is the midpoint of line segment PQ.

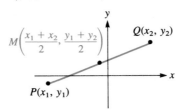

Figure 7-9

Example 6   Find the midpoint of the line segment PQ if P is $(-2, 3)$ and Q is $(3, -5)$.

Solution   The x-coordinate of M is

$$\frac{x_1 + x_2}{2} = \frac{-2 + 3}{2} = \frac{1}{2}$$

The y-coordinate of M is

$$\frac{y_1 + y_2}{2} = \frac{3 + (-5)}{2} = -1$$

Hence, the midpoint of segment PQ is point $M(\frac{1}{2}, -1)$.

■ EXERCISE 7.1

In Exercises 1–12, graph each equation.

1. $x + y = 4$	**2.** $x - y = 2$	**3.** $2x - y = 3$	**4.** $x + 2y = 5$
5. $3x + 4y = 12$	**6.** $4x - 3y = 12$	**7.** $y = -3x + 2$	**8.** $y = 2x - 3$
9. $3y = 6x - 9$		**10.** $2x = 4y - 10$	
11. $3x + 4y - 8 = 0$		**12.** $-2y - 3x + 9 = 0$	

In Exercises 13–22, find the distance between P and Q.

13. $Q(0, 0)$; $P(3, -4)$	**14.** $Q(0, 0)$; $P(-6, 8)$
15. $P(2, 4)$; $Q(5, 8)$	**16.** $P(5, 9)$; $Q(8, 13)$
17. $P(-2, -8)$; $Q(3, 4)$	**18.** $P(-5, -2)$; $Q(7, 3)$
19. $P(6, 8)$; $Q(12, 16)$	**20.** $P(10, 4)$; $Q(2, -2)$
21. $Q(-3, 5)$; $P(-5, -5)$	**22.** $Q(2, -3)$; $P(4, -8)$

In Exercises 23–36, find the midpoint of the line segment PQ.

23. $P(2, 4)$; $Q(5, 8)$	**24.** $P(5, 9)$; $Q(8, 13)$
25. $P(-2, -8)$; $Q(3, 4)$	**26.** $P(-5, -2)$; $Q(7, 3)$
27. $P(6, 8)$; $Q(12, 16)$	**28.** $P(10, 4)$; $Q(2, -2)$
29. $Q(-3, 5)$; $P(-5, -5)$	**30.** $Q(2, -3)$; $P(4, -8)$

31. $Q(0, 0)$; $P(3, -4)$

32. $Q(0, 0)$; $P(6, -8)$

33. $Q(a, b)$; $P(c, d)$

34. $Q(a + b, b)$; $P(c, c + d)$

35. $P(\sqrt{2}, \sqrt{3})$; $Q(\sqrt{3}, \sqrt{2})$

36. $P(\sqrt{5}, -\sqrt{3})$; $Q(-\sqrt{3}, 2\sqrt{5})$

37. Show that a triangle with vertices at $(-2, 4)$, $(2, 8)$, and $(6, 4)$ is isosceles (has two sides of equal length).

38. Show that a triangle with vertices at $(-2, 13)$, $(-8, 9)$, and $(-2, 5)$ is isosceles.

39. If $M(-2, 3)$ is the midpoint of segment PQ and the coordinates of P are $(-8, 5)$, find the coordinates of Q.

40. If $M(6, -5)$ is the midpoint of segment PQ and the coordinates of Q are $(-5, -8)$, find the coordinates of P.

41. Every point on the line CD in Illustration 1 is equidistant from points A and B. Use the distance formula to find the equation of line CD.

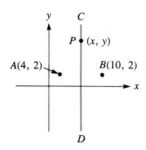

Illustration 1

42. Show that a triangle with vertices at $(2, 3)$, $(-3, 4)$, and $(1, -2)$ is a right triangle.

43. Find the coordinates of the two points on the x-axis that are $\sqrt{5}$ units away from the point $(5, 1)$.

44. The square shown in Illustration 2 has an area of 18 square units and its diagonals lie on the x- and y-axes. Find the coordinates of each corner of the square.

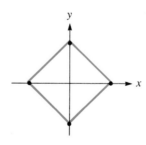

Illustration 2

7.2   SLOPE OF A NONVERTICAL LINE

The **slope of a nonvertical line** drawn in a rectangular coordinate system is a measure of its tilt or inclination. To develop a formula for computing the slope of a nonvertical line, we consider the line in Figure 7-10 that passes through the points $P(2, 3)$ and $Q(6, 9)$. If line RQ is parallel to the y-axis and line PR is parallel to the x-axis, then triangle PRQ is a right triangle and point R has coordinates $(6, 3)$. The distance from point R to point Q is the change in the y-coordinates of points R and Q, often denoted as Δy and read as "delta y." As we move from point R to point Q, $\Delta y = 9 - 3$ or 6 units. The distance from point P to point R is the change in the x-coordinates of points P and R, denoted as Δx and read as "delta x." As we move from point P to point R, $\Delta x = 6 - 2$ or 4 units.

The slope of line PQ is defined to be the distance from R to Q divided by the distance from P to R:

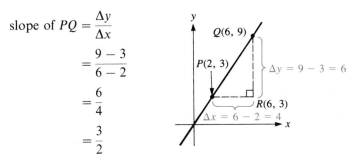

$$\text{slope of } PQ = \frac{\Delta y}{\Delta x}$$

$$= \frac{9 - 3}{6 - 2}$$

$$= \frac{6}{4}$$

$$= \frac{3}{2}$$

Figure 7-10

In general, the slope of a nonvertical line passing through the points $P(x_1, y_1)$ and $Q(x_2, y_2)$ is given by the following definition:

Definition. The **slope of a nonvertical line** passing through the points $P(x_1, y_1)$ and $Q(x_2, y_2)$ is the change in the y-coordinates divided by the corresponding change in the x-coordinates. See Figure 7-11.

In symbols, if m represents the slope of line PQ, then

$$m = \frac{\Delta y}{\Delta x} = \frac{y_2 - y_1}{x_2 - x_1} \qquad (x_2 \neq x_1)$$

Figure 7-11

Example 1 Point P in Figure 7-12 has coordinates $(-2, 5)$, and point Q has coordinates $(6, -7)$. Find the slope of the line passing through points P and Q.

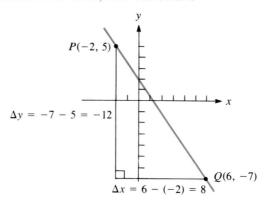

Figure 7-12

Solution   Let $P(x_1, y_1) = P(-2, 5)$ and $Q(x_2, y_2) = Q(6, -7)$. Then

$$m = \frac{\Delta y}{\Delta x} = \frac{y_2 - y_1}{x_2 - x_1}$$

$$= \frac{-7 - 5}{6 - (-2)}$$

$$= \frac{-12}{8}$$

$$= -\frac{3}{2}$$

The slope of the line passing through P and Q is $-\frac{3}{2}$.

Note that you would obtain the same result if you let $P(x_1, y_1) = P(6, -7)$ and $Q(x_2, y_2) = Q(-2, 5)$.   ∎

Example 2   The graph of the equation $3x - 4y = 12$ is a straight line. Find its slope.

Solution   Begin by finding the coordinates of two points on the line. If $x = 0$, for example, then $y = -3$. Thus, the point $(0, -3)$ is on the line. If $y = 0$, then $x = 4$. Thus, the point $(4, 0)$ is on the line.

Let $P(x_1, y_1) = P(0, -3)$ and $Q(x_2, y_2) = Q(4, 0)$, and substitute into the formula for slope to obtain

$$m = \frac{\Delta y}{\Delta x} = \frac{y_2 - y_1}{x_2 - x_1}$$

$$= \frac{0 - (-3)}{4 - 0}$$

$$= \frac{3}{4}$$

The slope of the line is $\frac{3}{4}$.   ∎

If a line has an equation of the form $y = k$, where k is a constant, it is a horizontal line and has a slope of 0. If a line has an equation of the form $x = k$, it is a vertical line. If $P(x_1, y_1)$ and $Q(x_2, y_2)$ are two points on a vertical line, then $x_1 = x_2$ and $x_2 - x_1 = 0$. Because the denominator of the fraction $\dfrac{y_2 - y_1}{x_2 - x_1}$ cannot be 0, a vertical line has no defined slope.

Example 3 Graph the equations **a.** $y = -2$ and **b.** $x = 5$.

Solution **a.** To graph the equation $y = -2$, note that any ordered pair with a y-coordinate of -2 satisfies the equation. Thus, such points as $(-2, -2)$, $(0, -2)$ and $(1, -2)$ satisfy the equation because each has a y-coordinate of -2. Make a table of values, plot the points, and draw the line.

b. To graph the equation $x = 5$, note that points such as $(5, -2)$ and $(5, 1)$ satisfy the equation because each has an x-coordinate of 5. Make a table of values, plot the points, and draw the line. See Figure 7-13.

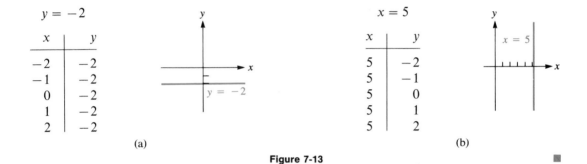

$y = -2$

x	y
-2	-2
-1	-2
0	-2
1	-2
2	-2

(a)

$x = 5$

x	y
5	-2
5	-1
5	0
5	1
5	2

(b)

Figure 7-13

If a line rises as we follow it from left to right, its slope is positive (see Figure 7-14a). If a line drops as we follow it from left to right, its slope is negative (see Figure 7-14b). If a line is horizontal, its slope is 0 (see Figure 7-14c). If a line is vertical, it has no defined slope (see Figure 7-14d).

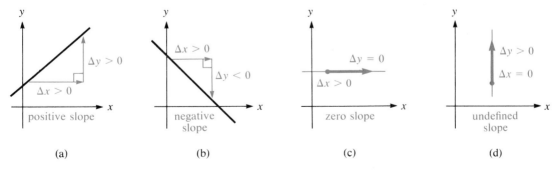

(a) positive slope

(b) negative slope

(c) zero slope

(d) undefined slope

Figure 7-14

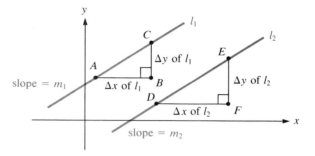

Figure 7-15

A theorem relates parallel lines to their slopes. To see this relationship, we consider the parallel lines l_1 and l_2 in Figure 7-15, with slopes of m_1 and m_2, respectively. Because the right triangles ABC and DFE are similar, it follows that

$$m_1 = \frac{\Delta y \text{ of } l_1}{\Delta x \text{ of } l_1}$$

$$= \frac{\Delta y \text{ of } l_2}{\Delta x \text{ of } l_2}$$

$$= m_2$$

Thus, if two nonvertical lines are parallel, they have the same slope. It is also true that if two lines have the same slope, they are parallel.

Theorem.   Nonvertical parallel lines have the same slope, and lines with the same slope are parallel.

Because vertical lines are parallel, two lines with undefined slope are parallel.

Example 4   Find y if the line passing through $P(3, -2)$ and $Q(-3, 4)$ is parallel to the line passing through $R(-2, 5)$ and $S(3, y)$.

Solution   Because the lines PQ and RS are parallel, they must have equal slopes. Find the slope of each line, set the slopes equal, and solve the resulting equation for y:

Slope of PQ Slope of RS

$$\frac{4 - (-2)}{-3 - 3} = \frac{y - 5}{3 - (-2)}$$

$$\frac{6}{-6} = \frac{y - 5}{5}$$

$$-1 = \frac{y - 5}{5}$$

$$-5 = y - 5$$

$$0 = y$$

Thus $y = 0$. The line passing through $P(3, -2)$ and $Q(-3, 4)$ is parallel to the line passing through $R(-2, 5)$ and $S(3, 0)$.   ■

Example 5   If graphed on a coordinate system, lines PQ and OR are distinct lines. Is the line passing through $P(-2, 5)$ and $Q(3, 9)$ parallel to the line passing through the origin O and the point $R(5, 4)$?

Solution   Find the slope of each line:

$$\text{Slope of } PQ = \frac{\Delta y}{\Delta x} = \frac{y_2 - y_1}{x_2 - x_1} = \frac{9 - 5}{3 - (-2)} = \frac{4}{5}$$

$$\text{Slope of } OR = \frac{\Delta y}{\Delta x} = \frac{y_2 - y_1}{x_2 - x_1} = \frac{4 - 0}{5 - 0} = \frac{4}{5}$$

Because the slopes are equal, the lines are parallel.   ■

If a and b are two numbers such that $ab = -1$, they are called **negative reciprocals**. For example,

$$-\frac{4}{3} \quad \text{and} \quad \frac{3}{4}$$

are negative reciprocals because $-\frac{4}{3}(\frac{3}{4}) = -1$.

The following theorem, accepted without proof, relates perpendicular lines to their slopes:

Theorem.   If two nonvertical lines are perpendicular, their slopes are negative reciprocals.

If the slopes of two lines are negative reciprocals, the lines are perpendicular.

Because a horizontal line is perpendicular to a vertical line, a line with a slope of 0 is perpendicular to a line with undefined slope.

Example 6   Two lines intersect at point $P(3, -4)$. One passes through the origin O and the other passes through point $Q(9, 4)$. Are the lines perpendicular?

Solution   See Figure 7-16. Find the slope of lines OP and PQ.

$$\text{Slope of } OP = \frac{\Delta y}{\Delta x} = \frac{y_2 - y_1}{x_2 - x_1} = \frac{-4 - 0}{3 - 0} = -\frac{4}{3}$$

$$\text{Slope of } PQ = \frac{\Delta y}{\Delta x} = \frac{y_2 - y_1}{x_2 - x_1} = \frac{4 - (-4)}{9 - 3} = \frac{8}{6} = \frac{4}{3}$$

Because the slopes are not negative reciprocals, the lines are not perpendicular.

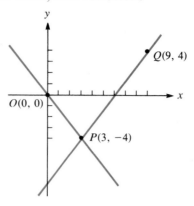

Figure 7-16

■ EXERCISE 7.2

In Exercises 1–10, find the slope of the line that passes through the given points. If the slope is undefined, so indicate.

1. $(0, 0)$ and $(3, 9)$

2. $(9, 6)$ and $(0, 0)$

3. $(-1, 8)$ and $(6, 1)$

4. $(-5, -8)$ and $(3, 8)$

5. $(3, -1)$ and $(-6, 2)$

6. $(0, -8)$ and $(-5, 0)$

7. $(7, 5)$ and $(-9, 5)$

8. $(2, -8)$ and $(3, -8)$

9. $(-7, -5)$ and $(-7, -2)$

10. $(3, -5)$ and $(3, 14)$

In Exercises 11–20, find the slope of the line determined by each equation.

11. $3x + 2y = 12$

12. $2x - y = 6$

13. $3x = 4y - 2$

14. $x = y$

15. $y = \dfrac{x - 4}{2}$

16. $x = \dfrac{3 - y}{4}$

17. $4y = 3(y + 2)$

18. $x + y = \dfrac{2 - 3y}{3}$

19. $x(y + 2) = y(x - 3) + 4$

20. $-y(x - 3) + 2 = x(4 - y)$

In Exercises 21–26, tell whether the slope of the line in each graph is positive, negative, 0, or undefined.

21.

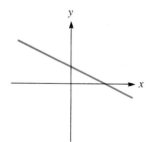

22.

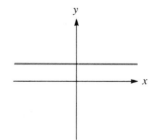

23.

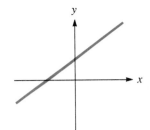

24.

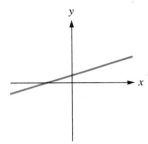

25.

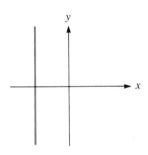

26.

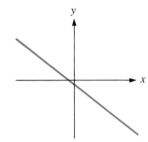

In Exercises 27–34, determine whether the lines with the given slopes are parallel, perpendicular, or neither.

27. $m_1 = 3; \quad m_2 = -\dfrac{1}{3}$

28. $m_1 = \dfrac{1}{4}; \quad m_2 = 4$

29. $m_1 = \sqrt{12}; \quad m_2 = 2\sqrt{3}$

30. $m_1 = \dfrac{\sqrt{3}}{3}; \quad m_2 = \dfrac{1}{\sqrt{3}}$

31. $m_1 = -\sqrt{7}; \quad m_2 = \dfrac{\sqrt{7}}{7}$

32. $m_1 = \sqrt{12}; \quad m_2 = -\dfrac{\sqrt{3}}{6}$

33. $m_1 = \dfrac{2\sqrt{7}}{7}; \quad m_2 = -\dfrac{2}{\sqrt{7}}$

34. $m_1 = 3 - \sqrt{2}; \quad m_2 = -\dfrac{3 + \sqrt{2}}{7}$

In Exercises 35–40, determine if the line passing through points P and Q is parallel or perpendicular to the line passing through R(2, −4) and S(−4, 8). If it is neither, so indicate.

35. $P(3, 4); \quad Q(4, 2)$

36. $P(6, 4); \quad Q(8, 5)$

37. $P(-2, 1); \quad Q(6, 5)$

38. $P(3, 4); \quad Q(-3, -5)$

39. $P(5, 4); \quad Q(6, 6)$

40. $P(-2, 3); \quad Q(4, -9)$

In Exercises 41–46, find the slope of lines PQ and PR and determine whether the points P, Q, and R lie on the same line. (Hint: Two lines with the same slope and a point in common must be the same line.)

41. $P(-2, 4); \quad Q(4, 8); \quad R(10, 12)$

42. $P(6, 10); \quad Q(0, 6); \quad R(3, 8)$

43. $P(-4, 10); \quad Q(-6, 0); \quad R(-1, 5)$

44. $P(-10, -13); \quad Q(-8, -10); \quad R(-12, -16)$

45. $P(-2, 4); \quad Q(0, 8); \quad R(2, 12)$

46. $P(8, -4); \quad Q(0, -12); \quad R(8, -20)$

47. What is the equation of the *x*-axis? What is its slope?

48. What is the equation of the *y*-axis? What is its slope, if any?

49. Show that points with coordinates of $(-3, 4)$, $(4, 1)$, and $(-1, -1)$ are vertices of a right triangle.

50. Show that a triangle with vertices at $(0, 0)$, $(12, 0)$, and $(12, 13)$ is a right triangle.

51. A square has vertices at points $A(a, 0)$, $B(0, a)$, $C(-a, 0)$, and $D(0, -a)$. Show that its adjacent sides are perpendicular.

52. Show that the points $A(2b, a)$, $B(b, b)$, and $C(a, 0)$ are vertices of a right triangle.

53. Show that the points $A(0, 0)$, $B(0, a)$, $C(b, c)$, and $D(b, a + c)$ are the vertices of a parallelogram. (*Hint:* Opposite sides of a **parallelogram** are parallel.)

54. Show that the points $A(0, 0)$, $B(0, b)$, $C(8, b + 2)$, and $D(12, 3)$ are the vertices of a trapezoid. (*Hint:* A **trapezoid** is a four-sided figure with exactly two sides parallel.)

55. The points $A(3, a)$, $B(5, 7)$, and $C(7, 10)$ lie on a line. Find a.

56. The line passing through points $A(1, 3)$ and $B(-2, 7)$ is perpendicular to the line passing through points $C(4, b)$ and $D(8, -1)$. Find b.

7.3   EQUATIONS OF LINES

Any two points on a nonvertical line can be used to determine the slope of the line. Suppose the nonvertical line l shown in Figure 7-17 has a slope of m and passes through the fixed point $P(x_1, y_1)$. Then, if $Q(x, y)$ is another point on line l, by the definition of slope we have

$$m = \frac{y - y_1}{x - x_1}$$

or

$$y - y_1 = m(x - x_1)$$

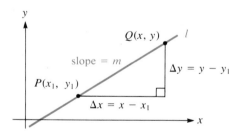

Figure 7-17

Because the equation $y - y_1 = m(x - x_1)$ displays both the coordinates of the fixed point (x_1, y_1) on the line and the line's slope m, it is called the **point–slope form** of the equation of a line.

> **The point–slope form of the equation of a line.**   The equation of the line passing through $P(x_1, y_1)$ with slope m is
>
> $$y - y_1 = m(x - x_1)$$

Example 1   Use the point–slope form to write the equation of a line with a slope of $-\frac{2}{3}$ passing through the point $P(-4, 5)$.

Solution   Substitute $-\frac{2}{3}$ for m, -4 for x_1, and 5 for y_1 in the point–slope form and simplify:

$$y - y_1 = m(x - x_1)$$

$$y - 5 = -\frac{2}{3}[x - (-4)]$$

$$y - 5 = -\frac{2}{3}(x + 4)$$

$$y = 5 - \frac{2}{3}x - \frac{8}{3} \qquad \text{Remove parentheses and add 5 to both sides.}$$

$$y = -\frac{2}{3}x + \frac{7}{3}$$

The desired equation is $y = -\frac{2}{3}x + \frac{7}{3}$.   ∎

Example 2   Use the point–slope form of the equation of a line to write the equation of the line passing through points $P(-5, 4)$ and $Q(8, -5)$.

Solution   First find the slope of the line:

$$m = \frac{y_2 - y_1}{x_2 - x_1} = \frac{-5 - 4}{8 - (-5)} = -\frac{9}{13}$$

Because the line passes through both P and Q, choose either point and substitute its coordinates into the equation for the point–slope form. For the purpose of illustration, choose $P(-5, 4)$. Then substitute -5 for x_1, 4 for y_1, and $-\frac{9}{13}$ for m and simplify.

$$y - y_1 = m(x - x_1)$$

$$y - 4 = -\frac{9}{13}[x - (-5)]$$

$$y - 4 = -\frac{9}{13}(x + 5)$$

$$y - 4 = -\frac{9}{13}x - \frac{45}{13}$$

$$y = -\frac{9}{13}x + \frac{7}{13} \qquad \text{Add 4 to both sides.}$$

The desired equation is $y = -\frac{9}{13}x + \frac{7}{13}$.   ∎

Slope–Intercept Form of the Equation of a Line

If the y-intercept of the line l with slope m shown in Figure 7-18 is b, the line intersects the y-axis at $P(0, b)$. We can write the equation of this line by sub-

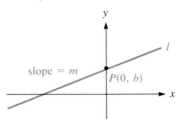

Figure 7-18

stituting 0 for x_1 and b for y_1 in the point–slope form and simplifying:

$$y - y_1 = m(x - x_1)$$
$$y - b = m(x - 0)$$
$$y - b = mx$$
$$y = mx + b$$

Because the equation $y = mx + b$ displays both the slope m and the y-intercept b of the line, it is called the **slope–intercept form** of the equation of a line.

The slope–intercept form of the equation of a line.   The equation of the line with slope m and y-intercept b is

$$y = mx + b$$

Example 3   Use the slope–intercept form to write the equation of the line with slope 4 that passes through the point $P(5, 9)$.

Solution   You are given that $m = 4$ and that the ordered pair $(5, 9)$ satisfies the equation. Substitute 5 for x, 9 for y, and 4 for m in the equation $y = mx + b$. Then solve for b, the y-intercept:

$$y = mx + b$$
$$9 = 4(5) + b$$
$$9 = 20 + b$$
$$-11 = b \qquad \text{Add } -20 \text{ to both sides.}$$

Because $m = 4$ and $b = -11$, the desired equation is $y = 4x - 11$.   ■

It is easy to graph an equation if it is written in slope–intercept form. For example, to graph the equation $y = \frac{4}{3}x - 2$, we first determine that the y-intercept is -2. Thus, the line passes through the point $P(0, -2)$. See Figure 7-19. Because the slope of the line is $\frac{4}{3}$, we can locate another point Q on the line by starting at point P and counting 3 units to the right and 4 units up. Note that the change in x (Δx) from point P to point Q is 3, and that the corre-

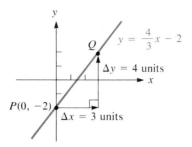

Figure 7-19

sponding change in y (Δy) is 4. The line joining points P and Q is the graph of the equation.

Example 4   Find the slope and the y-intercept of the line with equation $2(x - 3) = -3(y + 5)$.

Solution   Write the equation in the form $y = mx + b$ to determine the slope m and the y-intercept b:

$$2(x - 3) = -3(y + 5)$$
$$2x - 6 = -3y - 15$$
$$3y - 6 = -2x - 15 \qquad \text{Add } 3y - 2x \text{ to both sides.}$$
$$3y = -2x - 9 \qquad \text{Add } 6 \text{ to both sides.}$$
$$y = -\frac{2}{3}x - 3 \qquad \text{Divide both sides by 3.}$$

The slope of the line is $-\frac{2}{3}$ and the y-intercept is -3.   ∎

Example 5   Show that the lines represented by $4x + 8y = 10$ and $2x = 12 - 4y$ are parallel.

Solution   Solve each equation for y and observe that the lines are distinct and that their slopes are equal.

$$4x + 8y = 10 \qquad\qquad 2x = 12 - 4y$$
$$8y = -4x + 10 \qquad\qquad 4y = -2x + 12$$
$$y = -\frac{1}{2}x + \frac{5}{4} \qquad\qquad y = -\frac{1}{2}x + 3$$

Because the y-intercepts of the lines represented by these equations are different, the lines are distinct. Because the slope of each line is $-\frac{1}{2}$, they are parallel.   ∎

Example 6   Show that the lines represented by $4x + 8y = 10$ and $4x - 2y = 21$ are perpendicular.

Solution   Solve each equation for y and observe that their slopes are negative reciprocals:

$$4x + 8y = 10 \qquad\qquad 4x - 2y = 21$$
$$8y = -4x + 10 \qquad\qquad -2y = -4x + 21$$
$$y = -\frac{1}{2}x + \frac{5}{4} \qquad\qquad y = 2x - \frac{21}{2}$$

Because the slopes of these lines are $-\frac{1}{2}$ and 2 (which are negative reciprocals), the lines are perpendicular.   ∎

Example 7   Use the slope–intercept form to write the equation of the line passing through the point $P(-2, 5)$ and parallel to the line $y = 8x - 2$.

Solution   The line represented by the desired equation must have a slope of 8 because it is parallel to a line with slope of 8. Substitute -2 for x, **5** for y, and **8** for m in the slope–intercept form and solve for b:

$$y = mx + b$$
$$5 = 8(-2) + b$$
$$5 = -16 + b$$
$$21 = b \qquad\qquad \text{Add 16 to both sides.}$$

Because $m = 8$ and $b = 21$, the desired equation is $y = 8x + 21$.   ∎

Example 8   Use the point–slope form to write the equation of the line passing through $P(-2, 5)$ and perpendicular to the line $y = 8x - 2$.

Solution   The slope of the given line is 8. Thus, the slope of the desired line must be $-\frac{1}{8}$, which is the negative reciprocal of 8. Substitute -2 for x_1, **5** for y_1, and $-\frac{1}{8}$ for m into the point–slope form. Then simplify.

$$y - y_1 = m(x - x_1)$$
$$y - 5 = -\frac{1}{8}[x - (-2)]$$
$$y - 5 = -\frac{1}{8}(x + 2)$$
$$8y - 40 = -(x + 2) \qquad\qquad \text{Multiply both sides by 8.}$$
$$8y - 40 = -x - 2$$
$$x + 8y = 38 \qquad\qquad \text{Add } x + 40 \text{ to both sides.}$$

The equation of the line is $x + 8y = 38$.   ∎

General Form of the Equation of a Line

The final equation given in Example 8 was written in the form

$$x + 8y = 38$$

This is an example of a form called the **general form** of the equation of a line. Any equation that is written in the form $Ax + By = C$ is said to be written in general form.

> **Theorem.**   If A, B, and C are real numbers and $B \neq 0$, then the graph of the equation
>
> $$Ax + By = C$$
>
> is a nonvertical line with a slope of $-\dfrac{A}{B}$ and a y-intercept of $\dfrac{C}{B}$.

You will be asked to prove the previous theorem in the Exercises.

If $B = 0$ and $A \neq 0$, the equation $Ax + By = C$ represents a vertical line with x-intercept of C/A. Because the graph of any equation that can be written in the form $Ax + By = C$ is a straight line, the equation $Ax + By = C$ is called a **linear equation in x and y.**

Example 9   Show that the lines represented by $4x + 3y = 7$ and $3x - 4y = 12$ are perpendicular.

Solution   To show that the lines are perpendicular, show that their slopes are negative reciprocals. The first equation, $4x + 3y = 7$, is written in general form, with $A = 4$, $B = 3$, and $C = 7$. By the previous theorem, the slope of the line is

$$m_1 = -\frac{A}{B} = -\frac{4}{3}$$

Similarly, the second equation, $3x - 4y = 12$, is written in general form, with $A = 3$, $B = -4$, and $C = 12$. The slope of this line is

$$m_2 = -\frac{A}{B} = -\frac{3}{-4} = \frac{3}{4}$$

Because the slopes of the two lines are negative reciprocals, the lines are perpendicular.   ∎

The following box summarizes the various forms for the equation of a line:

General form of a linear equation	$Ax + By = C$ A and B cannot both be zero
Slope–intercept form of a linear equation	$y = mx + b$ slope is m and the y-intercept is b
Point–slope form of a linear equation	$y - y_1 = m(x - x_1)$ slope is m and the line passes through (x_1, y_1)
A horizontal line	$y = k$ slope of 0 and the y-intercept is k
A vertical line	$x = k$ no defined slope and the x-intercept is k

■ **EXERCISE 7.3** ■

In Exercises 1–6, use the point–slope form to write the equation of the line with the given properties. **Write each equation in general form.**

1. $m = 5$ and passing through $P(0, 7)$

2. $m = -8$ and passing through $P(0, -2)$

3. $m = -3$ and passing through $P(2, 0)$

4. $m = 4$ and passing through $P(-5, 0)$

5. $m = \dfrac{3}{2}$ and passing through $P(2, 5)$

6. $m = -\dfrac{2}{3}$ and passing through $P(-3, 2)$

In Exercises 7–12, use the point–slope form to write the equation of the line passing through the two given points. **Write each equation in the form $y = mx + b$.**

7. $P(0, 0);$ $Q(4, 4)$

8. $P(-5, -5);$ $Q(0, 0)$

9. $P(3, 4);$ $Q(0, -3)$

10. $P(4, 0);$ $Q(6, -8)$

11. $P(-2, 4);$ $Q(3, -5)$

12. $P(3, -5);$ $Q(-1, 12)$

In Exercises 13–20, use the slope–intercept form to write the equation of the line with the given properties. **Write each equation in slope–intercept form.**

13. $m = 3$ and $b = 17$

14. $m = -2$ and $b = 11$

15. $m = -7$ and passing through $P(7, 5)$

16. $m = 3$ and passing through $P(-2, -5)$

17. $m = 0$ and passing through $P(2, -4)$

18. $m = -7$ and passing through the origin

19. passing through $P(6, 8)$ and $Q(2, 10)$

20. passing through $P(-4, 5)$ and $Q(2, -6)$

In Exercises 21–26, write each equation in slope–intercept form to determine the slope and the y-intercept. Then use the method of Figure 7-19 to graph the equation.

21. $y + 1 = x$

22. $x + y = 2$

23. $x = \dfrac{3}{2}y - 3$

24. $x = -\dfrac{4}{5}y + 2$

25. $3(y - 4) = -2(x - 3)$

26. $-4(2x + 3) = 3(3y + 8)$

In Exercises 27–32, find the slope and the y-intercept of the line determined by the given equation.

27. $3x - 2y = 8$

28. $-2x + 4y = 12$

29. $-2(x + 3y) = 5$

30. $5(2x - 3y) = 4$

31. $x = \dfrac{2y - 4}{7}$

32. $3x + 4 = -\dfrac{2(y - 3)}{5}$

In Exercises 33–44, indicate whether the lines given by each pair of equations are parallel or perpendicular. If they are neither, so indicate.

33. $y = 3x + 4;$ $y = 3x - 7$

34. $y = 4x - 13;$ $y = \dfrac{1}{4}x + 13$

35. $x + y = 2;$ $y = x + 5$

36. $x = y + 2;$ $y = x + 3$

37. $y = 3x + 7;$ $2y = 6x - 9$

38. $2x + 3y = 9;$ $3x - 2y = 5$

39. $x = 3y + 8;$ $y = -3x + 7$

40. $3x + 6y = 1;$ $y = \dfrac{1}{2}x$

41. $y = 3;$ $x = 4$

42. $y = -3;$ $y = -7$

43. $x = \dfrac{y - 2}{3};$ $3(y - 3) + x = 0$

44. $2y = 8;$ $3(2 + y) = 3(x + 2)$

In Exercises 45–50, write the equation of the line that passes through the given point and is parallel to the given line.

45. $P(0, 0)$; $y = 4x - 7$

46. $P(0, 0)$; $x = -3y - 12$

47. $P(2, 5)$; $4x - y = 7$

48. $P(-6, 3)$; $y + 3x = -12$

49. $P(4, -2)$; $x = \dfrac{5}{4}y - 2$

50. $P(1, -5)$; $x = -\dfrac{3}{4}y + 5$

In Exercises 51–56, write the equation of the line that passes through the given point and is perpendicular to the given line.

51. $P(0, 0)$; $y = 4x - 7$

52. $P(0, 0)$; $x = -3y - 12$

53. $P(2, 5)$; $4x - y = 7$

54. $P(-6, 3)$; $y + 3x = -12$

55. $P(4, -2)$; $x = \dfrac{5}{4}y - 2$

56. $P(1, -5)$; $x = -\dfrac{3}{4}y + 5$

In Exercises 57–60, use the method of Example 9 to determine whether the graphs determined by each pair of equations are parallel, perpendicular, or neither.

57. $4x + 5y = 20$; $5x - 4y = 20$

58. $9x - 12y = 17$; $3x - 4y = 17$

59. $2x + 3y = 12$; $6x + 9y = 32$

60. $5x + 6y = 30$; $6x + 5y = 24$

61.   Find the equation of the line perpendicular to the line $y = 3$ and passing through the midpoint of the segment joining $(2, 4)$ and $(-6, 10)$.

62.   Find the equation of the line parallel to the line $y = -8$ and passing through the midpoint of the segment joining $(-4, 2)$ and $(-2, 8)$.

63.   Find the equation of the line parallel to the line $x = 3$ and passing through the midpoint of the segment joining $(2, -4)$ and $(8, 12)$.

64.   Find the equation of the line perpendicular to the line $x = 3$ and passing through the midpoint of the segment joining $(-2, 2)$ and $(4, -8)$.

65.   Solve the equation $Ax + By = C$ for y and thereby show that the slope of its graph is $-A/B$ and its y-intercept is C/B.

66.   Show that the x-intercept of the graph of $Ax + By = C$ is C/A.

7.4   FUNCTIONS AND FUNCTION NOTATION

The equations we have discussed so far have described a correspondence between two variables. The equation $y = -2x + 3$, for example, describes a correspondence in which each number x determines *exactly one* value of y. Such correspondences are called **functions**.

> **Definition.**   A **function** is a correspondence that assigns to each element x of some set X a single value y of a set Y.
>   The set X is called the **domain** of the function. The value y that corresponds to a particular x in the domain is called the **image** of x under the function. The set Y of all images of x is called the **range** of the function.

Because each value of y depends on some number x, we call y the **dependent variable**. The variable x is called the **independent variable**. Unless indicated otherwise, the domain of a function is its **implied domain**—the set of all real numbers for which the function is defined.

By definition, a function f is a correspondence from the elements of one set, X, to the elements of another set, Y. We can visualize this correspondence with the diagram in Figure 7-20. We represent the function f by drawing arrows from each element of X to the corresponding element of Y. If a function f assigns the element y_1 to the element x_1, we draw an arrow leaving x_1 and pointing to y_1. The set X, consisting of all elements from which arrows originate, is the domain of the function. The set Y, consisting of all elements to which arrows point, is the range.

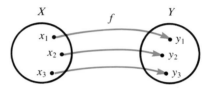

Figure 7-20

A correspondence is still a function if arrows point from several different elements of X to the same element y in the range. See Figure 7-21.

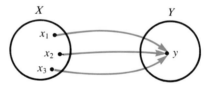

Figure 7-21

However, if arrows leave some element x in the domain and point to several elements in the range, the correspondence is *not* a function because more than one y value corresponds to the number x. See Figure 7-22.

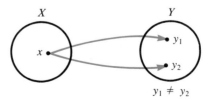

Figure 7-22

Such a correspondence is called a **relation**.

> **Definition.**   A **relation** is a correspondence that assigns to each number x in a set X one or more values of y in a set Y.
> The set X is called the **domain** of the relation. The **range** of the relation is the set Y of all values of y that correspond to some number x in the domain.

Because of the previous definitions, *a function is always a relation but a relation is not necessarily a function.*

Example 1   Does the equation $y = 4x - 5$ define y to be a function of x? If so, find its domain and range.

Solution   By the definition of function, each real number x must determine *exactly one* value of y. To determine y in this equation, 5 must be subtracted from the product of 4 and x. Because this arithmetic gives a single result, only one value of y is produced. Thus, the equation does determine a function.

Because x can be any real number, the implied domain of the function is the set of all real numbers.

Because y can be any real number, the range of the function also is the set of real numbers.   ■

Example 2   Does the equation $y^2 = 9x$ determine y to be a function of x? If not, give the domain and the range of the relation.

Solution   Let x be the number 4, for example. Then

$$y^2 = 9(4) = 36$$

and y could be either 6 or -6. Because more than one value of y corresponds to $x = 4$, this equation does not determine a function. However, it does determine a relation.

Because y^2 is always a nonnegative number, x must be a nonnegative number also. Thus, the domain of the relation is the set of nonnegative real numbers. Because y can be any real number, the range is the set of all real numbers.   ■

Example 3   Does the equation $y = \dfrac{12}{x - 3}$ determine y to be a function of x? If so, find its domain and range.

Solution   Since division by 0 is undefined, the denominator of the fraction cannot be 0. Thus, x cannot be 3. Because each real number x other than 3 does determine *exactly one* value of y, the equation does determine a function.

Since x cannot be 3, it follows that the domain of this function is the set of all real numbers except 3.

A fraction with a nonzero numerator cannot be 0. Thus, the range of this function is the set of all real numbers except 0.   ■

Function Notation

There is a notation that reinforces the idea that y is a function of x. If the variable y in a function is equal to some expression that involves the variable x, then y is often denoted as $f(x)$ and read as "f of x." The symbol $f(x)$ indicates the number to substitute for x when evaluating y. For example, if $y = f(x) = 4x + 2$, then $f(3)$ is the value of $4x + 2$ when $x = 3$:

$$f(3) = 4(3) + 2 = 14$$

Likewise,

$$f(-2) = 4(-2) + 2 = -6$$

Example 4   Suppose that y is the function of x determined by $y = f(x) = \dfrac{x}{2} + 1$. Find

    a. $f(4)$,   **b.** $f(8)$, and   **c.** $f(-12)$.

Solution   **a.** $f(4)$ is the value of y when $x = 4$:

$$f(4) = \frac{4}{2} + 1 = 2 + 1 = 3$$

    b. $f(8)$ is the value of y when $x = 8$:

$$f(8) = \frac{8}{2} + 1 = 4 + 1 = 5$$

    c. $f(-12)$ is the value of y when $x = -12$:

$$f(-12) = \frac{-12}{2} + 1 = -6 + 1 = -5$$ ■

Example 5   If $f(x) = x^2 + 3x + 1$, find   **a.** $f(z)$,   **b.** $f(-2)$, and   **c.** $f(z - 2)$.

Solution   **a.** $f(z) = z^2 + 3z + 1$

    b. $f(-2) = (-2)^2 + 3(-2) + 1$
$$= 4 - 6 + 1$$
$$= -1$$

    c. $f(z - 2) = (z - 2)^2 + 3(z - 2) + 1$
$$= z^2 - 4z + 4 + 3z - 6 + 1$$
$$= z^2 - z - 1$$

Because $f(z) + f(-2) = z^2 + 3z$ and $f(z - 2) = z^2 - z - 1$,

$$f(z) + f(-2) \neq f(z - 2)$$ ■

Composition of Functions

If $f(x)$ and $g(x)$ are both functions of x, then $f(g(x))$ and $g(f(x))$ are also functions of x. Such functions are called **composite functions**. The expression $f(g(x))$ is often denoted as $(f \circ g)(x)$ and read as "f composition g." The expression $g(f(x))$ is often denoted as $(g \circ f)(x)$. For example, if $f(x) = 4x$ and $g(x) = 3x + 2$, then

$$(f \circ g)(x) = f(g(x)) = f(3x + 2) = 4(3x + 2) = 12x + 8$$
$$(g \circ f)(x) = g(f(x)) = g(4x) = 3(4x) + 2 = 12x + 2$$

Note that $(f \circ g)(x) \neq (g \circ f)(x)$. Thus, the composition of functions is not commutative.

Example 6 Suppose that $f(x) = 2x + 1$ and $g(x) = x - 4$. Find   **a.** $(f \circ g)(8)$,
b. $(g \circ f)(-2)$, and   **c.** $(f \circ g)(x)$.

Solution **a.** $(f \circ g)(8)$ means $f(g(8))$. Because $g(8) = 8 - 4 = 4$, you have

$$(f \circ g)(8) = f(g(8)) = f(4) = 2(4) + 1 = 9$$

b. $(g \circ f)(-2)$ means $g(f(-2))$. Because $f(-2) = 2(-2) + 1 = -3$, you have

$$(g \circ f)(-2) = g(f(-2)) = g(-3) = -3 - 4 = -7$$

c. $(f \circ g)(x)$ means $f(g(x))$. Because $g(x) = x - 4$, you have

$$(f \circ g)(x) = f(g(x)) = f(x - 4) = 2(x - 4) + 1 = 2x - 7$$ ∎

Algebra of Functions

It is possible to add, subtract, multiply, and divide functions.

Definition. If the ranges of functions f and g are subsets of the real numbers, then

The **sum** of f and g, denoted as $f + g$, is defined by
$$(f + g)(x) = f(x) + g(x)$$
The **difference** of f and g, denoted as $f - g$, is defined by
$$(f - g)(x) = f(x) - g(x)$$
The **product** of f and g, denoted as $f \cdot g$, is defined by
$$(f \cdot g)(x) = f(x)g(x)$$
The **quotient** of f and g, denoted as f/g, is defined by
$$(f/g)(x) = \frac{f(x)}{g(x)} \qquad [g(x) \neq 0]$$

The domain of each of these functions is the set of all real numbers x that are in the domain of *both* f and g. In the case of the quotient, there is the further restriction that $g(x) \neq 0$.

Example 7   Let $f(x) = x^2$ and $g(x) = 2x + 4$, then

a. $(f + g)(x) = f(x) + g(x) = x^2 + 2x + 4$
    The domain of $f + g$ is the set of real numbers that are in the domain of
    both f and g. Since the domain of both f and g is the set of real numbers,
    the domain of $f + g$ is also the set of real numbers.

b. $(f - g)(x) = f(x) - g(x) = x^2 - (2x + 4) = x^2 - 2x - 4$
    The domain of $f - g$ is the set of real numbers.

c. $(f \cdot g)(x) = f(x)g(x) = x^2(2x + 4) = 2x^3 + 4x^2$
    The domain of $f \cdot g$ is the set of real numbers.

d. $(f/g)(x) = \dfrac{f(x)}{g(x)} = \dfrac{x^2}{2x + 4}$

    Because the denominator of $\dfrac{x^2}{2x + 4}$ cannot be 0, x cannot be -2. Thus,
    the domain of $(f/g)(x)$ is the set of real numbers except -2.   ■

■ EXERCISE 7.4

In Exercises 1–10, tell whether the given relation defines y to be a function of x. Assume that x and y represent real numbers in each ordered pair (x, y).

1. $y = 3x - 4$
2. $y = -\dfrac{1}{2}x + 7$
3. $3x + 2y = 4$
4. $x = 3y - 4$

5. $y^2 = 25x$
6. $y^2 = x$
7. $x = \dfrac{1}{y^2}$
8. $y = \dfrac{2}{x + 2}$

9. $x = \dfrac{y + 4}{3}$
10. $y^2 = \dfrac{x + 3}{5}$

In Exercises 11–20, give the domain and the range of each function. Assume that x and y represent real numbers in each ordered pair (x, y).

11. $y = 2x + 3$
12. $y = \dfrac{1}{2}x - 4$
13. $y = \dfrac{4}{2 - x}$
14. $y = \dfrac{x - 6}{5}$

15. $y = x^4$
16. $y = \dfrac{5}{x - 3}$
17. $y = 3$
18. $y = \dfrac{1}{2}x^2$

19. $y = \sqrt{x - 2}$
20. $y = \sqrt{x^2 - 4}$

In Exercises 21–28, assume that $y = f(x) = 5x + 2$. Evaluate each expression.

21. $f(1)$
22. $f(0)$
23. $f(-2)$
24. $f(-1)$

25. $f\left(-\dfrac{2}{5}\right)$
26. $f\left(\dfrac{2}{3}\right)$
27. $f(x + 2)$
28. $f(x - 1)$

In Exercises 29–36, assume that $y = g(x) = x^2 + 4$. Evaluate each expression.

29. $g(0)$
30. $g(1)$
31. $g(-1)$
32. $g(-2)$

33. $g\left(\dfrac{2}{3}\right)$
34. $g\left(-\dfrac{1}{2}\right)$
35. $g(x^2)$
36. $g\left(\dfrac{1}{x^2}\right)$

In Exercises 37–42, assume that $y = f(x) = 2x + 1$ and $g(x) = x^2 - 1$. Evaluate each expression.

37. $(f \circ g)(2)$ **38.** $(g \circ f)(2)$ **39.** $(g \circ f)(-3)$ **40.** $(f \circ g)(-3)$

41. $(f \circ g)(x)$ **42.** $(g \circ f)(x)$

In Exercises 43–48, assume that $f(x) = 3x - 2$ and $g(x) = x^2 + x$. Evaluate each expression.

43. $(f \circ g)(4)$ **44.** $(g \circ f)(4)$ **45.** $(g \circ f)(-3)$ **46.** $(f \circ g)(-3)$

47. $(g \circ f)(x)$ **48.** $(f \circ g)(x)$

49.   If $f(x) = x + 1$ and $g(x) = 2x - 5$, show that $(f \circ g)(x) \neq (g \circ f)(x)$.

50.   If $f(x) = x^2 + 1$ and $g(x) = 3x^2 - 2$, show that $(f \circ g)(x) \neq (g \circ f)(x)$.

51.   If $f(x) = x^2 + 2x - 3$, find $f(a)$, $f(h)$, and $f(a + h)$. Then show that $f(a + h) \neq f(a) + f(h)$.

52.   If $g(x) = 2x^2 + 10$, find $g(a)$, $g(h)$, and $g(a + h)$. Then show that $g(a + h) \neq g(a) + g(h)$.

53.   If $f(x) = x^2 + 2$, find $\dfrac{f(x + h) - f(x)}{h}$.   **54.**   If $f(x) = x^3 - 1$, find $\dfrac{f(x + h) - f(x)}{h}$.

In Exercises 55–58, let $f(x) = 2x + 1$ and $g(x) = x - 3$. Find each function and determine its domain.

55. $f + g$ **56.** $f - g$ **57.** $f \cdot g$ **58.** f/g

In Exercises 59–62, let $f(x) = 3x - 2$ and $g(x) = 2x^2 + 1$. Find each function and determine its domain.

59. $f - g$ **60.** $f + g$ **61.** f/g **62.** $f \cdot g$

In Exercises 63–66, let $f(x) = x^2 - 1$ and $g(x) = x^2 - 4$. Find each function and determine its domain.

63. $f - g$ **64.** $f + g$ **65.** g/f **66.** $f \cdot g$

7.5   LINEAR FUNCTIONS AND THEIR INVERSES

A relation defines a set of ordered pairs (x, y), where x is an element in the domain of the relation and y is the value that corresponds to x. The graph of a relation is the graph of all these ordered pairs on a rectangular coordinate system. If the relation is a function defined by the equation $y = f(x)$, the graph of the function consists of those points and only those points in the xy-plane with coordinates $(x, y) = (x, f(x))$.

A linear function is a function whose graph is a straight line. More formally, we define a linear function as follows:

> **Definition.**   A **linear function** is a function defined by an equation that can be written in the form
>
> $$y = mx + b$$
>
> where m is the slope of its straight-line graph and b is its y-intercept.

A linear function is often written in the form $f(x) = mx + b$, where $f(x)$ is just another symbol for y. Thus, $y = f(x)$.

Example 1   Solve the equation $3x + 2y = 10$ for y to show that it defines a linear function. Then graph the line.

Solution   Solve the equation for y as follows:

$$3x + 2y = 10$$
$$2y = -3x + 10 \qquad \text{Add } -3x \text{ to both sides.}$$
$$y = -\frac{3}{2}x + 5 \qquad \text{Divide both sides by 2.}$$

Because the equation can be written in the form $y = mx + b$, it defines a linear function. Because the form $y = mx + b$ is the slope–intercept form of the equation of a line, the slope of its graph is $-\frac{3}{2}$ and its y-intercept is 5. The graph appears in Figure 7-23.

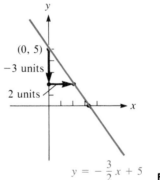

$$y = -\frac{3}{2}x + 5 \qquad \textbf{Figure 7-23}$$

A nonempty set of ordered pairs (x, y) defines a relation because it determines a correspondence between values of x and y. For example, a relation R is defined by the set

$$R = \{(1, 10), (2, 20), (3, 30)\}$$

The domain of R is the set of first components of the ordered pairs, and its range is the set of second components. Thus, the domain of R is $\{1, 2, 3\}$ and the range is $\{10, 20, 30\}$. Because only one second component y corresponds to each first component x of an ordered pair, the relation R is a function. However, the relation Q, where

$$Q = \{(5, 15), (5, 20), (6, 30), (7, 40)\}$$

with domain of $\{5, 6, 7\}$ and range $\{15, 20, 30, 40\}$ is not a function because two values of y (15 and 20) correspond to the first component 5.

Example 2   Let R be a relation defined by ordered pairs (x, y) such that $R = \{(2, 3), (5, 7),$ $(6, 8), (9, 8)\}$. Give the domain and range of R and tell whether R is a function.

Solution   The domain of R is the set of first components of the ordered pairs. Thus, the domain is $\{2, 5, 6, 9\}$.

The range of R is the set of second components of the ordered pairs. Thus, the range is $\{3, 7, 8\}$.

Because only one value of y corresponds to each number x, the relation R is a function.   ∎

Inverse Functions

If the components of each ordered pair in any given relation are interchanged, a new relation is formed called the **inverse relation**. For example, if R is the relation

$$R = \{(1, 10), (2, 20), \ (3, 30)\}$$

then the inverse relation of R, denoted as R^{-1}, is

$$R^{-1} = \{(10, 1), (20, 2), (30, 3)\}$$

The -1 in the notation R^{-1} is not an exponent. It refers to the inverse of the relation. The symbol R^{-1} is read as "the inverse relation of R" or just "R inverse."

The domain of R and the range of R^{-1} is $\{1, 2, 3\}$. The range of R and the domain of R^{-1} is $\{10, 20, 30\}$. In general, we have the following definition:

Definition.   If R is any relation and R^{-1} is the relation obtained from R by interchanging the components of each ordered pair of R, then R^{-1} is called the **inverse relation** of R.

The domain of R^{-1} is the range of R, and the range of R^{-1} is the domain of R.

An equation in x and y determines a set of ordered pairs. For example, if $x = 1, 2, 3, 4,$ and 5, the equation $y = 6x$ determines the following set of ordered pairs:

$$T = \{(x, y) : (1, 6), (2, 12), (3, 18), (4, 24), (5, 30)\}$$

To form T^{-1}, we simply interchange the x and y coordinates to obtain

$$T^{-1} = \{(x, y) : (6, 1), (12, 2), (18, 3), (24, 4), (30, 5)\}$$

Example 3   The set of all pairs (x, y) determined by the equation $y = 4x + 2$ determines a function, f. Find the inverse relation of $y = 4x + 2$ and tell whether the inverse is a function.

Solution   To find the inverse relation of $y = 4x + 2$, interchange the variables x and y to obtain

1. $x = 4y + 2$

To decide whether this inverse relation is a function, solve Equation 1 for y:

$$x = 4y + 2$$

$$x - 2 = 4y \qquad \text{Add } -2 \text{ to both sides.}$$

$$y = \frac{x - 2}{4} \qquad \begin{array}{l}\text{Divide both sides by 4 and apply the symmetric}\\ \text{property of equality.}\end{array}$$

Because each number x that is substituted into this equation gives a single value y, the inverse relation is a function. ∎

In Example 3 the inverse relation of the function $y = 4x + 2$ was found to be the function $y = \dfrac{x - 2}{4}$. In function notation, this inverse function can be denoted as

$$f^{-1}(x) = \frac{x - 2}{4} \qquad \text{Read as "}f \text{ inverse of } x \text{ is } \frac{x-2}{4}\text{."}$$

Example 4   The set of all pairs (x, y) determined by the equation $3x + 2y = 6$ is a function. Find the inverse function and graph both functions on a single coordinate system.

Solution   To find the inverse function of $3x + 2y = 6$, interchange the x and y to obtain

$$3y + 2x = 6$$

Then solve the equation for y:

$$3y + 2x = 6$$

$$3y = -2x + 6$$

$$y = -\frac{2}{3}x + 2$$

The graphs of the equations $3x + 2y = 6$ and $y = -\dfrac{2}{3}x + 2$ appear in Figure 7-24.

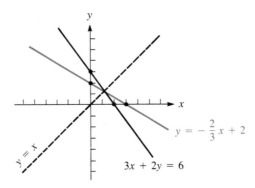

Figure 7-24

In Example 4, the graphs of the equations $3x + 2y = 6$ and $y = -\frac{2}{3}x + 2$ are symmetric about the line $y = x$. This is always the case because, if the coordinates (a, b) satisfy an equation, the coordinates (b, a) will satisfy its inverse.

■ EXERCISE 7.5

In Exercises 1–10, indicate whether the given equation determines a linear function.

1. $3x + 2y = 5$

2. $x - 3y = 8$

3. $3(x + 2) = 5y$

4. $-2(y - 3) = x$

5. $y = \dfrac{3}{x + 2}$

6. $y = \dfrac{x + 2}{3}$

7. $y = 9$

8. $x = 8$

9. $y^2 = x$

10. $x(x - y) = y$

In Exercises 11—16, give the domain and range of each relation. If the relation is a function, so indicate.

11. $\{(1, 2), (3, 4), (5, 9), (5, 12)\}$

12. $\{(3, 4), (-5, 3), (8, 9), (3, 6)\}$

13. $\{(1, 2), (-1, 3), (4, 4), (-4, 5)\}$

14. $\{(4, 0), (3, 0), (2, 0), (1, 0)\}$

15. $\{(5, 8), (6, 9), (5, 10)\}$

16. $\{(-2, 0), (-1, 1), (0, 2), (1, 3)\}$

In Exercises 17–22, find the inverse relation of each set of ordered pairs (x, y) and tell whether the inverse relation is a function.

17. $\{(3, 2), (2, 1), (1, 0)\}$

18. $\{(4, 1), (5, 1), (6, 1), (7, 1)\}$

19. $\{(1, 2), (2, 3), (1, 3), (1, 5)\}$

20. $\{(-1, -1), (0, 0), (1, 1), (2, 2)\}$

21. $\{(1, 1), (2, 4), (3, 9), (4, 16)\}$

22. $\{(1, 1), (2, 2), (3, 3), (4, 3)\}$

In Exercises 23–30, find the inverse relation of each set of ordered pairs (x, y) determined by the given equation and tell whether that inverse relation is a function. If the inverse relation is a linear function, express it in the form $f^{-1}(x) = mx + b$.

23. $y = 3x + 1$

24. $y + 1 = 5x$

25. $x + 4 = 5y$

26. $x = 3y + 1$

27. $y = \dfrac{x - 4}{5}$

28. $y^2 = \dfrac{2x + 6}{3}$

29. $4x - 5y^2 = 20$

30. $3x + 5y = 15$

In Exercises 31–40, find the inverse of each linear function. Then graph both the function and its inverse on a single coordinate system. What is the equation of the line of symmetry?

31. $y = 2x - 3$

32. $x = 3y - 1$

33. $x = \dfrac{y - 2}{3}$

34. $y = \dfrac{x + 3}{4}$

35. $3x - y = 5$

36. $2x + 3y = 18$

37. $3(x + y) = 2x + 4$

38. $-4(y - 1) + x = 2$

39. $\dfrac{3}{2}x = \dfrac{18 - 3y}{3}$

40. $2y - \dfrac{6y}{5} = \dfrac{2(3 - x)}{5}$

41. A high-contrast photographic developer uses 5.0 grams of potassium bromide for each liter of developer solution. Find an equation that describes the linear relation between g, the number of grams of potassium bromide, and l, the number of liters of solution.

42. A dentist charges \$30, plus \$15 for each tooth he fills. Find an equation that describes the relation between n, the number of teeth filled, and c, the amount the dentist charges.

43. To wash windows, Bill charges \$10, plus \$1 for each window. Find an equation that describes the linear relation between n, the number of windows washed, and c, the amount Bill charges.

44. A 20-minute long-distance telephone call costs \$3.15. If that call had lasted 30 minutes, it would have cost \$4.35. Find an equation that describes the linear relation between t, the time of the telephone call, and c, its cost.

45. The boiling point of water is 212°F or 100°C. The freezing point of water is 32°F or 0°C. Find an equation that describes the linear relation between temperature as measured on the Fahrenheit and Celsius scales.

46. The population of a town has been growing at the rate of 850 people per year since a census ten years ago. The current population of the town is 38,500 people. Find an equation that describes the linear relation between p (the population) and n (the number of years elapsed since the census year). Use this equation to compute the expected population ten years from now.

7.6 PROPORTION AND VARIATION

An indicated quotient of two numbers is often called a **ratio**. For example, the fraction $\frac{2}{3}$ can be read as "the ratio of 2 to 3." An equation indicating that two ratios are equal is called a **proportion**. Two examples of proportions are

$$\frac{1}{4} = \frac{2}{8} \quad \text{and} \quad \frac{4}{7} = \frac{12}{21}$$

In the proportion $\dfrac{a}{b} = \dfrac{c}{d}$, the terms a and d are called the **extremes** of the proportion and the terms b and c are called the **means**.

To develop a fundamental property of proportions, we suppose that

$$\frac{a}{b} = \frac{c}{d}$$

is a proportion and multiply both sides by bd to obtain

$$(bd)\frac{a}{b} = (bd)\frac{c}{d}$$

$$\frac{\cancel{b}da}{\cancel{b}} = \frac{b\cancel{d}c}{\cancel{d}}$$

$$ad = bc$$

Thus, in a proportion, *the product of the extremes equals the product of the means.*

Example 1 Solve the proportion $\dfrac{x + 1}{x} = \dfrac{x}{x + 2}$ for x.

Solution

$$\frac{x+1}{x} = \frac{x}{x+2}$$

$(x + 1)(x + 2) = x \cdot x$ The product of the extremes equals the product of the means.

$x^2 + 3x + 2 = x^2$

$3x + 2 = 0$ Add $-x^2$ to both sides.

$x = -\dfrac{2}{3}$ Add -2 to both sides and divide by 3.   ∎

Variation

Consider the formula

$$C = \pi D$$

for the circumference of a circle, where C is the circumference, D is the diameter, and $\pi \approx 3.14159$. If we double the diameter of a circle, we determine another circle with a larger circumference C_1 such that

$$C_1 = \pi(2D) = 2\pi D = 2C$$

Thus, doubling the diameter results in doubling the circumference. Likewise, if we triple the diameter, we triple the circumference. In this formula, we say that the variables C and D **vary directly**, or that they are **directly proportional**. That is, as one variable gets larger, so does the other, and in a predictable way. The constant π is called the **constant of variation**, or the **constant of proportionality**.

Direct variation.   The words "**y varies directly as x**," or "**y is directly proportional to x**" mean that $y = kx$ for some constant k.
   k is called the **constant of variation** or the **constant of proportionality**.

An example of direct variation is Hooke's law from physics. Hooke's law states that the distance a spring will stretch varies directly with the force that is applied to it. If d represents a distance and f represents a force, Hooke's law can be expressed mathematically as

$$d = kf$$

where k is the constant of variation. If the spring stretches 10 inches when a force of 6 pounds is attached, k can be computed for this spring as follows:

$$d = kf$$
$$10 = k(6)$$
$$\frac{5}{3} = k$$

To find the force required to stretch the spring by a distance of 35 inches, we can solve the equation $d = kf$ for f, with $d = 35$ and $k = \frac{5}{3}$:

$$d = kf$$

$$35 = \frac{5}{3}f$$

$$105 = 5f \qquad \text{Multiply both sides by 3.}$$

$$21 = f \qquad \text{Divide both sides by 5.}$$

Thus, the force required to stretch the spring by a distance of 35 inches is 21 pounds.

Example 2   The frequency of a vibrating string varies directly as the square root of its tension. If a spring is vibrating at a frequency of 144 hertz due to a tension of 3 pounds, what would the frequency be if the tension were 12 pounds?

Solution   The words "frequency varies directly as the square root of its tension" can be expressed mathematically as

$$f = k\sqrt{t}$$

where k is the constant of variation. To find k, substitute **144** for f and **3** for t and solve for k:

$$144 = k\sqrt{3}$$

$$k = \frac{144}{\sqrt{3}}$$

$$= \frac{144\sqrt{3}}{3}$$

$$= 48\sqrt{3}$$

To find the frequency when the tension is 12 pounds, substitute $48\sqrt{3}$ for k and **12** for t into the formula $f = k\sqrt{t}$. Then simplify.

$$f = k\sqrt{t}$$

$$f = 48\sqrt{3} \cdot \sqrt{12}$$

$$= 48\sqrt{36}$$

$$= 288$$

The frequency is 288 hertz.   ■

In the formula $w = \frac{12}{l}$, w gets smaller as l gets larger, and w gets larger as l gets smaller. Since these variables vary in opposite directions in a predictable way, we say that they **vary inversely**, or that they are **inversely proportional**. The constant 12 is the **constant of variation**.

> **Inverse variation.**   The words "**y varies inversely with x**," or "**y is inversely proportional to x**" mean that $y = \dfrac{k}{x}$ for some constant k.
>
> k is called the **constant of variation** or the **constant of proportionality**.

Because of gravity, an object in space is attracted to the earth. The force of this attraction varies inversely with the square of the object's distance from the center of the earth. If f represents the force and d represents the distance, this information can be expressed mathematically with the equation

$$f = \frac{k}{d^2}$$

If we know that an object 4000 miles from the center of the earth is attracted to the earth with a force of 90 pounds, we can compute the constant of variation k:

$$f = \frac{k}{d^2}$$

$$90 = \frac{k}{4000^2}$$

$$k = 90(4000^2)$$

$$= 1.44 \times 10^9$$

To find the force of attraction when the object is 5000 miles from the center of the earth, we substitute **5000** for d and **1.44×10^9** for k and simplify:

$$f = \frac{k}{d^2}$$

$$f = \frac{1.44 \times 10^9}{5000^2}$$

$$= 57.6$$

The object will be attracted to the earth with a force of 57.6 pounds when it is 5000 miles from the earth's center.

Example 3   The intensity I of light received by an object from a light source varies inversely with the square of the object's distance from the light source. If the intensity from a light source 4 feet from an object is 8 candelas, what is the intensity at a distance of 2 feet?

Solution   The words "intensity varies inversely with the square of the distance d" can be expressed mathematically as

$$I = \frac{k}{d^2}$$

To find k, substitute 8 for I and 4 for d and solve for k:

$$I = \frac{k}{d^2}$$

$$8 = \frac{k}{4^2}$$

$$128 = k$$

To find the intensity when the object is 2 feet from the light source, substitute 2 for d and 128 for k and simplify.

$$I = \frac{k}{d^2}$$

$$I = \frac{128}{2^2}$$

$$= 32$$

The intensity at 2 feet is 32 candelas. ■

There are many occasions when one variable varies with the product of several variables. For example, the area of a triangle varies directly with the product of its base and height:

$$A = \frac{1}{2} bh$$

Such variation is called **joint variation**.

Joint variation.   If one variable varies directly with the product of two or more variables, the relationship is called **joint variation**; if y varies jointly with x and z, then $y = kxz$. The constant k is called the **constant of variation**.

Example 4   The volume V of a cone varies jointly with its height h and the area of its base B. Express this relationship as an equation and as a proportion.

Solution   "V varies jointly with h and B" means that "V varies directly as the product of h and B." Hence,

$$V = khB$$

To write this equation as a proportion, divide both sides of the equation by hB and express k as $k/1$.

$$\frac{V}{hB} = \frac{k}{1}$$

Note that this relationship is often read as "V is directly proportional to the product of h and B." ∎

Many applied problems involve a combination of direct and inverse variation. Such variation is called **combined variation**.

Example 5 The time that it takes to build a highway varies directly with the length of the road, but inversely with the number of workers. If it takes 100 workers 4 weeks to build 2 miles of roadway, how long will it take 80 workers to build 10 miles of roadway?

Solution Let T represent time in weeks, L represent length in miles, and W represent the number of workers. The relationship between these variables is expressed by the equation

$$T = \frac{kL}{W}$$

Substitute $T = 4$, $W = 100$, and $L = 2$ to find the constant of variation:

$$4 = \frac{k(2)}{100}$$

$$400 = 2k \qquad \text{Multiply both sides by 100.}$$

$$200 = k \qquad \text{Divide both sides by 2.}$$

Now, substitute 80 for W, 10 for L, and 200 for k in the equation $T = \frac{kL}{W}$, and simplify:

$$T = \frac{kL}{W}$$

$$T = \frac{200(10)}{80}$$

$$= 25 \qquad \text{Simplify.}$$

It will take 25 weeks for 80 workers to build 10 miles of roadway. ∎

■ EXERCISE 7.6

In Exercises 1–12, solve each proportion for the variable, if possible.

1. $\dfrac{x}{5} = \dfrac{15}{25}$
2. $\dfrac{4}{y} = \dfrac{6}{27}$
3. $\dfrac{r-2}{3} = \dfrac{r}{5}$
4. $\dfrac{2}{c} = \dfrac{c-3}{2}$

5. $\dfrac{y}{4} = \dfrac{4}{y}$
6. $\dfrac{2}{3x} = \dfrac{12x}{36}$
7. $\dfrac{3}{n} = \dfrac{2}{n+1}$
8. $\dfrac{4}{x+3} = \dfrac{3}{5}$

9. $\dfrac{x + 1}{x - 1} = \dfrac{6}{4}$

10. $\dfrac{5}{5z + 3} = \dfrac{2z}{2z^2 + 6}$

11. $\dfrac{9t + 6}{t(t + 3)} = \dfrac{7}{t + 3}$

12. $\dfrac{(x - 7)(x + 2)}{2} = \dfrac{(x + 3)(x + 2)}{4}$

In Exercises 13–22, express each sentence as a formula.

13. A varies directly with the square of p.

14. z varies inversely with the cube root of t.

15. v varies inversely with the fourth root of r.

16. r varies directly with the square root of s.

17. B varies jointly with m and n.

18. C varies jointly with x, y, and z.

19. P varies directly with the square of a, and inversely with the cube of j.

20. M varies inversely with the cube of n, and jointly with x and the square of z.

21. The force of attraction F between two masses m_1 and m_2 varies directly with the product of m_1 and m_2, and inversely with the square of the distance between them.

22. The force of wind on a vertical surface varies jointly with the area of the surface and the square of the velocity of the wind.

In Exercises 23–30, express each formula in words. In each formula, k is the constant of variation.

23. $L = kmn$

24. $P = \dfrac{km}{n}$

25. $E = kab^2$

26. $U = krs^2t$

27. $X = \dfrac{kx^2}{y^2}$

28. $Z = \dfrac{kw}{xy}$

29. $R = \dfrac{kL}{d^2}$

30. $e = \dfrac{kPL}{A}$

31. The area of a circle varies directly with the square of its radius with the constant of variation equal to π. Find the area of a circle with a radius of 6 inches.

32. An object in free fall travels a distance s that is directly proportional to the square of the time t. If an object falls 1024 feet in 8 seconds, how far will it fall in 10 seconds?

33. The distance that a car can travel is directly proportional to the number of gallons of gasoline that it consumes. If a car can go 288 miles on 12 gallons of gasoline, how far can it go on a full tank of 18 gallons?

34. A farmer's harvest in bushels varies directly as the number of acres planted. If 144 bushels can be reaped from 8 acres, how many acres are required to produce 1152 bushels?

35. The length of time that a given number of bushels of corn will last when feeding cattle varies inversely with the number of animals. If x bushels will feed 25 cows for 10 days, how long will the feed last for 10 cows?

36. For a fixed area, the length of a rectangle is inversely proportional to its width. A rectangle has a width of 18 feet and a length of 12 feet. If the length is increased to 16 feet, how wide is the rectangle?

37. Under constant temperature, the volume occupied by a gas is inversely proportional to the pressure applied. If the gas occupies a volume of 20 cubic inches under 6 pounds per square inch of pressure, what is the volume of the gas when it is subjected to 10 pounds per square inch of pressure?

38. Assume that the value of a car varies inversely with its age. If a car is worth $7000 when it is 3 years old, how much will it be worth when it is 7 years old?

39. The frequency of vibration of air in an organ pipe is inversely proportional to the length of the pipe. If a pipe 2 feet long vibrates 256 times per second, how many times per second will a 6-foot pipe vibrate?

40. The area of a rectangle varies jointly with its length and width. If both the length and width are tripled, by what factor is the area multiplied?

41. The volume of a rectangular solid varies jointly with its length, width, and height. If the length is doubled, the width is tripled, and the height is doubled, by what factor is the volume multiplied?

42. When you go shopping, the cost of like items varies jointly with the number of units purchased and the price per unit. If 15 items cost $105, what is the cost of 35 items?

43. The number of gallons of oil that can be stored in a cylindrical tank varies jointly with the height of the tank and the square of the radius of its base. The constant of proportionality is **23.5**. Find the number of gallons of oil that can be stored in a cylindrical tank with a height of 20 feet and a circular base with a diameter of 15 feet.

44. The quantity L varies jointly with x and y and inversely with z. The value of L is 30 when $x = 15$, $y = 5$, and $z = 10$. Evaluate the constant of variation and express this relationship with an equation.

45. The voltage measured across an electrical component called a resistor is directly proportional to the current flowing through the resistor. The voltage is measured in volts, the current in amperes, and the constant of proportionality is called the **resistance**, measured in ohms. If 6 volts is measured across a resistor through which a current of 2 amperes flows, what is the resistance?

46. The power lost (usually in the form of heat) in a resistor is directly proportional to the square of the current passing through it. The constant of proportionality is the resistance, measured in ohms. What power is lost in a 5-ohm resistor carrying a 3-ampere current?

47. The deflection of a beam is inversely proportional to its width and the cube of its depth. If the deflection of a 4-inch by 4-inch beam is 1.1 inches, calculate the deflection of a 2-inch by 8-inch beam in each of its two orientations (on its side; on its edge). See Illustration 1.

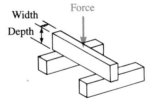

Width
Depth
Force

Illustration 1

48. The pressure of a certain amount of gas is directly proportional to the temperature (measured in degrees Kelvin), and inversely proportional to the volume. A sample of gas at a pressure of 1 atmosphere occupies a volume of 1 cubic meter at a temperature of 273 Kelvin (about 0° Celsius). When heated, the gas expands to twice its volume, but the pressure remains constant. To what temperature was it heated?

7.7 GRAPHS OF LINEAR INEQUALITIES IN TWO VARIABLES

The **graph of an inequality** in x and y is the graph of all ordered pairs (x, y) that satisfy the inequality. In this section we consider graphs of **linear inequalities**— inequalities that can be expressed in a form such as $Ax + By < C$, $Ax + By > C$, $Ax + By \leq C$, or $Ax + By \geq C$.

To graph the inequality $y > 3x + 2$, for example, we first note that exactly one of the following statements is true:

$$y < 3x + 2, \qquad y = 3x + 2, \qquad \text{or} \qquad y > 3x + 2$$

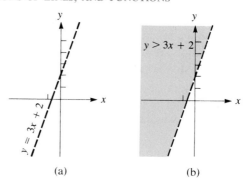

Figure 7-25

Because the equation $y = 3x + 2$ determines a linear function, its graph is a straight line. See Figure 7-25a. The graph of each inequality is a half-plane, one on each side of that line. Thus, we can think of the graph of $y = 3x + 2$ as a boundary separating the two half-planes. The graph of $y = 3x + 2$ is drawn with a broken line to indicate that it is not part of the desired graph for $y > 3x + 2$. To determine which half-plane is the graph of $y > 3x + 2$, we can substitute the coordinates of the origin, $(0, 0)$, into the inequality and simplify:

$$y > 3x + 2$$
$$0 > 3(0) + 2$$
$$0 \not> 2 \qquad \text{Read as "0 is not greater than 2."}$$

Because the coordinates of the origin do not satisfy the inequality, the origin is not in the half-plane that is the graph of $y > 3x + 2$. Thus, the half-plane on the other side of the broken line is the graph. The graph of the inequality $y > 3x + 2$ is shown in Figure 7-25b.

Example 1   Graph the inequality $2x - 3y \leq 6$.

Solution   This inequality is the combination of the inequality $2x - 3y < 6$ and the equation $2x - 3y = 6$. Begin by graphing the linear equation $2x - 3y = 6$ to establish the boundary that separates the two half-planes. However, this time draw a solid line because equality is permitted. See Figure 7-26a. To decide which half-plane represents $2x - 3y < 6$, check to see whether the coordinates of the origin satisfy the inequality:

$$2x - 3y < 6$$
$$2(0) - 3(0) < 6$$
$$0 < 6$$

In this case the origin is in the half-plane that is the graph of $2x - 3y < 6$. The complete graph is shown in Figure 7-26b.

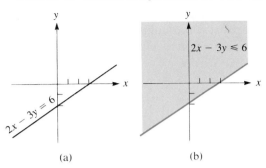

Figure 7-26

Example 2   Graph the inequality $y < 2x$.

Solution   Begin by graphing the equation $y = 2x$. Because it is not part of the graph, use a broken line as in Figure 7-27a. To decide which half-plane represents $y < 2x$, check to see whether the coordinates of some fixed point satisfies the inequality. This time, however, the origin cannot be used as a test point because the boundary line passes through the origin. Choose some other point, say (3, 1), for a test point:

$$y < 2x$$
$$1 < 2(3)$$
$$1 < 6$$

Because $1 < 6$, the point (3, 1) is in the graph. Thus, the graph is as shown in Figure 7-27b.

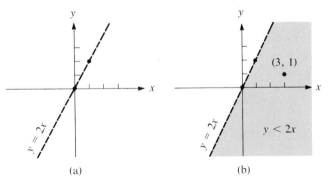

Figure 7-27

Example 3   Graph the inequality $2 < x \le 5$.

Solution   The double inequality $2 < x \le 5$ is equivalent to the inequalities

$$2 < x \qquad \text{and} \qquad x \le 5$$

Thus, the graph of $2 < x \le 5$ must contain all points in the plane that satisfy the inequalities $2 < x$ and $x \le 5$ simultaneously. These points are in the shaded region of Figure 7-28.

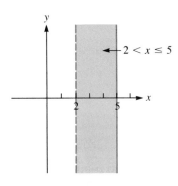

Figure 7-28

EXERCISE 7.7

In Exercises 1–18, graph each inequality.

1. $y > x + 1$ **2.** $y < 2x - 1$ **3.** $y \ge x$ **4.** $y \le 2x$

5. $2x + y \le 6$ **6.** $x - 2y \ge 4$ **7.** $3x \ge -y + 3$ **8.** $2x \le -3y - 12$

9. $y \ge 1 - \dfrac{3}{2}x$ **10.** $y < \dfrac{1}{3}x - 1$ **11.** $x < 4$ **12.** $y \ge -2$

13. $-2 \le x < 0$ **14.** $0 < y \le 5$

15. $y < -2$ or $y > 3$ **16.** $-x \le 1$ or $x \ge 2$

17. $-3 < y \le -1$ **18.** $-5 \ge x > -8$

In Exercises 19–28, find the equation of the boundary line or lines. Then give the inequality whose graph is shown.

19.

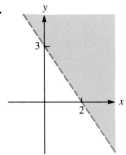

20.

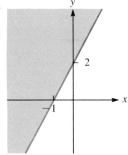

21.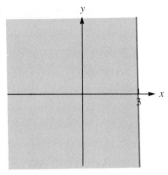

22.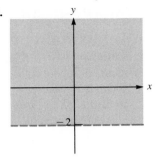

23.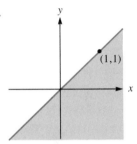

24.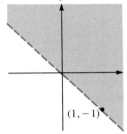

25.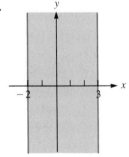

26.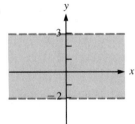

27.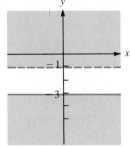

28.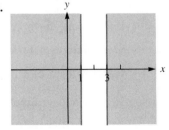

CHAPTER SUMMARY

Key Words

abscissa (7.1)
Cartesian coordinate system (7.1)
composite functions (7.4)
constant of variation (7.6)
dependent variable (7.4)
direct variation (7.6)
domain (7.4)
extremes of a proportion (7.6)
function (7.4)
graph of an equation (7.1)
graph of an inequality (7.7)
image (7.4)
independent variable (7.4)
inverse relation (7.5)
inverse variation (7.6)
joint variation (7.6)
linear equation in x and y (7.2)

linear inequalities (7.7)
means of a proportion (7.6)
ordered pairs (7.1)
ordinate (7.1)
origin (7.1)
proportion (7.6)
ratio (7.6)
rectangular coordinate system (7.1)
relation (7.4)
slope of a nonvertical line (7.2)
x-axis (7.1)
x-coordinate (7.1)
x-intercept (7.1)
y-axis (7.1)
y-coordinate (7.1)
y-intercept (7.1)

Key Ideas

(7.1) The **distance formula**: $d(PQ) = \sqrt{(x_2 - x_1)^2 + (y_2 - y_1)^2}$

The **midpoint formula**: If $P(x_1, y_1)$ and $Q(x_2, y_2)$ are two points on a line, then the midpoint of the line segment PQ is point M, where the coordinates of M are

$$\left(\frac{x_1 + x_2}{2}, \frac{y_1 + y_2}{2}\right)$$

(7.2) The **slope of a nonvertical line**: $m = \dfrac{\Delta y}{\Delta x} = \dfrac{y_2 - y_1}{x_2 - x_1}$

Nonvertical parallel lines have the same slope, and lines with the same slope are parallel.

If two nonvertical lines are perpendicular, their slopes are negative reciprocals.

If the slopes of two lines are negative reciprocals, the lines are perpendicular.

(7.3) **Point–slope form** of a linear equation: $y - y_1 = m(x - x_1)$

Slope–intercept form of a linear equation: $y = mx + b$

General form of a linear equation: $Ax + By = C$

(7.4) If $y = f(x)$ is a function, then $f(a)$ represents the value of y when $x = a$.

$$(f \circ g)(x) = f(g(x)) \qquad (f + g)(x) = f(x) + g(x) \qquad (f - g)(x) = f(x) - g(x)$$

$$(f \cdot g)(x) = f(x)g(x) \qquad (f/g)(x) = \frac{f(x)}{g(x)} \qquad [g(x) \neq 0]$$

(7.5)   A **linear function** is defined by an equation that can be written in the form $y = f(x) = mx + b$, where m and b are constants.

If R is a relation, then R^{-1} is the relation formed by interchanging the components of each ordered pair of R.

(7.6)   In a proportion, the product of the extremes is equal to the product of the means.

If $y = kx$ and k is a constant, then x and y **vary directly**.

If $y = k/x$ and k is a constant, then x and y **vary inversely**.

If $y = kxz$ and k is a constant, then y **varies jointly** with x and z.

The expression $y = kx/z$ (k is a constant) indicates **combined variation**, with y and x varying directly and y and z varying inversely.

(7.7)   To graph an inequality such as $y > ax + b$, first graph the linear function defined by $y = ax + b$. The line graph will determine two half-planes, one on each side of the line. Then determine which half-plane represents the graph of $y > ax + b$.

REVIEW EXERCISES

In Review Exercises 1–4, graph each equation.

1. $x + y = 4$ 2. $2x - y = 8$ 3. $y = 3x + 4$ 4. $x = 4 - 2y$

In Review Exercises 5–8, find the distance between each pair of points.

5. $P(2, 6)$; $Q(5, 10)$ 6. $P(-2, 5)$; $Q(3, 17)$
7. $P(-2, -5)$; $Q(6, 8)$ 8. $P(4, -5)$; $Q(-6, 12)$

In Review Exercises 9–12, find the midpoint of each line segment PQ.

9. $P(2, 6)$; $Q(2, 12)$ 10. $P(8, -2)$; $Q(-6, -2)$
11. $P(2, -6)$; $Q(5, 10)$ 12. $P(-3, -7)$; $Q(10, -6)$

In Review Exercises 13–16, find the slope of the line passing through points P and Q.

13. $P(2, 5)$; $Q(5, 8)$ 14. $P(-3, -2)$; $Q(6, 12)$
15. $P(-3, 4)$; $Q(-5, -6)$ 16. $P(5, -4)$; $Q(-6, -9)$

17.   Use the point–slope form to write the equation of the line with slope of $-\frac{3}{2}$ that passes through $P(-2, 5)$.
18.   Use the slope–intercept form to write the equation of the line with slope of $-\frac{3}{2}$ that passes through $P(-2, 5)$.

In Review Exercises 19–22, write the equation of the line with the given properties. **Write each answer in general form.**

19.   Slope of 3 and passing through $P(-8, 5)$ 20.   Passing through the points $(-2, 4)$ and $(6, -9)$
21.   Passing through the point $(-3, -5)$ and parallel to the graph of $3x - 2y = 7$
22.   Passing through the point $(-3, -5)$ and perpendicular to the graph of $3x - 2y = 7$

In Review Exercises 23–28, find the domain and range of each relation. Assume that all ordered pairs (x, y) consist of real numbers. If a relation is a function, so indicate.

23. $y = 4x - 1$ **24.** $x = 3y - 10$ **25.** $y = 3x^2 + 1$ **26.** $y = \dfrac{4}{2 - x}$

27. $x = \dfrac{y + 3}{2}$ **28.** $y^2 = 4x$

In Review Exercises 29–34, let $f(x) = 3x + 2$ and $g(x) = x^2 - 4$. Find each indicated value.

29. $f(-3)$ **30.** $g(8)$ **31.** $(g \circ f)(-2)$ **32.** $(f \circ g)(3)$

33. $(f \circ g)(x)$ **34.** $(g \circ f)(x)$

In Review Exercises 35–38, let $f(x) = 2x + 1$ and $g(x) = 3x - 2$. Find each function and determine its domain.

35. $f + g$ **36.** $f - g$ **37.** $f \cdot g$ **38.** f/g

In Review Exercises 39–40, find the inverse relation of each function defined by ordered pairs (x, y). If the inverse relation is a function, so indicate.

39. $y = 7x + 2$ **40.** $5 = \dfrac{3y + 4}{x - 2}$

In Review Exercises 41–42, solve each proportion.

41. $\dfrac{x + 1}{8} = \dfrac{4x - 2}{24}$ **42.** $\dfrac{1}{x + 6} = \dfrac{x + 10}{12}$

43. Assume that x varies directly with y. If $x = 12$ when $y = 2$, what is the value of x when $y = 12$?

44. Assume that x varies inversely with y. If $x = 24$ when $y = 3$, what is the value of y when $x = 12$?

45. Assume that x varies jointly with y and z. What is the constant of variation if $x = 24$ when $y = 3$ and $z = 4$?

46. Assume that x varies directly with t and inversely with y. Find the constant of variation if $x = 2$ when $t = 8$ and $y = 64$.

In Review Exercises 47–50, graph each inequality.

47. $2x + 3y > 6$ **48.** $y < 4 - x$

49. $-2 < x < 4$ **50.** $y \le -2$ or $y > 1$

CHAPTER SEVEN TEST

1. Graph the equation $2x - 5y = 10$.

2. Find the x- and y-intercepts of the graph of $y = \dfrac{x - 3}{5}$.

In Problems 3–5, consider points $P(-5, -6)$ and $Q(5, -2)$.

3. Find the distance between P and Q.   **4.** Find the midpoint of line segment PQ.

5. Find the slope of the line that passes through points P and Q.

6. Find the slope of the straight line graph of the equation $x = \dfrac{3y - 8}{2}$.

7. Write the equation of the line with slope of $\frac{2}{3}$ that passes through the point $P(4, -5)$. Give the answer in slope–intercept form.

8. Write the equation of the line that passes through $P(-2, 6)$ and $Q(-4, -10)$. Give the answer in general form.

9. Find the slope and the y-intercept of the graph of $-2(x - 3) = 3(2y + 5)$.

10. Determine whether the graphs of $4x - y = 12$ and $y = \frac{1}{4}x + 3$ are parallel, perpendicular, or neither.

11. Determine whether the graphs of $y = -\frac{2}{3}x + 4$ and $2y = 3x - 3$ are parallel, perpendicular, or neither.

12. Write the equation of the line that passes through the origin and is parallel to the graph of $y = \frac{3}{2}x - 7$.

13. Write the equation of the line that passes through the point $P(-3, 6)$ and is perpendicular to the graph of $y = -\frac{2}{3}x - 7$.

14. Does $|y| = x$ determine y to be a function of x? Explain.

15. Find the domain and the range of the function determined by $y = \dfrac{9}{x - 2}$.

In Problems 16–22, let $f(x) = 3x + 1$ and $g(x) = x^2 - 2$. Find each quantity.

16. $f(3)$ 17. $g(0)$ 18. $f(g(-2))$ 19. $g(f(x))$

20. $(f + g)(x)$ 21. $(f \cdot g)(x)$ 22. $(f/g)(x)$

23. Does the equation $y(x + 3) + 4 = x(y - 2)$ determine a linear function?

24. Find the inverse of the linear function $y = -\frac{1}{2}x + 5$. Give the answer using $f^{-1}(x)$ notation.

25. Solve the proportion $\dfrac{3}{x - 2} = \dfrac{x + 3}{2x}$.

26. The force of attraction F between two masses m_1 and m_2 varies directly with the product of m_1 and m_2 and inversely with the square of the distance between them. If k is the constant of variation express this relationship with an equation.

27. Assume that x varies directly with y. If $x = 30$ when $y = 4$, find x when $y = 9$.

28. Assume that V varies inversely with t. If $V = 55$ when $t = 20$, find t when $V = 75$.

29. Graph the inequality $3x + 2y \geq 6$.

30. Graph the inequality $-2 \leq y < 5$.

ECONOMIST Economists study the way a society uses resources such as land, labor, raw materials, and machinery to provide goods and services. They analyze the results of their research to determine the costs and benefits of making, distributing, and using resources in a particular way. Some economists are theoreticians who use mathematical models to explain the causes of recession and inflation. Most economists, however, are concerned with practical applications of economic policy in a particular area.

Qualifications Economists must thoroughly understand economic theory, mathematical methods of economic analysis, and basic statistical procedures. Training in computer science is highly recommended.

Job outlook Employment of economists is expected to grow faster than the average for all occupations through the mid-1990s. Opportunities should be best for economists in business and industry, research organizations, and consulting firms.

Example application An electronics firm manufactures tape recorders, receiving $120 for each unit they make. If x represents the number of recorders produced, then the income received is determined by the *revenue function*, given by the linear equation

$$R(x) = 120x$$

The manufacturer has determined that the *fixed costs* for advertising, insurance, utilities, and so on, are $12,000 per month, and the *variable cost* for materials is $57.50 for each machine produced. Thus, the *cost function* is given by the linear equation

$$C(x) = variable\ cost + fixed\ costs$$
$$= 57.50x + 12,000$$

The company's profit is the amount by which their revenue exceeds their costs. It is determined by the *profit function*, given by the equation

$$Profit = revenue - costs$$
$$P(x) = (R - C)(x)$$
$$= R(x) - C(x)$$
$$= 120x - (57.50x + 12,000)$$
$$= 62.50x - 12,000$$

If $P(x) > 0$, the company is making money. If $P(x) < 0$, they are operating at a loss. How many recorders must the company manufacture to break even?

Solution Graph the profit function $y = P(x) = 62.50x - 12,000$. The break-even point is that value of x that gives a profit of zero. It is the x-intercept of the graph of the profit function. See Illustration 1. To find it, set $P(x)$ equal to 0 and solve for x.

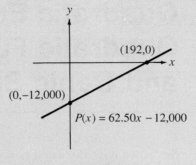

Illustration 1

$$P(x) = 0$$
$$62.50x - 12,000 = 0$$
$$62.50x = 12,000 \qquad \text{Add 12,000 to both sides.}$$
$$x = 192 \qquad \text{Divide both sides by 62.50.}$$

The company must manufacture and sell 192 tape recorders each month to break even.

Exercises

1. Find the revenue and the cost of manufacturing 192 units, and verify that the revenue and the cost are equal.
2. Determine the company's profit if they manufacture 150 units each month.
3. Determine the company's profit if they manufacture 400 units each month.
4. How many units must be manufactured each month to produce a total profit of $47,375?

(Answers: **1.** $23,040 revenue and cost   **2.** $2,625 loss
3. $13,000 profit   **4.** 950 units)

Quadratic Equations, Quadratic Functions, and Conic Sections

So far we have discussed how to solve linear equations and certain quadratic equations in which the quadratic expression was factorable. In this chapter we will discuss more general methods for solving quadratic equations and will consider the graphs of some nonlinear functions and relations.

8.1 COMPLETING THE SQUARE AND THE QUADRATIC FORMULA

Recall that a quadratic equation is an equation of the form $ax^2 + bx + c = 0$, where a, b, and c are real numbers and $a \neq 0$. In Chapter 4 we discussed how to solve quadratic equations by factoring. For example, to solve the equation $6x^2 - 7x - 3 = 0$, we first factor the quadratic trinomial and then apply the zero-factor theorem.

$$6x^2 - 7x - 3 = 0$$
$$(2x - 3)(3x + 1) = 0$$

$$2x - 3 = 0 \qquad \text{or} \qquad 3x + 1 = 0$$

$$x = \frac{3}{2} \qquad\qquad x = -\frac{1}{3}$$

A check will show that both solutions satisfy the original equation.

If the quadratic expression in a quadratic equation factors easily, the factoring method for solving quadratic equations is very convenient. Unfortunately, quadratic expressions do not always factor easily. For example, it would be difficult to factor the left-hand side of the equation $2x^2 + 4x + 1 = 0$ because it cannot be factored by using integers only.

To develop methods for solving all quadratic equations, we first consider the equation $x^2 = c$. If c is positive, the two real solutions of $x^2 = c$ can be found by adding $-c$ to both sides, factoring the binomial $x^2 - c$, setting each factor equal to 0, and solving for x:

$$x^2 = c$$
$$x^2 - c = 0$$
$$x^2 - (\sqrt{c})^2 = 0$$
$$(x + \sqrt{c})(x - \sqrt{c}) = 0$$

$$x + \sqrt{c} = 0 \qquad \text{or} \qquad x - \sqrt{c} = 0$$

$$x = -\sqrt{c} \qquad\qquad x = \sqrt{c}$$

Thus, the solutions of the equation $x^2 = c$ are $x = \sqrt{c}$ and $x = -\sqrt{c}$. This result is often called the **square root property**.

The Square Root Property.   If $c > 0$, then the equation $x^2 = c$ has two real solutions. They are

$$x = \sqrt{c} \qquad \text{and} \qquad x = -\sqrt{c}$$

Example 1   Solve the equation $x^2 - 12 = 0$.

Solution   Write the equation as $x^2 = 12$ and use the square root property:

$$x^2 - 12 = 0$$
$$x^2 = 12 \qquad\qquad \text{Add 12 to both sides.}$$
$$x = \sqrt{12} \quad \text{or} \quad x = -\sqrt{12} \qquad \text{Use the square root property.}$$
$$x = 2\sqrt{3} \qquad\qquad x = -2\sqrt{3} \qquad \text{Simplify each radical.}$$

Verify that both solutions check.   ∎

Example 2   Solve the equation $(x - 3)^2 = 16$.

Solution   Use the square root property:

$$(x - 3)^2 = 16$$
$$x - 3 = \sqrt{16} \quad \text{or} \quad x - 3 = -\sqrt{16}$$
$$x - 3 = 4 \qquad\qquad x - 3 = -4$$
$$x = 3 + 4 \qquad\qquad x = 3 - 4$$
$$x = 7 \qquad\qquad\qquad x = -1$$

Verify that both solutions check.   ∎

Completing the Square

All quadratic equations can be solved by a method called **completing the square**. This method is based on the special products

$$x^2 + 2yx + y^2 = (x + y)^2$$

and

$$x^2 - 2yx + y^2 = (x - y)^2$$

The trinomials $x^2 + 2yx + y^2$ and $x^2 - 2yx + y^2$ are both perfect trinomial squares because both factor as the square of a binomial. In each case, the coefficient of the first term is 1 and, if we take one-half of the coefficient of x in the middle term and square it, we obtain the third term:

$$\left[\frac{1}{2}(2y)\right]^2 = y^2$$

$$\left[\frac{1}{2}(-2y)\right]^2 = (-y)^2 = y^2$$

Thus, to make the expression $x^2 + 10x$ a perfect square, we must take one-half of 10 to get 5, square 5 to get 25, and add 25 to $x^2 + 10x$:

$$x^2 + 10x + \left[\frac{1}{2}(10)\right]^2 = x^2 + 10x + (5)^2$$

$$= x^2 + 10x + 25$$

Note that $x^2 + 10x + 25 = (x + 5)^2$.

To make the expression $x^2 - 6x$ a perfect square, we must take one-half of -6 to get -3, square -3 to get 9, and add 9 to $x^2 - 6x$:

$$x^2 - 6x + \left[\frac{1}{2}(-6)\right]^2 = x^2 - 6x + (-3)^2$$

$$= x^2 - 6x + 9$$

Note that $x^2 - 6x + 9 = (x - 3)^2$.

Example 3   Use the method of completing the square to solve the equation $x^2 + 8x + 7 = 0$.

Solution   Note that the coefficient of x^2 is 1. This condition is necessary before the square can be completed. Begin by adding -7 to both sides to get the constant on the right-hand side of the equals sign:

$$x^2 + 8x + 7 = 0$$
$$x^2 + 8x = -7$$

To complete the square, you must add a number k to both sides of the equation so that $x^2 + 8x + k$ is a perfect trinomial square. To determine k, take one-half of the coefficient of x (one-half of 8 is 4), square it (4^2 is 16), and add 16 to both sides of the equation:

$$x^2 + 8x + 16 = 16 - 7$$

The left-hand side of the previous equation is a perfect trinomial square because $x^2 + 8x + 16 = (x + 4)^2$. Now factor the left-hand side and combine terms on the right-hand side to obtain

$$(x + 4)^2 = 9$$

Solve this equation using the square root property:

$$(x + 4)^2 = 9$$

$x + 4 = \sqrt{9}$ or $x + 4 = -\sqrt{9}$ Use the square root property.

$x + 4 = 3$ $x + 4 = -3$

$x = -1$ $x = -7$

Verify that both solutions check. ■

Example 4   Solve $6x^2 + 5x - 6 = 0$.

Solution   Begin by dividing both sides of the equation by 6 to make the coefficient of x^2 equal to 1. Then add 1 to both sides.

$$6x^2 + 5x - 6 = 0$$

$$x^2 + \frac{5}{6}x - 1 = 0$$

$$x^2 + \frac{5}{6}x = 1$$

To complete the square on x, take one-half of $\frac{5}{6}$ to get $\frac{5}{12}$, square $\frac{5}{12}$ to get $\frac{25}{144}$, and add $\frac{25}{144}$ to both sides of the equation:

$$x^2 + \frac{5}{6}x + \frac{25}{144} = \frac{25}{144} + 1$$

$$\left(x + \frac{5}{12}\right)^2 = \frac{169}{144} \qquad \text{Factor and add the fractions.}$$

Apply the square root property:

$x + \dfrac{5}{12} = \sqrt{\dfrac{169}{144}}$ or $x + \dfrac{5}{12} = -\sqrt{\dfrac{169}{144}}$

$x + \dfrac{5}{12} = \dfrac{13}{12}$ $x + \dfrac{5}{12} = -\dfrac{13}{12}$

$x = -\dfrac{5}{12} + \dfrac{13}{12}$ $x = -\dfrac{5}{12} - \dfrac{13}{12}$

$x = \dfrac{8}{12}$ $x = -\dfrac{18}{12}$

$x = \dfrac{2}{3}$ $x = -\dfrac{3}{2}$

Verify that both solutions check. ■

    The solution used in Example 4 suggests a list of steps to follow when using the method of completing the square to solve quadratic equations of the form $ax^2 + bx + c = 0$.

---

1. Make sure that the coefficient of x^2 is 1. If it is not, make it 1 by dividing both sides of the equation by the coefficient of x^2.
2. If necessary, add a number to both sides of the equation to get the constant terms on the right-hand side of the equals sign.
3. Complete the square:
   a. Identify the coefficient of x.
   b. Find one-half of the coefficient of x and square it.
   c. Add that square to both sides of the equation.
4. Factor the trinomial square and combine terms.
5. Solve the resulting equation by applying the square root property.

Example 5   Solve $2x^2 + 4x + 1 = 0$.

Solution   Use the method of completing the square:

$$2x^2 + 4x + 1 = 0$$

$$x^2 + 2x + \frac{1}{2} = 0 \qquad \text{Divide both sides by 2 to make the coefficient of } x^2 \text{ equal to 1.}$$

$$x^2 + 2x = -\frac{1}{2} \qquad \text{Add } -\frac{1}{2} \text{ to both sides.}$$

$$x^2 + 2x + 1 = 1 - \frac{1}{2} \qquad \text{Square half the coefficient of } x, \text{ and add it to both sides.}$$

$$(x + 1)^2 = \frac{1}{2} \qquad \text{Factor and combine terms.}$$

$$x + 1 = \sqrt{\frac{1}{2}} \qquad \text{or} \qquad x + 1 = -\sqrt{\frac{1}{2}} \qquad \text{Use the square root property.}$$

$$x + 1 = \frac{\sqrt{2}}{2} \qquad\qquad x + 1 = -\frac{\sqrt{2}}{2}$$

$$x = -1 + \frac{\sqrt{2}}{2} \qquad\qquad x = -1 - \frac{\sqrt{2}}{2}$$

$$x = \frac{-2 + \sqrt{2}}{2} \qquad\qquad x = \frac{-2 - \sqrt{2}}{2}$$

Both values check.   ∎

The Quadratic Formula

All quadratic equations can be solved by the method of completing the square. However, the method is sometimes tedious. Fortunately, there is a formula, called the *quadratic formula*, that gives the solutions of any quadratic equation with less effort. To develop the quadratic formula, we will use the method of completing the square.

Recall that a quadratic equation is any equation that can be written in the form $ax^2 + bx + c = 0$, with $a \neq 0$. If we solve this general form of a quadratic equation by completing the square, we are solving every possible quadratic equation at one time. The solution of the general quadratic equation results in the quadratic formula.

To solve the equation $ax^2 + bx + c = 0$, with $a \neq 0$, we first divide both sides by a and then proceed as follows:

$$ax^2 + bx + c = 0$$

$$x^2 + \frac{bx}{a} + \frac{c}{a} = \frac{0}{a}$$

$$x^2 + \frac{bx}{a} = -\frac{c}{a} \qquad \text{Simplify } \frac{0}{a} \text{ and add } -\frac{c}{a} \text{ to both sides.}$$

$$x^2 + \frac{b}{a}x + \left(\frac{b}{2a}\right)^2 = \left(\frac{b}{2a}\right)^2 - \frac{c}{a} \qquad \text{Complete the square on } x.$$

$$x^2 + \frac{b}{a}x + \frac{b^2}{4a^2} = \frac{b^2}{4a^2} - \frac{4ac}{4aa} \qquad \begin{array}{l}\text{Remove parentheses and get a common}\\\text{denominator on the right-hand side.}\end{array}$$

1. $\quad \left(x + \frac{b}{2a}\right)^2 = \frac{b^2 - 4ac}{4a^2} \qquad \begin{array}{l}\text{Factor the left-hand side and add the}\\\text{fractions on the right-hand side.}\end{array}$

We can solve Equation 1 by using the square root property:

$$x + \frac{b}{2a} = \sqrt{\frac{b^2 - 4ac}{4a^2}} \qquad \text{or} \qquad x + \frac{b}{2a} = -\sqrt{\frac{b^2 - 4ac}{4a^2}}$$

$$x + \frac{b}{2a} = \frac{\sqrt{b^2 - 4ac}}{2a} \qquad\qquad x + \frac{b}{2a} = -\frac{\sqrt{b^2 - 4ac}}{2a}$$

$$x = -\frac{b}{2a} + \frac{\sqrt{b^2 - 4ac}}{2a} \qquad\qquad x = -\frac{b}{2a} - \frac{\sqrt{b^2 - 4ac}}{2a}$$

$$x = \frac{-b + \sqrt{b^2 - 4ac}}{2a} \qquad\qquad x = \frac{-b - \sqrt{b^2 - 4ac}}{2a}$$

These two values of x are the solutions to the equation $ax^2 + bx + c = 0$, with $a \neq 0$. They are usually written as a single expression called the **quadratic formula**. Read the symbol $\pm$ as "plus or minus."

> **The Quadratic Formula.** If $a \neq 0$, then the solutions of $ax^2 + bx + c = 0$ are given by the formula
>
> $$x = \frac{-b \pm \sqrt{b^2 - 4ac}}{2a}$$

Example 6   Solve the equation $2x^2 - 3x - 5 = 0$.

Solution   In this equation $a = 2$, $b = -3$, and $c = -5$. Substitute these values into the quadratic formula, and simplify:

$$x = \frac{-b \pm \sqrt{b^2 - 4ac}}{2a}$$

$$= \frac{-(-3) \pm \sqrt{(-3)^2 - 4(2)(-5)}}{2(2)}$$

$$= \frac{3 \pm \sqrt{9 + 40}}{4}$$

$$= \frac{3 \pm \sqrt{49}}{4}$$

$$= \frac{3 \pm 7}{4}$$

$x = \dfrac{3 + 7}{4}$	or	$x = \dfrac{3 - 7}{4}$
$= \dfrac{10}{4}$		$= \dfrac{-4}{4}$
$= \dfrac{5}{2}$		$= -1$

Verify that both solutions check.   ■

Example 7   Solve the equation $2x^2 + 4x + 1 = 0$.

Solution   In this equation $a = 2$, $b = 4$ and $c = 1$. Substitute these values, and simplify.

$$x = \frac{-b \pm \sqrt{b^2 - 4ac}}{2a}$$

$$= \frac{-4 \pm \sqrt{4^2 - 4(2)(1)}}{2(2)}$$

$$= \frac{-4 \pm \sqrt{16 - 8}}{4}$$

$$= \frac{-4 \pm \sqrt{8}}{4}$$

$$= \frac{-4 \pm 2\sqrt{2}}{4}$$

$$= \frac{-2 \pm \sqrt{2}}{2}$$

$x = \dfrac{-2 + \sqrt{2}}{2}$	or	$x = \dfrac{-2 - \sqrt{2}}{2}$

The solutions in this example are irrational numbers and the check is difficult. Try it anyhow. ∎

Example 8   The length of a rectangle is 12 centimeters more than its width, and the area of the rectangle is 253 square centimeters. Find the dimensions of the rectangle.

Solution   Let w represent the width of the rectangle. Then, $w + 12$ represents its length. Because the area of a rectangle is found by multiplying its length and width, and because that area is given to be 253 square centimeters, you can form the equation

$$w(w + 12) = 253$$

Solve this equation for w as follows:

$$w(w + 12) = 253$$
$$w^2 + 12w = 253 \qquad \text{Remove parentheses.}$$
$$w^2 + 12w - 253 = 0 \qquad \text{Add } -253 \text{ to both sides.}$$

Solution by factoring

$$(w - 11)(w + 23) = 0$$
$$w - 11 = 0 \quad \text{or} \quad w + 23 = 0$$
$$w = 11 \quad \mid \quad w = -23$$

Solution by formula

$$w = \frac{-12 \pm \sqrt{12^2 - 4(1)(-253)}}{2(1)}$$
$$= \frac{-12 \pm \sqrt{144 + 1012}}{2}$$
$$= \frac{-12 \pm \sqrt{1156}}{2}$$
$$= \frac{-12 \pm 34}{2}$$
$$w = 11 \quad \text{or} \quad w = -23$$

Because a rectangle cannot have a side with a negative width, the solution -23 must be discarded. Thus, the dimensions of the rectangle are 11 centimeters by $(11 + 12)$ centimeters, or 11 centimeters by 23 centimeters.

Note that 23 is 12 more than 11, and that the area of a rectangle 11 centimeters by 23 centimeters is 253 square centimeters. ∎

◼ EXERCISE 8.1

In Exercises 1–12, solve each quadratic equation by factoring.

1. $6x^2 + 12x = 0$
2. $5x^2 + 11x = 0$
3. $2y^2 - 50 = 0$
4. $4y^2 - 64 = 0$
5. $r^2 + 6r + 8 = 0$
6. $x^2 + 9x + 20 = 0$
7. $x^2 - 7x + 6 = 0$
8. $t^2 - 5t + 6 = 0$
9. $2z^2 - 5z + 2 = 0$
10. $2x^2 - x - 1 = 0$
11. $6s^2 + 11s - 10 = 0$
12. $3x^2 + 10x - 8 = 0$

In Exercises 13–24, solve each equation by using the square root property.

13. $x^2 = 36$ **14.** $x^2 = 144$ **15.** $z^2 = 5$ **16.** $u^2 = 24$

17. $3x^2 - 16 = 0$ **18.** $5x^2 - 49 = 0$ **19.** $(x + 1)^2 = 1$ **20.** $(x - 1)^2 = 4$

21. $(s - 7)^2 - 9 = 0$ **22.** $(t + 4)^2 = 16$ **23.** $(x + 5)^2 - 3 = 0$ **24.** $(x + 3)^2 - 7 = 0$

In Exercises 25–38, solve each equation by completing the square.

25. $x^2 + 2x - 8 = 0$ **26.** $x^2 + 6x + 5 = 0$

27. $x^2 - 6x + 8 = 0$ **28.** $x^2 + 8x + 15 = 0$

29. $x^2 + 5x + 4 = 0$ **30.** $x^2 - 11x + 30 = 0$

31. $2x^2 - x - 1 = 0$ **32.** $2x^2 - 5x + 2 = 0$ **33.** $6x^2 + 11x + 3 = 0$ **34.** $6x^2 + x - 2 = 0$

35. $8r^2 + 6r = 9$ **36.** $3w^2 - 11w = -10$ **37.** $\dfrac{7x + 1}{5} = -x^2$ **38.** $\dfrac{3x^2}{8} = \dfrac{1}{8} - x$

In Exercises 39–50, solve each equation by using the quadratic formula.

39. $x^2 + 3x + 2 = 0$ **40.** $x^2 - 3x + 2 = 0$ **41.** $x^2 + 12x = -36$ **42.** $y^2 - 18y = -81$

43. $5x^2 + 5x + 1 = 0$ **44.** $4w^2 + 6w + 1 = 0$ **45.** $8u = -4u^2 - 3$ **46.** $4t + 3 = 4t^2$

47. $16y^2 + 8y - 3 = 0$ **48.** $16x^2 + 16x + 3 = 0$

49. $\dfrac{x^2}{2} + \dfrac{5}{2}x = -1$ **50.** $-3x = \dfrac{x^2}{2} + 2$

In Exercises 51–72, solve each word problem.

51. The product of two consecutive even integers is 288. Find the integers.

52. The product of two consecutive odd integers is 143. Find the integers.

53. The sum of the squares of two consecutive positive integers is 85. What are the integers?

54. The sum of the squares of three consecutive positive integers is 77. What are the integers?

55. One side of a rectangle is 4 feet longer than it is wide. The area of the rectangle is 96 square feet. Find the dimensions of the rectangle.

56. One side of a rectangle is 3 times another. The area of the rectangle is 147 square meters. Find its dimensions.

57. The area of a certain square is numerically equal to its perimeter. What is the length of one side of the square?

58. A rectangle is 2 inches longer than it is wide. Numerically, its area exceeds its perimeter by 11. What is the perimeter of the rectangle?

59. The width of a rectangle is 5 feet less than its length, and its area is 50 square feet. Find the dimensions of the rectangle.

60. A rectangle is 7 feet longer than it is wide. Find the perimeter of the rectangle if its area is 60 square feet.

61. The height of a triangle is 5 centimeters longer than three times its base. Find the base of the triangle if its area is 6 square centimeters.

62. The height of a triangle is 4 meters longer than twice its base. Find the height of the triangle if its area is 15 square meters.

63. June drives 150 miles at r miles per hour. She could have gone the same distance in 2 hours less time if she had increased her speed by 20 miles per hour. Find r.

64. Jeff bicycles 160 miles at r miles per hour. The same trip would have taken 2 hours longer if he had decreased his speed by 4 miles per hour. Find r.

65. Movie tickets at a certain theater cost $4, and the average attendance is 300 persons. It is projected that, for every 10¢ increase in ticket price, the average attendance will decrease by 5. At what ticket price will nightly receipts be $1248?

66. A bus company has 3000 passengers daily, paying a 25¢ fare. For each nickel increase in fare, the company estimates that it will lose 80 passengers. What increase in fare will produce $994 in daily revenue?

67. The *Gazette's* profit is $20 per year from each of its 3000 subscribers. The management estimates that the profit per subscriber would increase by 1¢ for each additional subscriber over the current 3000. How many subscribers will bring a total profit of $120,000?

68. A woman deposits $1000 in a savings bank where interest is compounded annually at a rate r. After 1 year she deposits an additional $2000. After 2 years the balance in the account is

$$1000(1 + r)^2 + 2000(1 + r)$$

dollars. If this amount is $3368.10, what is the interest rate r?

69. The frame surrounding a 10-inch by 12-inch photograph has a constant width. The area of the frame equals the area of the photograph. How wide is the frame? (*Hint:* You will need to use the quadratic formula.)

70. Find a quadratic equation with solutions of 4 and 6.

71. Find a quadratic equation with solutions of -4 and 6.

72. Find a third-degree equation with solutions of 2, 3, and -4.

8.2 COMPLEX NUMBERS

So far, all of the work with quadratic expressions has involved real numbers only. The solutions of some quadratic equations are not real numbers. Consider the following example.

Example 1 Solve the quadratic equation $x^2 + x + 1 = 0$.

Solution Because the factoring method does not work conveniently, use the quadratic formula, with $a = 1$, $b = 1$, and $c = 1$

$$x = \frac{-b \pm \sqrt{b^2 - 4ac}}{2a}$$

$$= \frac{-1 \pm \sqrt{1^2 - 4(1)(1)}}{2(1)}$$

$$= \frac{-1 \pm \sqrt{1 - 4}}{2}$$

$$= \frac{-1 \pm \sqrt{-3}}{2}$$

$$x = \frac{-1 + \sqrt{-3}}{2} \quad \text{or} \quad x = \frac{-1 - \sqrt{-3}}{2}$$ ∎

Each solution in Example 1 involves the number $\sqrt{-3}$. This number is not a real number, because the square of no real number equals -3; the square of any real number is never negative. For years, mathematicians believed that numbers like $\sqrt{-3}$, $\sqrt{-1}$, and $\sqrt{-9}$ were nonsense, as illegal as division by zero. Even the great English mathematician Sir Isaac Newton (1642–1727) called them impossible. In the 17th century, these numbers were named *imaginary numbers* by René Descartes (1596–1650). That term is still used today.

Mathematicians no longer think of imaginary numbers as being fictitious or ridiculous. In fact, imaginary numbers have important uses such as describing the behavior of alternating current in electronics.

The imaginary number $\sqrt{-1}$ occurs often enough to warrant a special symbol; the letter i is used to denote $\sqrt{-1}$. Because i represents the square root of -1, it follows that

$$i^2 = -1$$

The powers of the imaginary number i produce an interesting pattern:

$$i = \sqrt{-1} = i$$
$$i^2 = \sqrt{-1}\sqrt{-1} = -1$$
$$i^3 = i^2 \cdot i = -1 \cdot i = -i$$
$$i^4 = i^2 \cdot i^2 = (-1)(-1) = 1$$

$$i^5 = i^4 \cdot i = 1 \cdot i = i$$
$$i^6 = i^4 \cdot i^2 = 1(-1) = -1$$
$$i^7 = i^4 \cdot i^3 = 1(-i) = -i$$
$$i^8 = i^4 \cdot i^4 = (1)(1) = 1$$

The pattern continues: $i, -1, -i, 1, \ldots$.

If we assume that multiplication of imaginary numbers is commutative and associative, then

$$(2i)^2 = 2^2 i^2 = 4(-1) = -4$$

Because $(2i)^2 = -4$, it follows that $2i$ is a square root of -4 and we write

$$\sqrt{-4} = 2i$$

Note that this result could have been obtained by the following process:

$$\sqrt{-4} = \sqrt{4(-1)}$$
$$= \sqrt{4}\sqrt{-1}$$
$$= 2i$$

Similarly, we have

$$\sqrt{-25} = \sqrt{25(-1)} = \sqrt{25}\sqrt{-1} = 5i$$
$$\sqrt{-\frac{1}{9}} = \sqrt{\frac{1}{9}(-1)} = \sqrt{\frac{1}{9}}\sqrt{-1} = \frac{1}{3}i$$

and

$$\sqrt{\frac{-100}{49}} = \sqrt{\frac{100}{49}(-1)} = \frac{\sqrt{100}}{\sqrt{49}}\sqrt{-1} = \frac{10}{7}i$$

The previous examples illustrate the following rule.

If at least one of a and b is a nonnegative real number and if there are no divisions by 0, then

$$\sqrt{ab} = \sqrt{a}\sqrt{b} \quad \text{and} \quad \sqrt{\frac{a}{b}} = \frac{\sqrt{a}}{\sqrt{b}}$$

Imaginary numbers such as $\sqrt{-3}, \sqrt{-1}$, and $\sqrt{-9}$ form a subset of a broader set of numbers called **complex numbers**.

Definition.   A **complex number** is any number that can be written in the form $a + bi$, where a and b are real numbers, and $i = \sqrt{-1}$. The number a is called the **real part** and the number b is called the **imaginary part** of the complex number $a + bi$.

If $b = 0$, the complex number $a + bi$ is a real number. If $b \neq 0$ and $a = 0$, the complex number $0 + bi$ (or just bi) is an imaginary number. Any imaginary number such as $\sqrt{-3}, \sqrt{-1}$, and $\sqrt{-9}$ can be expressed in bi form:

$$\sqrt{-3} = \sqrt{3(-1)} = \sqrt{3}\sqrt{-1} = \sqrt{3}i$$

$$\sqrt{-1} = i$$

$$\sqrt{-9} = \sqrt{9(-1)} = \sqrt{9}\sqrt{-1} = 3i$$

We now discuss some properties of complex numbers.

Definition.   The complex numbers $a + bi$ and $c + di$ are equal if and only if $a = c$ and $b = d$.

Example 2   **a.** $2 + 3i = \sqrt{4} + \frac{6}{2}i$, because $2 = \sqrt{4}$ and $3 = \frac{6}{2}$.

b. $4 - 5i = \frac{12}{3} - \sqrt{25}i$, because $4 = \frac{12}{3}$ and $-5 = -\sqrt{25}$.

c. $x + yi = 4 + 7i$ if and only if $x = 4$ and $y = 7$.   ■

Definition.   Complex numbers are added as if they were binomials:

$$(a + bi) + (c + di) = (a + c) + (b + d)i$$

Example 3   **a.** $(8 + 4i) + (12 + 8i) = 8 + 4i + 12 + 8i$
$$= 20 + 12i$$

b. $(7 - 4i) + (9 + 2i) = 7 - 4i + 9 + 2i$
$$= 16 - 2i$$

c. $(-6 + i) - (3 - 4i) = -6 + i - 3 + 4i$
$$= -9 + 5i$$

d. $(2 - 4i) - (-4 + 3i) = 2 - 4i + 4 - 3i$
$$= 6 - 7i$$ ∎

To multiply a complex number by an imaginary number, we use the distributive property to remove parentheses and then simplify. For example,

$$-5i(4 - 8i) = -5i(4) - (-5i)(8i)$$
$$= -20i + 40i^2$$
$$= -40 - 20i \qquad \text{Remember } i^2 = -1.$$

To multiply two complex numbers, we use the following definition:

> **Definition.**   Complex numbers are multiplied as if they were binomials, with $i^2 = -1$:
>
> $$(a + bi)(c + di) = ac + adi + bci + bdi^2$$
> $$= (ac - bd) + (ad + bc)i$$

Example 4   **a.** $(2 + 3i)(3 - 2i) = 6 - 4i + 9i - 6i^2$
$$= 6 + 5i + 6$$
$$= 12 + 5i$$

b. $(3 + i)(1 + 2i) = 3 + 6i + i + 2i^2$
$$= 3 + 7i - 2$$
$$= 1 + 7i$$

c. $(-4 + 2i)(2 + i) = -8 - 4i + 4i + 2i^2$
$$= -8 - 2$$
$$= -10$$ ∎

The next example shows how to write several complex numbers in $a + bi$ form. When writing answers, it is common practice to accept the form $a - bi$ as a substitute for the form $a + (-b)i$.

Example 5 **a.** $7 = 7 + 0i$

b. $3i = 0 + 3i$

c. $4 - \sqrt{-16} = 4 - \sqrt{-1(16)} = 4 - \sqrt{16}\sqrt{-1} = 4 - 4i$

d. $5 + \sqrt{-11} = 5 + \sqrt{-1(11)} = 5 + \sqrt{11}\sqrt{-1} = 5 + \sqrt{11}\,i$

e. $2i^2 + 4i^3 = 2(-1) + 4(-i) = -2 - 4i$

f. $\dfrac{3}{2i} = \dfrac{3}{2i} \cdot \dfrac{i}{i} = \dfrac{3i}{2i^2} = \dfrac{3i}{2(-1)} = \dfrac{3i}{-2} = 0 - \dfrac{3}{2}i$

g. $-\dfrac{5}{i} = -\dfrac{5}{i} \cdot \dfrac{i^3}{i^3} = -\dfrac{5(-i)}{1} = 5i = 0 + 5i$ ∎

We must rationalize denominators to write complex numbers such as

$$\frac{1}{3 + i}, \quad \frac{3 - i}{2 + i}, \quad \text{and} \quad \frac{5 + i}{5 - i}$$

in $a + bi$ form. To this end, we make the following definition.

Definition.   The complex numbers $a + bi$ and $a - bi$ are called **complex conjugates** of each other.

For example,

$3 + 4i$ and $3 - 4i$ are complex conjugates.

$5 - 7i$ and $5 + 7i$ are complex conjugates.

$8 + 17i$ and $8 - 17i$ are complex conjugates.

Example 6   Find the product of the complex number $3 + i$ and its complex conjugate.

Solution   The complex conjugate of $3 + i$ is $3 - i$. Find the product of these two binomials as follows:

$$
\begin{aligned}
(3 + i)(3 - i) &= 9 - 3i + 3i - i^2 \\
&= 9 - i^2 \\
&= 9 - (-1) \qquad \text{Because } i^2 = -1. \\
&= 10
\end{aligned}
$$
   ∎

In general, the product of the complex number $a + bi$ and its complex conjugate $a - bi$ is the real number $a^2 + b^2$, as the following work shows:

$$
\begin{aligned}
(a + bi)(a - bi) &= a^2 - abi + abi - b^2i^2 \\
&= a^2 - b^2(-1) \\
&= a^2 + b^2
\end{aligned}
$$

Thus, we have

$$(a + bi)(a - bi) = a^2 + b^2$$

Example 7   Write $\dfrac{1}{3 + i}$ in $a + bi$ form.

Solution   Because the product of $3 + i$ and its conjugate is a real number, rationalize the denominator by multiplying both the numerator and the denominator of the fraction by the complex conjugate of the denominator, and simplify.

$$\frac{1}{3 + i} = \frac{1}{3 + i} \cdot \frac{3 - i}{3 - i}$$

$$= \frac{3 - i}{9 - 3i + 3i - i^2}$$

$$= \frac{3 - i}{9 - (-1)}$$

$$= \frac{3 - i}{10}$$

$$= \frac{3}{10} - \frac{1}{10}i$$

Example 8   Write $\dfrac{3 - i}{2 + i}$ in $a + bi$ form.

Solution   Rationalize the denominator by multiplying both the numerator and the denominator of the fraction by the complex conjugate of the denominator, and simplify:

$$\frac{3 - i}{2 + i} = \frac{3 - i}{2 + i} \cdot \frac{2 - i}{2 - i}$$

$$= \frac{6 - 3i - 2i + i^2}{4 - 2i + 2i - i^2}$$

$$= \frac{5 - 5i}{4 - (-1)}$$

$$= \frac{5(1 - i)}{5} \qquad \text{Factor out 5 in the numerator.}$$

$$= 1 - i \qquad \text{Simplify.}$$

Example 9   Divide $5 + i$ by $5 - i$ and express the quotient in $a + bi$ form.

Solution   The quotient obtained when dividing $5 + i$ by $5 - i$ can be expressed as the fraction $\dfrac{5 + i}{5 - i}$. To express this quotient in $a + bi$ form, rationalize the denominator by multiplying both the numerator and the denominator by the complex conjugate of the denominator. Then simplify.

$$\frac{5+i}{5-i} = \frac{5+i}{5-i} \cdot \frac{5+i}{5+i}$$

$$= \frac{25 + 5i + 5i + i^2}{25 + 5i - 5i - i^2}$$

$$= \frac{25 + 10i - 1}{25 - (-1)}$$

$$= \frac{24 + 10i}{26}$$

$$= \frac{2(12 + 5i)}{26} \qquad \text{Factor out 2 in the numerator.}$$

$$= \frac{12 + 5i}{13} \qquad \text{Simplify.}$$

$$= \frac{12}{13} + \frac{5}{13}i \qquad \blacksquare$$

In most cases, the complex numbers that you encounter will not be in $a + bi$ form. To avoid mistakes, always put complex numbers in $a + bi$ form before doing any arithmetic involving the numbers.

Example 10   Write $\dfrac{4 + \sqrt{-16}}{2 + \sqrt{-4}}$ in $a + bi$ form.

Solution

$$\frac{4 + \sqrt{-16}}{2 + \sqrt{-4}} = \frac{4 + 4i}{2 + 2i}$$

$$= \frac{2(2 + 2i)}{2 + 2i} \qquad \begin{array}{l}\text{Factor out 2 in the}\\ \text{numerator, and simplify.}\end{array}$$

$$= 2 + 0i \qquad \blacksquare$$

Definition.   The **absolute value** of the complex number $a + bi$ is $\sqrt{a^2 + b^2}$. In symbols,

$$|a + bi| = \sqrt{a^2 + b^2}$$

Example 11   **a.** $|3 + 4i| = \sqrt{3^2 + 4^2} = \sqrt{9 + 16} = \sqrt{25} = 5$

b. $|5 - 12i| = \sqrt{5^2 + (-12)^2} = \sqrt{25 + 144} = \sqrt{169} = 13$

c. $|1 + i| = \sqrt{1^2 + 1^2} = \sqrt{1 + 1} = \sqrt{2}$

d. $|a + 0i| = \sqrt{a^2 + 0^2} = \sqrt{a^2} = |a|$

Note that the absolute value of any complex number is a nonnegative real number. Note also that the result of part d is consistent with the definition of the absolute value of a real number. ∎

Example 12 If a and b are both negative numbers, is the formula $\sqrt{a}\sqrt{b} = \sqrt{ab}$ still true?

Solution Let $a = -4$ and $b = -1$. Then, compute $\sqrt{a}\sqrt{b}$ and $\sqrt{ab}$ to see if their values are equal:

$$\sqrt{a}\sqrt{b} = \sqrt{-4}\sqrt{-1}$$
$$= 2i \cdot i$$
$$= 2i^2$$
$$= -2$$

On the other hand, you have
$$\sqrt{ab} = \sqrt{(-4)(-1)}$$
$$= \sqrt{4}$$
$$= 2$$

Because their values are different, the formula $\sqrt{ab} = \sqrt{a}\sqrt{b}$ is *not* true if both a and b are negative. ∎

■ EXERCISE 8.2

In Exercises 1–10, solve each quadratic equation. Write all roots in bi or a + bi form.

1. $x^2 + 9 = 0$ **2.** $x^2 + 16 = 0$ **3.** $3x^2 = -16$ **4.** $2x^2 = -25$

5. $x^2 + 2x + 2 = 0$ **6.** $x^2 + 3x + 3 = 0$ **7.** $2x^2 + x + 1 = 0$ **8.** $3x^2 + 2x + 1 = 0$

9. $3x^2 - 4x + 2 = 0$ **10.** $2x^2 - 3x + 2 = 0$

In Exercises 11–18, simplify each expression.

11. i^{21} **12.** i^{19} **13.** i^{27} **14.** i^{22}

15. i^{100} **16.** i^{42} **17.** i^{97} **18.** i^{200}

*In Exercises 19–60, express each number in a + bi form, if necessary, and perform the indicated operations. **Give all answers in a + bi form.***

19. $(3 + 4i) + (5 - 6i)$ **20.** $(5 + 3i) - (6 - 9i)$

21. $(7 - 3i) - (4 + 2i)$ **22.** $(8 + 3i) + (-7 - 2i)$

23. $(8 + \sqrt{-25}) + (7 + \sqrt{-4})$ **24.** $(-7 + \sqrt{-81}) - (-2 - \sqrt{-64})$

25. $(-8 - \sqrt{-3}) - (7 - \sqrt{-27})$ **26.** $(2 + \sqrt{-8}) + (-3 - \sqrt{-2})$

27. $3i(2 - i)$

28. $-4i(3 + 4i)$

29. $(2 + i)(3 - i)$

30. $(4 - i)(2 + i)$

31. $(2 - 4i)(3 + 2i)$

32. $(3 - 2i)(4 - 3i)$

33. $(2 + \sqrt{-2})(3 - \sqrt{-2})$

34. $(5 + \sqrt{-3})(2 - \sqrt{-3})$

35. $(-2 - \sqrt{-16})(1 + \sqrt{-4})$

36. $(-3 - \sqrt{-81})(-2 + \sqrt{-9})$

37. $(2 + \sqrt{-3})(3 - \sqrt{-2})$

38. $(1 + \sqrt{-5})(2 - \sqrt{-3})$

39. $(8 - \sqrt{-5})(-2 - \sqrt{-7})$

40. $(-1 + \sqrt{-6})(2 - \sqrt{-3})$

41. $\dfrac{1}{i}$

42. $\dfrac{1}{i^3}$

43. $\dfrac{4}{5i^3}$

44. $\dfrac{3}{2i}$

45. $\dfrac{3i}{8\sqrt{-9}}$

46. $\dfrac{5i^3}{2\sqrt{-4}}$

47. $\dfrac{-3}{5i^5}$

48. $\dfrac{-4}{6i^7}$

49. $\dfrac{-6}{\sqrt{-32}}$

50. $\dfrac{5}{\sqrt{-125}}$

51. $\dfrac{3}{5 + i}$

52. $\dfrac{-2}{2 - i}$

53. $\dfrac{-12}{7 - \sqrt{-1}}$

54. $\dfrac{4}{3 + \sqrt{-1}}$

55. $\dfrac{5i}{6 + 2i}$

56. $\dfrac{-4i}{2 - 6i}$

57. $\dfrac{3 - 2i}{3 + 2i}$

58. $\dfrac{2 + 3i}{2 - 3i}$

59. $\dfrac{3 + \sqrt{-2}}{2 + \sqrt{-5}}$

60. $\dfrac{2 - \sqrt{-5}}{3 + \sqrt{-7}}$

In Exercises 61–70, find each indicated value.

61. $|6 + 8i|$

62. $|12 + 5i|$

63. $|12 - 5i|$

64. $|3 - 4i|$

65. $|5 + 7i|$

66. $|6 - 5i|$

67. $|4 + \sqrt{-2}|$

68. $|3 + \sqrt{-3}|$

69. $|8 + \sqrt{-5}|$

70. $|7 - \sqrt{-6}|$

71. Show that $1 - 5i$ is a solution of $x^2 - 2x + 26 = 0$.

72. Show that $3 - 2i$ is a solution of $x^2 - 6x + 13 = 0$.

73. Show that i is a solution of $x^4 - 3x^2 - 4 = 0$.

74. Show that $2 + i$ is *not* a solution of $x^2 + x + 1 = 0$.

8.3   MORE ON QUADRATIC EQUATIONS

It is possible to discover what types of numbers will be solutions of a given quadratic equation without actually solving the equation. For example, suppose that the coefficients a, b, and c in the quadratic equation $ax^2 + bx + c = 0$ are real numbers. The two solutions of this equation are given by the quadratic formula

$$x = \frac{-b \pm \sqrt{b^2 - 4ac}}{2a} \qquad (a \neq 0)$$

The expression under the radical, $b^2 - 4ac$, is called the **discriminant**. If $b^2 - 4ac \geq 0$, then the solutions of the equation are real numbers. On the other hand, if $b^2 - 4ac < 0$, then the solutions are nonreal complex numbers. Thus, the value of the expression $b^2 - 4ac$ determines the nature of the solutions of any quadratic equation.

If a, b, and c are real numbers and	
if $b^2 - 4ac$ is · · ·	**the solutions are · · ·**
positive	real numbers and unequal.
0	real numbers and equal.
negative	nonreal complex numbers and complex conjugates.
If a, b, and c are rational numbers and	
if $b^2 - 4ac$ is · · ·	**the solutions are · · ·**
0	rational numbers and equal.
a nonzero perfect square	rational numbers and unequal.
positive and not a perfect square	irrational numbers and unequal

Example 1   Determine the type of solutions for the quadratic equation $x^2 + x + 1 = 0$.

Solution   Calculate the discriminant:

$$b^2 - 4ac = 1^2 - 4(1)(1) \qquad a = 1, b = 1, \text{ and } c = 1.$$
$$= -3$$

Because $b^2 - 4ac < 0$, the solutions of the equation are nonreal complex numbers.   ∎

Example 2   Determine the type of solutions for the quadratic equation $3x^2 + 5x + 2 = 0$.

Solution   Calculate the discriminant:

$$b^2 - 4ac = 5^2 - 4(3)(2) \qquad a = 3, b = 5, \text{ and } c = 2.$$
$$= 25 - 24$$
$$= 1$$

Because a, b, and c are rational numbers and $b^2 - 4ac$ is a perfect square, the solutions of the equation are rational and unequal.   ∎

Example 3   For what value of k will the solutions of the equation $kx^2 - 12x + 9 = 0$ be equal?

Solution   Calculate the discriminant:

$$b^2 - 4ac = (-12)^2 - 4(k)(9) \qquad a = k, b = -12, \text{ and } c = 9.$$
$$= -36k + 144$$

Because the solutions of the quadratic equation are to be equal, set $-36k + 144$ equal to 0 and solve for k:

$$-36k + 144 = 0$$
$$-36k = -144$$
$$k = 4$$

Thus, if $k = 4$, the solutions of the equation $kx^2 - 12x + 9 = 0$ will be equal. Verify this statement by solving the equation $4x^2 - 12x + 9 = 0$ and showing that it has two equal solutions. ∎

There are many types of equations that, while not quadratic equations, can be put into quadratic form. These equations can then be solved by using techniques for solving quadratic equations. For example, to solve $x^4 - 5x^2 + 4 = 0$, we can proceed as follows:

$$x^4 - 5x^2 + 4 = 0$$
$$(x^2)^2 - 5(x^2) + 4 = 0$$
$$y^2 - 5y + 4 = 0 \qquad \text{Let } y = x^2.$$
$$(y - 4)(y - 1) = 0 \qquad \text{Factor } y^2 - 5y + 4.$$
$$y - 4 = 0 \quad \text{or} \quad y - 1 = 0$$
$$y = 4 \qquad \qquad y = 1$$

Because $x^2 = y$, it follows that $x^2 = 4$ or $x^2 = 1$. Thus,

$$x^2 = 4 \qquad \text{or} \qquad x^2 = 1$$
$$x = 2 \quad \text{or} \quad x = -2 \qquad x = 1 \quad \text{or} \quad x = -1$$

This equation has four solutions. Verify that each one satisfies the original equation. Note that this equation could be solved directly by factoring.

Example 4   Solve the equation $x - 7x^{1/2} + 12 = 0$.

Solution   This equation is not a quadratic equation. However, if y^2 is substituted for x and y is substituted for $x^{1/2}$, the equation

$$x - 7x^{1/2} + 12 = 0$$

becomes a quadratic equation that can be solved by factoring:

$$y^2 - 7y + 12 = 0$$
$$(y - 3)(y - 4) = 0 \qquad \text{Factor } y^2 - 7y + 12.$$
$$y - 3 = 0 \quad \text{or} \quad y - 4 = 0$$
$$y = 3 \qquad \qquad y = 4$$

Because $x = y^2$, it follows that

$$x = 3^2 = 9 \qquad \text{or} \qquad x = 4^2 = 16$$

Verify that both solutions satisfy the original equation. ∎

Example 5   Solve the equation $\dfrac{24}{x} + \dfrac{12}{x+1} = 11$.

Solution   Because the denominator of a fraction cannot be 0, x cannot be 0 or -1. If either 0 or -1 appears as a suspected solution, it is extraneous and must be discarded. Solve the equation as follows:

$$\frac{24}{x} + \frac{12}{x+1} = 11$$

$$x(x+1)\left(\frac{24}{x} + \frac{12}{x+1}\right) = x(x+1)11 \qquad \text{Multiply both sides by } x(x+1).$$

$$24(x+1) + 12x = (x^2 + x)11$$

$$24x + 24 + 12x = 11x^2 + 11x$$

$$36x + 24 = 11x^2 + 11x$$

$$0 = 11x^2 - 25x - 24 \qquad \text{Add } -36x \text{ and } -24 \text{ to both sides.}$$

$$0 = (11x + 8)(x - 3) \qquad \text{Factor } 11x^2 - 25x - 24.$$

$$11x + 8 = 0 \qquad \text{or} \qquad x - 3 = 0$$

$$x = -\frac{8}{11} \qquad\qquad\qquad x = 3$$

Because both $-\frac{8}{11}$ and 3 are in the domain of x, each is a solution of the original equation. Verify this by checking each solution.   ∎

Example 6   Solve the equation $s = 16t^2 - 32$ for t.

Solution   Proceed as follows:

$$s = 16t^2 - 32$$

$$s + 32 = 16t^2 \qquad \text{Add 32 to both sides.}$$

$$\frac{s+32}{16} = t^2 \qquad \text{Divide both sides by 16.}$$

$$t^2 = \frac{s+32}{16} \qquad \text{Apply the symmetric property of equality.}$$

$$t = \pm\sqrt{\frac{s+32}{16}} \qquad \text{Apply the square root property.}$$

$$t = \frac{\pm\sqrt{s+32}}{\sqrt{16}} \qquad \sqrt{\frac{a}{b}} = \frac{\sqrt{a}}{\sqrt{b}}.$$

$$t = \frac{\pm\sqrt{s+32}}{4} \qquad\qquad\qquad\qquad\qquad ∎$$

Solutions of a Quadratic Equation

The solutions of a quadratic equation have some interesting properties. For example, if r_1 and r_2 are the solutions of the quadratic equation $ax^2 + bx + c = 0$, with $a \neq 0$, then $r_1 + r_2 = -b/a$ and $r_1 r_2 = c/a$. To prove this fact, we note that the solutions to the equation are given by the quadratic formula

$$r_1 = \frac{-b + \sqrt{b^2 - 4ac}}{2a} \quad \text{and} \quad r_2 = \frac{-b - \sqrt{b^2 - 4ac}}{2a}$$

Thus,

$$r_1 + r_2 = \frac{-b + \sqrt{b^2 - 4ac}}{2a} + \frac{-b - \sqrt{b^2 - 4ac}}{2a}$$

$$= \frac{-b + \sqrt{b^2 - 4ac} - b - \sqrt{b^2 - 4ac}}{2a}$$

$$= -\frac{2b}{2a}$$

$$= -\frac{b}{a}$$

and

$$r_1 r_2 = \frac{-b + \sqrt{b^2 - 4ac}}{2a} \cdot \frac{-b - \sqrt{b^2 - 4ac}}{2a}$$

$$= \frac{b^2 - (b^2 - 4ac)}{4a^2}$$

$$= \frac{b^2 - b^2 + 4ac}{4a^2}$$

$$= \frac{4ac}{4a^2}$$

$$= \frac{c}{a}$$

It is also true that, if the sum of two numbers r_1 and r_2 is $-a/b$ and if the product of the same two numbers is c/a, then r_1 and r_2 are the solutions of the quadratic equation $ax^2 + bx + c = 0$. This fact can be used to check the solutions of quadratic equations.

Example 7   Show that $\frac{3}{2}$ and $-\frac{1}{3}$ are solutions of the quadratic equation $6x^2 - 7x - 3 = 0$.

Solution   Note that, in the quadratic equation $6x^2 - 7x - 3 = 0$, $a = 6$, $b = -7$, and $c = -3$. If $\frac{3}{2}$ and $-\frac{1}{3}$ are two numbers such that

$$\frac{3}{2} + \left(-\frac{1}{3}\right) = -\frac{b}{a} = -\left(\frac{-7}{6}\right) = \frac{7}{6}$$

and

$$\frac{3}{2}\left(-\frac{1}{3}\right) = \frac{c}{a} = \frac{-3}{6} = -\frac{1}{2}$$

then $\frac{3}{2}$ and $-\frac{1}{3}$ are solutions of the given equation. Do the arithmetic to verify that these results are true:

$$\frac{3}{2} + \left(-\frac{1}{3}\right) = \frac{9}{6} - \frac{2}{6} = \frac{7}{6}$$

$$\frac{3}{2}\left(-\frac{1}{3}\right) = -\frac{3}{6} = -\frac{1}{2}$$

■ EXERCISE 8.3

In Exercises 1–8, use the discriminant to determine what type of solutions exist for each quadratic equation. **Do not solve the equation.**

1. $4x^2 - 4x + 1 = 0$ **2.** $6x^2 - 5x - 6 = 0$

3. $5x^2 + x + 2 = 0$ **4.** $3x^2 + 10x - 2 = 0$

5. $2x^2 = 4x - 1$ **6.** $9x^2 = 12x - 4$

7. $x(2x - 3) = 20$ **8.** $x(x - 3) = -10$

In Exercises 9–16, find the value(s) of k that will make the solutions of each given quadratic equation equal.

9. $x^2 + kx + 9 = 0$ **10.** $kx^2 - 12x + 4 = 0$

11. $9x^2 + 4 = -kx$ **12.** $9x^2 - kx + 25 = 0$

13. $(k - 1)x^2 + (k - 1)x + 1 = 0$ **14.** $(k + 3)x^2 + 2kx + 4 = 0$

15. $(k + 4)x^2 + 2kx + 9 = 0$ **16.** $(k + 15)x^2 + (k - 30)x + 4 = 0$

17. Use the discriminant to determine whether the solutions of $1492x^2 + 1776x - 1984 = 0$ are real numbers.

18. Use the discriminant to determine whether the solutions of $1776x^2 - 1492x + 1984 = 0$ are real numbers.

19. Determine k so that the solutions of $3x^2 + 4x = k$ are nonreal complex numbers.

20. Determine k so that the solutions of $kx^2 - 4x = 7$ are nonreal complex numbers.

In Exercises 21–46, solve each equation.

21. $x^4 - 17x^2 + 16 = 0$ **22.** $x^4 - 10x^2 + 9 = 0$

23. $x^4 - 3x^2 = -2$ **24.** $x^4 - 29x^2 = -100$ **25.** $x^4 = 6x^2 - 5$ **26.** $x^4 = 8x^2 - 7$

27. $2x^4 - 10x^2 = -8$ **28.** $2x^4 + 24 = 26x^2$ **29.** $2x + x^{1/2} - 3 = 0$ **30.** $2x - x^{1/2} - 1 = 0$

31. $3x + 5x^{1/2} + 2 = 0$ **32.** $3x - 4x^{1/2} + 1 = 0$

33. $x^{2/3} + 5x^{1/3} + 6 = 0$ **34.** $x^{2/3} - 7x^{1/3} + 12 = 0$

35. $x^{2/3} - 2x^{1/3} - 3 = 0$ **36.** $x^{2/3} + 4x^{1/3} - 5 = 0$

37. $x + 5 + \dfrac{4}{x} = 0$ **38.** $x - 4 + \dfrac{3}{x} = 0$ **39.** $x + 1 = \dfrac{20}{x}$ **40.** $x + \dfrac{15}{x} = 8$

41. $\dfrac{1}{x - 1} + \dfrac{3}{x + 1} = 2$ **42.** $\dfrac{6}{x - 2} - \dfrac{12}{x - 1} = -1$ **43.** $\dfrac{1}{x + 2} + \dfrac{24}{x + 3} = 13$ **44.** $\dfrac{36}{x} + \dfrac{24}{x + 1} = 17$

45. $x^{-4} - 2x^{-2} + 1 = 0$ **46.** $4x^{-4} + 1 = 5x^{-2}$

In Exercises 47–54, solve each equation for the indicated variable.

47. $x^2 + y^2 = r^2$ for x

48. $x^2 + y^2 = r^2$ for y

49. $I = \dfrac{k}{d^2}$ for d

50. $V = \dfrac{1}{3}\pi r^2 h$ for r

51. $xy^2 + 3xy + 7 = 0$ for y

52. $kx = ay - x^2$ for x

53. $\sigma = \sqrt{\dfrac{\Sigma x^2}{N} - \mu^2}$ for μ^2

54. $\sigma = \sqrt{\dfrac{\Sigma x^2}{N} - \mu^2}$ for N

In Exercises 55–60, solve each quadratic equation and verify that the sum of the solutions is $-b/a$ and that the product of the solutions is c/a.

55. $12x^2 - 5x - 2 = 0$

56. $8x^2 - 2x - 3 = 0$

57. $2x^2 + 5x + 1 = 0$

58. $3x^2 + 9x + 1 = 0$

59. $3x^2 - 2x + 4 = 0$

60. $2x^2 - x + 4 = 0$

8.4   GRAPHS OF QUADRATIC FUNCTIONS

We have defined a linear function as a correspondence between x and y determined by a first-degree equation that can be written in the form $y = mx + b$. We now discuss another important function called a **quadratic function**.

> **Definition.**   A **quadratic function** is a function defined by a second-degree polynomial equation of the form
>
> $$y = ax^2 + bx + c$$
>
> where a, b, and c are real numbers and $a \neq 0$.

To graph the quadratic function determined by $y = x^2 - 3$, for example, we calculate several ordered pairs that satisfy the equation, plot each point, and join them with a smooth curve, as shown in Figure 8-1. This curve is called a **parabola**.

$y = x^2 - 3$

x	y
-3	6
-2	1
-1	-2
0	-3
1	-2
2	1
3	6

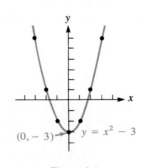

$(0, -3)$ $y = x^2 - 3$

Figure 8-1

Example 1   Graph the quadratic function determined by the equation $y = -x^2 + 2x + 1$.

Solution   Plot several points whose coordinates satisfy the equation and join them with a smooth curve to obtain the parabola shown in Figure 8-2.

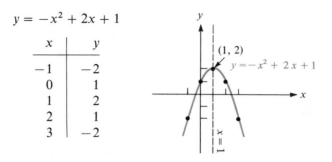

$$y = -x^2 + 2x + 1$$

x	y
-1	-2
0	1
1	2
2	1
3	-2

Figure 8-2

The graph of all functions determined by equations of the form $y = ax^2 + bx + c$, where $a \neq 0$, are parabolas. They open upward when $a > 0$ (as in Figure 8-1) and downward when $a < 0$ (as in Figure 8-2). The bottom point of a parabola that opens upward or the top of a parabola that opens downward is called the **vertex** of the parabola. The vertex of the parabola shown in Figure 8-1 is the point $(0, -3)$. The vertex of the parabola in Figure 8-2 is the point $(1, 2)$.

The vertical line that passes through the vertex of a parabola is called the **axis of symmetry** because it divides the parabola into two congruent halves. The axis of symmetry in Figure 8-1 is the y-axis. The axis of symmetry in Figure 8-2 is the line $x = 1$.

If a, h, and k are constants and $a \neq 0$, then the equation

$$y = a(x - h)^2 + k$$

also determines a quadratic function because this equation takes on the form $y = ax^2 + bx + c$ when the right-hand side is expanded and simplified. The graph of the equation $y = a(x - h)^2 + k$ is a parabola that opens upward when $a > 0$ and downward when $a < 0$. This form of the quadratic equation is useful because it displays the coordinates of the vertex of its parabolic graph as the following discussion shows.

Suppose that $a > 0$. Then the graph of

$$y = a(x - h)^2 + k$$

is a parabola opening upward, as in Figure 8-3. The vertex of the parabola is the point on the graph that has the smallest possible value for its y-coordinate. Because $a > 0$, the smallest possible y value occurs when the nonnegative quantity $a(x - h)^2$ on the right-hand side of the equation is 0. This occurs when $x = h$. When $x = h$, the value of $a(x - h)^2$ is 0 and the value of y is k. Thus,

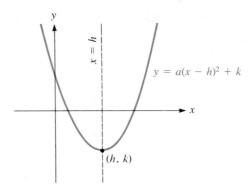

Figure 8-3

the vertex of the parabola is at the point with coordinates (h, k). A similar argument holds when $a < 0$.

The previous discussion leads to the following theorem:

Theorem.   The graph of the equation

$$y = a(x - h)^2 + k$$

where $a \neq 0$, is a parabola with vertex at (h, k). The parabola opens upward when $a > 0$ and downward when $a < 0$. The axis of symmetry is the line $x = h$.

Example 2   Find the vertex of the parabola determined by the quadratic equation $y = 2x^2 + 6x - 3$ and graph the parabola.

Solution   Proceed as follows to complete the square on the right-hand side of the given equation:

$$y = 2x^2 + 6x - 3$$

$$= 2(x^2 + 3x) - 3 \qquad \text{Factor out 2 from } 2x^2 + 6x \text{ to make the coefficient of } x^2 \text{ equal to 1.}$$

$$= 2\left(x^2 + 3x + \frac{9}{4}\right) - 3 - \frac{9}{2} \qquad \text{Add and subtract } \tfrac{9}{2}.$$

$$= 2\left(x + \frac{3}{2}\right)^2 - \frac{15}{2} \qquad \text{Factor } x^2 + 3x + \tfrac{9}{4}.$$

$$= 2\left[x - \left(-\frac{3}{2}\right)\right]^2 + \left(-\frac{15}{2}\right) \qquad \begin{array}{l}\text{To write the equation in the form} \\ y = a(x - h)^2 + k, \text{ write } x + \tfrac{3}{2} \\ \text{as } x - (-\tfrac{3}{2}), \text{ and } -\tfrac{15}{2} \text{ as } +(-\tfrac{15}{2}).\end{array}$$

Thus, the vertex is the point with coordinates $(-\frac{3}{2}, -\frac{15}{2})$. The graph of $y = 2x^2 + 6x - 3$ appears in Figure 8-4. Note that the parabola has the line $x = -\frac{3}{2}$ as an axis of symmetry.

$$y = 2x^2 + 6x - 3$$

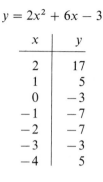

x	y
2	17
1	5
0	-3
-1	-7
-2	-7
-3	-3
-4	5

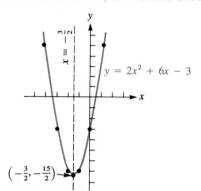

Figure 8-4   ■

Example 3   Suppose that a ball is thrown straight up in the air with an initial velocity of 128 feet per second. The quadratic function $s = 128t - 16t^2$ gives the relation between s and t, where s represents the number of feet the ball is above the ground and t represents the time measured in seconds. How high did the ball travel?

Solution   The equation $s = -16t^2 + 128t$ represents a quadratic function, so its graph is a parabola. The maximum height attained by the ball is given by the s-coordinate of the vertex of the parabola. Find the coordinates of the vertex by completing the square:

$$s = -16t^2 + 128t$$
$$= -16(t^2 - 8t) \qquad \text{Factor out } -16.$$
$$= -16(t^2 - 8t + 16) + 256 \qquad \text{Subtract and add 256.}$$
$$= -16(t - 4)^2 + 256 \qquad \text{Factor } t^2 - 8t + 16.$$

Thus, the coordinates of the vertex are (4, 256). Because $t = 4$ and $s = 256$ are the coordinates of the vertex, the ball reaches a maximum height of 256 feet in 4 seconds.

Because a parabola is symmetric, it will take an additional 4 seconds for the ball to return to earth. The total time of the flight is 8 seconds. Note that if $t = 8$, the value of s in the equation $s = -16t^2 + 128t$ is zero, and the ball is back to earth. Although this quadratic function describes the height of the ball in relation to time, it does not describe the path traveled by the ball. The ball went straight up and came back straight down.   ■

Example 4   A man wants to build a rectangular pen to house his dog. To save fencing, he intends to use one side of his garage as one side of the pen. Find the maximum area that he can enclose with 80 feet of fencing.

Solution   Let w be the width of the pen. Then the length is represented by $80 - 2w$ (see Figure 8-5).

The area of the pen is given by the product of the length and the width. Thus, you have

$$A = w(80 - 2w)$$

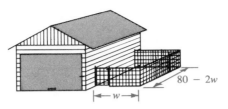

$80 - 2w$

$\longleftarrow w \longrightarrow$

Figure 8-5

Find the maximum value of A as follows:

$$A = w(80 - 2w)$$
$$= 80w - 2w^2 \qquad\qquad \text{Remove parentheses.}$$
$$= -2(w^2 - 40w) \qquad\qquad \text{Factor out } -2 \text{ and rearrange terms.}$$
$$= -2(w^2 - 40w + 400) + 800 \qquad \text{Subtract and add 800.}$$
$$= -2(w - 20)^2 + 800 \qquad\qquad \text{Factor } w^2 - 40w + 400.$$

Thus, the coordinates of the vertex of the graph of the quadratic function are (20, 800) and the maximum area is 800 square feet. ∎

It is easy to show that the vertex of the parabola determined by the equation $y = ax^2 + bx + c$ is the point with coordinates

$$\left(-\frac{b}{2a}, c - \frac{b^2}{4a}\right)$$

To do so, we use the method of completing the square to write the equation $y = ax^2 + bx + c$ in the form $y = a(x - h)^2 + k$ to determine the coordinates of the vertex:

$$y = ax^2 + bx + c$$
$$= a\left(x^2 + \frac{b}{a}x\right) + c \qquad\qquad\qquad \text{Factor out } a.$$
$$= a\left(x^2 + \frac{b}{a}x + \frac{b^2}{4a^2}\right) + c - \frac{b^2}{4a} \qquad \text{Add and subtract } \frac{b^2}{4a}.$$
$$= a\left(x + \frac{b}{2a}\right)^2 + c - \frac{b^2}{4a} \qquad\qquad \text{Factor the trinomial.}$$
$$= a\left[x - \left(-\frac{b}{2a}\right)\right]^2 + \left(c - \frac{b^2}{4a}\right)$$

Thus, the vertex is the point with coordinates of $\left(-\dfrac{b}{2a}, c - \dfrac{b^2}{4a}\right)$.

■ EXERCISE 8.4 ▬▬▬▬▬▬▬▬▬▬

In Exercises 1–12, graph each function determined by the given quadratic equation.

1. $y = x^2$
2. $y = -x^2$
3. $y = x^2 + 2$
4. $y = x^2 - 3$
5. $y = -(x - 2)^2$
6. $y = (x + 2)^2$
7. $y = -3x^2 + 2x$
8. $y = 5x + x^2$

9. $y = x^2 + x - 6$

10. $y = x^2 - x - 6$

11. $y = 6x^2 + 6x - 12$

12. $y = -4x^2 + 8x + 12$

In Exercises 13–24, find the coordinates of the vertex and the axis of symmetry of the graph of each equation. If necessary, complete the square on x to write the equation in the form $y = a(x - h)^2 + k$. **Do not graph the equation.**

13. $y = (x - 1)^2 + 2$

14. $y = 2(x - 2)^2 - 1$

15. $y = 2(x + 3)^2 - 4$

16. $y = -3(x + 1)^2 + 3$

17. $y = -3x^2$

18. $y = 3x^2 - 3$

19. $y = 2x^2 - 4x$

20. $y = 3x^2 + 6x$

21. $y = -4x^2 + 16x + 5$

22. $y = 5x^2 + 20x + 25$

23. $y - 7 = 6x^2 - 5x$

24. $y - 2 = 3x^2 + 4x$

25. The equation $y - k = (x - h)^2$ represents a quadratic function whose graph is a parabola. Find its vertex.

26. Show that $y = ax^2$, where $a \neq 0$, represents a quadratic function whose vertex is at the origin.

27. The sum of two numbers is 50 and their product is maximum. Find the numbers.

28. The sum of two numbers is 10 and the sum of their squares is minimum. Find the numbers.

29. If a ball is thrown straight up with an initial velocity of 48 feet per second, its height after t seconds is given by the equation $s = 48t - 16t^2$. Find the maximum height attained by the ball and the time it takes for the ball to return to earth.

30. From the top of a building 48 feet tall, a ball is thrown straight upward with an initial velocity of 32 feet per second. The equation $s = -16t^2 + 32t + 48$ gives the height of the ball t seconds after it was thrown. Find the maximum height reached by the ball and find the time it will take for the ball to hit the ground.

31. Find the dimensions of the rectangle of maximum area that can be constructed with 200 feet of fencing. What is the maximum area?

32. A farmer wants to fence in three sides of a rectangular field with 1000 feet of fencing. The fourth side of the rectangle is to be a river. If the enclosed area is to be maximum, find the dimensions of the field.

33. When priced at $30 each, the annual sales of a toy total 4000 units. The manufacturer estimates that each dollar increase in cost will decrease sales by 100 units. What unit price will maximize total revenue? (*Hint:*   Total revenue = price • the number of units sold.)

34. When priced at $57, the annual sales of a radio total 525 units. For each dollar the radio is reduced in price, the radio is expected to sell an additional 75 units. What unit price will maximize total revenue? (*Hint:*   Total revenue = price • the number of units sold.)

8.5 GRAPHS OF OTHER FUNCTIONS AND RELATIONS AND MORE ON INVERSE FUNCTIONS

In this section we continue the discussion of graphs of functions and relations.

Example 1   Graph the equation $y = |x|$ and tell whether it determines a function.

Solution   Make a table of values as in Figure 8-6. Plot each ordered pair and join the points as in the figure to obtain the graph of $y = |x|$. This equation does deter-

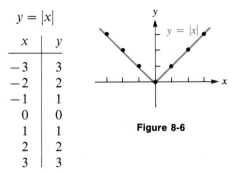

$y = |x|$

x	y
-3	3
-2	2
-1	1
0	0
1	1
2	2
3	3

Figure 8-6

mine a function because, for each number x, there corresponds exactly one value of y. For example, to the number $x = -2$, there corresponds only the value $y = 2$.   ■

The Vertical Line Test

A test called the **vertical line test** can be used to determine whether a relation represented by a graph is a function. If a vertical line intersects the graph more than once, the relation represented by the graph cannot be a function because to one number x there would correspond more than one value of y. The graph in Figure 8-7a represents a function because every vertical line that intersects the graph does so exactly once. However, the graph in Figure 8-7b does not represent a function because some vertical lines intersect the graph more than once.

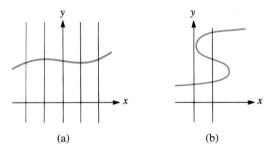

(a)   (b)

Figure 8-7

In Example 1, every vertical line that intersects the graph of $y = |x|$ will do so exactly once. Thus, by the vertical line test, the equation $y = |x|$ does determine a function.

Example 2   Graph the equation $xy = 6$ and use the vertical line test to decide whether the equation determines a function.

Solution   Make a table of values as in Figure 8-8a, plot the points, and draw the graph. Because neither x nor y can be 0, the curve will never cross the x- or y-axis. The curve shown in Figure 8-8a is called a **hyperbola**. To see whether the equation determines a function, draw several vertical lines that intersect the hyperbola, as in Figure 8-8b. Because each vertical line that intersects it does so exactly once, the equation does determine a function.

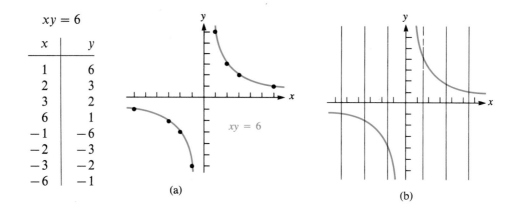

$xy = 6$

x	y
1	6
2	3
3	2
6	1
-1	-6
-2	-3
-3	-2
-6	-1

(a)   (b)

Figure 8-8

Example 3   Graph the equation $y^2 = -4x$ and tell whether it determines a function.

Solution   A table of values is shown in Figure 8-9a. Plot each ordered pair and join the points with a smooth curve. The graph of $y^2 = -4x$ is a parabola that opens to the left. Because some vertical lines intersect the parabola twice, the equation $y^2 = -4x$ does not determine a function. See Figure 8-9b.

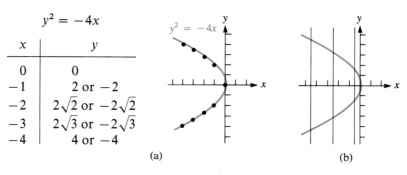

$y^2 = -4x$

x	y
0	0
-1	2 or -2
-2	$2\sqrt{2}$ or $-2\sqrt{2}$
-3	$2\sqrt{3}$ or $-2\sqrt{3}$
-4	4 or -4

(a)   (b)

Figure 8-9

There is a function, called the **greatest integer function**, that is used occasionally in mathematics and often in computer programming. The ordered pairs (x, y) of this function are determined by the equation $y = [x]$. The square brackets around x are read "the greatest integer in x." The value of y that corresponds to x is the greatest integer that is less than or equal to the number x. For example, the greatest integer that is less than or equal to π (3.14159 . . .) is 3. Thus, $[\pi] = 3$. Study each of the following equalities to understand the behavior of the greatest integer function:

$[3.7] = 3$ $[-3.7] = -4$

$[5] = 5$ $[-5] = -5$

$[0.99999999999] = 0$ $[-0.9999999999] = -1$

$[0] = 0$ $\left[\dfrac{13}{4}\right] = 3$

Example 4   Graph the function determined by $y = [x]$.

Solution   For values of x that are between two consecutive integers, the corresponding value of y is the smaller of the two integers. For an integer value of x, the corresponding value of y is the integer x. A table of values and the graph of the function appear in Figure 8-10. Note the positions of the solid circles and the open circles on the graph.

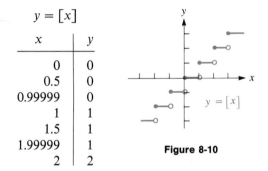

$y = [x]$

x	y
0	0
0.5	0
0.99999	0
1	1
1.5	1
1.99999	1
2	2

Figure 8-10

Inverses of Quadratic Functions

Recall that the inverse relation of a function is found by interchanging the variables x and y. Thus, to find the inverse relation of the function determined by $y = x^2$, we proceed as follows:

$y = x^2$

$x = y^2$ Interchange the x and y variables.

$y = \pm\sqrt{x}$ Apply the symmetric property of equality and the square root property.

If we graph both $y = x^2$ and its inverse relation $y = \pm\sqrt{x}$ on the same set of coordinate axes, we obtain the graphs in Figure 8-11. Once again, we see that the graphs of a function and its inverse are symmetric about the line $y = x$.

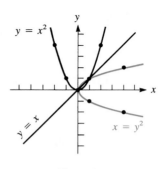

$$x = y^2$$
$$\text{or}$$
$$y = x^2 \qquad y = \pm\sqrt{x}$$

x	y	x	y
0	0	0	0
1	1	1	1
-1	1	1	-1
2	4	4	2
-2	4	4	-2

Figure 8-11

Example 5   Find the inverse relation of the function determined by $y = x^2$ with $x \geq 0$, and decide whether the inverse relation is a function. Graph both the function and its inverse on a single set of coordinate axes.

Solution   The inverse relation of the function $y = x^2$ with $x \geq 0$ is

$$x = y^2 \quad \text{with } y \geq 0$$

This equation can be written in the form

$$y = \pm\sqrt{x} \quad \text{with } y \geq 0$$

Because y must be greater than or equal to 0, each value of x gives only one value of y: $y = +\sqrt{x}$. Hence, the inverse relation of $y = x^2$ with $x \geq 0$ is a function. The graphs of the two functions appear in Figure 8-12. The line $y = x$ is included for reference.

$$y = x^2 \text{ and } x \geq 0 \qquad x = y^2 \text{ and } y \geq 0$$

x	y	x	y
0	0	0	0
1	1	1	1
2	4	4	2
3	9	9	3

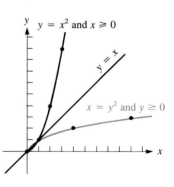

Figure 8-12   ■

If $y = f(x)$ represents a function of x, and if its inverse relation is also a function, the inverse function often is denoted as $y = f^{-1}(x)$. This equation is read as "y is the inverse function of x" or "y equals f inverse of x." This notation is used in the following example.

Example 6   The set of all pairs (x, y) such that $y = f(x) = x^3 + 2$ is a function. Show that its inverse is a function and express it using $f^{-1}(x)$ notation.

Solution   To find the inverse function, proceed as follows:

$$y = x^3 + 2$$
$$x = y^3 + 2 \qquad \text{Interchange the } x \text{ and } y \text{ variables.}$$
$$x - 2 = y^3 \qquad \text{Add } -2 \text{ to both sides.}$$
$$\sqrt[3]{x - 2} = y \qquad \text{Take the cube root of both sides.}$$

Note that, to each number x, there corresponds a single cube root. Thus, the equation $y = \sqrt[3]{x - 2}$ represents a function. In $f^{-1}(x)$ notation, you have

$$y = f^{-1}(x) = \sqrt[3]{x - 2} \qquad\qquad \blacksquare$$

If the inverse of a given function is also a function, the given function is called **one-to-one**. Stated another way, we have the following definition.

Definition.   A function is called **one-to-one** if and only if each element in the range of the function corresponds to only one element in the domain of the function.

The Horizontal Line Test

A test called the **horizontal line test** can be used to decide whether a function is one-to-one. If any horizontal line that intersects the graph of a function does so in at most one point, the function is one-to-one. Otherwise, the function is *not* one-to-one.

Example 7   Use the horizontal line test to decide whether the function defined by $y = x^3$ is one-to-one.

Solution   Graph the equation as in Figure 8-13. Because each horizontal line that intersects the graph does so exactly once, each value of y corresponds to only one number x. Thus, the function defined by $y = x^3$ *is* one-to-one.

$$y = x^3$$

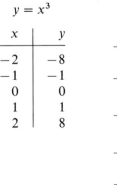

x	y
-2	-8
-1	-1
0	0
1	1
2	8

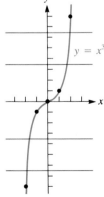

Figure 8-13 $\blacksquare$

Example 8   Use the horizontal line test to decide whether the function defined by $y = x^2 - 4$ is one-to-one.

Solution   Graph the equation as in Figure 8-14. Because many horizontal lines that intersect the graph do so twice, some values of y correspond to two numbers x. Thus, this function *is not* one-to-one.

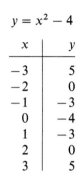

$$y = x^2 - 4$$

x	y
-3	5
-2	0
-1	-3
0	-4
1	-3
2	0
3	5

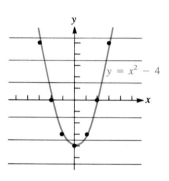

Figure 8-14

EXERCISE 8.5

In Exercises 1–18, graph each equation and use the vertical line test to decide whether the equation determines a function.

1. $x = |y|$

2. $y = |x - 2|$

3. $y = \dfrac{1}{2}|x + 4|$

4. $y = |x| - 1$

5. $|x| - y = 4$

6. $x + |y| = 1$

7. $x = y^2 + 4$

8. $x = 2y^2 - 2$

9. $x = -y^2 - 2$

10. $x = -\dfrac{1}{2}y^2$

11. $xy = 12$

12. $xy = 24$

13. $xy = -24$

14. $-xy = 12$

15. $y = x^3$

16. $x = y^3$

17. $x = -y^3$

18. $y = -x^3$

In Exercises 19–22, graph each function. The square brackets denote the greatest integer function.

19. $y = [2x]$

20. $y = [3x]$

21. $y = [-x]$

22. $y = \left[\dfrac{x}{2}\right]$

23.   Computer programmers use a function called the **signum function**, denoted by $y = sgn\ x$, that is defined in the following way:

If $x < 0$, then $y = -1$.

If $x = 0$, then $y = 0$.

If $x > 0$, then $y = 1$.

Graph this function.

24. The **Heaviside unit step function**, used in calculus, is defined by

$$y = H(x) = \begin{cases} 1 & \text{if } x > 0 \\ 0 & \text{if } x < 0 \end{cases}$$

$H(0)$ is undefined. Graph the equation $y = H(x) \cdot H(x - 1)$.

In Exercises 25–34, find the inverse relation of the relation determined by each equation. Tell whether the inverse relation is a function.

25. $y = x^2 + 4$ **26.** $x = y^2 - 2$ **27.** $x = y^2 - 4$ **28.** $y = x^2 + 5$

29. $y = x^3$ **30.** $xy = 4$ **31.** $y = \sqrt{x}; \ y \geq 0$ **32.** $y = \sqrt[3]{x}$

33. $x + 1 = \sqrt{y}$ **34.** $4y^2 = x - 3$

In Exercises 35–36, show that the inverse of the function determined by each equation is also a function. Express the inverse using $f^{-1}(x)$ notation.

35. $y = 2x^3 - 3$ **36.** $y = \dfrac{3}{x^3} - 1$

In Exercises 37–40, graph each equation and its inverse on one set of coordinate axes. What is the axis of symmetry?

37. $y = x^2 + 1$ **38.** $y = \dfrac{1}{4}x^2 - 3$ **39.** $y = -\sqrt{x}; \ y \leq 0$ **40.** $y = |x|$

In Exercises 41–50, each equation represents a function. Graph each equation and use the horizontal line test to decide whether each function is one-to-one.

41. $y = 3x + 2$ **42.** $y = 5 - 3x$ **43.** $y = \dfrac{x + 7}{8}$ **44.** $y = \dfrac{5 - x}{2}$

45. $y = 3x^2 + 2$ **46.** $y = 5 - x^2$ **47.** $y = \sqrt[3]{x}$ **48.** $y = \sqrt{x}$

49. $y = x^3 - x$ **50.** $y = -x^4 + x^2$

8.6   QUADRATIC AND OTHER NONLINEAR INEQUALITIES

We now turn our attention to solving quadratic inequalities. To introduce this topic we consider the quadratic inequality

$$x^2 + x - 6 < 0$$

To find the values of x that make this inequality true is to *solve the inequality.* We begin by factoring the trinomial to obtain

$$(x + 3)(x - 2) < 0$$

Here we have a product of two quantities that is less than 0, which can happen only when the values of $x + 3$ and $x - 2$ are opposite in sign. The best way to keep track of the signs of these factors is to construct a sign chart as in Figure 8-15.

This chart indicates that the binomial $x - 2$ is 0 when $x = 2$, is negative when $x < 2$, and is positive when $x > 2$. It also indicates that the binomial $x + 3$

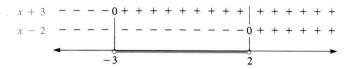

Figure 8-15

is 0 when $x = -3$, is negative when $x < -3$, and is positive when $x > -3$. The signs of the two binomials are opposite only in the interval between -3 and 2. This interval, which can be written as $-3 < x < 2$, is the solution to the inequality.

The graph of the solution set is shown on the number line in the sign chart in Figure 8-15.

Example 1   Solve the inequality $(x - 1)(x + 3) \geq 0$.

Solution   Construct a sign chart as in Figure 8-16. The binomial $x - 1$ is 0 when $x = 1$, is negative when $x < 1$, and is positive when $x > 1$. The binomial $x + 3$ is 0 when $x = -3$, is negative when $x < -3$, and is positive when $x > -3$. The product of $x - 1$ and $x + 3$ is greater than 0 when the signs of the binomials are the same, which occurs when $x < -3$ or $x > 1$. The numbers -3 and 1 are included because they make the product equal to 0. The graph of the solution set is shown on the number line in the sign chart in Figure 8-16. The solution set can be written as $\{x : x \leq -3 \text{ or } x \geq 1\}$.

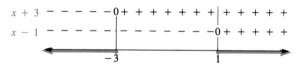

Figure 8-16   ■

Making a sign chart is useful for solving many inequalities that are neither linear nor quadratic.

Example 2   Solve the inequality $\dfrac{1}{x} < 6$.

Solution   Add -6 to both sides, find a common denominator, and add the fractions:

$$\frac{1}{x} < 6$$

$$\frac{1}{x} - 6 < 0$$

$$\frac{1}{x} - \frac{6x}{x} < 0$$

$$\frac{1 - 6x}{x} < 0$$

Now make a sign chart as in Figure 8-17. The denominator, x, is 0 when $x = 0$, is positive when $x > 0$, and is negative when $x < 0$. The numerator, $1 - 6x$, is 0 when $x = \frac{1}{6}$, is positive when $x < \frac{1}{6}$, and is negative when $x > \frac{1}{6}$. The quotient $\frac{1 - 6x}{x}$ is less than 0 when the binomial $1 - 6x$ and the monomial x are opposite in sign which occurs when $x < 0$ or when $x > \frac{1}{6}$. The graph of the solution set, $\{x : x < 0 \text{ or } x > \frac{1}{6}\}$, is shown on the number line in the sign chart in Figure 8-17.

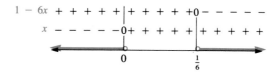

Figure 8-17

Example 3 Solve the inequality $\dfrac{x^2 - 3x + 2}{x - 3} \geq 0$.

Solution Rewrite the fraction by factoring the numerator:

$$\frac{(x - 2)(x - 1)}{x - 3} \geq 0$$

This time the signs of three binomials must be considered. To do so, construct a sign chart as in Figure 8-18.

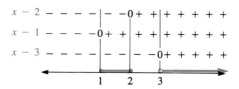

Figure 8-18

The fraction will be positive in the intervals where all factors are positive, or where two factors are negative. The numbers 1 and 2 are included because they make the numerator, and thus the fraction, equal to 0. The number 3 is not included because it leads to a 0 in the denominator. The graph of the solution set $\{x : x > 3 \text{ or } 1 \leq x \leq 2\}$ appears on the number line in the sign chart in Figure 8-18.

Example 4 Solve the inequality $\dfrac{3}{x - 1} < \dfrac{2}{x}$.

Solution   Add $-\dfrac{2}{x}$ to both sides, find a common denominator, and add the fractions:

$$\frac{3}{x-1} < \frac{2}{x}$$

$$\frac{3}{x-1} - \frac{2}{x} < 0$$

$$\frac{3x}{(x-1)x} - \frac{2(x-1)}{x(x-1)} < 0$$

$$\frac{3x - 2x + 2}{x(x-1)} < 0$$

$$\frac{x+2}{x(x-1)} < 0$$

Keep track of the signs of the three polynomials with a sign chart as in Figure 8-19. The fraction will be negative in those regions with either one or three negative factors. The number -2 is not included because it does not satisfy the inequality. The numbers 0 and 1 are not included because they lead to a 0 in the denominator. The graph of $\{x : x < -2 \text{ or } 0 < x < 1\}$ is shown on the number line in the sign chart.

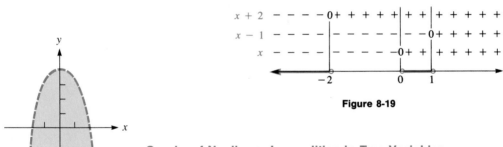

Figure 8-19   ■

Graphs of Nonlinear Inequalities in Two Variables

In Section 7.7 we saw how to graph linear inequalities in two variables. We now consider the graphs of nonlinear inequalities in two variables.

Figure 8-20

$y < -x^2 + 4$

Example 5   Graph $y < -x^2 + 4$.

Solution   In this example, the graph of the equation $y = -x^2 + 4$ can be thought of as a "fence" separating the region represented by $y < -x^2 + 4$ and the region represented by $y > -x^2 + 4$. Graph the quadratic function $y = -x^2 + 4$ using a broken curve because equality is not permitted. Note that the coordinates of the origin, $(0, 0)$, satisfy the inequality $y < -x^2 + 4$. Hence, the origin is in the graph of $y < -x^2 + 4$. The graph is shown in Figure 8-20.   ■

Example 6   Graph $x \le |y|$.

Solution   As in the previous example, first graph the equality $x = |y|$. Use a solid line because equality is permitted. See Figure 8-21a. This time the coordinates of the origin cannot be used to establish the proper region because the origin is on the graph of $x = |y|$. However, any other convenient point, such as $(1, 0)$, will do. Substitute 1 for x and 0 for y into the inequality $x \leq |y|$, and note that the result is a false statement. Because $(1, 0)$ does not satisfy $x \leq |y|$, it is *not* in the graph. Hence, the graph of $x \leq |y|$ is to the left of the graph of $x = |y|$. The complete graph of $x \leq |y|$ is shown in Figure 8-21b.

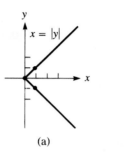

(a)

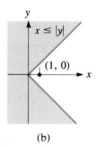

(b)

Figure 8-21

■ EXERCISE 8.6

In Exercises 1–40, solve each inequality and graph its solution set.

1. $x^2 - 5x + 4 < 0$

2. $x^2 - 3x - 4 > 0$

3. $x^2 - 8x + 15 > 0$

4. $x^2 + 2x - 8 < 0$

5. $x^2 + x - 12 \leq 0$

6. $x^2 + 7x + 12 \geq 0$

7. $x^2 + 2x \geq 15$

8. $x^2 - 8x \leq -15$

9. $x^2 + 8x < -16$

10. $x^2 - 6x > -9$

11. $x^2 \geq 9$

12. $x^2 \geq 16$

13. $2x^2 - 50 < 0$

14. $3x^2 - 240 < 0$

15. $\dfrac{1}{x} < 2$

16. $\dfrac{1}{x} > 3$

17. $\dfrac{4}{x} \geq 2$

18. $-\dfrac{6}{x} < 12$

19. $-\dfrac{5}{x} < 3$

20. $\dfrac{4}{x} \geq 8$

21. $\dfrac{x^2 - x - 12}{x - 1} < 0$

22. $\dfrac{x^2 + x - 6}{x - 4} \geq 0$

23. $\dfrac{x^2 + x - 20}{x + 2} \geq 0$

24. $\dfrac{x^2 - 10x + 25}{x + 5} < 0$

25. $\dfrac{2x^2 + x - 6}{x - 3} \leq 0$

26. $\dfrac{2x^2 - 5x + 2}{x + 2} > 0$

27. $\dfrac{6x^2 - 5x + 1}{2x + 1} > 0$

28. $\dfrac{6x^2 + 11x + 3}{3x - 1} < 0$

29. $\dfrac{3}{x - 2} < \dfrac{4}{x}$

30. $\dfrac{-6}{x + 1} \geq \dfrac{1}{x}$

31. $\dfrac{-5}{x + 2} \geq \dfrac{4}{2 - x}$

32. $\dfrac{-6}{x - 3} < \dfrac{5}{3 - x}$

33. $\dfrac{7}{x - 3} \geq \dfrac{2}{x + 4}$

34. $\dfrac{-5}{x - 4} < \dfrac{3}{x + 1}$

35. $\dfrac{x}{x + 4} \leq \dfrac{1}{x + 1}$

36. $\dfrac{x}{x + 9} \geq \dfrac{1}{x + 1}$

37. $\dfrac{x}{x + 16} > \dfrac{1}{x + 1}$

38. $\dfrac{x}{x + 25} < \dfrac{1}{x + 1}$

39. $(x + 2)^2 > 0$

40. $(x - 3)^2 < 0$

In Exercises 41–52, graph each inequality.

41. $y < x^2 + 1$ **42.** $y > x^2 - 3$ **43.** $y \leq x^2 + 5x + 6$ **44.** $y \geq x^2 + 5x + 4$

45. $x \geq y^2 - 3$ **46.** $x \leq y^2 + 1$

47. $-x^2 - y + 6 > -x$ **48.** $y > (x + 3)(x - 2)$

49. $y < |x + 4|$ **50.** $y \geq |x - 3|$ **51.** $y \leq -|x| + 2$ **52.** $y < |x| - 2$

8.7   THE CIRCLE AND THE PARABOLA

The graphs of second-degree equations in x and y represent figures that have interested mathematicians since the time of the ancient Greeks. However, the equations of these graphs were not studied carefully until the 17th century, when René Descartes (1596–1650) and Blaise Pascal (1623–1662) began investigating them.

Descartes discovered that the graphs of second-degree equations always fall into one of several categories: a pair of lines, a point, a circle, a parabola, an ellipse, a hyperbola, or no graph at all. Because all of these graphs can be formed by the intersection of a plane and a right circular cone, they are called **conic sections**. See Figure 8-22.

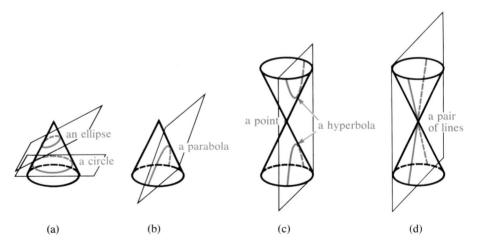

(a)   (b)   (c)   (d)

Figure 8-22

Conic sections have many applications. For example, a parabola can be rotated to generate a dish-shaped surface called a **paraboloid.** Any light or sound placed at a certain point, called the *focus* of the paraboloid, is reflected outward in parallel paths. This property makes parabolic surfaces ideal for flashlight and headlight reflectors. Using the same property in reverse makes parabolic surfaces good antennas because signals captured by such an antenna are concentrated at the focus. A parabolic mirror is capable of concentrating the rays of the sun at a single point and thereby generating tremendous heat. This fact is used in the design of certain solar furnaces. Any object that is thrown upward and outward travels in a parabolic path. In architecture many arches are parabolic in shape because of the strength possessed by such an arch; the cables that support a suspension bridge hang in the form of a parabola.

Ellipses have optical and acoustical properties that are useful in architecture and engineering. For example, many arches are portions of an ellipse because the shape is pleasing to the eye. To provide nonuniform motion, gears are often cut into elliptical shapes. The planets and some comets have elliptical orbits.

Hyperbolas serve as the basis of a navigational system known as LORAN (LOng RAnge Navigation). They are also used to find the source of a distress signal. The orbits of some comets may be one branch of a hyperbola.

In this section we discuss the circle and the parabola.

The Circle

> **Definition.**   A **circle** is the set of all points in a plane that are a fixed distance from a point called its **center**.
>
> The fixed distance is called the **radius** of the circle.

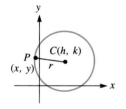

Figure 8-23

To develop the general equation of a circle, we must write the equation of a circle with a radius of r and with center at some point $C(h, k)$. See Figure 8-23. This task is equivalent to finding all points $P(x, y)$ such that the length of the line segment CP is r. We can use the distance formula to find r.

$$r = \sqrt{(x - h)^2 + (y - k)^2}$$

We then square both sides to obtain

1. $r^2 = (x - h)^2 + (y - k)^2$

Equation 1 is called the **standard form of the equation of a circle** with a radius of r and center at the point with coordinates (h, k). The previous discussion suggests the following theorem.

> **Theorem.**   Any equation that can be written in the form
>
> $$(x - h)^2 + (y - k)^2 = r^2$$
>
> has a graph that is a circle with radius r and center at point (h, k).

If $r = 0$, the graph reduces to a single point called a **point circle**. If $r < 0$, then a circle does not exist. If both coordinates of the center are 0, then the center of the circle is the origin.

> **Theorem.**   Any equation that can be written in the form
>
> $$x^2 + y^2 = r^2$$
>
> has a graph that is a circle with radius r and center at the origin.

Example 1   Graph the equation $x^2 + y^2 = 25$.

Solution   Because this equation can be written in the form $x^2 + y^2 = r^2$, its graph is a circle with center at the origin. Because $r^2 = 25 = 5^2$, the circle has a radius of 5. The graph appears in Figure 8-24.

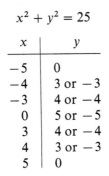

$$x^2 + y^2 = 25$$

x	y
-5	0
-4	3 or -3
-3	4 or -4
0	5 or -5
3	4 or -4
4	3 or -3
5	0

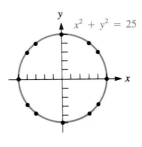

Figure 8-24

Example 2   Find the equation of the circle with radius 5 and center at $C(3, 2)$.

Solution   Substitute 5 for r, 3 for h, and 2 for k in the standard form and simplify:

$$(x - h)^2 + (y - k)^2 = r^2$$
$$(x - 3)^2 + (y - 2)^2 = 5^2$$
$$x^2 - 6x + 9 + y^2 - 4y + 4 = 25$$
$$x^2 + y^2 - 6x - 4y - 12 = 0$$

The equation of the circle is $x^2 + y^2 - 6x - 4y - 12 = 0$.

Example 3   Graph the circle $x^2 + y^2 - 4x + 2y = 20$.

Solution   Because the equation is not in standard form, the coordinates of the center and the length of the radius are not obvious. To put the equation in standard form, complete the square on both x and y as follows.

$$x^2 + y^2 - 4x + 2y = 20$$
$$x^2 - 4x + y^2 + 2y = 20$$
$$x^2 - 4x + 4 + y^2 + 2y + 1 = 20 + 4 + 1 \qquad \text{Add } 4 + 1 \text{ to both sides.}$$
$$(x - 2)^2 + (y + 1)^2 = 25 \qquad \text{Factor } x^2 - 4x + 4 \\ \text{and } y^2 + 2y + 1.$$

$$(x - 2)^2 + [y - (-1)]^2 = 5^2$$

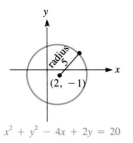

$x^2 + y^2 - 4x + 2y = 20$

Figure 8-25

Note that the radius of the circle is 5 and the coordinates of its center are $h = 2$ and $k = -1$. Plot the center of the circle and construct the circle with a radius of 5 units, as shown in Figure 8-25.   ∎

The Parabola

We have seen that equations of the form $y = a(x - h)^2 + k$, with $a \neq 0$, represent parabolas with vertex at the point (h, k). They open upward when $a > 0$ and downward when $a < 0$. Equations of the form $x = a(y - k)^2 + h$ also represent parabolas with vertex at the point (h, k); however, they open to the right when $a > 0$ and to the left when $a < 0$. Parabolas that open to the right or left do not represent functions because their graphs do not pass the vertical line test.

Several types of parabolas are summarized in the following chart. If $a > 0$, then

Parabola opening	Vertex at origin	Vertex at (h, k)
Up	$y = ax^2$	$y = a(x - h)^2 + k$
Down	$y = -ax^2$	$y = -a(x - h)^2 + k$
Right	$x = ay^2$	$x = a(y - k)^2 + h$
Left	$x = -ay^2$	$x = -a(y - k)^2 + h$

Example 4   Graph the equation $x = \dfrac{1}{2} y^2$.

Solution   Make a table of values, plot the points, and draw the parabola as in Figure 8-26. Because the equation is of the form $x = ay^2$, the parabola opens to the right and has its vertex at the origin.

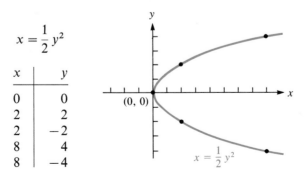

$$x = \frac{1}{2} y^2$$

x	y
0	0
2	2
2	-2
8	4
8	-4

Figure 8-26

Example 5   Graph the equation $x = -2(y - 2)^2 + 3$.

Solution   Make a table of values, plot the points, and draw the parabola as in Figure 8-27. Because the equation is of the form $x = -a(y - k)^2 + h$, the parabola opens to the left and has its vertex at the point with coordinates $(3, 2)$.

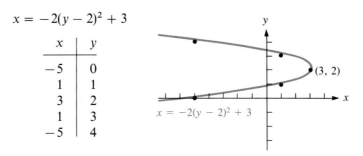

$$x = -2(y - 2)^2 + 3$$

x	y
-5	0
1	1
3	2
1	3
-5	4

$x = -2(y - 2)^2 + 3$

(3, 2)

Figure 8-27

Example 6   Graph the equation $y = -2x^2 + 12x - 15$.

Solution   Because the equation is not in standard form, the coordinates of its vertex are not obvious. To put the equation into standard form, complete the square on x:

$$y = -2x^2 + 12x - 15$$
$$y = -2(x^2 - 6x \quad\quad) - 15 \qquad\qquad \text{Factor out } -2 \text{ from } -2x^2 + 12x.$$
$$y = -2(x^2 - 6x + 9) - 15 + 18 \qquad \text{Subtract and add 18.}$$
$$y = -2(x - 3)^2 + 3$$

Because the equation is of the form $y = -a(x - h)^2 + k$, the parabola opens downward and has its vertex at $(3, 3)$. The graph of the equation appears in Figure 8-28.

$$y = -2x^2 + 12x - 15$$

x	y
1	-5
2	1
3	3
4	1
5	-5

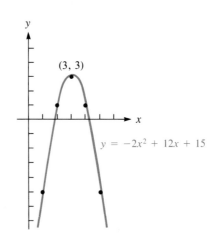

(3, 3)

$y = -2x^2 + 12x + 15$

Figure 8-28

■ EXERCISE 8.7

In Exercises 1–10, graph each equation.

1. $x^2 + y^2 = 9$

2. $x^2 + y^2 = 16$

3. $(x - 2)^2 + y^2 = 36$

4. $x^2 + (y - 3)^2 = 25$

5. $(x - 2)^2 + (y - 4)^2 = 4$

6. $(x - 3)^2 + (y - 2)^2 = 4$

7. $(x + 3)^2 + (y - 1)^2 = 16$

8. $(x - 1)^2 + (y + 4)^2 = 9$

9. $x^2 + (y + 3)^2 = 25$

10. $(x + 4)^2 + y^2 = 1$

In Exercises 11–18, write the equation of the circle with the given properties.

11. Center at the origin; radius of 1

12. Center at the origin; radius of 4

13. Center at (6, 8); radius of 5

14. Center at (5, 3); radius of 2

15. Center at (−2, 6); radius of 12

16. Center at (5, −4); radius of 6

17. Center at the origin; diameter of $2\sqrt{2}$

18. Center at the origin; diameter of $4\sqrt{3}$

In Exercises 19–26, graph each circle.

19. $x^2 + y^2 + 2x - 26 = 0$

20. $x^2 + y^2 - 4y = 12$

21. $9x^2 + 9y^2 - 12y = 5$

22. $4x^2 + 4y^2 + 8y = 12$

23. $x^2 + y^2 - 2x + 4y = 4$

24. $x^2 + y^2 + 4x + 2y = 4$

25. $x^2 + y^2 + 6x - 4y = 3$

26. $x^2 + y^2 + 8x + 2y = -13$

In Exercises 27–40, determine the vertex of each parabola and graph it.

27. $x = y^2$

28. $x = -y^2 + 1$

29. $x = -\dfrac{1}{4}y^2$

30. $x = 4y^2$

31. $y = x^2 + 4x + 5$

32. $y = -x^2 - 2x - 8$

33. $x = -y^2 - y + 6$

34. $x = \dfrac{1}{2}y^2 + 4y$

35. $y^2 + 4x - 6y = -1$

36. $x^2 - 2y - 2x = -7$

37. $y = 2(x - 1)^2 - 3$

38. $y = -2(x + 1)^2 + 2$

39. $x = -3(y + 2)^2 + 2$

40. $x = 2(y - 3)^2 - 4$

8.8   THE ELLIPSE AND THE HYPERBOLA

The graph of the equation

$$\frac{x^2}{36} + \frac{y^2}{9} = 1$$

is called an **ellipse**. To graph this ellipse, we can make a table of values, plot the points, and join them with a smooth curve, as shown in Figure 8-29. We note that the center of the ellipse is at the origin; it intersects the *x*-axis at points (6, 0) and (−6, 0) and intersects the *y*-axis at the points (0, 3) and (0, −3).

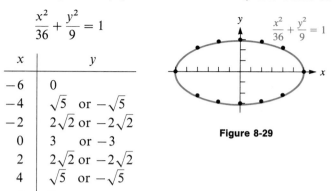

$$\frac{x^2}{36} + \frac{y^2}{9} = 1$$

x	y
-6	0
-4	$\sqrt{5}$ or $-\sqrt{5}$
-2	$2\sqrt{2}$ or $-2\sqrt{2}$
0	3 or -3
2	$2\sqrt{2}$ or $-2\sqrt{2}$
4	$\sqrt{5}$ or $-\sqrt{5}$
6	0

Figure 8-29

The previous discussion illustrates the following general theorem.

> **Theorem.** Any equation that can be written in the form
>
> $$\frac{x^2}{a^2} + \frac{y^2}{b^2} = 1$$
>
> has a graph that is an **ellipse** centered at the origin. It intersects the x-axis at $(a, 0)$ and $(-a, 0)$ and intersects the y-axis at $(0, b)$ and $(0, -b)$.

The following theorem gives the equation of an ellipse with center at the point with coordinates (h, k).

> **Theorem.** Any equation that can be written in the form
>
> $$\frac{(x - h)^2}{a^2} + \frac{(y - k)^2}{b^2} = 1$$
>
> is an ellipse with center at the point (h, k).

Example 1   Graph the ellipse $\dfrac{(x - 2)^2}{16} + \dfrac{(y + 3)^2}{25} = 1$.

Solution   Write the equation in the form

$$\frac{(x - 2)^2}{16} + \frac{[y - (-3)]^2}{25} = 1$$

to see that the center of the ellipse is at the point $(2, -3)$. To find some points on the ellipse, let $x = 2$ so that $x - 2 = 0$ and solve for y:

$$\frac{(x-2)^2}{16} + \frac{(y+3)^2}{25} = 1$$

$$\frac{(2-2)^2}{16} + \frac{(y+3)^2}{25} = 1$$

$$\frac{(y+3)^2}{25} = 1$$

$$(y+3)^2 = 25$$

$$y + 3 = 5 \quad \text{or} \quad y + 3 = -5 \qquad \text{Apply the square root theorem.}$$
$$y = 2 \quad \Big| \quad y = -8$$

Thus, the points with coordinates of (2, 2) and (2, −8) lie on the graph. Now let $y = -3$ and solve for x:

$$\frac{(x-2)^2}{16} + \frac{(y+3)^2}{25} = 1$$

$$\frac{(x-2)^2}{16} + \frac{(-3+3)^2}{25} = 1$$

$$(x-2)^2 = 16$$

$$x - 2 = 4 \quad \text{or} \quad x - 2 = -4$$
$$x = 6 \quad \Big| \quad x = -2$$

Thus, the points with coordinates of (6, −3) and (−2, −3) lie on the graph. The graph appears in Figure 8-30.

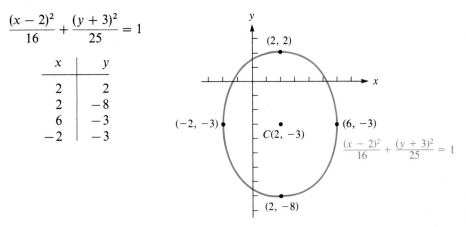

$$\frac{(x-2)^2}{16} + \frac{(y+3)^2}{25} = 1$$

x	y
2	2
2	−8
6	−3
−2	−3

Figure 8-30

The Hyperbola

The graph of the equation $\dfrac{x^2}{25} - \dfrac{y^2}{9} = 1$ is called a **hyperbola**. To graph this hyperbola, we can make a table of values, plot the points, and join them with a smooth curve, as in Figure 8-31.

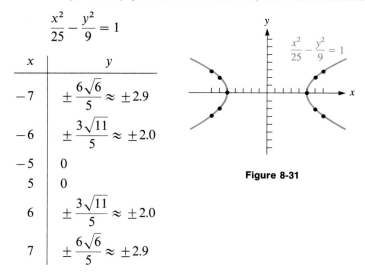

$$\frac{x^2}{25} - \frac{y^2}{9} = 1$$

x	y
-7	$\pm\dfrac{6\sqrt{6}}{5} \approx \pm 2.9$
-6	$\pm\dfrac{3\sqrt{11}}{5} \approx \pm 2.0$
-5	0
5	0
6	$\pm\dfrac{3\sqrt{11}}{5} \approx \pm 2.0$
7	$\pm\dfrac{6\sqrt{6}}{5} \approx \pm 2.9$

Figure 8-31

In general, we have the following theorem:

Theorem.   Any equation that can be written in the form

$$\frac{x^2}{a^2} - \frac{y^2}{b^2} = 1$$

has a graph that is a **hyperbola**. The graph is centered at the origin and intersects the x-axis at $(a, 0)$ and $(-a, 0)$.

The branches of the hyperbola in Figure 8-31 opened left and right. It is possible for hyperbolas to have different orientations with respect to the x- and y-axes. For example, the branches of a hyperbola can open upward and downward, in which case the following theorem applies.

Theorem.   Any equation that can be written in the form

$$\frac{y^2}{b^2} - \frac{x^2}{a^2} = 1$$

has a graph that is a **hyperbola**. The graph is centered at the origin and intersects the y-axis at $(0, b)$ and $(0, -b)$.

Example 2   Graph the equation $9y^2 - 4x^2 = 36$.

Solution   To write the equation in standard form, divide both sides by 36 and simplify to obtain

$$\frac{9y^2}{36} - \frac{4x^2}{36} = 1$$

$$\frac{y^2}{4} - \frac{x^2}{9} = 1$$

This equation represents a hyperbola centered at the origin and intersecting the y-axis at (0, 2) and (0, −2). To determine other points that lie on the graph, substitute several values for x in the equation $9y^2 - 4x^2 = 36$ and solve for y. The graph and a table of values appear in Figure 8-32.

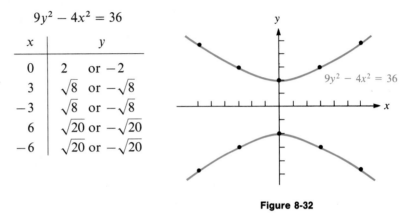

$9y^2 - 4x^2 = 36$

x	y
0	2 or -2
3	$\sqrt{8}$ or $-\sqrt{8}$
-3	$\sqrt{8}$ or $-\sqrt{8}$
6	$\sqrt{20}$ or $-\sqrt{20}$
-6	$\sqrt{20}$ or $-\sqrt{20}$

Figure 8-32

If a hyperbola is centered at the point with coordinates (h, k), the following theorem applies:

Theorem.   Any equation that can be written in the form

$$\frac{(x - h)^2}{a^2} - \frac{(y - k)^2}{b^2} = 1$$

is a hyperbola with center at (h, k) that opens left and right.
Any equation of the form

$$\frac{(y - k)^2}{b^2} - \frac{(x - h)^2}{a^2} = 1$$

is a hyperbola with center at (h, k) that opens up and down.

Example 3   Graph the hyperbola $\dfrac{(x - 3)^2}{16} - \dfrac{(y + 1)^2}{4} = 1.$

Solution Write the equation in the form

$$\frac{(x-3)^2}{16} - \frac{[y-(-1)]^2}{4} = 1$$

to see that the hyperbola is centered at the point $(h, k) = (3, -1)$. A table of values and the graph are shown in Figure 8-33.

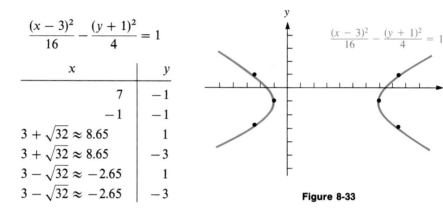

$$\frac{(x-3)^2}{16} - \frac{(y+1)^2}{4} = 1$$

x	y
7	-1
-1	-1
$3 + \sqrt{32} \approx 8.65$	1
$3 + \sqrt{32} \approx 8.65$	-3
$3 - \sqrt{32} \approx -2.65$	1
$3 - \sqrt{32} \approx -2.65$	-3

Figure 8-33

There is a special type of hyperbola also centered at the origin that does not intersect either the x- or y-axis. These hyperbolas have equations of the form $xy = k$, where k is a constant and $k \neq 0$.

Example 4 Graph the equation $xy = -8$.

Solution Make a table of values, plot the points, and join them with a smooth curve to obtain the hyperbola shown in Figure 8-34.

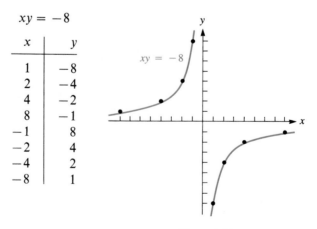

$$xy = -8$$

x	y
1	-8
2	-4
4	-2
8	-1
-1	8
-2	4
-4	2
-8	1

Figure 8-34

The result in Example 4 suggests the following theorem.

> **Theorem.** Any equation of the form $xy = k$, where k is a constant and $k \neq 0$, has a graph that is a **hyperbola** centered at the origin. This hyperbola does not intersect either the x- or y-axis.

■ EXERCISE 8.8

In Exercises 1–10, graph the ellipse represented by each equation.

1. $\dfrac{x^2}{4} + \dfrac{y^2}{9} = 1$ **2.** $x^2 + \dfrac{y^2}{9} = 1$

3. $x^2 + 9y^2 = 9$ **4.** $25x^2 + 9y^2 = 225$ **5.** $16x^2 + 4y^2 = 64$ **6.** $4x^2 + 9y^2 = 36$

7. $\dfrac{(x-2)^2}{9} + \dfrac{(y-1)^2}{4} = 1$ **8.** $\dfrac{(x-1)^2}{4} + \dfrac{(y-3)^2}{25} = 1$

9. $(x+1)^2 + 4(y+2)^2 = 4$ **10.** $4(x+4)^2 + 9(y-3)^2 = 36$

In Exercises 11–22, graph the hyperbola represented by each equation.

11. $\dfrac{x^2}{16} - \dfrac{y^2}{4} = 1$ **12.** $\dfrac{x^2}{36} - \dfrac{y^2}{25} = 1$ **13.** $\dfrac{y^2}{25} - \dfrac{x^2}{9} = 1$ **14.** $\dfrac{y^2}{4} - \dfrac{x^2}{64} = 1$

15. $25x^2 - y^2 = 25$ **16.** $9x^2 - 4y^2 = 36$

17. $\dfrac{(x-2)^2}{9} - \dfrac{y^2}{16} = 1$ **18.** $\dfrac{(x+2)^2}{16} - \dfrac{(y-3)^2}{25} = 1$

19. $4(x+3)^2 - (y-1)^2 = 4$ **20.** $(x+5)^2 - 16y^2 = 16$

21. $xy = 8$ **22.** $xy = -10$

CHAPTER SUMMARY

Key Words

axis of symmetry (8.4)
circle (8.7)
completing the square (8.1)
complex conjugates (8.2)
complex numbers (8.2)
conic sections (8.7)
discriminant (8.3)
ellipse (8.8)
greatest integer function (8.5)

horizontal line test (8.5)
hyperbola (8.8)
imaginary numbers (8.2)
one-to-one function (8.5)
parabola (8.4, 8.7)
quadratic function (8.4)
vertex of a parabola (8.4)
vertical line test (8.5)

Key Ideas

(8.1)   The **square root property:**   If $c > 0$, the equation $x^2 = c$ has two real solutions. They are $x = \sqrt{c}$ and $x = -\sqrt{c}$.

The **quadratic formula:** $x = \dfrac{-b \pm \sqrt{b^2 - 4ac}}{2a}$ $(a \neq 0)$

(8.2)   **Properties of complex numbers:**   If a, b, c, and d are real numbers and given that $i^2 = -1$, then

$$a + bi = c + di \quad \text{if and only if} \quad a = c \text{ and } b = d$$
$$(a + bi) + (c + di) = (a + c) + (b + d)i$$
$$(a + bi)(c + di) = (ac - bd) + (ad + bc)i$$
$$|a + bi| = \sqrt{a^2 + b^2}$$

(8.3)   The **discriminant,** $b^2 - 4ac$, is used to determine the nature of the solutions of the equation $ax^2 + bx + c = 0$, where a, b, and c are real numbers and $a \neq 0$.

If $b^2 - 4ac > 0$, the two solutions are unequal real numbers.

If $b^2 - 4ac = 0$, the two solutions are equal real numbers.

If $b^2 - 4ac < 0$, the two solutions are complex conjugates.

If r_1 and r_2 are the solutions of $ax^2 + bx + c = 0$, then

$$r_1 + r_2 = -\frac{b}{a} \quad \text{and} \quad r_1 r_2 = \frac{c}{a}$$

(8.4)   If $a \neq 0$, the graph of the equation $y = a(x - h)^2 + k$ is a parabola with vertex at (h, k). It opens upward when $a > 0$ and downward when $a < 0$.

(8.5)   If every vertical line that intersects a graph does so exactly once, the graph represents a function.

To find the inverse relation of a function, interchange the x and y variables and solve for y.

If every horizontal line that intersects the graph of a function does so exactly once, the function is one-to-one.

(8.6)   To graph a quadratic inequality in one variable, make a sign chart. To graph an inequality in two variables such as $y > 4x - 3$, first graph the equation $y = 4x - 3$. Then determine which half-plane represents the graph of $y > 4x - 3$.

(8.7)   The graph of $(x - h)^2 + (y - k)^2 = r^2$ is a circle centered at (h, k) with radius r.

The graph of $x^2 + y^2 = r^2$ is a circle centered at $(0, 0)$ with radius r.

Equations of parabolas: If $a > 0$,

Parabola opening	Vertex at origin	Vertex at (h, k)
Up	$y = ax^2$	$y = a(x - h)^2 + k$
Down	$y = -ax^2$	$y = -a(x - h)^2 + k$
Right	$x = ay^2$	$x = a(y - k)^2 + h$
Left	$x = -ay^2$	$x = -a(y - k)^2 + h$

(8.8) The graph of $\dfrac{x^2}{a^2} + \dfrac{y^2}{b^2} = 1$ is an ellipse centered at the origin.

The graph of $\dfrac{(x - h)^2}{a^2} + \dfrac{(y - k)^2}{b^2} = 1$ is an ellipse centered at (h, k).

The graph of $\dfrac{x^2}{a^2} - \dfrac{y^2}{b^2} = 1$ or $\dfrac{y^2}{b^2} - \dfrac{x^2}{a^2} = 1$ is a hyperbola centered at the origin.

The graph of $\dfrac{(x - h)^2}{a^2} - \dfrac{(y - k)^2}{b^2} = 1$ or $\dfrac{(y - k)^2}{b^2} - \dfrac{(x - h)^2}{a^2} = 1$ is a hyperbola centered at (h, k).

REVIEW EXERCISES

In Review Exercises 1–4, solve each equation by factoring or by using the square root property.

1. $12x^2 + x - 6 = 0$
2. $6x^2 + 17x + 5 = 0$
3. $15x^2 + 2x - 8 = 0$
4. $(x + 2)^2 = 36$

In Review Exercises 5–6, solve each equation by completing the square.

5. $x^2 + 6x + 8 = 0$
6. $2x^2 - 9x + 7 = 0$

In Review Exercises 7–10, solve each equation by using the quadratic formula.

7. $x^2 - 8x - 9 = 0$
8. $x^2 - 10x = 0$
9. $2x^2 + 13x - 7 = 0$
10. $3x^2 - 20x - 7 = 0$

In Review Exercises 11–28, perform the indicated operations. **Give all answers in a + bi form.**

11. $(5 + 4i) + (7 - 12i)$
12. $(-6 - 40i) - (-8 + 28i)$
13. $(-32 + \sqrt{-144}) - (64 + \sqrt{-81})$
14. $(-8 + \sqrt{-8}) + (6 - \sqrt{-32})$
15. $7i(-3 + 4i)$
16. $6i(2 + i)$
17. $(5 - 3i)(-6 + 2i)$
18. $(2 + \sqrt{-128})(3 - \sqrt{-98})$
19. $\dfrac{3}{4i}$
20. $\dfrac{-2}{5i^3}$
21. $\dfrac{6}{2 + i}$
22. $\dfrac{7}{3 - i}$
23. $\dfrac{4 + i}{4 - i}$
24. $\dfrac{3 - i}{3 + i}$
25. $\dfrac{3}{5 + \sqrt{-4}}$
26. $\dfrac{2}{3 - \sqrt{-9}}$
27. $|9 + 12i|$
28. $|24 - 10i|$

In Review Exercises 29–30, use the discriminant to determine what type of solutions exist for each equation.

29. $3x^2 + 4x - 3 = 0$ **30.** $4x^2 - 5x + 7 = 0$

31. Find the values of k that will make the solutions of $(k - 8)x^2 + (k + 16)x = -49$ equal.

32. Find values of k such that the solutions of $3x^2 + 4x = k + 1$ are real numbers.

33. A rectangle is 2 centimeters longer than it is wide. If both the length and width are doubled, its area is increased by 72 square centimeters. What are the dimensions of the original rectangle?

34. A rectangle is 1 foot longer than it is wide. If the length is tripled and the width is doubled, its area is increased by 30 square feet. What are the dimensions of the original rectangle?

In Review Exercises 35–38, solve each equation.

35. $x - 13x^{1/2} + 12 = 0$ **36.** $a^{2/3} + a^{1/3} - 6 = 0$

37. $\dfrac{1}{x + 1} - \dfrac{1}{x} = -\dfrac{1}{x + 1}$ **38.** $\dfrac{6}{x + 2} + \dfrac{6}{x + 1} = 5$

39. Find the sum of the solutions of the equation $3x^2 - 14x + 3 = 0$.

40. Find the product of the solutions of the equation $3x^2 - 14x + 3 = 0$.

In Review Exercises 41–44, graph each equation and give the coordinates of the vertex of the resulting parabola.

41. $y = 2x^2 - 3$ **42.** $y = -2x^2 - 1$

43. $y = -4(x - 2)^2 + 1$ **44.** $y = 5(x + 1)^2 - 6$

In Review Exercises 45–48, graph each equation and use the vertical line test to decide whether the graph determines a function.

45. $y = |2x + 4|$ **46.** $x = y^2 - 5$ **47.** $xy = -6$ **48.** $x = |y - 1|$

In Review Exercises 49–52, find the inverse of each function.

49. $y = 6x - 3$ **50.** $y = 4x^2 + 5$

51. $y = 2x^2 - 1$, with $x \geq 0$ **52.** $y = |x|$

In Review Exercises 53–54, each equation represents a function. Graph each equation and use the horizontal line test to decide whether the equation determines a one-to-one function.

53. $y = 2(x - 3)$ **54.** $y = x(2x - 3)$

In Review Exercises 55–58, solve each inequality and graph the solution set.

55. $x^2 + 2x - 35 > 0$ **56.** $x^2 + 7x - 18 < 0$ **57.** $\dfrac{3}{x} \leq 5$ **58.** $\dfrac{2x^2 - x - 28}{x - 1} > 0$

In Review Exercises 59–66, graph each equation.

59. $x^2 + y^2 = 36$ **60.** $(x - 3)^2 + (y + 2)^2 = 25$

61. $x = -3(y - 2)^2 + 5$ **62.** $x = 2(y + 1)^2 - 2$

63. $9x^2 + 16y^2 = 144$ **64.** $\dfrac{(x - 2)^2}{4} + \dfrac{(y - 4)^2}{9} = 1$

65. $xy = 15$ **66.** $9x^2 - y^2 = -9$

67. Graph the equation $x(x - y) = (x + 4)(x - 4)$.

68. Graph the equation $x(x + y) = (x + 3)(x - 3)$.

CHAPTER EIGHT TEST

In Problems 1–2, solve each equation.

1. $x^2 + 3x - 18 = 0$

2. $x(6x + 19) = -15$

In Problems 3–4, determine what number must be added to each binomial to make it a perfect square.

3. $x^2 + 24x$

4. $x^2 - 50x$

In Problems 5–6, solve each equation.

5. $x^2 + 4x + 1 = 0$

6. $x^2 - 5x - 3 = 0$

In Problems 7–11, perform the indicated operations. ***Give all answers in a + bi form.***

7. $(3 - \sqrt{-9}) - (-1 + \sqrt{-16})$

8. $2i(3 - 4i)$

9. $(3 + 2i)(-4 - i)$

10. $\dfrac{1}{i\sqrt{2}}$

11. $\dfrac{2 + i}{3 - i}$

12. Determine whether the solutions of $3x^2 + 5x + 17 = 0$ are real or nonreal.

13. For what value(s) of k are the solutions of $4x^2 - 2kx + k - 1 = 0$ equal?

14. One leg of a right triangle is 14 inches longer than the other, and the hypotenuse is 26 inches long. How long is the shortest side of the triangle?

15. Solve the equation $2y - 3y^{1/2} + 1 = 0$.

16. Graph the equation $y = \dfrac{1}{2}x^2 + 5$ and give the coordinates of its vertex.

17. Graph the equation $|y + 1| = x$ and use the vertical line test to decide whether the graph represents a function.

18. Find the inverse of $y = 3x^2 + 4$ with $x \le 0$. Is the inverse a function?

19. Graph the function represented by $y = \dfrac{1}{2}x^2 - 6$ and use the horizontal line test to decide whether the function is one-to-one.

In Problems 20–21, solve each inequality and graph its solution set.

20. $x^2 - 2x - 8 > 0$

21. $\dfrac{x - 2}{x + 3} \le 0$

In Problems 22–25, graph each equation.

22. $(x - 2)^2 + (y + 3)^2 = 4$

23. $9x^2 + 4y^2 = 36$

24. $-xy = 15$

25. $\dfrac{(x - 2)^2}{25} - y^2 = 1$

CHEMIST Chemists search for and put to practical use new knowledge about substances.

Over half of all chemists work in research and development. In basic research, chemists investigate the properties, composition, and structure of matter and the laws that govern the combination of elements and the reactions of substances. Their research has resulted in the development of a tremendous variety of synthetic materials, of ingredients that have improved other substances, and of processes that help save energy and reduce pollution. In applied research and development, they create new products or improve existing ones, often using knowledge gained from basic research.

Qualifications A bachelor's degree in chemistry or a related discipline is sufficient for many beginning jobs as a chemist. Graduate training, however, is required for most research jobs, and most college teaching jobs require a Ph.D. degree.

Job outlook The employment of chemists is expected to grow about as fast as the average for all occupations through the mid-1990s.

The majority of job openings are expected to be in private industry, primarily in the development of new products. In addition, industrial companies and government agencies will need chemists to help solve problems related to energy shortages, pollution control, and health care.

Example application A certain weak acid (0.1 M concentration) will break down into free cations (the hydrogen ion, H^+) and anions (A^-). When this acid dissociates, the following equilibrium equation is established.

1. $$\frac{[H^+][A^-]}{[HA]} = 4 \times 10^{-4}$$

where $[H^+]$, the hydrogen ion concentration, is equal to $[A^-]$, the anion concentration. $[HA]$ is the concentration of the undissociated acid itself. Find $[H^+]$ at equilibrium.

Solution Let x be the concentration of H^+. Then x is also $[A^-]$. From chemistry, it turns out that the concentration $[HA]$ of the undissociated acid is $0.1 - x$. Substituting these concentrations into Equation 1 gives a quadratic equation.

$$\frac{x^2}{0.1 - x} = 4 \times 10^{-4}$$

This equation can be solved as follows

$$\frac{x^2}{0.1 - x} = 0.0004 \qquad \text{Because } 4 \times 10^{-4} = 0.0004.$$

$$x^2 = 0.0004(0.1 - x) \qquad \text{Multiply both sides by } 0.1 - x.$$

$$x^2 = 0.00004 - 0.0004x \qquad \text{Remove parentheses.}$$

$$x^2 + 0.0004x - 0.00004 = 0 \qquad \text{Add } 0.0004x - 0.00004$$
$$\text{to both sides.}$$

Substitute into the quadratic formula with $a = 1$, $b = 0.0004$, and $c = -0.00004$. You only need to consider the positive root.

$$x = \frac{-b + \sqrt{b^2 - 4ac}}{2a}$$

$$x = \frac{-0.0004 + \sqrt{(0.0004)^2 - 4(1)(-0.00004)}}{2(1)}$$

$$x = \frac{-0.0004 + 0.01266}{2}$$

$$x = 0.00613$$

At equilibrium, $[H^+]$ is approximately 6.13×10^{-3} M.

Exercises

Use a scientific calculator to solve these problems.

1. A saturated solution of hydrogen sulfide (concentration 0.1 M) dissociates into cation H^+ and anion HS^-. When this solution dissociates, the following equilibrium equation is established.

    $$\frac{[H^+][HS^-]}{[HHS]} = 1.0 \times 10^{-7}$$

    Find $[H^+]$.

2. An HS^- anion dissociates into cation H^+ and anion S^{--} with equilibrium equation

    $$\frac{[H^+][S^{--}]}{[HS^-]} = 1.3 \times 10^{-13}$$

    Assume the concentration of HS^- to be 1×10^{-4} M. Find $[H^+]$.

3. Suppose the concentration of the acid of the example were 0.2 M. Find the hydrogen ion concentration, $[H^+]$, at equilibrium.

4. Show that the equation of the example has only one meaningful solution by showing that its other solution is negative.

(Answers: **1.** 9.995×10^{-5} **2.** 3.605×10^{-9} **3.** 8.75×10^{-3})

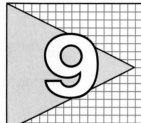

Systems of Equations and Inequalities

In the previous chapter, we considered many equations that contained two variables (usually x and y). We found that there was an unlimited number of ordered pairs (x, y) that satisfied each given equation. The graph of each equation was found by plotting points. We now consider **systems of equations**: either two equations, each with two variables; or three equations, each with three variables.

9.1 SOLUTION BY GRAPHING

Consider the pair of equations

$$\begin{cases} x + 2y = 4 \\ 2x - y = 3 \end{cases}$$

There are infinitely many ordered pairs (x, y) that satisfy the first equation, and there are infinitely many ordered pairs (x, y) that satisfy the second equation. However, there is only *one* ordered pair (x, y) that satisfies *both* equations simultaneously. The process of finding the ordered pair (x, y) that satisfies both of these equations simultaneously is called **solving the system of simultaneous equations**. The first example shows how to use a graphing method to solve this system.

Example 1 Solve the system $\begin{cases} x + 2y = 4 \\ 2x - y = 3 \end{cases}$ by graphing.

Solution Graph both equations on a single set of coordinate axes, as shown in Figure 9-1. Although an infinite number of pairs (x, y) satisfy the equation $x + 2y = 4$, and an infinite number of pairs (x, y) satisfy the equation $2x - y = 3$, only the coordinates of the point where the graphs intersect satisfy both equations simultaneously. The solution to this system is $x = 2$ and $y = 1$, or just $(2, 1)$. To check the answers, substitute 2 for x and 1 for y in each equation, and verify that $(2, 1)$ satisfies each equation.

$$\begin{array}{c|c} x + 2y \overset{?}{=} 4 & 2x - y \overset{?}{=} 3 \\ 2 + 2(1) \overset{?}{=} 4 & 2(2) - 1 \overset{?}{=} 3 \\ 4 = 4 & 3 = 3 \end{array}$$

$x + 2y = 4$ $2x - y = 3$

x	y
4	0
0	2
-2	3

x	y
1	-1
0	-3
-1	-5

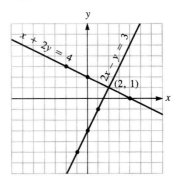

Figure 9-1

In Example 1, the graphs of the two equations were two distinct lines and the system had a solution. If two equations have distinct graphs, the equations are called **independent**. Otherwise, they are called **dependent**. If a system of equations has a solution, the system is called a **consistent system of equations**. If a system of equations does not have a solution, the system is called an **inconsistent system of equations**.

Example 2   Use the graphing method to solve the system $\begin{cases} 2x + 3y = 6 \\ 4x + 6y = 24 \end{cases}$.

Solution   Graph both equations on the same set of coordinate axes, as in Figure 9-2. The lines in the figure must be parallel because the slopes of the two lines are equal. Verify that the slope of each line is $-\frac{2}{3}$ by writing each equation in slope–intercept form. Because the graphs of these two equations are distinct lines, the two equations are independent equations. However, because they are parallel lines and parallel lines do not intersect, the system does not have a solution. Thus, this system is inconsistent.

$2x + 3y = 6$ $4x + 6y = 24$

x	y
3	0
0	2
-3	4

x	y
6	0
0	4
-3	6

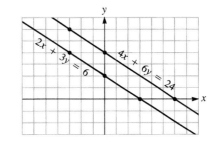

Figure 9-2

Example 3   Use the graphing method to solve the system $\begin{cases} 2y - x = 4 \\ 2x + 8 = 4y \end{cases}$.

Solution   Graph each equation on the same set of coordinate axes, as in Figure 9-3. In this example, the graphs of each equation coincide. This system is consistent because there are an infinite number of simultaneous solutions. Because the equations are essentially the same, any pair that satisfies one of the equations satisfies the other also. For this reason, these equations are dependent equations.

$$2y - x = 4 \qquad 2x + 8 = 4y$$

x	y
-4	0
0	2

x	y
-4	0
0	2

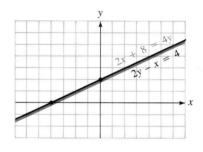

Figure 9-3

The following box summarizes the possibilities that can occur when two equations, each with two variables, are graphed.

If the	**Then**
lines are distinct and intersect,	the equations are independent and the system is consistent. One simultaneous solution exists.
lines are distinct and parallel,	the equations are independent and the system is inconsistent. No simultaneous solution exists.
lines coincide,	the equations are dependent and the system is consistent. An infinite number of simultaneous solutions exist.

Example 4   Use the graphing method to solve the system $\begin{cases} \dfrac{3}{2}x - y = \dfrac{5}{2} \\ x + \dfrac{1}{2}y = 4 \end{cases}$

Solution   Multiply both sides of the equation $\frac{3}{2}x - y = \frac{5}{2}$ by 2 to clear it of fractions and, thereby, obtain the equation $3x - 2y = 5$. Multiply both sides of the equation $x + \frac{1}{2}y = 4$ by 2 to clear it of fractions and, thereby, obtain the equation $2x + y = 8$. The new system

$$\begin{cases} 3x - 2y = 5 \\ 2x + y = 8 \end{cases}$$

STEVEN E. JABLONSKI

SERGEANT FIRST CLASS
U.S. ARMY NURSE RECRUITER

U.S. ARMY
NURSE RECRUITING STATION

4313 VALLEY GREEN MALL
ETTERS, PA 17319

OFFICE PHONE (717) 938-5531
HOME PHONE (717) 533-5039

has the same simultaneous solution as the given system of equations, but with no fractions it is easier to solve. Graph each equation in this new system as in Figure 9-4. The coordinates of the point of intersection of the two lines are (3, 2), and this pair is the solution to both the new and the original system. Verify that (3, 2) satisfies each of the equations in the original system. In this example, the equations are independent and the system is consistent.

$$3x - 2y = 5 \qquad 2x + y = 8$$

x	y
0	$-\dfrac{5}{2}$
$\dfrac{5}{3}$	0

x	y
4	0
1	6

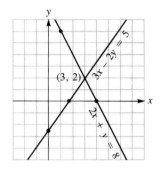

Figure 9-4

Example 5   Is the system $\begin{cases} y = 3x + 7 \\ 3x - y = 12 \end{cases}$ consistent?

Solution   Note that the slope of the graph of $y = 3x + 7$ is 3 and the slope of the graph of $3x - y = 12$ (or $y = 3x - 12$) is 3 also. Because both lines have the same slope, either the lines are parallel or they coincide. Because the y-intercept of the graph of $y = 3x + 7$ is 7 and the y-intercept of the graph of $3x - y = 12$ is -12, the graphs cannot coincide. Because the lines are distinct and parallel, the system is inconsistent.

EXERCISE 9.1

In Exercises 1–20, solve each system of equations by the graphing method. If a system is inconsistent or if the equations of a system are dependent, so indicate.

1. $\begin{cases} x + y = 6 \\ x - y = 2 \end{cases}$

2. $\begin{cases} x - y = 4 \\ 2x + y = 5 \end{cases}$

3. $\begin{cases} 2x + y = 1 \\ x - 2y = -7 \end{cases}$

4. $\begin{cases} 3x - y = -3 \\ 2x + y = -7 \end{cases}$

5. $\begin{cases} x = 13 - 4y \\ 3x = 4 + 2y \end{cases}$

6. $\begin{cases} 3x = 7 - 2y \\ 2x = 2 + 4y \end{cases}$

7. $\begin{cases} x = 3 - 2y \\ 2x + 4y - 6 = 0 \end{cases}$

8. $\begin{cases} 3x = 5 - 2y \\ 3x + 2y - 7 = 0 \end{cases}$

9. $\begin{cases} x = 5 \\ y = \dfrac{9 - x}{2} \end{cases}$

10. $\begin{cases} y = -2 \\ x = \dfrac{4 + 3y}{2} \end{cases}$

11. $\begin{cases} y = 5 \\ x = 2 \end{cases}$

12. $\begin{cases} 2x + 3y = -15 \\ 2x + y = -9 \end{cases}$

13. $\begin{cases} x = \dfrac{11 - 2y}{3} \\ y = \dfrac{11 - 6x}{4} \end{cases}$

14. $\begin{cases} x = \dfrac{1 - 3y}{4} \\ y = \dfrac{12 + 3x}{2} \end{cases}$

15. $\begin{cases} \dfrac{5}{2}x + y = \dfrac{1}{2} \\ 2x - \dfrac{3}{2}y = 5 \end{cases}$

16. $\begin{cases} \dfrac{5}{2}x + 3y = 6 \\ y = \dfrac{24 - 10x}{12} \end{cases}$

17. $\begin{cases} x = \dfrac{5y-4}{2} \\ x - \dfrac{5}{3}y + \dfrac{1}{3} = 0 \end{cases}$ **18.** $\begin{cases} 2x = 5y - 11 \\ 3x = 2y \end{cases}$ **19.** $\begin{cases} x = -\dfrac{3}{2}y \\ x = \dfrac{3}{2}y - 2 \end{cases}$ **20.** $\begin{cases} x = \dfrac{3y-1}{4} \\ y = \dfrac{4-8x}{3} \end{cases}$

21. Form an independent system of equations with the simultaneous solution $(-5, 2)$.

22. Form a dependent system of equations with one possible solution of $(-5, 2)$.

9.2 SOLUTION BY SUBSTITUTION AND ADDITION

The graphing method provides a nice way to visualize the process of solving systems of equations. However, it has two major deficiencies: The method gives exact answers only if the lines in a graph happen to intersect exactly at a point whose coordinates can be read accurately from the graph, and the method cannot be used to solve systems of higher order, such as three equations, each with three variables.

In this section, we will discuss two algebraic methods of solving systems of two equations in two variables. We begin with the **substitution method**.

Example 1 Use the substitution method to solve the system $\begin{cases} 4x + y = 13 \\ -2x + 3y = -17 \end{cases}$.

Solution There is a two-part strategy to the substitution method. First, solve one of the equations for one of its variables. Second, substitute that quantity for the same variable in the other equation. In this example, it is most convenient to solve the first equation for y, because y has a coefficient of 1 and no fractions are introduced. Then substitute that value of y for y in the second equation and thereby get one equation with one variable:

$$\begin{cases} 4x + y = 13 \Rightarrow y = \boxed{-4x + 13} \\ -2x + 3y = -17 \end{cases}$$

$$-2x + 3(-4x + 13) = -17$$

Solve this new equation for x as follows:

$$\begin{aligned} -2x + 3(-4x + 13) &= -17 \\ -2x - 12x + 39 &= -17 \qquad \text{Remove parentheses.} \\ -14x &= -56 \qquad \text{Combine terms and add } -39 \text{ to both sides.} \\ x &= 4 \qquad \text{Divide both sides by } -14. \end{aligned}$$

To find y, substitute 4 for x in the equation $y = -4x + 13$ and simplify:

$$\begin{aligned} y &= -4x + 13 \\ &= -4(4) + 13 \\ &= -3 \end{aligned}$$

The solution to this system is $x = 4$ and $y = -3$, or just $(4, -3)$. Verify that this solution satisfies each equation in the given system.   ∎

Example 2 Use the substitution method to solve the system $\begin{cases} \dfrac{4}{3}x + \dfrac{1}{2}y = -\dfrac{2}{3} \\ \dfrac{1}{2}x + \dfrac{2}{3}y = \dfrac{5}{3} \end{cases}$.

Solution First, find an equivalent system (one with the same solution) that has no fractions. Do this by multiplying each side of each equation by 6 to obtain the system

$$\begin{cases} 8x + 3y = -4 \\ 3x + 4y = 10 \end{cases}$$

Because no variable has a coefficient of 1, it is impossible to avoid fractions when solving either equation for a variable. For sake of argument, solve the second equation for x, and substitute for x in the first equation:

$$\begin{cases} 8x + 3y = -4 \\ 3x + 4y = 10 \end{cases} \Rightarrow x = \frac{10 - 4y}{3}$$

$$8\left(\frac{10 - 4y}{3}\right) + 3y = -4$$

$$\frac{8}{3}(10 - 4y) + 3y = -4$$

Clear this equation of fractions by multiplying each side by 3, and then solve for y:

$$8(10 - 4y) + 9y = -12$$
$$80 - 32y + 9y = -12 \qquad \text{Remove parentheses.}$$
$$-23y = -92 \qquad \text{Combine terms and add } -80 \text{ to both sides.}$$
$$y = 4 \qquad \text{Divide both sides by } -23.$$

Find x by substituting 4 for y in the equation $x = \dfrac{10 - 4y}{3}$ and simplifying:

$$x = \frac{10 - 4y}{3}$$
$$= \frac{10 - 4(4)}{3}$$
$$= \frac{-6}{3}$$
$$= -2$$

The solution to this system is the pair $(-2, 4)$. Verify that this solution satisfies each equation in the original system. ∎

We now consider a second algebraic method of solving systems of equations, called **solution by addition**.

Example 3   Use the addition method to solve the system $\begin{cases} 4x + y = 13 \\ -2x + 3y = -17 \end{cases}$.

Solution   Note that this system is repeated from Example 1. The strategy of the addition method is to adjust the equations so that, if you add their left-hand sides and add their right-hand sides, one of the variables drops out. It is then possible to solve for the remaining variable. In this example, it is convenient to multiply the second equation by 2 to obtain the system

$$\begin{cases} 4x + y = 13 \\ -4x + 6y = -34 \end{cases}$$

When these equations are added, the terms involving x drop out, and you get

$$7y = -21$$
$$y = -3 \qquad \text{Divide both sides by 7.}$$

To solve for x, get the terms involving y to drop out by multiplying the first equation of the original system by -3. This gives the system

$$\begin{cases} -12x - 3y = -39 \\ -2x + 3y = -17 \end{cases}$$

When these equations are added, the terms involving y drop out, and you get

$$-14x = -56$$
$$x = 4 \qquad \text{Divide both sides by } -14.$$

The solution to this system is $x = 4$ and $y = -3$, or just $(4, -3)$. You have already verified, in Example 1, that this solution satisfies each of the original equations. ∎

Example 4   Use the addition method to solve the system $\begin{cases} \dfrac{4}{3}x + \dfrac{1}{2}y = -\dfrac{2}{3} \\ \dfrac{1}{2}x + \dfrac{2}{3}y = \dfrac{5}{3} \end{cases}$.

Solution   Note that this system is repeated from Example 2. As in that example, begin by finding an equivalent system that has no fractions. Again, multiply each side of each equation by 6 to obtain the system

1. $\begin{cases} 8x + 3y = -4 \\ \textbf{2.} \quad 3x + 4y = 10 \end{cases}$

To solve for x, get the terms involving y to drop out by multiplying each side of Equation 1 by 4, and each side of Equation 2 by -3. This produces the system

$$\begin{cases} 32x + 12y = -16 \\ -9x - 12y = -30 \end{cases}$$

When these equations are added, the terms involving y drop out, and you get

$$23x = -46$$
$$x = -2 \qquad \text{Divide both sides by 23.}$$

To solve for y, you could get the terms involving x to drop out by multiplying each side of Equation 1 by 3, and each side of Equation 2 by -8 and adding the equations. However, it is easier to substitute -2 for x in either Equation 1 or Equation 2 and solve for y. For sake of argument, substitute -2 for x in Equation 2 and solve for y:

$$3x + 4y = 10$$
$$3(-2) + 4y = 10$$
$$-6 + 4y = 10 \qquad \text{Simplify.}$$
$$4y = 16 \qquad \text{Add 6 to both sides.}$$
$$y = 4 \qquad \text{Divide both sides by 4.}$$

The solution to this system is $(-2, 4)$. You verified that this pair of values satisfied each of the equations in the given system in Example 2. ■

Example 5   Solve the system $\begin{cases} y = 2x + 4 \\ 8x - 4y = 7 \end{cases}$.

Solution   Because the first equation in this system is already solved for y, use the substitution method and substitute $2x + 4$ for y in the second equation:

$$\begin{cases} y = \boxed{2x + 4} \\ 8x - 4y = 7 \end{cases}$$
$$8x - 4(2x + 4) = 7$$

Solve this equation as follows:

$$8x - 8x - 16 = 7 \qquad \text{Remove parentheses.}$$
$$-16 \neq 7 \qquad \text{Combine terms.}$$

Of course, -16 is not equal to 7. This impossible result indicates that the equations in the given system are independent and that the system is inconsistent. If each equation in this system were graphed, the lines would be parallel. There is no solution to this system. ■

Example 6   Solve the system $\begin{cases} 4x + 6y = 12 \\ -2x - 3y = -6 \end{cases}$.

Solution   Use the addition method. Multiply each side of the second equation by 2 to get the system

$$\begin{cases} 4x + 6y = 12 \\ -4x - 6y = -12 \end{cases}$$

Add the left-hand sides and the right-hand sides of these equations to obtain the equation

$$0x + 0y = 0$$

In this case, both the x and y terms drop out. However, the statement $0 = 0$ is true. This indicates that the equations in this system are dependent and that the system is consistent. Note that the equations are equivalent because, when the second equation is multiplied by -2, it becomes the first equation. If both equations in this system were graphed, the two lines would coincide. Any ordered pair that satisfies one of the equations in the system satisfies the other also.   ■

Example 7   Hi-Fi Electronics sells two models of CB radios. One model sells for $67 and the other for $100. In one week, 36 radios were sold. If the receipts from the sale of the radios totaled $2940, how many of each model were sold?

Solution   Let x represent the number of radios sold for $67, and let y represent the number of radios sold for $100. Then the receipts for the sale of the less expensive model were $67x$, and the receipts for the sale of the more expensive model were $100y$. The information in the problem gives the following two equations:

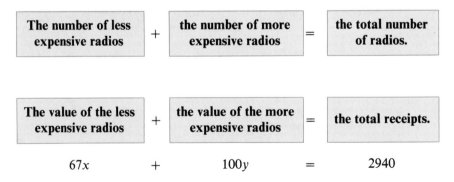

Now solve the following system of equations for x and y to find out how many of each model were sold:

$$\begin{cases} x + y = 36 \\ 67x + 100y = 2940 \end{cases}$$

Multiply both sides of the first equation by -100, add the resulting equation to the second equation, and solve for x:

$$\begin{array}{rcr} -100x - 100y &=& -3600 \\ 67x + 100y &=& 2940 \\ \hline -33x &=& -660 \\ x &=& 20 \end{array}$$ Divide both sides by -33.

To find y, substitute 20 for x in the equation $x + y = 36$, and solve for y:

$$\begin{aligned} x + y &= 36 \\ 20 + y &= 36 \\ y &= 16 \end{aligned}$$ Add -20 to both sides.

Twenty of the less expensive and 16 of the more expensive radios were sold.

Check: Note that, if 20 of one model were sold and 16 of the other model were sold, then a total of 36 radios were sold. Also note that, because the value of the less expensive radios is $20(\$67) = \$1340$ and the value of the more expensive radios is $16(\$100) = \$1600$, the total value is $\$2940$. ∎

Example 8   How many ounces of a 10% saline solution and how many ounces of a 20% saline solution must be mixed together to obtain 50 ounces of a 15% saline solution?

Solution   Let x represent the number of ounces of the 10% solution, and let y represent the number of ounces of the 20% solution that are to be mixed. The information given in the problem gives the following two equations:

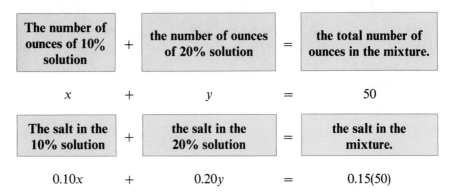

Thus, solve the following system of equations for x and y to find how many ounces of each are needed:

$$\begin{cases} x + \quad y = 50 \\ 0.10x + 0.20y = 0.15(50) \end{cases}$$

Multiply the second equation by 100 to eliminate the decimal fractions, and solve the resulting system as follows:

$$\begin{cases} x + y = 50 & \Rightarrow \quad y = \boxed{50 - x} \\ 10x + 20y = 15(50) \end{cases}$$

$$10x + 20(50 - x) = 750$$

$10x + 1000 - 20x = 750$ Remove parentheses.

$-10x + 1000 = 750$ Combine terms.

$-10x = -250$ Add -1000 to both sides.

$x = 25$ Divide both sides by -10.

Because $y = 50 - x$ and $x = 25$, it follows that

$$y = 50 - x$$
$$= 50 - 25$$
$$= 25$$

To obtain 50 ounces of a 15% solution, you must mix 25 ounces each of the 10% and 20% solutions.   ■

■ EXERCISE 9.2

In Exercises 1–10, solve each system of equations by the substitution method, if possible. If a system is inconsistent, or if the equations are dependent, so indicate.

1. $\begin{cases} y = x \\ x + y = 4 \end{cases}$
2. $\begin{cases} y = x + 2 \\ x + 2y = 16 \end{cases}$
3. $\begin{cases} x - y = 2 \\ 2x + y = 13 \end{cases}$
4. $\begin{cases} x - y = -4 \\ 3x - 2y = -5 \end{cases}$

5. $\begin{cases} x + 2y = 6 \\ 3x - y = -10 \end{cases}$
6. $\begin{cases} 2x - y = -21 \\ 4x + 5y = 7 \end{cases}$
7. $\begin{cases} 3x = 2y - 4 \\ 6x - 4y = -4 \end{cases}$
8. $\begin{cases} 8x = 4y + 10 \\ 4x - 2y - 5 = 0 \end{cases}$

9. $\begin{cases} 3x - 4y = 9 \\ x + 2y = 8 \end{cases}$
10. $\begin{cases} 3x - 2y = -10 \\ 6x + 5y = 25 \end{cases}$

In Exercises 11–20, solve each system of equations by the addition method. If a system is inconsistent, or if the equations are dependent, so indicate.

11. $\begin{cases} x - y = 3 \\ x + y = 7 \end{cases}$
12. $\begin{cases} x + y = 1 \\ x - y = 7 \end{cases}$
13. $\begin{cases} 2x + y = -10 \\ 2x - y = -6 \end{cases}$
14. $\begin{cases} x + 2y = -9 \\ x - 2y = -1 \end{cases}$

15. $\begin{cases} 2x + 3y = 8 \\ 3x - 2y = -1 \end{cases}$
16. $\begin{cases} 5x - 2y = 19 \\ 3x + 4y = 1 \end{cases}$
17. $\begin{cases} 4x + 9y = 8 \\ 2x - 6y = -3 \end{cases}$
18. $\begin{cases} 4x + 6y = 5 \\ 8x - 9y = 3 \end{cases}$

19. $\begin{cases} 8x - 4y = 16 \\ 2(x - 2) = y \end{cases}$
20. $\begin{cases} x = \dfrac{3}{2}y + 4 \\ 2x - 3y = 8 \end{cases}$

In Exercises 21–28, solve each system of equations by any method.

21. $\begin{cases} \dfrac{x}{2} + \dfrac{y}{2} = 6 \\ \dfrac{x}{2} - \dfrac{y}{2} = -2 \end{cases}$

22. $\begin{cases} \dfrac{x}{2} - \dfrac{y}{3} = -4 \\ \dfrac{x}{2} + \dfrac{y}{9} = 0 \end{cases}$

23. $\begin{cases} \dfrac{3}{4}x + \dfrac{2}{3}y = 7 \\ \dfrac{3}{5}x - \dfrac{1}{2}y = 18 \end{cases}$

24. $\begin{cases} \dfrac{2}{3}x - \dfrac{1}{4}y = -8 \\ \dfrac{1}{2}x - \dfrac{3}{8}y = -9 \end{cases}$

25. $\begin{cases} \dfrac{3x}{2} - \dfrac{2y}{3} = 0 \\ \dfrac{3x}{4} + \dfrac{4y}{3} = \dfrac{5}{2} \end{cases}$

26. $\begin{cases} \dfrac{3x}{5} + \dfrac{5y}{3} = 2 \\ \dfrac{6x}{5} - \dfrac{5y}{3} = 1 \end{cases}$

27. $\begin{cases} \dfrac{2}{5}x - \dfrac{1}{6}y = \dfrac{7}{10} \\ \dfrac{3}{4}x - \dfrac{2}{3}y = \dfrac{19}{8} \end{cases}$

28. $\begin{cases} \dfrac{5}{6}x + \dfrac{2}{3}y = \dfrac{7}{6} \\ \dfrac{10}{7}x - \dfrac{4}{9}y = \dfrac{17}{21} \end{cases}$

In Exercises 29–32, solve each system of equations for x and y. Consider solving for $\dfrac{1}{x}$ and $\dfrac{1}{y}$ first.

29. $\begin{cases} \dfrac{1}{x} + \dfrac{1}{y} = \dfrac{5}{6} \\ \dfrac{1}{x} - \dfrac{1}{y} = \dfrac{1}{6} \end{cases}$

30. $\begin{cases} \dfrac{1}{x} + \dfrac{1}{y} = \dfrac{9}{20} \\ \dfrac{1}{x} - \dfrac{1}{y} = \dfrac{1}{20} \end{cases}$

31. $\begin{cases} \dfrac{1}{x} + \dfrac{2}{y} = -1 \\ \dfrac{2}{x} - \dfrac{1}{y} = -7 \end{cases}$

32. $\begin{cases} \dfrac{3}{x} - \dfrac{2}{y} = -30 \\ \dfrac{2}{x} - \dfrac{3}{y} = -30 \end{cases}$

In Exercises 33–43, use two variables to solve each word problem.

33. The sum of two numbers is 49 and their difference is 7. Find the numbers.

34. The sum of the ages of two persons is 98 and the difference of their ages is 16. How old is each person?

35. The perimeter of a rectangle is 72 inches. Find the dimensions of the rectangle if twice the length added to three times the width is 88 inches.

36. A sporting goods salesperson sells 2 fishing reels and 5 rods for $270. The next day, the salesperson sells 4 reels and 2 rods for $220. How much does each cost?

37. In a certain right triangle, one acute angle is 15° greater than two times the other acute angle. Find the difference between the two acute angles.

38. Sam invested part of $8000 at 10% and the rest at 12% interest. His annual income from these investments was $900. How much did he invest at each rate?

39. How many ounces each of an 8% alcohol solution and a 15% alcohol solution must be mixed to obtain 100 ounces of a 12.2% solution?

40. How many pounds each of nuts that cost $2 per pound and $4 per pound must be mixed to obtain 60 pounds of nuts that are worth $3 per pound?

41. A car travels 50 miles in the same time that a plane travels 180 miles. The speed of the plane is 143 miles per hour faster than the speed of the car. Find the speed of the car.

42. In a certain two-digit number, the sum of the digits is 11. If the digits are reversed, the number is decreased by 45. What is the number? (*Hint:* Let t represent the tens digit and u the units digit. Then, $10t + u$ represents the number and $10u + t$ represents the number with its digits reversed.)

43. The manager of an apartment complex is also a tenant. He pays only three-quarters of the rent that each of the remaining 5 tenants pays. Each month, the landlord collects a total of $2070 from the 6 occupants. How much rent does the manager pay?

44. If $r_1 + r_2 = -b/a$, and if $r_1 r_2 = c/a$ ($a \neq 0$), show that r_1 and r_2 are solutions of the quadratic equation $ax^2 + bx + c = 0$.

9.3   SOLUTION BY DETERMINANTS

We now discuss a fourth method of solving certain systems of two equations in two variables. This method, named after the 18th-century mathematician Gabriel Cramer, involves a special number called a **determinant**, which is associated with a square array of numbers.

> **Definition.**   A **matrix** is any rectangular array of numbers.

The following rectangular arrays of numbers are examples of matrices:

$$A = \begin{bmatrix} 1 & 2 & 3 \\ 4 & 5 & 6 \end{bmatrix} \qquad B = \begin{bmatrix} 1 & 2 \\ 3 & 4 \\ 5 & 6 \end{bmatrix} \qquad C = \begin{bmatrix} 2 & 4 & 6 \\ 8 & 10 & 12 \\ 14 & 16 & 18 \end{bmatrix}$$

Because matrix A has two rows and three columns, it is called a 2 by 3 (read as "two by three" and denoted as 2×3) matrix. Matrix B is a 3×2 matrix, because the matrix has three rows and two columns. Matrix C is a 3×3 matrix (three rows and three columns). Any matrix that has the same number of rows as columns is called a **square matrix**. Thus, matrix C is an example of a square matrix.

There is a function, called the **determinant function**, that associates a numerical value with every square matrix. For any square matrix A, the symbol $\det(A)$ or the symbol $|A|$ represents the determinant of A.

> **Definition.**   If a, b, c, and d are numbers, then the **determinant** of the square matrix $A = \begin{bmatrix} a & b \\ c & d \end{bmatrix}$ is
>
> $$\det(A) = \begin{vmatrix} a & b \\ c & d \end{vmatrix} = ad - bc$$

Note that the determinant of a 2×2 matrix A is the *number* that is equal to the product of the entries on the major diagonal

$$\begin{vmatrix} a & b \\ c & d \end{vmatrix}$$

minus the product of the entries on the other diagonal

$$\begin{vmatrix} a & b \\ c & d \end{vmatrix}$$

Example 1   Find the values of the determinants associated with the matrices

a. $\begin{bmatrix} 3 & 2 \\ 6 & 9 \end{bmatrix}$,   **b.** $\begin{bmatrix} -2 & 7 \\ 8 & -5 \end{bmatrix}$, and   **c.** $\begin{bmatrix} -5 & \frac{1}{2} \\ -1 & 0 \end{bmatrix}$

Solution   **a.** $\det\left(\begin{bmatrix} 3 & 2 \\ 6 & 9 \end{bmatrix}\right) = \begin{vmatrix} 3 & 2 \\ 6 & 9 \end{vmatrix} = 3(9) - 2(6) = 27 - 12 = 15$

b. $\det\left(\begin{bmatrix} -2 & 7 \\ 8 & -5 \end{bmatrix}\right) = \begin{vmatrix} -2 & 7 \\ 8 & -5 \end{vmatrix} = (-2)(-5) - 7(8)$

$= 10 - 56 = -46$

c. $\det\left(\begin{bmatrix} -5 & \frac{1}{2} \\ -1 & 0 \end{bmatrix}\right) = \begin{vmatrix} -5 & \frac{1}{2} \\ -1 & 0 \end{vmatrix} = -5(0) - \frac{1}{2}(-1) = 0 + \frac{1}{2} = \frac{1}{2}$ ■

We now turn our attention to solving the general system of two equations in two variables. Consider the system

$$\begin{cases} ax + by = e \\ cx + dy = f \end{cases}$$

with variables x and y, and arbitrary constants a, b, c, d, e, and f. We will use the addition method to solve this system for x and y. We multiply each side of the first equation by d, each side of the second equation by $-b$, and add the equations to eliminate the variable y:

$$\begin{aligned} adx + bdy &= ed \\ -bcx - bdy &= -bf \\ \hline adx - bcx &= ed - bf \end{aligned}$$

We factor out the x on the left-hand side and divide each side by $ad - bc$ to solve for x:

$$(ad - bc)x = ed - bf$$

$$x = \frac{ed - bf}{ad - bc} \qquad \text{provided that } ad - bc \neq 0$$

The value of y can be found in a similar manner. After eliminating the x variable, we get

$$y = \frac{af - ec}{ad - bc} \qquad \text{provided that } ad - bc \neq 0$$

These formulas for x and y are not easy to remember because they involve so many quantities. However, determinants provide an easy way of remembering these formulas. Note that the denominator for the values of both x and y is

$$\begin{vmatrix} a & b \\ c & d \end{vmatrix} = ad - bc$$

The numerators can be expressed as determinants also:

$$x = \frac{ed - bf}{ad - bc} = \frac{\begin{vmatrix} e & b \\ f & d \end{vmatrix}}{\begin{vmatrix} a & b \\ c & d \end{vmatrix}} \quad \text{and} \quad y = \frac{af - ec}{ad - bc} = \frac{\begin{vmatrix} a & e \\ c & f \end{vmatrix}}{\begin{vmatrix} a & b \\ c & d \end{vmatrix}}$$

Compare these formulas with the original system of equations:

$$\begin{cases} ax + by = e \\ cx + dy = f \end{cases}$$

Note that, in the expressions for x and y above, the denominator determinant is formed by using the coefficients a, b, c, and d of the variables in the equations. The numerator determinants are similar to the denominator determinant. However, the column of coefficients of the variable for which we are solving is replaced with the column of constants e and f. Thus, to form the numerator determinant used in solving for x, the coefficients of x (a and c) in the matrix $\begin{bmatrix} a & b \\ c & d \end{bmatrix}$ are replaced by the constants e and f. Similarly, when solving for y, the coefficients of y (b and d) are replaced by the constants e and f:

$$x = \frac{\begin{vmatrix} e & b \\ f & d \end{vmatrix}}{\begin{vmatrix} a & b \\ c & d \end{vmatrix}} \qquad y = \frac{\begin{vmatrix} a & e \\ c & f \end{vmatrix}}{\begin{vmatrix} a & b \\ c & d \end{vmatrix}}$$

Example 2 Solve the system $\begin{cases} 4x - 3y = 6 \\ -2x + 5y = 4 \end{cases}$ by using determinants.

Solution The solution for x is the quotient of two determinants. The denominator determinant involves the matrix of the four coefficients of x and y. To solve for x, form a numerator determinant from that denominator determinant by replacing its first column (the coefficients of x) with the column of constants. The second column remains unchanged.

$$x = \frac{\begin{vmatrix} 6 & -3 \\ 4 & 5 \end{vmatrix}}{\begin{vmatrix} 4 & -3 \\ -2 & 5 \end{vmatrix}}$$

To solve for y, replace the second column (the coefficients of y) of the denominator determinant with the column of constants to form the numerator determinant. The first column remains unchanged.

$$y = \frac{\begin{vmatrix} 4 & 6 \\ -2 & 4 \end{vmatrix}}{\begin{vmatrix} 4 & -3 \\ -2 & 5 \end{vmatrix}}$$

There are three determinants to evaluate: $\begin{vmatrix} 4 & -3 \\ -2 & 5 \end{vmatrix}$, $\begin{vmatrix} 6 & -3 \\ 4 & 5 \end{vmatrix}$, and

$\begin{vmatrix} 4 & 6 \\ -2 & 4 \end{vmatrix}$. Always begin by evaluating the denominator determinant because, if it is 0, the system is a special case and the other determinants might not have to be evaluated. If the determinant in the denominator equals 0, then either the equations are dependent (there are an infinite number of solutions) or the system is inconsistent (there are no solutions). Evaluate the denominator determinant as follows:

$$\begin{vmatrix} 4 & -3 \\ -2 & 5 \end{vmatrix} = 4(5) - (-3)(-2) = 20 - 6 = 14$$

Because the denominator determinant is not 0, the equations in this system are independent and the system is consistent. Evaluate the other two determinants to complete the solutions for x and y:

$$\begin{vmatrix} 6 & -3 \\ 4 & 5 \end{vmatrix} = 6(5) - (-3)(4) = 30 + 12 = 42$$

$$\begin{vmatrix} 4 & 6 \\ -2 & 4 \end{vmatrix} = 4(4) - 6(-2) = 16 + 12 = 28$$

Hence, you have

$$x = \frac{\begin{vmatrix} 6 & -3 \\ 4 & 5 \end{vmatrix}}{\begin{vmatrix} 4 & -3 \\ -2 & 5 \end{vmatrix}} = \frac{42}{14} = 3 \quad \text{and} \quad y = \frac{\begin{vmatrix} 4 & 6 \\ -2 & 4 \end{vmatrix}}{\begin{vmatrix} 4 & -3 \\ -2 & 5 \end{vmatrix}} = \frac{28}{14} = 2$$

The solution to this system is (3, 2). Verify that $x = 3$ and $y = 2$ satisfy each equation in the given system. ■

The method used in Example 2 is called **Cramer's Rule**.

Example 3   Use Cramer's Rule to solve the system $\begin{cases} 7x = 8 - 4y \\ 2y = 3 - \dfrac{7}{2}x \end{cases}$.

Solution   Multiply each side of the second equation by 2 to clear the fractions and rewrite the system in the form

$$\begin{cases} 7x + 4y = 8 \\ 7x + 4y = 6 \end{cases}$$

Because the two equations are different (independent) and because the determinant in the denominator is 0,

$$\begin{vmatrix} 7 & 4 \\ 7 & 4 \end{vmatrix} = 7(4) - 4(7) = 0$$

this system is inconsistent. It has no solutions. ■

Cramer's Rule for Two Equations in Two Variables.

If the system

$$\begin{cases} ax + by = e \\ cx + dy = f \end{cases}$$

has a unique solution, it is given by

$$x = \frac{D_x}{D} \quad \text{and} \quad y = \frac{D_y}{D}$$

where

$$D = \begin{vmatrix} a & b \\ c & d \end{vmatrix}, D_x = \begin{vmatrix} e & b \\ f & d \end{vmatrix}, \text{ and } D_y = \begin{vmatrix} a & e \\ c & f \end{vmatrix}.$$

If the denominators and the numerators of these fractions are *all* 0, the system is consistent, but the equations are dependent.

If the denominators are 0 and one of the numerators is *not* 0, the system is inconsistent.

■ EXERCISE 9.3 ■

In Exercises 1–10, evaluate each determinant.

1. $\begin{vmatrix} 2 & 3 \\ -2 & 1 \end{vmatrix}$
2. $\begin{vmatrix} 3 & -2 \\ -2 & 4 \end{vmatrix}$
3. $\begin{vmatrix} -1 & 2 \\ 3 & -4 \end{vmatrix}$
4. $\begin{vmatrix} -1 & -2 \\ -3 & -4 \end{vmatrix}$

5. $\begin{vmatrix} x & y \\ y & x \end{vmatrix}$
6. $\begin{vmatrix} -x & 2 \\ x & 2 \end{vmatrix}$
7. $\begin{vmatrix} x & 2x \\ 3x & 4x \end{vmatrix}$
8. $\begin{vmatrix} x & y \\ x & y \end{vmatrix}$

9. $\begin{vmatrix} x+1 & x+1 \\ x+1 & x-1 \end{vmatrix}$
10. $\begin{vmatrix} x+y & x-y \\ x-y & x+y \end{vmatrix}$

In Exercises 11–24, use determinants to solve each system of equations for x and y, if possible. If a system is inconsistent or if the equations are dependent, so indicate.

11. $\begin{cases} x + y = 6 \\ x - y = 2 \end{cases}$
12. $\begin{cases} x - y = 4 \\ 2x + y = 5 \end{cases}$
13. $\begin{cases} 2x + y = 1 \\ x - 2y = -7 \end{cases}$
14. $\begin{cases} 3x - y = -3 \\ 2x + y = -7 \end{cases}$

15. $\begin{cases} 2x + 3y = 0 \\ 4x - 6y = -4 \end{cases}$
16. $\begin{cases} 4x - 3y = -1 \\ 8x + 3y = 4 \end{cases}$
17. $\begin{cases} y = \dfrac{-2x+1}{3} \\ 3x - 2y = 8 \end{cases}$
18. $\begin{cases} 2x + 3y = -1 \\ x = \dfrac{y-9}{4} \end{cases}$

19. $\begin{cases} y = \dfrac{11 - 3x}{2} \\ x = \dfrac{11 - 4y}{6} \end{cases}$
20. $\begin{cases} x = \dfrac{12 - 6y}{5} \\ y = \dfrac{24 - 10x}{12} \end{cases}$
21. $\begin{cases} x = \dfrac{5y - 4}{2} \\ y = \dfrac{3x - 1}{5} \end{cases}$
22. $\begin{cases} y = \dfrac{1 - 5x}{2} \\ x = \dfrac{3y + 10}{4} \end{cases}$

23. $\begin{cases} ax + y = k \\ x + dy = q \end{cases}$
24. $\begin{cases} rx + sy = t \\ x + ry = s \end{cases}$

In Exercises 25–28, evaluate each determinant and solve the resulting equation.

25. $\begin{vmatrix} x & 1 \\ 3 & 2 \end{vmatrix} = 1$

26. $\begin{vmatrix} x & -x \\ 2 & -3 \end{vmatrix} = -5$

27. $\begin{vmatrix} x & -2 \\ 3 & 1 \end{vmatrix} = \begin{vmatrix} 4 & 2 \\ x & 3 \end{vmatrix}$

28. $\begin{vmatrix} x & 3 \\ x & 2 \end{vmatrix} = \begin{vmatrix} 3 & 2 \\ 1 & 1 \end{vmatrix}$

9.4 SOLUTIONS OF THREE EQUATIONS IN THREE VARIABLES

We have previously shown that a solution to the system of equations

$$\begin{cases} ax + by = e \\ cx + dy = f \end{cases}$$

is an ordered pair of real numbers (x, y) that satisfies both of the given equations simultaneously. Likewise, a solution to the system

$$\begin{cases} ax + by + cz = j \\ dx + ey + fz = k \\ gx + hy + iz = l \end{cases}$$

is an ordered triple of numbers (x, y, z) that satisfies each of the three given equations simultaneously.

A linear equation in two variables has a graph that is a straight line. A system of two linear equations in two variables is consistent or inconsistent depending on whether a pair of lines intersect or are parallel.

The graph of an equation in three variables of the form $ax + by + cz = j$ is a flat surface called a **plane**. A system of three equations in three variables is consistent or inconsistent depending on how the three planes corresponding to the three equations intersect. The drawings in Figure 9-5 illustrate some of the possibilities.

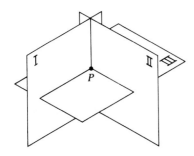

The three planes intersect at a single point P:
One solution
(a)

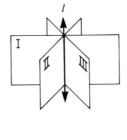

The three planes have a line l in common:
An infinite number of solutions
(b)

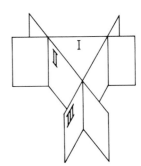

The three planes have no point in common:
No solutions
(c)

Figure 9-5

Example 1 discusses a consistent system of three equations in three variables. Example 2 discusses a system that is inconsistent.

Example 1 Solve the system $\begin{cases} 2x + y + 4z = 12 \\ x + 2y + 2z = 9 \\ 3x - 3y - 2z = 1 \end{cases}$.

Solution You are given the following system of equations in three variables:

1. $\begin{cases} 2x + y + 4z = 12 \\ x + 2y + 2z = 9 \\ 3x - 3y - 2z = 1 \end{cases}$
2.
3.

Use the addition method to eliminate the variable z and, thereby, obtain a system of two equations in two variables. If Equations 2 and 3 are added, the variable z is eliminated:

$$\begin{array}{ll} 2. & x + 2y + 2z = 9 \\ 3. & 3x - 3y - 2z = 1 \\ \hline 4. & 4x - y \quad\quad = 10 \end{array}$$

Now pick a different pair of equations and eliminate the variable z again. If each side of Equation 3 is multiplied by 2 and the resulting equation is added to Equation 1, the variable z is eliminated again:

$$\begin{array}{ll} 1. & 2x + y + 4z = 12 \\ & 6x - 6y - 4z = 2 \\ \hline 5. & 8x - 5y \quad\quad = 14 \end{array}$$

Equations 4 and 5 form a system of two equations in two variables:

$$\begin{array}{l} 4. \\ 5. \end{array} \begin{cases} 4x - y = 10 \\ 8x - 5y = 14 \end{cases}$$

To solve this system, multiply Equation 4 by -5, add the resulting equation to Equation 5 to eliminate the variable y, and solve for x:

$$\begin{array}{ll} & -20x + 5y = -50 \\ 5. & 8x - 5y = 14 \\ \hline & -12x \quad\quad = -36 \\ & x = 3 \quad\quad \text{Divide both sides by } -12. \end{array}$$

To find the variable y, substitute 3 for x in an equation containing the variables x and y, such as Equation 5, and solve for y:

$$\begin{array}{ll} 5. & 8x - 5y = 14 \\ & 8(3) - 5y = 14 \\ & 24 - 5y = 14 \quad\quad \text{Simplify.} \\ & -5y = -10 \quad\quad \text{Add } -24 \text{ to both sides.} \\ & y = 2 \quad\quad\quad \text{Divide both sides by } -5. \end{array}$$

To find the variable z, substitute 3 for x and 2 for y in an equation containing the variables x, y, and z, such as Equation 1, and solve for z:

$$
\begin{array}{lll}
\textbf{1.} & 2x + y + 4z = 12 & \\
& 2(3) + 2 + 4z = 12 & \\
& 8 + 4z = 12 & \text{Simplify.} \\
& 4z = 4 & \text{Add } -8 \text{ to both sides.} \\
& z = 1 & \text{Divide both sides by 4.}
\end{array}
$$

The solution of this system is $(x, y, z) = (3, 2, 1)$. Verify that these values satisfy each of the equations in the system.   ■

Example 2   Solve the system $\begin{cases} 2x + y - 3z = -3 \\ 3x - 2y + 4z = 2 \\ 4x + 2y - 6z = -7 \end{cases}$.

Solution   You are given the following system of equations:

$$
\begin{array}{ll}
\textbf{1.} & 2x + y - 3z = -3 \\
\textbf{2.} & 3x - 2y + 4z = 2 \\
\textbf{3.} & 4x + 2y - 6z = -7
\end{array}
$$

Begin by multiplying Equation 1 by 2 and adding the resulting equation to Equation 2 to eliminate the variable y:

$$
\begin{array}{ll}
& 4x + 2y - 6z = -6 \\
\textbf{2.} & 3x - 2y + 4z = 2 \\
\hline
\textbf{4.} & 7x - 2z = -4
\end{array}
$$

Now add Equations 2 and 3 to eliminate the variable y again:

$$
\begin{array}{ll}
\textbf{2.} & 3x - 2y + 4z = 2 \\
\textbf{3.} & 4x + 2y - 6z = -7 \\
\hline
\textbf{5.} & 7x - 2z = -5
\end{array}
$$

Equations 4 and 5 form the system

$$
\begin{array}{ll}
\textbf{4.} & 7x - 2z = -4 \\
\textbf{5.} & 7x - 2z = -5
\end{array}
$$

Because $7x - 2z$ cannot equal both -4 and -5, this system must be inconsistent. Thus, the original system has no solutions, either; it is inconsistent.   ■

Example 3   The sum of three integers is 2. The third integer is 2 greater than the second and 17 greater than the first. Find the three integers.

Solution   Let a, b, and c represent the three integers. Because their sum is 2, you know that

$$a + b + c = 2$$

Because the third integer is 2 greater than the second and 17 greater than the first, you know that

$$c - b = 2$$
$$c - a = 17$$

Put these three equations together to form a system of three equations in three variables:

1.
2. $\begin{cases} a + b + c = 2 \\ -b + c = 2 \\ -a + c = 17 \end{cases}$
3.

Add Equations 1 and 2 to get Equation 4:

4. $a + 2c = 4$

Equations 3 and 4 form a system of two equations in two variables:

3.
4. $\begin{cases} -a + c = 17 \\ a + 2c = 4 \end{cases}$

Add Equations 3 and 4 to get the equation

$$3c = 21$$
$$c = 7$$

Substitute 7 for c in Equation 4 to find a:

4. $a + 2c = 4$

$$a + 2(7) = 4$$
$$a + 14 = 4 \qquad \text{Simplify.}$$
$$a = -10 \qquad \text{Add } -14 \text{ to both sides.}$$

Substitute 7 for c in Equation 2 to find b:

2. $-b + c = 2$

$$-b + 7 = 2$$
$$-b = -5 \qquad \text{Add } -7 \text{ to both sides.}$$
$$b = 5 \qquad \text{Divide both sides by } -1.$$

Thus, the three integers are -10, 5, and 7. Note that these three integers have a sum of 2, that 7 is 2 greater than 5, and that 7 is 17 greater than -10. ∎

■ EXERCISE 9.4

In Exercises 1–12, solve each system of equations. If a system of equations is inconsistent, or if the equations are dependent, so indicate.

1. $\begin{cases} x + y + z = 4 \\ 2x + y - z = 1 \\ 2x - 3y + z = 1 \end{cases}$

2. $\begin{cases} x + y + z = 4 \\ x - y + z = 2 \\ x - y - z = 0 \end{cases}$

3. $\begin{cases} 2x + 2y + 3z = 10 \\ 3x + y - z = 0 \\ x + y + 2z = 6 \end{cases}$

4. $\begin{cases} x - y + z = 4 \\ x + 2y - z = -1 \\ x + y - 3z = -2 \end{cases}$

5. $\begin{cases} x + y + 2z = 7 \\ x + 2y + z = 8 \\ 2x + y + z = 9 \end{cases}$

6. $\begin{cases} x + 2y + 2z = 10 \\ 2x + y + 2z = 9 \\ 2x + 2y + z = 11 \end{cases}$

7. $\begin{cases} 2x + y - z = 1 \\ x + 2y + 2z = 2 \\ 4x + 5y + 3z = 3 \end{cases}$

8. $\begin{cases} 4x + 3z = 4 \\ 2y - 6z = -1 \\ 8x + 4y + 3z = 9 \end{cases}$

9. $\begin{cases} 2x + 3y + 4z - 6 = 0 \\ 2x - 3y - 4z + 4 = 0 \\ 4x + 6y + 8z - 12 = 0 \end{cases}$

10. $\begin{cases} x - 3y + 4z - 2 = 0 \\ 2x + y + 2z - 3 = 0 \\ 4x - 5y + 10z - 7 = 0 \end{cases}$

11. $\begin{cases} x + \frac{1}{3}y + z = 13 \\ \frac{1}{2}x - y + \frac{1}{3}z = -2 \\ x + \frac{1}{2}y - \frac{1}{3}z = 2 \end{cases}$

12. $\begin{cases} x - \frac{1}{5}y - z = 9 \\ \frac{1}{4}x + \frac{1}{5}y - \frac{1}{2}z = 5 \\ 2x + y + \frac{1}{6}z = 12 \end{cases}$

In Exercises 13–22, solve each word problem.

13. The sum of three numbers is 18. The third number is four times the second, and the second number is 6 more than the first. Find the numbers.

14. The sum of three numbers is 48. If the first number is doubled, the sum is 60. If the second number is doubled, the sum is 63. Find the numbers.

15. Three numbers have a sum of 30. The third number is 8 less than the sum of the first and second, and the second number is half the sum of the first and third. Find the numbers.

16. The sum of the three angles in any triangle is 180°. In triangle ABC, angle A is 100° less than the sum of angles B and C, and angle C is 40° less than twice angle B. Find each angle.

17. A collection of 17 nickels, dimes, and quarters has a value of $1.50. There are twice as many nickels as dimes. How many of each kind are there?

18. A unit of food contains 1 gram of fat, 1 gram of carbohydrate, and 2 grams of protein. A second contains 2 grams of fat, 1 gram of carbohydrate, and 1 gram of protein. A third contains 2 grams of fat, 1 gram of carbohydrate, and 2 grams of protein. How many units of each must be used to provide exactly 11 grams of fat, 6 grams of carbohydrate, and 10 grams of protein?

19. A factory manufactures three types of footballs at a monthly cost of $2425 for 1125 footballs. The manufacturing costs for the three types of footballs are $4, $3, and $2. These footballs sell for $16, $12, and $10, respectively. How many of each type are manufactured if the monthly profit is $9275? (*Hint:* Profit = income − cost)

20. A retailer purchased 105 radios from sources A, B, and C. Five fewer units were purchased from C than from A and B combined. If twice as many had been purchased from A, the total would have been 130. Find the number purchased from each source.

21. Tickets for a concert cost $5, $3, and $2. Twice as many $5 tickets were sold as $2 tickets. The receipts for 750 tickets were $2625. How many of each price ticket were sold?

22. The owner of a candy store wants to mix some peanuts worth $3 per pound, some cashews worth $9 per pound, and some brazil nuts worth $9 per pound to get 50 pounds of a mixture that will sell for $6 per pound. She used 15 fewer pounds of cashews than peanuts. How many pounds of each did she use?

9.5 SOLUTIONS OF THREE EQUATIONS IN THREE VARIABLES BY DETERMINANTS

The determinant of a 3×3 matrix may be defined so that Cramer's Rule can be used to solve many systems of three equations in three variables. Finding the value of the determinant of a 3×3 matrix requires a process known as **expanding a determinant by minors**.

Definition.   If a is the element in the ith row and the jth column of a 3×3 matrix, the **minor** of a is the determinant of the 2×2 matrix formed by those elements in the 3×3 matrix that do not lie in the ith row or the jth column.

Example 1   Find the minor of the element 3 in the matrix $\begin{bmatrix} 1 & 2 & 7 \\ 3 & 5 & 9 \\ -2 & -1 & 4 \end{bmatrix}$.

Solution   The number 3 is in the 2nd row and the 1st column. The minor of 3 is the determinant of the 2×2 matrix formed by those elements that are not in the 2nd row and are not in the 1st column:

$$\begin{bmatrix} 1 & 2 & 7 \\ 3 & 5 & 9 \\ -2 & -1 & 4 \end{bmatrix}$$

The minor of 3 is $\det\left(\begin{bmatrix} 2 & 7 \\ -1 & 4 \end{bmatrix}\right)$ or $\begin{vmatrix} 2 & 7 \\ -1 & 4 \end{vmatrix}$.   ∎

Example 2   Find the minor of the element -7 in the matrix $\begin{bmatrix} 3 & -2 & 5 \\ 0 & -5 & 4 \\ 1 & -7 & 6 \end{bmatrix}$.

Solution   Because -7 is in the 3rd row and the 2nd column, cross out the 3rd row and the 2nd column to find the minor of -7:

$$\begin{bmatrix} 3 & -2 & 5 \\ 0 & -5 & 4 \\ 1 & -7 & 6 \end{bmatrix}$$

The minor of -7 is the determinant of the remaining 2×2 matrix:

$$\begin{vmatrix} 3 & 5 \\ 0 & 4 \end{vmatrix}$$ ■

Definition. If a is an element in the ith row and jth column of a 3×3 matrix, then the **cofactor** of a is the minor of a if $i + j$ is even and the negative of the minor of a if $i + j$ is odd.

Example 3 Find the cofactor of **a.** 3 and **b.** 7 in the 3×3 matrix

$$\begin{bmatrix} 2 & 3 & 5 \\ 1 & 7 & 4 \\ 9 & -3 & 8 \end{bmatrix}.$$

Solution **a.** The minor of the element 3 is the 2×2 determinant

$$\begin{vmatrix} 1 & 4 \\ 9 & 8 \end{vmatrix}$$

Because 3 is in the 1st row and 2nd column, and $1 + 2$ equals 3, which is odd, the cofactor of 3 is the negative of this minor. The cofactor of 3 is

$$-\begin{vmatrix} 1 & 4 \\ 9 & 8 \end{vmatrix}$$

b. The minor of 7 is the 2×2 determinant

$$\begin{vmatrix} 2 & 5 \\ 9 & 8 \end{vmatrix}$$

Because 7 is in the 2nd row and the 2nd column, and $2 + 2$ equals 4, which is even, the cofactor of 7 is equal to its minor. The cofactor of 7 is

$$\begin{vmatrix} 2 & 5 \\ 9 & 8 \end{vmatrix}$$ ■

We accept the following theorem without proof.

Theorem. The value of the determinant of a 3×3 matrix is the sum of the products of each of the elements of any chosen row or column and the cofactors of those elements.

Example 4 Use the method of expanding by minors to evaluate the determinant

$$\begin{vmatrix} 1 & 3 & -2 \\ 2 & 1 & 3 \\ 1 & 2 & 3 \end{vmatrix}.$$

Solution The method of expansion by minors works for expansion along any row or column. For sake of argument, expand along the second row with its elements of 2, 1, and 3. The cofactor of 2 is

$$-\begin{vmatrix} 3 & -2 \\ 2 & 3 \end{vmatrix}$$

The cofactor of 1 is

$$\begin{vmatrix} 1 & -2 \\ 1 & 3 \end{vmatrix}$$

The cofactor of 3 is

$$-\begin{vmatrix} 1 & 3 \\ 1 & 2 \end{vmatrix}$$

The value of the determinant is the sum of the products of these elements with their cofactors. Hence, you have

$$\begin{vmatrix} 1 & 3 & -2 \\ 2 & 1 & 3 \\ 1 & 2 & 3 \end{vmatrix} = 2\left(-\begin{vmatrix} 3 & -2 \\ 2 & 3 \end{vmatrix}\right) + 1\left(\begin{vmatrix} 1 & -2 \\ 1 & 3 \end{vmatrix}\right) + 3\left(-\begin{vmatrix} 1 & 3 \\ 1 & 2 \end{vmatrix}\right)$$

$$= -2\begin{vmatrix} 3 & -2 \\ 2 & 3 \end{vmatrix} + \begin{vmatrix} 1 & -2 \\ 1 & 3 \end{vmatrix} - 3\begin{vmatrix} 1 & 3 \\ 1 & 2 \end{vmatrix}$$

$$= -2(9 + 4) + (3 + 2) - 3(2 - 3)$$

$$= -26 + 5 + 3$$

$$= -18$$

Example 5 Evaluate $\begin{vmatrix} 1 & 3 & -2 \\ 2 & 1 & 3 \\ 1 & 2 & 3 \end{vmatrix}$ by expanding the determinant by minors along the second column.

Solution The second column contains the elements 3, 1, and 2. The cofactor of 3 is

$$-\begin{vmatrix} 2 & 3 \\ 1 & 3 \end{vmatrix}$$

The cofactor of 1 is

$$\begin{vmatrix} 1 & -2 \\ 1 & 3 \end{vmatrix}$$

The cofactor of 2 is

$$-\begin{vmatrix} 1 & -2 \\ 2 & 3 \end{vmatrix}$$

The value of the determinant is the sum of the products of these elements with their cofactors. Hence, you have

$$\begin{vmatrix} 1 & 3 & -2 \\ 2 & 1 & 3 \\ 1 & 2 & 3 \end{vmatrix} = 3\left(-\begin{vmatrix} 2 & 3 \\ 1 & 3 \end{vmatrix}\right) + 1\left(\begin{vmatrix} 1 & -2 \\ 1 & 3 \end{vmatrix}\right) + 2\left(-\begin{vmatrix} 1 & -2 \\ 2 & 3 \end{vmatrix}\right)$$

$$= -3\begin{vmatrix} 2 & 3 \\ 1 & 3 \end{vmatrix} + \begin{vmatrix} 1 & -2 \\ 1 & 3 \end{vmatrix} - 2\begin{vmatrix} 1 & -2 \\ 2 & 3 \end{vmatrix}$$

$$= -3(6 - 3) + (3 + 2) - 2(3 + 4)$$

$$= -9 + 5 - 14$$

$$= -18 \qquad \blacksquare$$

Note that the values obtained in Examples 4 and 5 are equal. This illustrates that either rows or columns can be used to expand determinants.

If a row or column of a determinant contains one or more 0s, it is a good idea to expand the determinant along that row or column. This will reduce the amount of arithmetic required because 0 times its cofactor is 0.

We can now solve a system of three equations in three variables by using determinants.

Example 6   Use Cramer's Rule to solve the system $\begin{cases} 2x + y + 4z = 12 \\ x + 2y + 2z = 9 \\ 3x - 3y - 2z = 1 \end{cases}$.

Solution   Follow the same procedure as for the 2×2 case: The denominator determinant is the determinant of the matrix formed by the coefficients of the variables, and the numerator determinants are formed by replacing the coefficients of the variable being solved for by the column of constants. Form the quotients for x, y, and z, and evaluate the determinants:

$$x = \frac{\begin{vmatrix} 12 & 1 & 4 \\ 9 & 2 & 2 \\ 1 & -3 & -2 \end{vmatrix}}{\begin{vmatrix} 2 & 1 & 4 \\ 1 & 2 & 2 \\ 3 & -3 & -2 \end{vmatrix}} = \frac{-72}{-24} = 3$$

$$y = \frac{\begin{vmatrix} 2 & 12 & 4 \\ 1 & 9 & 2 \\ 3 & 1 & -2 \end{vmatrix}}{\begin{vmatrix} 2 & 1 & 4 \\ 1 & 2 & 2 \\ 3 & -3 & -2 \end{vmatrix}} = \frac{-48}{-24} = 2$$

$$z = \dfrac{\begin{vmatrix} 2 & 1 & 12 \\ 1 & 2 & 9 \\ 3 & -3 & 1 \end{vmatrix}}{\begin{vmatrix} 2 & 1 & 4 \\ 1 & 2 & 2 \\ 3 & -3 & -2 \end{vmatrix}} = \dfrac{-24}{-24} = 1$$

The solution to this system is (3, 2, 1). ∎

Cramer's Rule for Three Equations in Three Variables.   If the system

$$\begin{cases} ax + by + cz = j \\ dx + ey + fz = k \\ gx + hy + iz = l \end{cases}$$

has a unique solution, it is given by

$$x = \frac{D_x}{D}, \quad y = \frac{D_y}{D}, \quad \text{and} \quad z = \frac{D_z}{D}$$

where

$$D = \begin{vmatrix} a & b & c \\ d & e & f \\ g & h & i \end{vmatrix}$$

and

$$D_x = \begin{vmatrix} j & b & c \\ k & e & f \\ l & h & i \end{vmatrix}, \quad D_y = \begin{vmatrix} a & j & c \\ d & k & f \\ g & l & i \end{vmatrix}, \quad D_z = \begin{vmatrix} a & b & j \\ d & e & k \\ g & h & l \end{vmatrix}$$

If the denominators and the numerators of these fractions are *all* 0, the system is consistent, but the equations are dependent.

If the denominators are 0 and at least one of the numerators is *not* 0, the system is inconsistent.

▬ EXERCISE 9.5 ▬▬▬▬▬▬▬▬▬▬

In Exercises 1–10, evaluate each determinant by expanding the determinant by minors.

1. $\begin{vmatrix} 1 & 0 & 1 \\ 0 & 1 & 0 \\ 1 & 1 & 1 \end{vmatrix}$ **2.** $\begin{vmatrix} 1 & 2 & 0 \\ 0 & 1 & 2 \\ 0 & 0 & 1 \end{vmatrix}$ **3.** $\begin{vmatrix} -1 & 2 & 1 \\ 2 & 1 & -3 \\ 1 & 1 & 1 \end{vmatrix}$ **4.** $\begin{vmatrix} 1 & 2 & 3 \\ 1 & 2 & 3 \\ 1 & 2 & 3 \end{vmatrix}$

5. $\begin{vmatrix} 1 & -2 & 3 \\ -2 & 1 & 1 \\ -3 & -2 & 1 \end{vmatrix}$ 　　**6.** $\begin{vmatrix} 1 & 1 & 2 \\ 2 & 1 & -2 \\ 3 & 1 & 3 \end{vmatrix}$ 　　**7.** $\begin{vmatrix} 1 & 2 & 3 \\ 4 & 5 & 6 \\ 7 & 8 & 9 \end{vmatrix}$ 　　**8.** $\begin{vmatrix} 1 & 4 & 7 \\ 2 & 5 & 8 \\ 3 & 6 & 9 \end{vmatrix}$

9. $\begin{vmatrix} a & 2a & -a \\ 2 & -1 & 3 \\ 1 & 2 & -3 \end{vmatrix}$ 　　　　　　　　　**10.** $\begin{vmatrix} 1 & 2b & -3 \\ 2 & -b & 2 \\ 1 & 3b & 1 \end{vmatrix}$

In Exercises 11–24, use Cramer's Rule to solve each system of equations. If Cramer's Rule fails, so indicate.

11. $\begin{cases} x + y + z = 4 \\ x + y - z = 0 \\ x - y + z = 2 \end{cases}$ 　　　　　　　**12.** $\begin{cases} x + y + z = 4 \\ x - y + z = 2 \\ x - y - z = 0 \end{cases}$

13. $\begin{cases} x + y + 2z = 7 \\ x + 2y + z = 8 \\ 2x + y + z = 9 \end{cases}$ 　　　　　　　**14.** $\begin{cases} x + 2y + 2z = 10 \\ 2x + y + 2z = 9 \\ 2x + 2y + z = 1 \end{cases}$

15. $\begin{cases} 2x + y - z = 1 \\ x + 2y + 2z = 2 \\ 4x + 5y + 3z = 3 \end{cases}$ 　　　　　　**16.** $\begin{cases} 4x + 3z = 4 \\ 2y - 6z = -1 \\ 8x + 4y + 3z = 9 \end{cases}$

17. $\begin{cases} 2x + y + z = 5 \\ x - 2y + 3z = 10 \\ x + y - 4z = -3 \end{cases}$ 　　　　　　**18.** $\begin{cases} 3x + 2y - z = -8 \\ 2x - y + 7z = 10 \\ 2x + 2y - 3z = -10 \end{cases}$

19. $\begin{cases} 2x + 3y + 4z - 6 = 0 \\ 2x - 3y - 4z + 4 = 0 \\ 4x + 6y + 8z - 12 = 0 \end{cases}$ 　　　**20.** $\begin{cases} x - 3y + 4z - 2 = 0 \\ 2x + y + 2z - 3 = 0 \\ 4x - 5y + 10z - 7 = 0 \end{cases}$

21. $\begin{cases} x + y = 1 \\ \dfrac{y}{2} + z = \dfrac{5}{2} \\ x - z = -3 \end{cases}$ 　　　　　**22.** $\begin{cases} 3x + 4y + 14z = 7 \\ -\dfrac{x}{2} - y + 2z = \dfrac{3}{2} \\ x + \dfrac{3}{2}y + \dfrac{5}{2}z = 1 \end{cases}$

23. $\begin{cases} 4z - y + 2x + 2 = 0 \\ 5x + 7z + 8y = -8 \\ 3y + x + z + 3 = 0 \end{cases}$ 　　　**24.** $\begin{cases} \dfrac{1}{2}x + y + z + \dfrac{3}{2} = 0 \\ x + \dfrac{1}{2}y + z - \dfrac{1}{2} = 0 \\ x + y + \dfrac{1}{2}z + \dfrac{1}{2} = 0 \end{cases}$

25. Show that $\begin{vmatrix} x & y & 1 \\ 2 & 3 & 1 \\ 4 & 5 & 1 \end{vmatrix} = 0$ represents the equation of a line passing through (2, 3) and (4, 5).

26. Show that $\begin{vmatrix} a & a & d \\ b & b & e \\ c & c & f \end{vmatrix} = 0.$

9.6 SOLUTION BY MATRICES

Recall that a **matrix** is a rectangular array of numbers. The arrays

$$\begin{bmatrix} 2 & 1 \\ -3 & 2 \end{bmatrix}, \quad \begin{bmatrix} 2 & 1 & 3 \\ 4 & -2 & -1 \end{bmatrix}, \quad \begin{bmatrix} 1 & 2 \\ 3 & 4 \\ 5 & 6 \end{bmatrix}$$

are all examples of matrices. Because the matrix on the left has two rows and two columns, it is a 2×2 matrix. Because the matrix in the center has two rows and three columns, it is a 2×3 matrix. The matrix on the right is a 3×2 matrix.

We can use matrices to solve systems of linear equations. We begin the discussion by considering the system of equations

$$\begin{cases} x - 2y - z = 6 \\ 2x + 2y - z = 1 \\ -x - y + 2z = 1 \end{cases}$$

This system of equations can be represented by the following matrix, called an **augmented matrix**:

$$\begin{bmatrix} 1 & -2 & -1 & \vdots & 6 \\ 2 & 2 & -1 & \vdots & 1 \\ -1 & -1 & 2 & \vdots & 1 \end{bmatrix}$$

The 3×3 matrix to the left of the dashed line, called the **coefficient matrix**, is determined by the coefficients of x, y, and z in the equations of the system. The 3×1 matrix to the right of the dashed line is determined by the constants in the equations of the system. Note that each row of the augmented matrix represents exactly one equation of the system:

$$\begin{bmatrix} 1 & -2 & -1 & \vdots & 6 \\ 2 & 2 & -1 & \vdots & 1 \\ -1 & -1 & 2 & \vdots & 1 \end{bmatrix} \begin{matrix} \leftrightarrow \\ \leftrightarrow \\ \leftrightarrow \end{matrix} \begin{cases} x - 2y - z = 6 \\ 2x + 2y - z = 1 \\ -x - y + 2z = 1 \end{cases}$$

To solve the given system we shall use a method called **Gaussian elimination**. The strategy is to transform the augmented matrix into the following form, called **triangular form**,

$$\begin{bmatrix} a & b & c & \vdots & d \\ 0 & e & f & \vdots & g \\ 0 & 0 & h & \vdots & i \end{bmatrix} \quad (a, b, c, \ldots, i \text{ are real numbers})$$

by using three operations called **elementary row operations**.

Elementary Row Operations.

1. Any two rows of a matrix can be interchanged.
2. Any row of a matrix can be multiplied by a nonzero constant.
3. Any row of a matrix can be changed by adding to it a constant multiple of another row.

After we have written the matrix in triangular form, we can solve the corresponding system of equations by a substitution process. Note that a type 1 row operation corresponds to interchanging two equations of the system, a type 2 row operation corresponds to multiplying both sides of an equation by a non-zero constant, and a type 3 row operation corresponds to adding a multiple of one equation to another. None of these operations will change the solution of the given system of equations.

The first example shows how to solve the previous system of equations by matrix methods.

Example 1   Solve the system
$$\begin{cases} x - 2y - z = 6 \\ 2x + 2y - z = 1. \\ -x - y + 2z = 1 \end{cases}$$

Solution   First represent the system by an augmented matrix:

$$\left[\begin{array}{ccc|c} 1 & -2 & -1 & 6 \\ 2 & 2 & -1 & 1 \\ -1 & -1 & 2 & 1 \end{array}\right]$$

To get 0s under the 1 in the 1st column, use a type 3 row operation twice:

Multiply row 1 by -2 Multiply row 1 by 1
and add to row 2.   and add to row 3.

$$\left[\begin{array}{ccc|c} 1 & -2 & -1 & 6 \\ 2 & 2 & -1 & 1 \\ -1 & -1 & 2 & 1 \end{array}\right] \approx \left[\begin{array}{ccc|c} 1 & -2 & -1 & 6 \\ 0 & 6 & 1 & -11 \\ -1 & -1 & 2 & 1 \end{array}\right] \approx \left[\begin{array}{ccc|c} 1 & -2 & -1 & 6 \\ 0 & 6 & 1 & -11 \\ 0 & -3 & 1 & 7 \end{array}\right]$$

The symbol "$\approx$" is read as "is row equivalent to." Each of the above matrices represents a system of equations, and they are all equivalent.

To get a 0 under the -2 and 6 in the second column of the last matrix, use another type 3 row operation:

Multiply row 2 by $\frac{1}{2}$
and add to row 3.

$$\left[\begin{array}{ccc|c} 1 & -2 & -1 & 6 \\ 0 & 6 & 1 & -11 \\ 0 & -3 & 1 & 7 \end{array}\right] \approx \left[\begin{array}{ccc|c} 1 & -2 & -1 & 6 \\ 0 & 6 & 1 & -11 \\ 0 & 0 & \frac{3}{2} & \frac{3}{2} \end{array}\right]$$

Finally, use a type 2 row operation:

Multiply row 3 by $\frac{2}{3}$.

$$\left[\begin{array}{ccc|c} 1 & -2 & -1 & 6 \\ 0 & 6 & 1 & -11 \\ 0 & 0 & \frac{3}{2} & \frac{3}{2} \end{array}\right] \approx \left[\begin{array}{ccc|c} 1 & -2 & -1 & 6 \\ 0 & 6 & 1 & -11 \\ 0 & 0 & 1 & 1 \end{array}\right]$$

The final matrix represents the system of equations

$$\begin{array}{rl} \textbf{1.} & \left\{ \begin{array}{l} x - 2y - z = \quad 6 \\ \textbf{2.} \quad 0x + 6y + z = -11 \\ \textbf{3.} \quad 0x + 0y + z = \quad 1 \end{array} \right. \end{array}$$

From Equation 3, you can read that $z = 1$. To find y, substitute $\mathbf{1}$ for z in Equation 2 and solve for y:

$$\begin{array}{rl} \textbf{2.} & 6y + z = -11 \\ & 6y + \mathbf{1} = -11 \\ & 6y = -12 \\ & y = -2 \end{array}$$

Thus, $y = -2$. To find x, substitute $\mathbf{1}$ for z and -2 for y in Equation 1 and solve for x:

$$\begin{array}{rl} \textbf{1.} & x - 2y - z = 6 \\ & x - 2(-2) - \mathbf{1} = 6 \\ & x + 4 - 1 = 6 \\ & x + 3 = 6 \\ & x = 3 \end{array}$$

Thus, $x = 3$. The solution to the given system is $(3, -2, 1)$. Verify that this triple satisfies each equation of the original system. ∎

We can use matrices to solve systems of equations that have more equations than variables.

Example 2   Solve the system $\left\{ \begin{array}{r} x + y = -1 \\ 2x - y = 7 \\ -x + 2y = -8 \end{array} \right.$

Solution   This system can be represented by a 3×3 augmented matrix:

$$\left[\begin{array}{rr|r} 1 & 1 & -1 \\ 2 & -1 & 7 \\ -1 & 2 & -8 \end{array} \right]$$

To get 0s under the $\mathbf{1}$ in the first column, perform a type 3 row operation twice:

Multiply row 1 by -2 and add to row 2.   Multiply row 1 by 1 and add to row 3.

$$\left[\begin{array}{rr|r} 1 & 1 & -1 \\ 2 & -1 & 7 \\ -1 & 2 & -8 \end{array} \right] \approx \left[\begin{array}{rr|r} 1 & 1 & -1 \\ 0 & -3 & 9 \\ -1 & 2 & -8 \end{array} \right] \approx \left[\begin{array}{rr|r} 1 & 1 & -1 \\ 0 & -3 & 9 \\ 0 & 3 & -9 \end{array} \right]$$

Perform other row operations to get

$$\begin{array}{ccc} \text{Multiply row 3 by 1} & \text{Interchange} & \text{Multiply row 2} \\ \text{and add to row 2.} & \text{row 2 and row 3.} & \text{by } \frac{1}{3}. \end{array}$$

$$\left[\begin{array}{cc|c} 1 & 1 & -1 \\ 0 & -3 & 9 \\ 0 & 3 & -9 \end{array}\right] \approx \left[\begin{array}{cc|c} 1 & 1 & -1 \\ 0 & 0 & 0 \\ 0 & 3 & -9 \end{array}\right] \approx \left[\begin{array}{cc|c} 1 & 1 & -1 \\ 0 & 3 & -9 \\ 0 & 0 & 0 \end{array}\right] \approx \left[\begin{array}{cc|c} 1 & 1 & -1 \\ 0 & 1 & -3 \\ 0 & 0 & 0 \end{array}\right]$$

The final matrix represents the system

$$\begin{cases} x + y = -1 \\ 0x + y = -3 \\ 0x + 0y = 0 \end{cases}$$

The third equation may be discarded because $0x + 0y = 0$ for all x and y. From the second equation, you can read that $y = -3$. To find x, substitute -3 for y in the first equation and solve for x:

$$x + y = -1$$
$$x - 3 = -1$$
$$x = 2$$

The solution to the original system is $(2, -3)$. Verify that this solution satisfies all three equations of the original system. ∎

If the last row of the final matrix of Example 2 had been of the form $0x + 0y = k$, where $k \neq 0$, the system could have no solution. No values of x and y could make the expression $0x + 0y$ equal to a nonzero constant.

Example 3 Solve the system $\begin{cases} x + y - 2z = -1 \\ 2x - y + z = -3 \end{cases}$.

Solution In this example, there are more variables than equations. The system can be represented by the 2×4 augmented matrix

$$\left[\begin{array}{ccc|c} 1 & 1 & -2 & -1 \\ 2 & -1 & 1 & -3 \end{array}\right]$$

To get a 0 under the 1 in the first column, perform a type 3 row operation:

$$\begin{array}{c} \text{Multiply row 1 by } -2 \\ \text{and add to row 2.} \end{array}$$

$$\left[\begin{array}{ccc|c} 1 & 1 & -2 & -1 \\ 2 & -1 & 1 & -3 \end{array}\right] \approx \left[\begin{array}{ccc|c} 1 & 1 & -2 & -1 \\ 0 & -3 & 5 & -1 \end{array}\right]$$

Then perform a type 2 row operation:

$$\text{Multiply row 2 by } -\tfrac{1}{3}.$$

$$\left[\begin{array}{ccc|c} 1 & 1 & -2 & -1 \\ 0 & -3 & 5 & -1 \end{array}\right] \approx \left[\begin{array}{ccc|c} 1 & 1 & -2 & -1 \\ 0 & 1 & -\frac{5}{3} & \frac{1}{3} \end{array}\right]$$

The final matrix represents the system

$$\begin{cases} x + y - 2z = -1 \\ y - \dfrac{5}{3}z = \dfrac{1}{3} \end{cases}$$

Add $\frac{5}{3}z$ to both sides of the second equation to obtain

$$y = \frac{1}{3} + \frac{5}{3}z$$

Substitute $\dfrac{1}{3} + \dfrac{5}{3}z$ for y in the first equation and simplify to get

$$x + y - 2z = -1$$
$$x + \frac{1}{3} + \frac{5}{3}z - 2z = -1$$
$$x + \frac{1}{3} - \frac{1}{3}z = -1$$
$$x - \frac{1}{3}z = -\frac{4}{3}$$
$$x = -\frac{4}{3} + \frac{1}{3}z$$

A solution to this system must have the form

$$\left(-\frac{4}{3} + \frac{1}{3}z, \ \frac{1}{3} + \frac{5}{3}z, \ z \right)$$

for all values of z. Thus, if $z = 0$, then the corresponding solution is $(-\frac{4}{3}, \frac{1}{3}, 0)$. If $z = 1$, the corresponding solution is $(-1, 2, 1)$. Verify that both of these solutions satisfy each equation of the given system.

This system has an infinite number of solutions, a different one for each value of z. The equations of this system are dependent. ∎

EXERCISE 9.6

In Exercises 1–12, use matrices to solve each system of equations. Each system has one solution.

1. $\begin{cases} x + y = 2 \\ x - y = 0 \end{cases}$ **2.** $\begin{cases} x + y = 3 \\ x - y = -1 \end{cases}$ **3.** $\begin{cases} x + 2y = -4 \\ 2x + y = 1 \end{cases}$ **4.** $\begin{cases} 2x - 3y = 16 \\ -4x + y = -22 \end{cases}$

5. $\begin{cases} 3x + 4z = -12 \\ 9x - 2z = 6 \end{cases}$ **6.** $\begin{cases} 5x - 4y = 10 \\ x - 7y = 2 \end{cases}$ **7.** $\begin{cases} x + y + z = 6 \\ x + 2y + z = 8 \\ x + y + 2z = 9 \end{cases}$ **8.** $\begin{cases} x - y + z = 2 \\ x + 2y - z = 6 \\ 2x - y - z = 3 \end{cases}$

9. $\begin{cases} 2x + y + 3z = 3 \\ -2x - y + z = 5 \\ 4x - 2y + 2z = 2 \end{cases}$

10. $\begin{cases} 3x + 2y + z = 8 \\ 6x - y + 2z = 16 \\ -9x + y - z = -20 \end{cases}$

11. $\begin{cases} 3x - 2y + 4z = 4 \\ x + y + z = 3 \\ 6x - 2y - 3z = 10 \end{cases}$

12. $\begin{cases} 2x + 3y - z = -8 \\ x - y - z = -2 \\ -4x + 3y + z = 6 \end{cases}$

In Exercises 13–18, use matrices to solve each system of equations. If a system has no solution, so indicate.

13. $\begin{cases} x + y = 3 \\ 3x - y = 1 \\ 2x + y = 4 \end{cases}$

14. $\begin{cases} x - y = -5 \\ 2x + 3y = 5 \\ x + y = 1 \end{cases}$

15. $\begin{cases} 2x - y = 4 \\ x + 3y = 2 \\ -x - 4y = -2 \end{cases}$

16. $\begin{cases} 3x - 2y = 5 \\ x + 2y = 7 \\ -3x - y = -11 \end{cases}$

17. $\begin{cases} 2x + y = 7 \\ x - y = 2 \\ -x + 3y = -2 \end{cases}$

18. $\begin{cases} 3x - y = 2 \\ -6x + 3y = 0 \\ -x + 2y = -4 \end{cases}$

In Exercises 19–22, use matrices to solve each system of equations. The equations of each system are dependent.

19. $\begin{cases} x + 2y + 3z = -2 \\ -x - y - 2z = 4 \end{cases}$

20. $\begin{cases} 2x - 4y + 3z = 6 \\ -4x + 6y + 4z = -6 \end{cases}$

21. $\begin{cases} x - y = 1 \\ y + z = 1 \\ x + z = 2 \end{cases}$

22. $\begin{cases} x + z = 1 \\ x + y = 2 \\ 2x + y + z = 3 \end{cases}$

9.7   SYSTEMS OF INEQUALITIES

In a previous section, we considered graphs of inequalities containing two variables. We now consider the graphs of systems of such inequalities.

Example 1   Graph the solution set of the system $\begin{cases} x + y \leq 1 \\ 2x - y > 2 \end{cases}$.

Solution   On the same set of coordinate axes, graph each inequality. See Figure 9-6. The graph of the inequality $x + y \leq 1$ includes the line graph of the equation

$x + y = 1$ $2x - y = 2$

x	y
0	1
1	0

x	y
0	-2
1	0

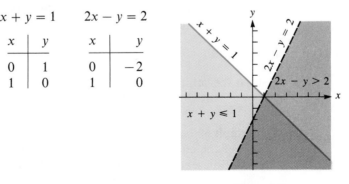

Figure 9-6

$x + y = 1$ and all points below it. Because the boundary line is included, it is drawn as a solid line. The graph of the inequality $2x - y > 2$ contains only those points below the line graph of the equation $2x - y = 2$. Because the boundary line is not included, it is drawn as a broken line. The area that is shaded twice represents the simultaneous solutions of the given system of inequalities. Any point in the doubly shaded region has coordinates that satisfy both inequalities in the system.   ∎

Example 2   Graph the solution set of the system $\begin{cases} y < x^2 \\ y > \dfrac{x^2}{4} - 2 \end{cases}$.

Solution   The graph of the equation $y = x^2$ is a parabola opening upward with vertex at the origin. See Figure 9-7. The points with coordinates that satisfy the inequality $y < x^2$ are those points below the parabola.

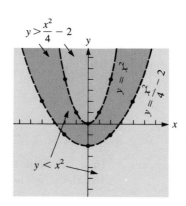

$y = x^2$

x	y
0	0
1	1
-1	1
2	4
-2	4

$y = \dfrac{x^2}{4} - 2$

x	y
0	-2
2	-1
-2	-1
4	2
-4	2

Figure 9-7

The graph of $y = x^2/4 - 2$ is a parabola opening upward with vertex at $(0, -2)$. However, this time the points with coordinates that satisfy the inequality are those points above the parabola. Thus, the graph of the solution set of this system is the shaded area between the two parabolas.   ∎

Example 3   Graph the solution set of the system $\begin{cases} x \ge 1 \\ y \ge x \\ 4x + 5y < 20 \end{cases}$.

Solution   The graph of the solution set of the inequality $x \ge 1$ includes those points on the graph of the equation $x = 1$ and to the right. See Figure 9-8a. The graph of the solution set of the inequality $y \ge x$ includes those points on the graph of the equation $y = x$ and above it. See Figure 9-8b. The graph of the solution set of

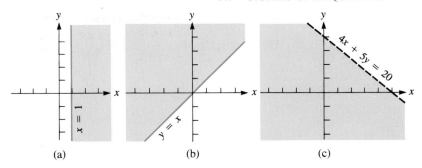

Figure 9-8

the inequality $4x + 5y < 20$ includes those points below the line graph of the equation $4x + 5y = 20$. See Figure 9-8c.

If these three graphs are merged onto a single set of coordinate axes, the graph of the original system of inequalities includes those points within the shaded triangle together with the points on the sides of the triangle drawn as solid lines. See Figure 9-9.

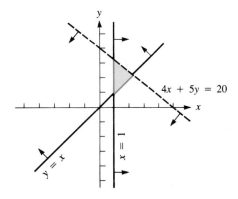

Figure 9-9

▬ EXERCISE 9.7 ▬

Graph the solution set of each system of inequalities.

1. $\begin{cases} y < 3x + 2 \\ y < -2x + 3 \end{cases}$

2. $\begin{cases} y \le x - 2 \\ y \ge 2x + 1 \end{cases}$

3. $\begin{cases} 3x + 2y > 6 \\ x + 3y \le 2 \end{cases}$

4. $\begin{cases} x + y < 2 \\ x + y \le 1 \end{cases}$

5. $\begin{cases} 3x + y \le 1 \\ -x + 2y \ge 9 \end{cases}$

6. $\begin{cases} x + 2y < 3 \\ 2x + 4y < 8 \end{cases}$

7. $\begin{cases} 2x - y > 4 \\ y < -x^2 + 2 \end{cases}$

8. $\begin{cases} x \le y^2 \\ y \ge x \end{cases}$

9. $\begin{cases} y > x^2 - 4 \\ y < -x^2 + 4 \end{cases}$

10. $\begin{cases} x \ge y^2 \\ y \ge x^2 \end{cases}$

11. $\begin{cases} 2x + 3y \le 5 \\ 3x + y \le 1 \\ x \le 0 \end{cases}$

12. $\begin{cases} 2x + y \le 2 \\ y \ge x \\ x \ge 0 \end{cases}$

13. $\begin{cases} x - y < 4 \\ y \geq 0 \\ xy = 12 \end{cases}$

14. $\begin{cases} xy \leq 1 \\ x \geq 0 \\ y \geq 0 \end{cases}$

15. $\begin{cases} x \geq 0 \\ y \geq 0 \\ 9x + 3y \leq 18 \\ 3x + 6y \leq 18 \end{cases}$

16. $\begin{cases} x + y \geq 1 \\ x - y \leq 1 \\ x - y \geq 0 \\ x \leq 2 \end{cases}$

CHAPTER SUMMARY

Key Words

augmented matrix (9.6)
coefficient matrix (9.6)
cofactor (9.5)
consistent system of equations (9.1)
Cramer's Rule (9.3, 9.5)
dependent equations (9.1)
determinant (9.3)
determinant function (9.3)
elementary row operations (9.6)

Gaussian elimination (9.6)
inconsistent system of
  equations (9.1)
independent equations (9.1)
matrix (9.3, 9.6)
minor (9.5)
square matrix (9.3)
systems of equations (9.1)
triangular form of a matrix (9.6)

Key Ideas

(9.1) Systems of two linear equations in two variables can be solved by graphing.

If a system of equations has at least one solution, the system is a **consistent system**. Otherwise, the system is an **inconsistent system**.

If the graphs of the equations of a system are distinct, the equations are **independent equations**. Otherwise, the equations are **dependent equations**.

(9.2) Systems of two linear equations in two variables can be solved by the **substitution method** or the **addition method**.

(9.3) A **matrix** is any rectangular array of numbers.

A **determinant of a square matrix** is a real number:

$$\det\left(\begin{bmatrix} a & b \\ c & d \end{bmatrix}\right) = \begin{vmatrix} a & b \\ c & d \end{vmatrix} = ad - bc$$

Many systems of two linear equations in two variables can be solved by using Cramer's Rule.

(9.4) A system of three equations in three variables can be solved by using the addition method.

(9.5) The value of the determinant of a 3 × 3 matrix is the sum of the products of the elements of any chosen row or column and the **cofactors** of those elements.

Many systems of three linear equations in three variables can be solved by using Cramer's Rule.

(9.6) Many systems of linear equations can be solved by using matrices and the method of Gaussian elimination.

(9.7) Systems of inequalities can be solved by graphing.

REVIEW EXERCISES

In Review Exercises 1–4, solve each system of equations by the graphing method. If a system is inconsistent or if the equations of a system are dependent, so indicate.

1. $\begin{cases} 2x + y = 11 \\ -x + 2y = 7 \end{cases}$

2. $\begin{cases} 3x + 2y = 0 \\ 2x - 3y = -13 \end{cases}$

3. $\begin{cases} \dfrac{1}{2}x + \dfrac{1}{3}y = 2 \\ y = 6 - \dfrac{3}{2}x \end{cases}$

4. $\begin{cases} \dfrac{1}{3}x - \dfrac{1}{2}y = 1 \\ 6x - 9y = 2 \end{cases}$

In Review Exercises 5–8, solve each system of equations by substitution.

5. $\begin{cases} y = x + 4 \\ 2x + 3y = 7 \end{cases}$

6. $\begin{cases} y = 2x + 5 \\ 3x - 5y = -4 \end{cases}$

7. $\begin{cases} x + 2y = 11 \\ 2x - y = 2 \end{cases}$

8. $\begin{cases} 2x + 3y = -2 \\ 3x + 5y = -2 \end{cases}$

In Review Exercises 9–14, solve each system of equations by addition.

9. $\begin{cases} x + y = -2 \\ 2x + 3y = -3 \end{cases}$

10. $\begin{cases} 3x + 2y = 1 \\ 2x - 3y = 5 \end{cases}$

11. $\begin{cases} x + \dfrac{1}{2}y = 7 \\ -2x = 3y - 6 \end{cases}$

12. $\begin{cases} y = \dfrac{x - 3}{2} \\ x = \dfrac{2y + 7}{2} \end{cases}$

13. $\begin{cases} x + y + z = 6 \\ x - y - z = -4 \\ -x + y - z = -2 \end{cases}$

14. $\begin{cases} 2x + 3y + z = -5 \\ -x + 2y - z = -6 \\ 3x + y + 2z = 4 \end{cases}$

In Review Exercises 15–18, evaluate each determinant.

15. $\begin{vmatrix} 2 & 3 \\ -4 & 3 \end{vmatrix}$

16. $\begin{vmatrix} -3 & -4 \\ 5 & -6 \end{vmatrix}$

17. $\begin{vmatrix} -1 & 2 & -1 \\ 2 & -1 & 3 \\ 1 & -2 & 2 \end{vmatrix}$

18. $\begin{vmatrix} 3 & -2 & 2 \\ 1 & -2 & -2 \\ 2 & 1 & -1 \end{vmatrix}$

In Review Exercises 19–22, use Cramer's Rule to solve each system of equations.

19. $\begin{cases} 3x + 4y = 10 \\ 2x - 3y = 1 \end{cases}$

20. $\begin{cases} 2x - 5y = -17 \\ 3x + 2y = 3 \end{cases}$

21. $\begin{cases} x + 2y + z = 0 \\ 2x + y + z = 3 \\ x + y + 2z = 5 \end{cases}$

22. $\begin{cases} 2x + 3y + z = 2 \\ x + 3y + 2z = 7 \\ x - y - z = -7 \end{cases}$

In Review Exercises 23–24, solve each system of equations by using matrices.

23. $\begin{cases} x + 2y = 4 \\ 2x - y = 3 \end{cases}$

24. $\begin{cases} x + y + z = 6 \\ 2x - y + z = 1 \\ 4x + y - z = 5 \end{cases}$

In Review Exercises 25–26, graph the solution set of each system of inequalities.

25. $\begin{cases} y \geq x + 1 \\ 3x + 2y < 6 \end{cases}$

26. $\begin{cases} y \geq x^2 - 4 \\ y < x + 3 \end{cases}$

CHAPTER NINE TEST

1. Solve the system $\begin{cases} 2x + y = 5 \\ y = 2x - 3 \end{cases}$ by graphing.

In Problems 2–3, consider the following system of equations:

$$\begin{cases} 3(x + y) = x - 3 \\ -y = \dfrac{2x + 3}{3} \end{cases}$$

2. Are the equations of the given system dependent or independent?

3. Is the system consistent or inconsistent?

4. Use any method to solve the following system for x:

$$\begin{cases} 2x - 4y = 14 \\ x = -2y + 7 \end{cases}$$

5. Use any method to solve the following system for y:

$$\begin{cases} \dfrac{x}{2} - \dfrac{y}{4} = -4 \\ x + y = -2 \end{cases}$$

6. Evaluate $\begin{vmatrix} 2 & -3 \\ 4 & 5 \end{vmatrix}$.

7. Evaluate $\begin{vmatrix} -3 & -4 \\ -2 & 3 \end{vmatrix}$.

In Problems 8–10, consider the system $\begin{cases} x - y = -6 \\ 3x + y = -6 \end{cases}$, *which is to be solved with Cramer's Rule.*

8. When solving for x, what is the numerator determinant? **(Don't evaluate it.)**

9. When solving for y, what is the denominator determinant? **(Don't evaluate it.)**

10. Solve the system for y.

In Problems 11–13, consider the system $\begin{cases} x + y + z = 4 \\ x + y - z = 6 \\ 2x - 3y + z = -1 \end{cases}$

11. Solve for x.

12. Solve for y.

13. Solve for z.

14. Evaluate $\begin{vmatrix} 3 & 0 & -2 \\ 3 & 2 & 0 \\ 1 & -2 & -1 \end{vmatrix}$.

15. Write the augmented matrix that represents the system of equations given in Problems 11–13.

16. Use the method of graphing to solve the following system of inequalities:

$$\begin{cases} y \geq x^2 \\ y < x + 3 \end{cases}$$

CUMULATIVE REVIEW EXERCISES (CHAPTERS 7–9)

1. Graph the equation $2x - 3y = 6$ and tell whether it represents a function.

2. Find the distance between $P(-2, 5)$ and $Q(8, -9)$.

In Problems 3–5, consider the points $P(-2, 5)$ and $Q(8, -9)$.

3. Find the midpoint of PQ.

4. Find the slope of PQ.

5. Write the equation of line PQ.

6. Are the lines represented by $3x + 2y = 12$ and $2x - 3y = 5$ parallel or perpendicular?

In Exercises 7–10, assume that $f(x) = 3x^2 + 2$ and $g(x) = 2x - 1$. Evaluate each expression.

7. $f(-1)$

8. $(g \circ f)(2)$

9. $(f \circ g)(x)$

10. $(g \circ f)(x)$

11. Find the inverse relation of the function $y = \dfrac{x + 3}{3}$ and tell whether it is a function.

12. Solve the proportion $\dfrac{x + 3}{2x} = \dfrac{3x}{6x + 5}$.

13. Express the following statement as a formula: "y varies directly with the product of x and z but inversely with the product of r and s."

14. The volume of a cylindrical tank varies jointly with the height of the tank and the square of the radius of its circular base. The volume is 4π cubic feet when $h = 4$ feet and $r = 1$ foot. Find the height when $V = 8\pi$ cubic feet and $r = 2$ feet.

15. Graph the inequality $2x - 3y \leq 12$.

16. Use the method of completing the square to solve the equation $2x^2 + x - 3 = 0$.

17. Use the quadratic formula to solve the equation $3x^2 + 4x - 1 = 0$.

18. Write in $a + bi$ form: $(3 - 2i) - (4 + i)^2$.

19. Write in $a + bi$ form: $\dfrac{1}{3 - i} + |3 + \sqrt{-4}|$.

20. Solve $x^4 - 13x^2 + 12 = 0$.

21. Solve $x - x^{1/2} - 12 = 0$.

22. For what values of k will the solutions of $2x^2 + 4x = k$ be equal?

23. Graph the equation $y = \dfrac{1}{2}x^2 - x + 1$ and find the coordinates of its vertex.

24. Graph $2y = |x|$ and tell whether the graph represents a function.

25. Is the function determined by $y = x^3 + 2$ one-to-one?

26. Solve $x^2 - x - 6 > 0$.

27. Graph the equation $(x - 2)^2 + (y + 1)^2 = 9$.

28. Solve the system $\begin{cases} 2x + y = 5 \\ x - 2y = 0 \end{cases}$ by graphing.

29. Solve the system $\begin{cases} 3x + y = 4 \\ 2x - 3y = -1 \end{cases}$ by substitution.

30. Solve the system $\begin{cases} x + 2y = -2 \\ 2x - y = 6 \end{cases}$ by addition.

31. Solve the system $\begin{cases} \dfrac{x}{10} + \dfrac{y}{5} = \dfrac{1}{2} \\ \dfrac{x}{2} - \dfrac{y}{5} = \dfrac{13}{10} \end{cases}$ by any method.

32. Evaluate $\begin{vmatrix} 3 & -2 \\ 1 & -1 \end{vmatrix}$

33. Solve the system $\begin{cases} 4x - 3y = -1 \\ 3x + 4y = -7 \end{cases}$ for y only by Cramer's Rule.

34. Solve the system $\begin{cases} x + y + z = 1 \\ 2x - y - z = -4 \\ x - 2y + z = 4 \end{cases}$ by any method.

35. Solve the system $\begin{cases} x + 2y + 3z = 6 \\ 3x + 2y + z = 6 \\ 2x + 3y + z = 6 \end{cases}$ for z only by Cramer's Rule.

36. Use matrix methods to solve the system $\begin{cases} x + y + z = 6 \\ x - z = -2. \\ x - y + z = 2 \end{cases}$

37. Solve the system $\begin{cases} 3x - 2y < 6 \\ y < -x + 2 \end{cases}$ by graphing.

38. Solve the system $\begin{cases} y < x + 2 \\ 3x + y \le 6 \end{cases}$ by graphing.

CAREERS AND MATHEMATICS

ELECTRICAL/ ELECTRONIC ENGINEER

Electrical engineers design, develop, test, and supervise the manufacture of electrical and electronic equipment. Electrical engineers who work with electronic equipment are often called **electronic engineers**.

Electrical engineers generally specialize in a major area—such as power distributing equipment, integrated circuits, computers, electrical equipment manufacturing, or communications. Besides manufacturing and research, development, and design, many are employed in administration and management, technical sales, or teaching.

Qualifications

A bachelor's degree in engineering is generally acceptable for beginning engineering jobs. College graduates with a degree in a natural science or mathematics also may qualify for some jobs.

Engineers should be able to work as part of a team and should have creativity, an analytical mind, and a capacity for detail. In addition, engineers should be able to express themselves well—both orally and in writing.

Job outlook

Employment of electrical engineers is expected to increase faster than the average for all occupations through the mid-1990s. Although increased demand for computers, communications equipment, and military electronics is expected to be the major contributor to this growth, demand for electrical and electronic consumer goods, along with increased research and development in new types of power generation, should create additional jobs.

Example application

In a radio, an inductor and a capacitor are used in a resonant circuit to select a desired station at frequency f, and reject all others. The inductance L and the capacitance C determine the inductive reactance X_L and the capacitive reactance X_C of that circuit, where

$$X_L = 2\pi f L \quad \text{and} \quad X_C = \frac{1}{2\pi f C}$$

The radio station that will be selected will be at the frequency f for which the inductive reactance and capacitive reactance are equal. To find that frequency, solve for f in terms of L and C.

Solution

Since X_L and X_C are to be equal, you can solve the system of equations

$$\begin{cases} X_L = 2\pi f L \\ X_C = \dfrac{1}{2\pi f C} \\ X_L = X_C \end{cases}$$

to express f in terms of L and C.

$$X_L = X_C$$

$$2\pi fL = \frac{1}{2\pi fC}$$

$$(2\pi fL)(2\pi fC) = 1 \qquad \text{Multiply both sides by } 2\pi fC.$$

$$4\pi^2 f^2 LC = 1 \qquad \text{Simplify.}$$

$$f^2 = \frac{1}{4\pi^2 LC} \qquad \text{Divide both sides by } 4\pi^2 LC.$$

$$f = \sqrt{\frac{1}{4\pi^2 LC}} \qquad \text{Take the positive square root of both sides.}$$

$$f = \frac{1}{2\pi \sqrt{LC}} \qquad \text{Simplify.}$$

Exercises

1. At what frequency will a 0.0001 farad capacitor and a 0.005 henry inductor resonate? (*Hint:* $C = 0.0001$ and $L = 0.005$)
2. If the inductor and the capacitor in Exercise 1 were both doubled, what would the resonant frequency become?
3. At what frequency will a 0.0008 farad capacitor and a 0.002 henry inductor resonate?
4. If the inductor of Exercise 3 were doubled, and the capacitance were reduced by one-half, what would the resonant frequency be?

(Answers: **1.**   225 hertz   **2.**   113 hertz   **3.**   126 hertz
4.   126 hertz)

10

Exponential and Logarithmic Functions

In the late 16th century, logarithms were invented to simplify computations in astronomy. They proved to be so successful that they were soon used to simplify calculations in many other fields. Today, the widespread use of calculators has eliminated the need for logarithms as a computational aid. However, logarithms have retained their importance in mathematics, science, engineering, and economics in other ways such as expressing the acidity of a solution, expressing the voltage gain of an amplifier, or computing the intensity of an earthquake. Logarithms are related to functions that involve exponential expressions. For this reason we begin by considering exponential functions.

10.1 EXPONENTIAL FUNCTIONS

We have previously defined exponential expressions such as b^x, where x is a rational number. We now give meaning to the expression b^x, where x is an irrational number. To do so we consider the expression $3^{\sqrt{2}}$, where $\sqrt{2}$ is an irrational number with a decimal value of $1.414213562\ldots$ and, consequently, $1 < \sqrt{2} < 2$. Because 3 is a number greater than 1, it can be shown that

$$3^1 < 3^{\sqrt{2}} < 3^2$$

Similarly, because $1.4 < \sqrt{2} < 1.5$,

$$3^{1.4} < 3^{\sqrt{2}} < 3^{1.5}$$

Because the numbers 1.4 and 1.5 are rational numbers, the exponential expressions $3^{1.4}$ and $3^{1.5}$ are defined. If we continue in this manner, $3^{\sqrt{2}}$ becomes fenced in by two values, one that is greater than $3^{\sqrt{2}}$ and one that is less than $3^{\sqrt{2}}$. As the list continues we can see that the value of $3^{\sqrt{2}}$ gets squeezed into a smaller and smaller interval:

$$3^1 = 3 \qquad < 3^{\sqrt{2}} < 9 \qquad = 3^2$$
$$3^{1.4} \approx 4.656 \quad < 3^{\sqrt{2}} < 5.196 \quad \approx 3^{1.5}$$
$$3^{1.41} \approx 4.7070 \quad < 3^{\sqrt{2}} < 4.7590 \quad \approx 3^{1.42}$$
$$3^{1.414} \approx 4.727695 < 3^{\sqrt{2}} < 4.732892 \approx 3^{1.415}$$

In calculus classes it is shown that $3^{\sqrt{2}}$ has a fixed value greater than any of the increasing values on the left of the previous list, but less than any of the decreasing values on the right. To find an approximation for $3^{\sqrt{2}}$, we can press these keys on a calculator:

$$\boxed{3} \quad \boxed{y^x} \quad \boxed{2} \quad \boxed{\sqrt{}} \quad \boxed{=}$$

The display will show 4.728804388. Thus,

$$3^{\sqrt{2}} \approx 4.728804388 \ldots$$

The previous discussion suggests this important fact: *If b is a positive real number and x is a real number (either rational or irrational), the exponential expression b^x represents a single real number.*

The function defined by the equation $y = f(x) = b^x$ is called an **exponential function**.

Definition.   The **exponential function with base b** is defined by the equation

$$y = b^x$$

where $b > 0$, $b \neq 1$, and x is a real number.

The **domain** of the exponential function is the set of real numbers, and the **range** is the set of positive real numbers.

Because the exponential function defined by $y = f(x) = b^x$ has the real number set for its domain and the positive real number set for its range, we can graph the function on a Cartesian coordinate system.

Example 1   Graph the exponential functions defined by   **a.** $y = 2^x$ and   **b.** $y = 4^x$.

Solution   **a.**   To graph the function defined by the equation $y = 2^x$, calculate several pairs of points (x, y) that satisfy the equation, plot the points, and join them with a smooth curve. The graph appears in Figure 10-1a.

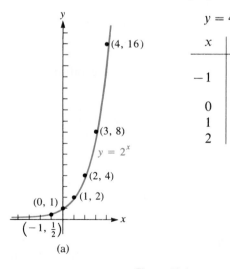

$y = 2^x$

x	y
-1	$\dfrac{1}{2}$
0	1
1	2
2	4
3	8

(a)

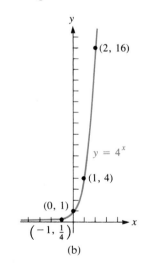

$y = 4^x$

x	y
-1	$\dfrac{1}{4}$
0	1
1	4
2	16

(b)

Figure 10-1

b.   To graph the function defined by the equation $y = 4^x$, proceed as in part **a**. The graph appears in Figure 10-1b.   ■

In each graph of Example 1, the values of y increase as the values of x increase; thus, the functions defined by the equations $y = 2^x$ and $y = 4^x$ are called **increasing functions**. Note that each graph approaches the x-axis as x gets smaller and that each graph passes through the point $(0, 1)$. Note also that the graph of $y = 2^x$ passes through the point $(1, 2)$ and that the graph of $y = 4^x$ passes through the point $(1, 4)$. In general, the graph of $y = b^x$ passes through the points $(0, 1)$ and $(1, b)$.

Example 2   Graph the exponential functions $y = \left(\dfrac{1}{2}\right)^x$ and $y = \left(\dfrac{1}{4}\right)^x$.

Solution   Calculate and plot several pairs (x, y) that satisfy each equation. The graph of $y = (\tfrac{1}{2})^x$ appears in Figure 10-2a, and the graph of $y = (\tfrac{1}{4})^x$ appears in Figure 10-2b.

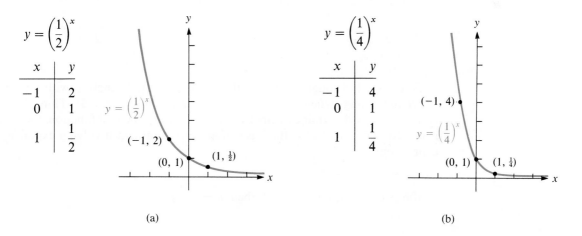

$y = \left(\dfrac{1}{2}\right)^x$

x	y
-1	2
0	1
1	$\dfrac{1}{2}$

$y = \left(\dfrac{1}{4}\right)^x$

x	y
-1	4
0	1
1	$\dfrac{1}{4}$

(a)   (b)

Figure 10-2   ■

In each graph of Example 2, the values of y decrease as the values of x increase; thus, the functions defined by the equations $y = (\tfrac{1}{2})^x$ and $y = (\tfrac{1}{4})^x$ are called **decreasing functions**. Each graph approaches the x-axis as x gets larger and each graph passes through the point $(0, 1)$. Note that the graph of $y = (\tfrac{1}{2})^x$ passes through the point $(1, \tfrac{1}{2})$ and that the graph of $y = (\tfrac{1}{4})^x$ passes through the point $(1, \tfrac{1}{4})$. In general, the graph of $y = (1/b)^x$ passes through the points $(0, 1)$ and $(1, 1/b)$.

Example 3   On the same set of coordinate axes, graph the exponential functions defined by
a. $y = (\tfrac{3}{2})^x$ and   **b.** $y = (\tfrac{2}{3})^x$.

Solution   Calculate and plot several pairs (x, y) that satisfy each equation. The graphs appear in Figure 10-3. Note that the equation $y = (\frac{3}{2})^x$ represents an increasing function and that the equation $y = (\frac{2}{3})^x$ represents a decreasing function. Note also that the bases of the functions are reciprocals of each other and that the graphs are reflections of each other in the y-axis.

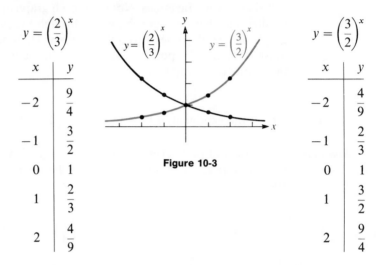

$y = \left(\dfrac{2}{3}\right)^x$

x	y
-2	$\dfrac{9}{4}$
-1	$\dfrac{3}{2}$
0	1
1	$\dfrac{2}{3}$
2	$\dfrac{4}{9}$

Figure 10-3

$y = \left(\dfrac{3}{2}\right)^x$

x	y
-2	$\dfrac{4}{9}$
-1	$\dfrac{2}{3}$
0	1
1	$\dfrac{3}{2}$
2	$\dfrac{9}{4}$

■

Examples 1 through 3 suggest that an exponential function with base b is either increasing (when $b > 1$) or decreasing (when $0 < b < 1$). Thus, distinct real numbers x determine distinct values b^x. The exponential function, therefore, is one-to-one. This fact is the basis of an important fact involving exponential expressions.

> If $b > 0$, $b \neq 1$, and $b^m = b^n$, then $m = n$.

Because exponential functions are one-to-one, they all have inverse functions. The next example shows how to find the inverse of an exponential function.

Example 4   Find the inverse function of the exponential function defined by $y = 3^x$. Graph both functions on the same set of coordinate axes.

Solution   The graph of $y = 3^x$ appears in Figure 10-4. The equation that defines the inverse function of $y = 3^x$ is found by interchanging x and y. Thus, the inverse of $y = 3^x$ is

$$x = 3^y$$

The graph of $x = 3^y$ also appears in the figure. As expected, the two graphs are symmetric with respect to the line $y = x$.

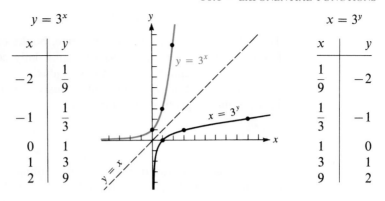

$$y = 3^x$$

x	y
-2	$\dfrac{1}{9}$
-1	$\dfrac{1}{3}$
0	1
1	3
2	9

$$x = 3^y$$

x	y
$\dfrac{1}{9}$	-2
$\dfrac{1}{3}$	-1
1	0
3	1
9	2

Figure 10-4

Applications of Exponential Functions

A mathematical description of an observed event is called a **model** of that event. Many observed events can be modeled by functions defined by equations of the form $y = ab^{kx}$. For example, the amount of money in a savings account that earns **compound interest** is modeled by the formula

Formula for
Compound Interest
$$A = A_0 \left(1 + \frac{r}{k} \right)^{kt}$$

where A represents the amount in the account after t years, with interest paid k times a year at an annual rate of $r\%$ on an initial deposit of A_0.

Example 5   If \$1500 is deposited in a bank account that pays 10% annual interest, compounded quarterly, how much will be in the account after 20 years?

Solution   Calculate A using the formula

$$A = A_0 \left(1 + \frac{r}{k} \right)^{kt}$$

with $A_0 = 1500$, $r = 0.10$, and $t = 20$. Because quarterly compounding pays interest four times a year, $k = 4$.

$$A = A_0 \left(1 + \frac{r}{k} \right)^{kt}$$

$$A = 1500 \left(1 + \frac{0.10}{4} \right)^{4(20)}$$

$$= 1500(1.025)^{80}$$

$$= 10,814.35$$

Use a calculator and press these keys:
1.025 [y^x] 80 [=] [×] 1500 [=].

After 20 years, the account will contain \$10,814.35.

Base-e Exponential Functions

To make their services more attractive, banks often compound interest more frequently than quarterly. Monthly and daily compounding is common, but some banks compound interest *continuously*.

If interest is compounded continuously, the amount A contained in a savings account is given by the formula

Formula for Continuous Compounding of Interest

$$A = A_0 e^{rt}$$

where A_0 is the initial deposit, r is the annual interest rate, t is the length of time in years, and e is an irrational number with a value of

$$e \approx 2.718281828$$

To compute the amount to which $1500 will grow if invested for 10 years at 10% annual interest compounded continuously, we substitute **1500** for A_0, **0.10** for r, and **10** for t in the formula $A = A_0 e^{rt}$ and simplify:

$$A = A_0 e^{rt}$$
$$A = 1500(e)^{0.10(10)}$$
$$= 1500(e)^1$$
$$\approx 4077.42$$

Use a calculator and press these keys:

1 $\boxed{e^x}$ $\boxed{\times}$ 1500 $\boxed{=}$.

After 10 years, the account will contain $4077.42.

If your calculator does not have an $\boxed{e^x}$ key, try pressing $\boxed{\text{INV}}$ $\boxed{\ln x}$ in place of $\boxed{e^x}$ or consult your owner's manual.

In mathematical models of natural events, the number e appears often as the base of an exponential function. For example, in the **Malthusian model for population growth**, the current population of a colony is given by the formula

Formula for Population Growth

$$P = P_0 e^{kt}$$

where P_0 is the population at $t = 0$, and k is the difference between the birth and death rates. If t is measured in years, k is called the **annual growth rate**.

Example 6   The annual growth rate within a city is 2%. If this growth rate remains constant and if the current population of the city is 15,000, what will the population be in 100 years?

Solution   Use the Malthusian model for population growth,

$$P = P_0 e^{kt}$$

with $P_0 = 15{,}000$, $k = 0.02$, and $t = 100$.

$$P = P_0 e^{kt}$$
$$P = 15{,}000 e^{0.02(100)}$$
$$= 15{,}000 e^2$$
$$\approx 110{,}835.84$$

Use a calculator and press these keys:

$$2 \boxed{e^x} \boxed{\times} 15000 \boxed{=}.$$

In 100 years the population will be 110,836.   ■

■ **EXERCISE 10.1**

In Exercises 1–12, graph the function defined by each equation.

1. $y = 3^x$

2. $y = 5^x$

3. $y = \left(\dfrac{1}{5}\right)^x$

4. $y = \left(\dfrac{1}{3}\right)^x$

5. $y = \left(\dfrac{3}{4}\right)^x$

6. $y = \left(\dfrac{5}{3}\right)^x$

7. $y = 2 + 2^x$

8. $y = -2 + 3^x$

9. $y = 3(2^x)$

10. $y = 2(3^x)$

11. $y = 2^{-x}$

12. $y = 3^{-x}$

In Exercises 13–16, graph the exponential function defined by each equation and graph its inverse on the same set of coordinate axes.

13. $y = 2^x$

14. $y = 4^x$

15. $y = \left(\dfrac{1}{4}\right)^x$

16. $y = \left(\dfrac{1}{2}\right)^x$

In Exercises 17–22, find the value of b, if any, that would cause the graph of $y = b^x$ to look like the graph indicated.

17.

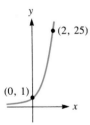

18.

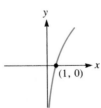

19.

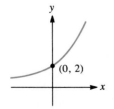

20.

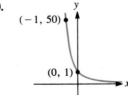

21.

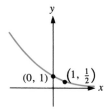

22.

23. An initial deposit of $10,000 earns 8% annual interest compounded quarterly. How much will be in the account after 10 years?

24. An initial deposit of $10,000 earns 8% annual interest compounded monthly. How much will be in the account after 10 years?

25. An initial deposit of $20,000 earns 7% annual interest compounded monthly. How much will be in the account after 25 years?

26. An initial deposit of $20,000 earns 7% annual interest compounded daily. How much will be in the account after 25 years?

27. An initial deposit of $8000 earns 6% annual interest compounded continuously. How much will be in the account after 10 years?

28. An initial deposit of $22,000 earns 9% annual interest compounded continuously. How much will be in the account after 20 years?

29. A population of fish is growing according to the Malthusian model. How many fish will there be in 10 years if the annual growth rate is 3% and the initial population is 2700 fish?

30. The population of a small town is 1350. The town is expected to grow according to the Malthusian model with an annual growth rate of 6%. What will be the population of the town in 20 years?

31. The population of the world is approximately 5 billion. If the world population is growing according to the Malthusian model with an annual growth rate of 1.8%, what will be the population of the world in 30 years?

32. The population of the world is approximately 5 billion. If the world population is growing according to the Malthusian model with an annual growth rate of 1.8%, what will be the population of the world in 60 years?

33. A colony of 6 million bacteria is growing in a culture medium. The population P after t hours is given by the formula $P = (6 \times 10^6)(2.3)^t$. What is the population after 4 hours?

34. The population of North Rivers is growing exponentially according to the formula $P = 375(1.3)^t$, where t is measured in years from the present date. What will be the population of North Rivers in 6 years and 9 months?

35. The charge remaining in a battery decreases as the battery slowly discharges. The charge C after t days is given by the formula $C = (3 \times 10^{-4})(0.7)^t$. What is the charge after 5 days?

36. A radioactive material decays according to the formula $A = A_0(\frac{2}{3})^t$, where A_0 is the initial amount present and t is measured in years. What amount will be present in 5 years?

10.2 LOGARITHMIC FUNCTIONS

The exponential function defined by the equation $y = b^x$ is one-to-one and, therefore, possesses an inverse function. The equation that defines its one-to-one inverse is obtained by interchanging x and y. Thus, the inverse of the function $y = b^x$ is defined by the equation $x = b^y$. To express the equation $x = b^y$ in a form that is solved for y, we need the following definition:

Definition.   The **logarithmic function with base b** is a one-to-one function defined by the equation

$$y = \log_b x$$

where $b > 0$ and $b \neq 1$. This equation is read as "y equals log base b of x," and is equivalent to the exponential equation $x = b^y$.

The **domain** of the logarithmic function is the set of positive real numbers. The **range** is the set of real numbers.

The previous definition points out that any pair of values (x, y) that satisfies the equation $y = \log_b x$ also satisfies the equation $x = b^y$. From this definition, it follows that

$\log_{10} x = y$ means $10^y = x$

$\log_7 1 = 0$ means $7^0 = 1$

$\log_3 m = n$ means $3^n = m$

$\log_4 16 = 2$ means $4^2 = 16$

$\log_2 32 = 5$ means $2^5 = 32$

$\log_5 \dfrac{1}{25} = -2$ means $5^{-2} = \dfrac{1}{25}$

In each case the logarithm of a number appears as the exponent in the corresponding exponential expression. This is no coincidence, because the logarithm of any number is always an exponent. In fact, $\log_b x$ is the exponent to which a base b must be raised in order to obtain x:

$$b^{\log_b x} = x$$

Because there is no real number x such that $2^x = -8$, for example, there is no meaning to the expression $\log_2(-8)$. In general, *there are no logarithms for nonpositive numbers*.

Because base-10 logarithms appear so often in mathematics and science, it is a good idea to learn the following base-10 logarithms:

$\log_{10} \dfrac{1}{100} = -2$ because $10^{-2} = \dfrac{1}{100}$

$\log_{10} \dfrac{1}{10} = -1$ because $10^{-1} = \dfrac{1}{10}$

$\log_{10} 1 = 0$ because $10^0 = 1$

$\log_{10} 10 = 1$ because $10^1 = 10$

$\log_{10} 100 = 2$ because $10^2 = 100$

$\log_{10} 1000 = 3$ because $10^3 = 1000$

In general, we have

$$\log_{10} 10^x = x$$

Example 1   Find the value of x in each equation: **a.** $\log_3 81 = x$,   **b.** $\log_x 125 = 3$, and   **c.** $\log_4 x = 3$

Solution   **a.** $\log_3 81 = x$ is equivalent to the equation $3^x = 81$. Because $3^4 = 81$, it follows that $x = 4$.

b. $\log_x 125 = 3$ is equivalent to the equation $x^3 = 125$. Because $5^3 = 125$, it follows that $x = 5$.

c. $\log_4 x = 3$ is equivalent to the equation $4^3 = x$ Because $4^3 = 64$, it follows that $x = 64$.   ■

Example 2   Find the value of x in each equation: **a.** $\log_{1/3} x = 2$,   **b.** $\log_{1/3} x = -2$, and   **c.** $\log_{1/3} \dfrac{1}{27} = x$.

Solution   **a.** $\log_{1/3} x = 2$ is equivalent to $\left(\dfrac{1}{3}\right)^2 = x$. Thus, $x = \dfrac{1}{9}$.

b. $\log_{1/3} x = -2$ is equivalent to $\left(\dfrac{1}{3}\right)^{-2} = x$. Thus,

$$x = \left(\frac{1}{3}\right)^{-2} = 3^2 = 9$$

c. $\log_{1/3} \dfrac{1}{27} = x$ is equivalent to $\left(\dfrac{1}{3}\right)^x = \dfrac{1}{27}$. Because $\left(\dfrac{1}{3}\right)^3 = \dfrac{1}{27}$, it follows that $x = 3$.   ■

Because both the domain and the range of the logarithmic function are subsets of the real numbers, it is possible to graph logarithmic functions on the Cartesian coordinate system.

Example 3   Graph the logarithmic function defined by $y = \log_2 x$.

Solution   The equation $y = \log_2 x$ is equivalent to the equation $x = 2^y$. Calculate a table of values (x, y) that satisfy the equation $x = 2^y$, plot the corresponding points, and connect them with a smooth curve. The graph appears in Figure 10-5.

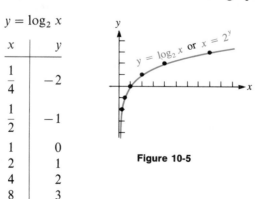

$y = \log_2 x$

x	y
$\dfrac{1}{4}$	-2
$\dfrac{1}{2}$	-1
1	0
2	1
4	2
8	3

Figure 10-5

■

Example 4   Graph the logarithmic function defined by $y = \log_{1/2} x$.

Solution   The equation $y = \log_{1/2} x$ is equivalent to the equation $x = \left(\frac{1}{2}\right)^y$. Calculate a table of values (x, y) that satisfy the equation $x = \left(\frac{1}{2}\right)^y$, plot the corresponding points, and connect them with a smooth curve. The graph appears in Figure 10-6.

$y = \log_{1/2} x$

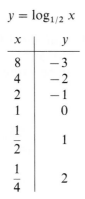

x	y
8	-3
4	-2
2	-1
1	0
$\dfrac{1}{2}$	1
$\dfrac{1}{4}$	2

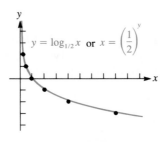

Figure 10-6

If the graphs of Examples 3 and 4 are merged onto a single set of coordinate axes as in Figure 10-7, it is apparent that they are reflections of each other in the x-axis. Note that both graphs pass through the point with coordinates of $(1, 0)$, and that both curves approach but never touch the y-axis.

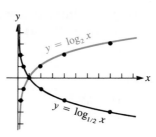

Figure 10-7

Properties of Logarithms

Because logarithmic equations can be written as exponential equations, it is not surprising that the laws of exponents have counterparts in logarithmic notation. We will use this fact to develop many of the following important properties of logarithms.

Properties of Logarithms.   If M, N, p, and b are positive numbers and $b \neq 1$, then

1. $\log_b 1 = 0$ 2. $\log_b b = 1$
3. $\log_b b^x = x$ 4. $b^{\log_b x} = x$
5. $\log_b MN = \log_b M + \log_b N$ 6. $\log_b \dfrac{M}{N} = \log_b M - \log_b N$
7. $\log_b M^p = p \log_b M$ 8. If $\log_b x = \log_b y$, then $x = y$.

Properties 1 through 4 follow directly from the definition of logarithm.

To prove Property 5, we let $x = \log_b M$ and $y = \log_b N$. Because of the definition of logarithm, these equations can be written in the form

$$M = b^x \quad \text{and} \quad N = b^y$$

Then,

$$MN = b^x b^y$$

and the properties of exponents give

$$MN = b^{x+y}$$

Using the definition of logarithms in reverse gives

$$\log_b MN = x + y$$

Substituting the values of x and y completes the proof:

$$\log_b MN = \log_b M + \log_b N$$

To prove Property 6, we again let $x = \log_b M$ and $y = \log_b N$. These equations can be written as

$$M = b^x \quad \text{and} \quad N = b^y$$

Then,

$$\frac{M}{N} = \frac{b^x}{b^y}$$

and the properties of exponents give

$$\frac{M}{N} = b^{x-y}$$

Using the definition of logarithms in reverse gives

$$\log_b \frac{M}{N} = x - y$$

Substituting the values of x and y completes the proof:

$$\log_b \frac{M}{N} = \log_b M - \log_b N \qquad \square$$

To prove Property 7, we let $x = \log_b M$, write the expression in exponential form, and raise both sides to the pth power:

$$M = b^x$$
$$(M)^p = (b^x)^p$$
$$M^p = b^{px}$$

Using the definition of logarithms in reverse gives

$$\log_b M^p = px$$

Substituting the value for x completes the proof:

$$\log_b M^p = p \log_b M \qquad \square$$

Property 8 follows from the fact that the logarithmic function is a one-to-one function.

Property 5 of logarithms asserts that the logarithm of the *product* of two numbers is equal to the *sum* of their logarithms. The logarithm of a *sum* or a *difference* usually does not simplify. In general,

$$\log_b(M + N) \neq \log_b M + \log_b N$$

Likewise,

$$\log_b(M - N) \neq \log_b M - \log_b N$$

Property 6 of logarithms asserts that the logarithm of the *quotient* of two numbers is equal to the *difference* of their logarithms. The logarithm of a quotient is not the quotient of the logarithms:

$$\log_b \frac{M}{N} \neq \frac{\log_b M}{\log_b N}$$

Example 5   Simplify each expression: **a.** $\log_3 1$,   **b.** $\log_4 4$,   **c.** $\log_7 7^3$, and   **d.** $b^{\log_b 3}$.

Solution   **a.**   By Property 1, $\log_3 1 = 0$.

b.   By Property 2, $\log_4 4 = 1$.

c.   By Property 3, $\log_7 7^3 = 3$.

d.   By Property 4, $b^{\log_b 3} = 3$.   ∎

The properties of logarithms are often used to expand or condense a logarithmic expression as in the following two examples.

Example 6   Assume x, y, z, and b are positive numbers ($b \neq 1$). Use properties of logarithms to write each expression in terms of the logarithms of x, y, and z.

a. $\log_b \dfrac{xy}{z}$,   b. $\log_b(x^3y^2z)$, and   c. $\log_b \dfrac{y^2\sqrt{z}}{x}$

Solution   a. $\log_b \dfrac{xy}{z} = \log_b(xy) - \log_b z$ Use Property 6.

$= \log_b x + \log_b y - \log_b z$ Use Property 5.

b. $\log_b(x^3y^2z) = \log_b x^3 + \log_b y^2 + \log_b z$ Use Property 5 twice.

$= 3 \log_b x + 2 \log_b y + \log_b z$ Use Property 7 twice.

c. $\log_b \dfrac{y^2\sqrt{z}}{x} = \log_b(y^2\sqrt{z}) - \log_b x$ Use Property 6.

$= \log_b y^2 + \log_b \sqrt{z} - \log_b x$ Use Property 5.

$= \log_b y^2 + \log_b z^{1/2} - \log_b x$ Write $\sqrt{z}$ as $z^{1/2}$.

$= 2 \log_b y + \dfrac{1}{2} \log_b z - \log_b x$ Use Property 7 twice.   ■

Example 7   Assume x, y, z, and b are positive numbers ($b \neq 1$). Use properties of logarithms to write each expression as the logarithm of a single quantity.

a. $2 \log_b x + \dfrac{1}{3} \log_b y$ and   b. $\dfrac{1}{2} \log_b(x - 2) - \log_b y + 3 \log_b z$

Solution   a. $2 \log_b x + \dfrac{1}{3} \log_b y = \log_b x^2 + \log_b y^{1/3}$ Use Property 7 twice.

$= \log_b(x^2 y^{1/3})$ Use Property 5.

$= \log_b(x^2 \sqrt[3]{y})$ Write $y^{1/3}$ as $\sqrt[3]{y}$.

b. $\dfrac{1}{2} \log_b(x - 2) - \log_b y + 3 \log_b z$

$= \log_b(x - 2)^{1/2} - \log_b y + \log_b z^3$ Use Property 7 twice.

$= \log_b \dfrac{(x - 2)^{1/2}}{y} + \log_b z^3$ Use Property 6.

$= \log_b \dfrac{z^3 \sqrt{x - 2}}{y}$ Use Property 5 and write $(x - 2)^{1/2}$ as $\sqrt{x - 2}$.

Example 8   Given that $\log_{10} 2 \approx 0.3010$ and $\log_{10} 3 \approx 0.4771$, find the approximate values of   a. $\log_{10} 6$,   b. $\log_{10} 9$,   c. $\log_{10} 5$,   d. $\log_{10} \sqrt{5}$,   and   e. $\log_{10} 1.5$.

Solution   a. $\log_{10} 6 = \log_{10}(2 \cdot 3)$ Factor 6 as $2 \cdot 3$.

$= \log_{10} 2 + \log_{10} 3$ Use Property 5 of logarithms.

$\approx 0.3010 + 0.4771$ Substitute the value of each logarithm.

≈ 0.7781

b. $\log_{10} 9 = \log_{10} 3^2$ Write 9 as 3^2.

$\qquad = 2 \log_{10} 3$ Use Property 7 of logarithms.

$\qquad \approx 2(0.4771)$ Substitute the value of the logarithm of 3.

$\qquad \approx 0.9542$

c. $\log_{10} 5 = \log_{10} \dfrac{10}{2}$ Write 5 as $\dfrac{10}{2}$.

$\qquad = \log_{10} 10 - \log_{10} 2$ Use Property 6 of logarithms.

$\qquad = 1 - \log_{10} 2$ Use Property 2 of logarithms ($\log_{10} 10 = 1$).

$\qquad \approx 1 - 0.3010$ Substitute the value of the logarithm of 2.

$\qquad \approx 0.6990$

d. $\log_{10} \sqrt{5} = \log_{10} 5^{1/2}$ Write $\sqrt{5}$ as $5^{1/2}$.

$\qquad = \dfrac{1}{2} \log_{10} 5$ Use Property 7 of logarithms.

$\qquad \approx \dfrac{1}{2} (0.6990)$ Use the answer from part **c**.

$\qquad \approx 0.3495$

e. $\log_{10} 1.5 = \log_{10} \dfrac{3}{2}$ Write 1.5 as $\dfrac{3}{2}$.

$\qquad = \log_{10} 3 - \log_{10} 2$ Use Property 6 of logarithms.

$\qquad \approx 0.4771 - 0.3010$ Substitute the value of each logarithm.

$\qquad \approx 0.1761$ ∎

■ EXERCISE 10.2

In Exercises 1–8, write each equation in exponential form.

1. $\log_3 81 = 4$ **2.** $\log_7 7 = 1$ **3.** $\log_{1/2} \dfrac{1}{8} = 3$ **4.** $\log_{1/5} 1 = 0$

5. $\log_4 \dfrac{1}{64} = -3$ **6.** $\log_6 \dfrac{1}{36} = -2$ **7.** $\log_x y = z$ **8.** $\log_m n = \dfrac{1}{2}$

In Exercises 9–16, write each equation in logarithmic form.

9. $8^2 = 64$ **10.** $10^3 = 1000$ **11.** $4^{-2} = \dfrac{1}{16}$ **12.** $3^{-4} = \dfrac{1}{81}$

13. $\left(\dfrac{1}{2}\right)^{-5} = 32$ **14.** $\left(\dfrac{1}{3}\right)^{-3} = 27$ **15.** $x^y = z$ **16.** $m^n = p$

In Exercises 17–24, find each value.

17. $\log_7 1$ **18.** $\log_2 4$ **19.** $\log_5 25$ **20.** $\log_9 81$

21. $\log_{1/8} \dfrac{1}{8}$ **22.** $\log_{1/6} \dfrac{1}{36}$ **23.** $\log_{2/5} \dfrac{8}{125}$ **24.** $\log_{3/4} \dfrac{81}{256}$

In Exercises 25–50, find each value of x.

25. $\log_2 8 = x$ **26.** $\log_3 27 = x$ **27.** $\log_5 125 = x$ **28.** $\log_3 81 = x$

29. $\log_4 x = 2$ **30.** $\log_5 x = 3$ **31.** $\log_3 x = 4$ **32.** $\log_2 x = 6$

33. $\log_x 16 = 4$ **34.** $\log_x 9 = 2$ **35.** $\log_x 32 = 5$ **36.** $\log_x 125 = 3$

37. $\log_{1/5} \dfrac{1}{25} = x$ **38.** $\log_{1/5} x = -3$ **39.** $\log_x \dfrac{1}{9} = -2$ **40.** $\log_{1/4} x = -4$

41. $\log_x 9 = \dfrac{1}{2}$ **42.** $\log_x 8 = \dfrac{1}{3}$ **43.** $\log_{125} x = \dfrac{2}{3}$ **44.** $\log_{100} \dfrac{1}{1000} = x$

45. $\log_{5/2} \dfrac{4}{25} = x$ **46.** $\log_x \dfrac{9}{4} = 2$ **47.** $\log_x \dfrac{\sqrt{3}}{3} = \dfrac{1}{2}$ **48.** $\log_{1/2} x = 0$

49. $\log_{\sqrt{2}} 4 = x$ **50.** $\log_{\sqrt{3}} 9 = x$

In Exercises 51–58, graph each logarithmic function.

51. $y = \log_3 x$ **52.** $y = \log_4 x$ **53.** $y = \log_{1/4} x$ **54.** $y = \log_{1/3} x$

55. $y = 1 + \log_2 x$ **56.** $y = -2 + \log_5 x$ **57.** $y = 3\log_3 x$ **58.** $y = \log_2(3x)$

In Exercises 59–60, graph each pair of functions on a single set of coordinate axes. Note the line of symmetry.

59. $y = 4^x$ and $y = \log_4 x$ **60.** $y = 3^x$ and $y = \log_3 x$

In Exercises 61–66, find the value of b, if any, that would cause the graph of $y = \log_b x$ to look like the graph shown.

61.

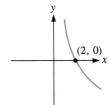

62.

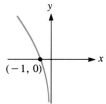

63.

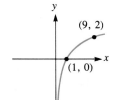

64.

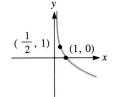

65.

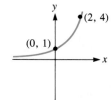

66.

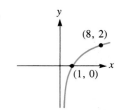

In Exercises 67–74, assume that x, y, z, and b are positive numbers. Use the properties of logarithms to write each expression in terms of the logarithms of x, y, and z.

67. $\log_b xyz$

68. $\log_b \dfrac{x}{yz}$

69. $\log_b \left(\dfrac{x}{y}\right)^2$

70. $\log_b (xz)^{1/3}$

71. $\log_b x\sqrt{x-z}$

72. $\log_b \dfrac{\sqrt[3]{x}}{\sqrt[3]{y-z}}$

73. $\log_b \sqrt[4]{\dfrac{x^3 y^2}{z^4}}$

74. $\log_b x \sqrt{\dfrac{\sqrt{y}}{z}}$

In Exercises 75–82, assume that x, y, and z are positive numbers. Use the properties of logarithms to write each expression as the logarithm of a single quantity.

75. $\log_b(x+1) - \log_b x$

76. $\log_b x + \log_b(x+2) - \log_b 8$

77. $2\log_b x + \dfrac{1}{3}\log_b y$

78. $-2\log_b x - 3\log_b y + \log_b z$

79. $-3\log_b x - 2\log_b y + \dfrac{1}{2}\log_b z$

80. $3\log_b(x+1) - 2\log_b(x+2) + \log_b x$

81. $\log_b\left(\dfrac{x}{z}+x\right) - \log_b\left(\dfrac{y}{z}+y\right)$

82. $\log_b(xy+y^2) - \log_b(xz+yz) + \log_b x$

In Exercises 83–98, tell whether each statement is true. If not, so indicate.

83. $\log_b 0 = 1$

84. $7^{\log_7 7} = 7$

85. $\log_b 1 = 0$

86. $\log_b 1 = 1$

87. $\log_b 1 = b$

88. $\log_b b = 1$

89. $\log_7 7^7 = 7$

90. $\dfrac{\log_b A}{\log_b B} = \log_b A - \log_b B$

91. $\log_b ab = 1 + \log_b a$

92. $\log_b \dfrac{1}{a} = -\log_b a$

93. $\log_b 2 = \log_2 b$

94. A logarithm cannot be negative.

95. $\log_b xy = (\log_b x)(\log_b y)$

96. $\dfrac{1}{3}\log_b a^3 = \log_b a$

97. If $\log_a b = c$, then $\log_b a = c$.

98. If $\log_b a = c$, then $\log_b a^p = pc$.

In Exercises 99–112, assume that $\log_{10} 4 \approx 0.6021$, $\log_{10} 7 \approx 0.8451$, *and* $\log_{10} 9 \approx 0.9542$. *Use these values and the properties of logarithms to find the approximate value of each quantity.*

99. $\log_{10} 28$ **100.** $\log_{10} \dfrac{7}{4}$ **101.** $\log_{10} 2.25$ **102.** $\log_{10} 36$

103. $\log_{10} \dfrac{63}{4}$ **104.** $\log_{10} \dfrac{4}{63}$ **105.** $\log_{10} 252$ **106.** $\log_{10} 49$

107. $\log_{10} 2$ **108.** $\log_{10} 3$ **109.** $\log_{10} 6$ **110.** $\log_{10} 81$

111. $\log_{10} 50$ **112.** $\log_{10} 0.5$

10.3 APPLICATIONS OF BASE-10 AND BASE-*e* LOGARITHMS

Logarithms are necessary to solve many applied problems. In the past mathematicians had to use extensive tables to find logarithms of numbers, but today logarithms are easy to find with a calculator.

For computational purposes, base-10 logarithms, often called **common logarithms**, are the most convenient. In this book, if the base b is not given in the notation $\log_b x$, we will always assume that b is 10; thus,

$$\log x \quad \text{means} \quad \log_{10} x$$

To use a calculator to find log 2.34, for example, we enter the number 2.34 into a calculator and press the $\boxed{\text{log}}$ key. (*Caution:* Some calculators require pressing a $\boxed{\text{2nd}}$ function key first.) The display on the calculator will read $\boxed{.3692158574}$. Thus, $\log 2.34 \approx 0.3692$. Note that this statement is equivalent to the statement $10^{0.3692} \approx 2.34$.

Example 1   Use a calculator to find the value of x in each equation: **a.** $\log 8.75 = x$, **b.** $\log 379 = x$, and   **c.** $\log x = -2.1180$.

Solution   **a.**   To find log 8.75, enter the number 8.75 and press the $\boxed{\text{log}}$ key. The display will read $\boxed{0.942008053}$. Thus, $\log 8.75 \approx 0.9420$.

b.   To find log 379, enter the number 379 and press the $\boxed{\text{log}}$ key. The display will read $\boxed{2.57863921}$. Thus, $\log 379 \approx 2.5786$.

c.   To find x in the equation $\log x = -2.1180$, change the equation to exponential form:

$$10^{-2.1180} = x$$

To evaluate $10^{-2.1180}$, enter the number 10, press the $\boxed{y^x}$ key, and enter the number 2.1180. Then press the $\boxed{+/-}$ key to change 2.1180 to -2.1180, and press the $\boxed{=}$ key. The display will read $\boxed{.0076207901}$. Thus, $x = 10^{-2.1180} \approx 0.0076$.

If your calculator has a $\boxed{10^x}$ key, you can enter 2.1180, press the $\boxed{+/-}$ key, and then press the $\boxed{10^x}$ key.   ∎

Applications of Base-10 Logarithms

Common logarithms are used in chemistry to express the acidity of solutions. The more acidic a solution, the greater the concentration of hydrogen ions. This hydrogen ion concentration is indicated indirectly by the **pH scale**, defined by the equation

Formula for
pH Scale

$$pH = -\log[H^+]$$

where $[H^+]$ is the hydrogen ion concentration in gram-ions per liter. Pure water, for example, has some hydrogen ions—approximately 10^{-7} gram-ions per liter. Thus, the pH of pure water is

$$pH = -\log 10^{-7}$$
$$= -(-7)\log 10 \qquad \text{Use Property 7 of logarithms.}$$
$$= 7 \qquad\qquad \text{Recall that } \log 10 = 1.$$

Example 2   Seawater has a pH of approximately 8.5. Find its hydrogen ion concentration.

Solution   To find the hydrogen ion concentration of seawater, solve the equation $8.5 = -\log[H^+]$ for $[H^+]$:

$$8.5 = -\log[H^+]$$
$$-8.5 = \log[H^+] \qquad \text{Multiply both sides by } -1.$$
$$[H^+] = 10^{-8.5} \qquad \text{Change the equation to exponential form.}$$
$$[H^+] \approx 3.2 \times 10^{-9} \qquad \text{Use a calculator.}$$

Thus, seawater has a hydrogen ion concentration of approximately 3.2×10^{-9} gram-ions per liter.   ∎

Common logarithms are used in electrical engineering to express the voltage gain (or loss) of an electronic device such as an amplifier. The unit of gain (or loss), called the **decibel**, is defined by a logarithmic relation. If E_O is the output voltage of a device and E_I is the input voltage, the decibel voltage gain is given by the formula

Formula for
Decibel Voltage Gain

$$\textbf{Decibel voltage gain} = 20 \log \frac{E_O}{E_I}$$

Example 3   If the input to an amplifier is 0.5 volt and the output is 40 volts, find the decibel voltage gain of the amplifier.

Solution   The decibel voltage gain is calculated by substituting 0.5 for E_I and 40 for E_O in the formula and simplifying:

$$\text{Decibel voltage gain} = 20 \log \frac{E_O}{E_I}$$

$$= 20 \log \frac{40}{0.5}$$

$$= 20 \log 80$$

$$\approx 38 \qquad \text{Use a calculator.}$$

The decibel voltage gain is approximately 38. ■

The intensity of earthquakes is measured on the **Richter scale**, which is based on the formula

Richter Scale $$R = \log \frac{A}{P}$$

where A is the amplitude of the tremor (measured in micrometers) and P is the period of the tremor (measured in seconds).

Example 4   Find the measure on the Richter scale of an earthquake with an amplitude of 10,000 micrometers and a period of 0.1 second.

Solution   Substitute $10,000$ for A and 0.1 for P in the formula defining the Richter scale and proceed as follows:

$$R = \log \frac{A}{P}$$

$$R = \log \frac{10,000}{0.1}$$

$$= \log 100,000$$

$$= 5$$

The earthquake measures 5 on the Richter scale. ■

Natural Logarithms

A second important base for logarithms is the number e. Logarithms with a base of e are called **natural logarithms** and are denoted by the symbol $\ln x$ (read as "the natural logarithm of x," rather than the symbol $\log_e x$. Thus, in this book

$$\ln x \quad \text{means} \quad \log_e x$$

Natural logarithms are found by using the $\boxed{\ln x}$ key on a calculator. For example, to find the value of $\ln 25.3$, enter 25.3 and press the $\boxed{\ln x}$ key. The display will read $\boxed{3.230804396}$. Thus, $\ln 25.3 \approx 3.2308$.

Example 5 Use a calculator to find the value of x in each equation: **a.** $\ln 8.75 = x$, **b.** $\ln 379 = x$, and **c.** $\ln x = -2.1180$.

Solution **a.** To find $\ln 8.75$, enter the number 8.75 and press the $\boxed{\ln x}$ key. The display will read $\boxed{2.1690537}$. Thus, $x = \ln 8.75 \approx 2.1691$.

b. To find $\ln 379$, enter the number 379 and press the $\boxed{\ln x}$ key. The display will read $\boxed{5.937536205}$. Thus, $x = \ln 379 \approx 5.9375$.

c. To find x in the equation $\ln x = -2.1180$, write the equation in exponential form:

$$e^{-2.1180} = x$$

To evaluate $e^{-2.1180}$, enter 2.1180, press the $\boxed{+/-}$ key, and then press the $\boxed{e^x}$ key. The display will read $\boxed{0.120271932}$. Thus, $x = e^{-2.1180} \approx 0.1203$. ■

Applications of Natural Logarithms

The length of time t that it takes for money to double if left in an account where interest is compounded continuously is given by the formula

$$t = \frac{\ln 2}{r}$$

where r is the annual rate of interest.

Example 6 How long will it take \$1000 to double if it is invested at an annual interest rate of 8% compounded continuously?

Solution Substitute 0.08 for r and simplify:

$$t = \frac{\ln 2}{r}$$

$$t = \frac{\ln 2}{0.08}$$

$$\approx 8.6643 \qquad \text{Use a calculator.}$$

It will take approximately $8\frac{2}{3}$ years for the money to double. ■

The loudness of a sound is not directly proportional to the intensity of the sound. Experiments in physiology suggest that the relationship between loudness and intensity is a logarithmic one known as the **Weber–Fechner Law**. This law states that the apparent loudness L of a sound is directly proportional to the natural logarithm of its intensity I. In symbols,

Weber–Fechner Law $\qquad L = k \ln I$

where k is the constant of proportionality.

Example 7   Find the increase in intensity that will cause the apparent loudness of a sound to double.

Solution   If the original loudness L_0 is caused by an actual intensity I_0, then

$$1. \quad L_0 = k \ln I_0$$

To double the apparent loudness, multiply both sides of Equation 1 by 2 and apply Property 7 of logarithms:

$$2L_0 = 2k \ln I_0$$
$$= k \ln(I_0)^2$$

Thus, to double the apparent loudness of the sound, the intensity must be squared.   ■

■ EXERCISE 10.3

In Exercises 1–18, use a calculator to find the value of the variable. Express all answers to four decimal places.

1. $\log 3.25 = x$ **2.** $\log 32.1 = x$ **3.** $\log \dfrac{17}{7} = x$ **4.** $\log \dfrac{3}{31} = x$

5. $\log N = 3.29$ **6.** $\log M = -9.17$ **7.** $\log M = -\dfrac{21}{5}$ **8.** $\log N = \dfrac{4}{13}$

9. $\ln 3.9 = y$ **10.** $\ln 0.087 = z$ **11.** $\ln \dfrac{1}{37} = t$ **12.** $\ln 4300 = r$

13. $\ln M = -8.3$ **14.** $\ln N = 0.763$ **15.** $\ln N = 2.83$ **16.** $\ln M = -23.2$

17. $\ln A = \log 7$ **18.** $\log B = \ln 8$

In Exercises 19–32, solve each word problem.

19. Find the pH of a solution with a hydrogen ion concentration of 1.7×10^{-5} gram-ions per liter.

20. The hydrogen ion concentration of sour pickles is 6.31×10^{-4}. What is the pH?

21. What is the hydrogen ion concentration of a saturated solution of calcium hydroxide with pH $= 13.2$?

22. Find the hydrogen ion concentration of the water in a pond if the pH of the water is 6.5.

23. The safe pH for a freshwater tank containing tropical fish can range from 6.8 to 7.6. What is the corresponding range in hydrogen ion concentration?

24. The pH of apples ranges from 2.9 to 3.3. What is the corresponding range in hydrogen ion concentration?

25. The decibel voltage gain of an amplifier is 29. If the output is 20 volts, what is the input?

26. The decibel voltage gain of an amplifier is 35. If the input voltage is 0.05 volt, what is the output?

27. An amplifier produces an output of 30 volts when driven by an input of 0.1 volt. What is the decibel voltage gain of the amplifier?

28. An amplifier produces an output of 80 volts when driven by an input of 0.12 volt. What is the amplifier's decibel voltage gain?

29. An earthquake has an amplitude of 5000 micrometers and a period of 0.2 second. What does the earthquake measure on the Richter scale?

30. An earthquake with an amplitude of 8000 micrometers measures 6 on the Richter scale. What is its period?

31. An earthquake with a period of $\frac{1}{4}$ second measures 4 on the Richter scale. What is the earthquake's amplitude?

32. By what factor must the period of an earthquake change to increase its severity by 1 point on the Richter scale? Assume that the amplitude remains constant.

In Exercises 33–34, use the formula

$$t = \frac{\ln 3}{r}$$

which gives the length of time t for an amount of money to triple if left in an account paying an annual rate r, compounded continuously.

33. Find the length of time it will take $2500 to triple if invested in an account paying 9% annual interest compounded continuously.

34. Find the length of time it will take $15,500 to triple if invested in an account paying 15% annual interest compounded continuously.

In Exercises 35–36, use the formula

$$t = \frac{\ln 4}{r}$$

which gives the length of time t for an amount of money to quadruple if left in an account paying an annual rate r, compounded continuously.

35. Find the length of time it will take $1700 to quadruple if it is invested in an account that pays 6% annual interest compounded continuously.

36. Find the length of time it will take $50,500 to quadruple if it is invested in an account paying 9% annual interest compounded continuously.

In Exercises 37–40, solve each word problem.

37. If the intensity of a sound is doubled, what will be the apparent change in loudness?

38. If the intensity of a sound is tripled, what will be the apparent change in loudness?

39. What increase in intensity will cause an apparent tripling of the loudness of a sound?

40. What increase in the intensity of a sound will cause the apparent loudness to be multiplied by 4?

10.4 EXPONENTIAL AND LOGARITHMIC EQUATIONS

An **exponential equation** is an equation in which the variable appears in an exponent, and a **logarithmic equation** is an equation that involves the logarithm of an expression containing the variable. Some exponential and logarithmic equations are difficult to solve, but others can be solved by using the properties of exponents and logarithms and the usual equation-solving techniques.

Because the logarithmic function is one-to-one, it can be shown that logarithms of equal positive numbers are equal. This fact enables us to solve many equations by taking the logarithms of both sides of the equation and then applying Property 7 of logarithms to remove the variable from its position as an exponent. To solve the equation $3^x = 5$, for example, we take the common logarithm of both sides of the equation and use Property 7 of logarithms:

$$3^x = 5$$
$$\log 3^x = \log 5$$
$$x \log 3 = \log 5$$

We then divide both sides of the equation by log 3 to obtain

$$\textbf{1.} \quad x = \frac{\log 5}{\log 3}$$

To obtain a decimal approximation, we use a calculator to determine that $\log 5 \approx 0.6990$ and $\log 3 \approx 0.4771$ and then substitute these values into Equation 1 and simplify:

$$x \approx \frac{0.6990}{0.4771}$$
$$\approx 1.465$$

A careless reading of Equation 1 leads to a common error. Because $\log \dfrac{A}{B} = \log A - \log B$, many students believe that the expression $\dfrac{\log 5}{\log 3}$ also involves a subtraction. It does not. The expression $\dfrac{\log 5}{\log 3}$ indicates the quotient that is obtained when log 5 is divided by log 3.

Example 1   Solve the equation $6^{x-3} = 2$.

Solution   Take the common logarithm of both sides of the equation and proceed as follows:

$$6^{x-3} = 2$$
$$\log 6^{x-3} = \log 2$$
$$(x - 3) \log 6 = \log 2 \qquad \text{Use Property 7 of logarithms.}$$
$$x - 3 = \frac{\log 2}{\log 6} \qquad \text{Divide both sides by log 6.}$$
$$x = \frac{\log 2}{\log 6} + 3 \qquad \text{Add 3 to both sides.}$$

To obtain a decimal approximation, use a calculator and press 2 $\boxed{\log}$ $\boxed{\div}$ 6 $\boxed{\log}$ $\boxed{+}$ 3 $\boxed{=}$ to get the following result rounded to four decimal places:

$$x \approx 3.3869$$

If pressing these keys gives you a different result, consult the owner's manual for your calculator. ∎

Example 2   Solve the equation $\log x + \log(x - 3) = 1$.

Solution   Use Property 5 of logarithms to write the left-hand side of the equation as $\log x(x - 3)$ to obtain

$$\log [x(x - 3)] = 1$$

This equation is equivalent to the exponential equation

$$x(x - 3) = 10^1 \qquad \log_{10}[x(x - 3)] = 1 \text{ is equivalent to } 10^1 = x(x - 3).$$

which is a quadratic equation that can be solved by factoring:

$$x(x - 3) = 10^1$$
$$x^2 - 3x - 10 = 0$$
$$(x + 2)(x - 5) = 0$$
$$x = -2 \qquad \text{or} \qquad x = 5$$

Because logarithms of negative numbers are not defined, the number -2 is not a solution. However, 5 is a solution because it satisfies the original equation.

Check:

$$\log x + \log(x - 3) = 1$$
$$\log 5 + \log(5 - 3) \overset{?}{=} 1$$
$$\log 5 + \log 2 \overset{?}{=} 1$$
$$\log 10 \overset{?}{=} 1 \qquad \log 5 + \log 2 = \log(5 \cdot 2) = \log 10.$$
$$1 = 1 \qquad \text{Use Property 5 of logarithms.} \qquad ∎$$

Example 3   Solve the equation $\log_b(3x + 2) - \log_b(2x - 3) = 0$.

Solution   Proceed as follows:

$$\log_b(3x + 2) - \log_b(2x - 3) = 0$$
$$\log_b(3x + 2) = \log_b(2x - 3) \qquad \text{Add } \log_b(2x - 3) \text{ to both sides.}$$
$$3x + 2 = 2x - 3 \qquad \text{Use Property 8 of logarithms.}$$
$$x = -5$$

Check:

$$\log_b(3x + 2) - \log_b(2x - 3) = 0$$
$$\log_b[3(-5) + 2] - \log_b[2(-5) - 3] = 0$$
$$\log_b(-13) - \log_b(-13) = 0$$

Because there are no logarithms of negative numbers, the previous equation is meaningless. The original equation has no solutions. Its solution set is $\varnothing$. ∎

The Change-of-Base Formula

A formula for finding logarithms with different bases can be found by solving the equation $b^x = y$ for the variable x. To produce the desired result, we begin by taking the base-a logarithm of both sides:

$$b^x = y$$
$$\log_a b^x = \log_a y$$
$$x \log_a b = \log_a y \qquad \text{Use Property 7 of logarithms.}$$
$$\textbf{2.} \qquad x = \frac{\log_a y}{\log_a b}$$

Because the original equation $b^x = y$ is equivalent to the equation $x = \log_b y$, we can substitute $\log_b y$ for x in Equation 2 to obtain the **change-of-base formula:**

The Change-of-Base Formula.

$$\log_b y = \frac{\log_a y}{\log_a b}$$

If we know the logarithm of some number y to some base a (for example, $a = 10$), we can then use the change-of-base formula to find the logarithm of y to a different base b. To do so, we divide the base-a logarithm of y by the base-a logarithm of the new base b.

Example 4   Use the change-of-base formula to find $\log_3 5$.

Solution   Substitute 3 for b, 5 for y, and 10 for a in the change-of-base formula and simplify:

$$\log_b y = \frac{\log_a y}{\log_a b}$$
$$\log_3 5 = \frac{\log_{10} 5}{\log_{10} 3}$$
$$\approx \frac{0.6990}{0.4771} \qquad \log_{10} 5 \approx 0.6990 \text{ and } \log_{10} 3 \approx 0.4771.$$
$$\approx 1.465$$
∎

Applications of Exponential and Logarithmic Equations

The atomic structure of radioactive material changes as the material emits radiation. Uranium, for example, decays into thorium, then into radium, and eventually into lead. Experiments have determined the time for half of a given amount of radioactive material to decompose. This time, called the **half-life**, is constant for any given substance. The amount A of radioactive material present decays exponentially according to the model

Formula for Radioactive Decay

$$A = A_0 2^{-t/h}$$

where A_0 is the amount present at time $t = 0$ and h is the half-life of the material.

Example 5 How long will it take 1 gram of radium, with a half-life of 1600 years, to decompose to 0.75 gram?

Solution Substitute 1 for A_0, 0.75 for A, and 1600 for h in the formula for radioactive decay, and solve for t:

$$A = A_0 2^{-t/h}$$

$$0.75 = 1 \cdot 2^{-t/1600}$$

Take the common logarithm of both sides to obtain

$$\log 0.75 = \log 2^{-t/1600}$$

$$\log 0.75 = -\frac{t}{1600} \log 2 \qquad \text{Use Property 7 of logarithms.}$$

$$t = -1600 \frac{\log 0.75}{\log 2}$$

$$t \approx 664$$

It will take approximately 664 years for 1 gram of radium to decompose to 0.75 gram. ■

When a living organism dies, the oxygen/carbon dioxide cycle common to all living things stops, and carbon-14, a radioactive isotope of carbon that is present in the atmosphere, is no longer absorbed. Because the half-life of carbon-14 is about 5700 years, a wooden artifact with half of the radioactivity of wood from a living tree would be 5700 years old. Carbon dating is a common technique used by archeologists to date objects as old as 25,000 years.

Example 6 How old is a wooden statue if it contains $\frac{1}{3}$ of the carbon-14 of a living tree?

Solution Substitute $\frac{1}{3}A_0$ for A and 5700 for h in the formula for radioactive decay, and solve for t:

$$A = A_0 2^{-t/h}$$

$$\frac{1}{3}A_0 = A_0 2^{-t/5700}$$

$$\frac{1}{3} = 2^{-t/5700} \qquad \text{Divide both sides by } A_0.$$

Take the common logarithm of both sides of the equation, and solve for t:

$$\log \frac{1}{3} = \log 2^{-t/5700}$$

$$\log 1 - \log 3 = -\frac{t}{5700} \log 2 \qquad \text{Use Properties 6 and 7 of logarithms.}$$

$$-\log 3 = -\frac{t}{5700} \log 2 \qquad \log 1 = 0$$

$$t = 5700 \frac{\log 3}{\log 2}$$

$$\approx 9034.29$$

To the nearest thousand years, the statue is approximately 9000 years old.   ■

While there is a sufficient food supply and space to grow, populations of living things tend to increase according to the Malthusian model of population growth:

Formula for
Population Growth
$$P = P_0 e^{kt}$$

where P_0 is the initial population at time $t = 0$ and k depends on the rate of growth.

Example 7   A bacteria culture increased from an initial population of 500 to a population of 1500 in 3 hours. How long will it take the population to reach 10,000?

Solution   Substitute **1500** for P, **500** for P_0, and **3** for t in the formula for population growth:

$$P = P_0 e^{kt}$$
$$1500 = 500 e^{k \cdot 3}$$
$$3 = e^{3k}$$

Because the base of e^{3k} is e, take the natural logarithm of both sides and simplify to obtain

$$\ln 3 = \ln e^{3k}$$
$$\ln 3 = 3 k \ln e \qquad \text{Use Property 7 of logarithms.}$$
$$\ln 3 = 3 k \qquad \ln e = 1$$
$$k = \frac{\ln 3}{3}$$

To find when the population will reach 10,000, substitute **10,000** for P, **500** for P_0, and $\dfrac{\ln 3}{3}$ for k in the equation for population growth, and solve for t:

$$P = P_0 e^{kt}$$

$$10,000 = 500 e^{[(\ln 3)/3]t}$$

$$20 = e^{[(\ln 3)/3]t} \qquad \text{Divide both sides by 500.}$$

$$\ln 20 = \ln e^{[(\ln 3)/3]t} \qquad \text{Take the natural logarithm of both sides.}$$

$$\ln 20 = \frac{\ln 3}{3} t \ln e \qquad \text{Use Property 7 of logarithms.}$$

$$t = \frac{3 \ln 20}{\ln 3} \qquad \text{Remember that } \ln e = 1.$$

$$\approx 8.1805$$

In approximately 8 hours the culture will have a population of 10,000.   ■

■ EXERCISE 10.4

In Exercises 1–16, solve each equation for x.

1. $4^x = 5$ **2.** $7^x = 12$ **3.** $2^{x-1} = 10$ **4.** $3^{x+1} = 100$

5. $2^{x+1} = 2^{2x+3}$ **6.** $3^{3x+2} = 3^{2x+7}$ **7.** $10^{x^2} = 100$ **8.** $10^{x^2} = 1000$

9. $3^x = 4^x$ **10.** $3^{2x} = 4^x$ **11.** $3^{x+1} = 4^x$ **12.** $3^x - 4^{x+1} = 0$

13. $10^{\sqrt{x}} = 1000$ **14.** $100^{\sqrt{x}} = 10$ **15.** $10^{\sqrt[3]{x}} - 100 = 0$ **16.** $1000^{\sqrt[3]{x}} - 10 = 0$

In Exercises 17–38, solve each equation for x.

17. $\log 2x = \log 4$ **18.** $\log 3x = \log 9$

19. $\log(3x + 1) = \log(x + 7)$ **20.** $\log(x^2 + 4x) = \log(x^2 + 16)$

21. $\log(2x - 3) - \log(x + 4) = 0$ **22.** $\log(3x + 5) - \log(2x + 6) = 0$

23. $\log \dfrac{4x + 1}{2x + 9} = 0$ **24.** $\log \dfrac{5x + 2}{2(x + 7)} = 0$

25. $\log x^2 = 2$ **26.** $\log x^3 = 3$

27. $\log x + \log(x - 48) = 2$ **28.** $\log x + \log(x + 9) = 1$

29. $\log x + \log(x - 15) = 2$ **30.** $\log x + \log(x + 21) = 2$

31. $\log(x + 90) = 3 - \log x$ **32.** $\log(x - 90) = 3 - \log x$

33. $\log(x - 6) - \log(x - 2) = \log \dfrac{5}{x}$ **34.** $\log(x - 1) - \log 6 = \log(x - 2) - \log x$

35. $\log_7(2x - 3) - \log_7(x - 1) = 0$ **36.** $\log x^2 = (\log x)^2$

37. $\log(\log x) = 1$ **38.** $\log_3 x = \log_3 \dfrac{1}{x} + 4$

In Exercises 39–46, use the change-of-base formula to find each logarithm.

39. $\log_3 7$ **40.** $\log_7 3$ **41.** $\log_{1/3} 3$ **42.** $\log_{1/2} 6$

43. $\log_8 \sqrt{2}$ **44.** $\log_6 \sqrt{6}$ **45.** $\log_\pi \sqrt{2}$ **46.** $\log_{\sqrt{2}} \pi$

47. Show that $\log_{b^2} x = \frac{1}{2}\log_b x$.

48. Show that $\log_b b^x = x$.

49. Show that $b^{\log_b x} = x$.

50. Show that $e^{x \ln a} = a^x$.

In Exercises 51–62, solve each word problem.

51. The half-life of tritium is 12.4 years. How long will it take for 25% of a sample of tritium to decompose?

52. Twenty percent of a newly discovered radioactive element, mathematicum, decays in 2 years. What is its half-life?

53. An isotope of thorium, ^{227}Th, has a half-life of 18.4 days. How long will it take 80% of a sample to decompose?

54. An isotope of lead, ^{201}Pb, has a half-life of 8.4 hours. How many hours ago was there 30% more of the substance?

55. A parchment fragment is found in a newly discovered ancient tomb. It contains 60% of the carbon-14 it is assumed to have had initially. Approximately how old is the fragment?

56. Only 10% of the carbon-14 in a small wooden bowl remains. How old is the bowl?

57. A bacteria culture doubles in size every 24 hours. By how much will it have increased in 36 hours?

58. A bacteria culture triples in size every 3 days. By what factor will it have increased in 1 day?

59. The population of Waving Grass, Kansas, is presently 140 and is expected to triple every 15 years. When can the city planners expect the population to be double the present census?

60. The rodent population in a city is currently estimated at 20,000. It is expected to double every 5 years. When will the population reach 1 million?

61. An initial deposit of p dollars in a bank that pays 12% annual interest, compounded annually, will grow to an amount A in t years, where

$$A = p(1.12)^t$$

How long will it take $1000 to increase to $3000?

62. An initial deposit of p dollars in a bank that pays 12% annual interest, compounded monthly, will grow to an amount A in t years, where

$$A = p(1.01)^{12t}$$

How long will it take $1000 to increase to $3000?

10.5 LOGARITHMIC CALCULATIONS (OPTIONAL)

Before the widespread availability of calculators, logarithms provided the only reasonable way of performing certain long and involved calculations. To use logarithms, mathematicians had to rely on extensive tables of logarithmic values. In this section, we will show how logarithms can be used as a computational aid. We begin by discussing how to use Table B in Appendix II.

Example 1 Use Table B to find the base-10 logarithm of 2.71.

Solution To find log 2.71, run your finger down the left column of Table B (reproduced in Figure 10-8) until you reach 2.7. Then, slide your finger to the right until you reach entry .4330, which is in the column headed with 1. This value is the

N	0	1	2	3	4	5	6	7	8	9
2.6	.4150	.4166	.4183	.4200	.4216	.4232	.4249	.4265	.4281	.4298
2.7	.4314	.4330	.4346	.4362	.4378	.4393	.4409	.4425	.4440	.4456
2.8	.4472	.4487	.4502	.4518	.4533	.4548	.4564	.4579	.4594	.4609

Figure 10-8

logarithm of 2.71; that is, log 2.71 ≈ 0.4330. To verify that this is true, use a calculator to show that $10^{0.4330} \approx 2.71$. ∎

Example 2   Use Table B to find the base-10 logarithm of 2,710,000.

Solution   The logarithm of 2,710,000 cannot be found directly from the table. However, by using properties of logarithms, you can determine the logarithm of 2,710,000 as follows:

$2{,}710{,}000 = 2.71 \times 10^6$	Write 2,710,000 in scientific notation.
$\log 2{,}710{,}000 = \log(2.71 \times 10^6)$	Take the common logarithm of both sides.
$= \log 2.71 + \log 10^6$	Use Property 5 of logarithms.
$= \log 2.71 + 6 \log 10$	Use Property 7 of logarithms.
$= \log 2.71 + 6$	Use Property 2 of logarithms.
$\approx 0.4330 + 6$	Substitute 0.4330 for log 2.71.
≈ 6.4330	

Hence, log 2,710,000 ≈ 6.4330. To verify that this is true, use a calculator to show that $10^{6.4330} \approx 2{,}710{,}000$. ∎

If a common logarithm is written as the sum of an integer and a positive decimal between 0 and 1, the positive decimal is called the **mantissa**, and the integer part is called the **characteristic**. The characteristic of the value 0.4330 that was obtained in Example 1 is 0, and the mantissa is .4330. The characteristic of the value 6.4330 that was obtained in Example 2 is 6, and the mantissa is .4330.

Example 3   Find log 0.000271.

Solution   Proceed as in Example 2:

$0.000271 = 2.71 \times 10^{-4}$	Write 0.000271 in scientific notation.
$\log 0.000271 = \log(2.71 \times 10^{-4})$	Take the common logarithm of both sides.
$= \log 2.71 + \log 10^{-4}$	Use Property 5 of logarithms.
$= \log 2.71 - 4 \log 10$	Use Property 7 of logarithms.
$= \log 2.71 - 4$	Use Property 2 of logarithms.
$\approx 0.4330 - 4$	Substitute 0.4330 for log 2.71.

Hence, $\log 0.000271 \approx 0.4330 - 4$. The mantissa of $\log 0.000271$ is the positive decimal .4330 and the characteristic is -4. If the characteristic and the mantissa are combined, you have $\log 0.000271 \approx -3.5670$. To verify that this is true, use a calculator to show that $10^{-3.5670} \approx 0.000271$.   ■

The results of Examples 1, 2, and 3 indicate that

$$\log 2.71 = \log(2.71 \times 10^0) \approx 0.4330$$
$$\log 2,710,000 = \log(2.71 \times 10^6) \approx 6.4330$$

and

$$\log 0.000271 = \log (2.71 \times 10^{-4}) \approx 0.4330 - 4$$

From these results, it is apparent that the mantissa, .4330, is determined by the digits 271, and that the characteristic is determined by the location of the decimal point. In fact, the characteristic of the logarithm of a number *is* the exponent of 10 used when expressing that number in scientific notation.

Example 4   If $\log N = 0.4579$, find N.

Solution   In this example, you are given the logarithm 0.4579 and are asked to find N, called the **antilogarithm** of 0.4579. First, locate the mantissa .4579 in the body of Table B. (See Figure 10-8.) The number in the left column of the row that contains .4579 is 2.8, and the heading of the column that contains .4579 is 7. Hence, the antilogarithm of 0.4579 is approximately 2.87; that is, $N \approx 2.87$. To verify that this is true, use a calculator to show that $10^{0.4579} \approx 2.87$.   ■

Example 5   If $\log N = -2.1180$, find N.

Solution   Because $-2.1180 = -(2 + 0.1180) = -2 - 0.1180$ and the decimal -0.1180 is negative, the antilogarithm of -2.1180 cannot be found directly from the table. However, if you add and subtract 3 from -2.1180, you can establish a positive mantissa:

$$\log N = -2.1180 = (-2.1180 + 3) - 3$$
$$= 0.8820 - 3$$

The mantissa of this logarithm is the positive decimal .8820, and the characteristic is -3. The mantissa, .8820, can be found in Table B, and it determines the number 7.62. Note that $\log 7.62 \approx 0.8820$ and that $10^{0.8820} \approx 7.62$. To place the decimal point in the answer, write the equation $\log N = 0.8820 - 3$ in exponential form and simplify:

$$N = 10^{0.8820-3}$$
$$= 10^{0.8820} \cdot 10^{-3} \qquad \text{Use the product rule of exponents.}$$
$$\approx 7.62 \cdot 10^{-3} \qquad \text{Substitute 7.62 for } 10^{0.8820}$$
$$\approx 0.00762$$ ■

We will now show how logarithms can be used as a computational tool.

Example 6   Use logarithms to calculate $\sqrt[3]{25,300}$.

Solution Form the equation $N = \sqrt[3]{25,300}$, take the common logarithm of both sides, and proceed as follows:

$$N = \sqrt[3]{25,300}$$

$$\log N = \log \sqrt[3]{25,300} \qquad \text{Take the common logarithm of both sides.}$$

$$= \log(25,300)^{1/3} \qquad \text{Use the fact that } \sqrt[n]{x} = x^{1/n}.$$

$$= \frac{1}{3} \log(2.53 \times 10^4) \qquad \begin{array}{l} \text{Use Property 7 of logarithms and write} \\ \text{25,300 in scientific notation.} \end{array}$$

$$\approx \frac{1}{3}(4.4031) \qquad \text{Find } \log(2.53 \times 10^4).$$

$$\approx 1.4677$$

The characteristic of $\log N$ is 1, and the mantissa is .4677. Look up .4677 in the body of Table B. Because .4677 does not appear in the table, pick the mantissa closest to it and find that $.4677 \approx \log 2.94$ (or $10^{0.4677} \approx 2.94$). To place the decimal point in the answer, write the equation $\log N \approx 1.4677$ in exponential form and simplify:

$$N \approx 10^{1.4677}$$

$$\approx 10^{0.4677+1} \qquad \text{Separate the characteristic and mantissa.}$$

$$\approx 10^{0.4677} \cdot 10^1 \qquad \text{Use the product rule for exponents.}$$

$$\approx 2.94 \cdot 10^1 \qquad \text{Substitute 2.94 for } 10^{0.4677}.$$

$$\approx 29.4$$

Hence,

$$\sqrt[3]{25,300} \approx 29.4$$

∎

Example 7 Use logarithms to calculate $\dfrac{(3.14)^8}{\sqrt[7]{628}}$.

Solution Form the equation $N = \dfrac{(3.14)^8}{\sqrt[7]{628}}$, take the common logarithm of both sides, and proceed as follows:

$$N = \frac{(3.14)^8}{\sqrt[7]{628}}$$

$$\log N = \log \frac{(3.14)^8}{\sqrt[7]{628}} \qquad \text{Take the common logarithm of both sides.}$$

$$= \log(3.14)^8 - \log \sqrt[7]{628} \qquad \text{Use Property 6 of logarithms.}$$

$$= 8 \log 3.14 - \frac{1}{7} \log 628 \qquad \text{Use Property 7 of logarithms.}$$

$$\approx 8(0.4969) - \frac{1}{7}(2.7980) \qquad \text{Find } \log 3.14 \text{ and } \log 628.$$

$$\approx 3.9752 - 0.3997$$

$$\approx 3.5755$$

The characteristic of log N is 3 and the mantissa is .5755. Look up .5755 in the body of Table B, and find that $.5755 \approx \log 3.76$ (or $10^{0.5755} \approx 3.76$). To place the decimal point in the answer, write the equation $\log N \approx 3.5755$ in exponential form and simplify:

$$N \approx 10^{3.5755}$$
$$\approx 10^{0.5755+3} \qquad \text{Separate the characteristic and mantissa.}$$
$$\approx 10^{0.5755} \cdot 10^3 \qquad \text{Use the product rule for exponents.}$$
$$\approx 3.76 \cdot 10^3 \qquad \text{Substitute 3.76 for } 10^{0.5755}.$$
$$\approx 3760$$

Hence,

$$\frac{(3.14)^8}{\sqrt[7]{628}} \approx 3760 \qquad \blacksquare$$

Linear Interpolation

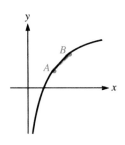

Figure 10-9

Table B in Appendix II gives values of N that are accurate to three significant digits, and values of the mantissas that are accurate to four significant digits. There is a process, called **linear interpolation**, that allows us to extend these accuracies by one additional significant digit.

The method of linear interpolation is based on the fact that the graph of the logarithmic function appears to be a straight line when we look at a small part of it. For example, the logarithmic curve in Figure 10-9 appears to be straight between points A and B. If line AB is assumed to be a straight line, we can set up proportions involving points on the line AB. The next example illustrates how.

Example 8   Use Table B to find log 3.974.

Solution   Table B does not give the value of log 3.974, but it does give the values for log 3.970 and log 3.980.

10 thousandths	4 thousandths	$\log 3.970 \approx 0.5988$ $\log 3.974 \approx ?$ $\log 3.980 \approx 0.5999$	x ten-thousandths	11 ten-thousandths

Note that the difference between 3.970 and 3.980 is 10 thousandths, and that the difference between 3.970 and 3.974 is 4 thousandths. Also note that the difference between 0.5988 and 0.5999 is 11 ten-thousandths. If you assume that the graph of $y = \log x$ is a straight line between $x = 3.97$ and $x = 3.98$, then the following proportion can be formed:

$$\frac{4 \text{ thousandths}}{10 \text{ thousandths}} = \frac{x \text{ ten-thousandths}}{11 \text{ ten-thousandths}}$$

or simply

$$\frac{4}{10} = \frac{x}{11}$$

You can solve this proportion for x as follows:

$$\frac{4}{10} = \frac{x}{11}$$

$$10x = 44$$

$$x = 4.4$$

$$x \approx 4$$

To get a good estimate of log 3.974, you must add 4 ten-thousandths to 0.5988. Hence,

$$\log 3.974 \approx 0.5988 + 0.0004$$

$$\approx 0.5992$$

Figure 10-10 shows the similar triangles that justify the previous proportion.

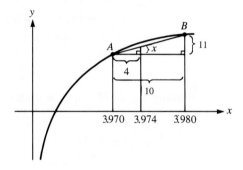

Figure 10-10

Example 9   Use linear interpolation to find N, where log $N = 0.4733$.

Solution   You cannot find .4733 in the body of Table B, but you can find two consecutive values, 0.4728 and 0.4742, that straddle 0.4733. Hence, you can set up the following chart.

10 thousandths	x thousandths	$\left[\begin{array}{c}\log 2.970 \approx 0.4728 \\ \log N \approx 0.4733 \\ \log 2.980 \approx 0.4742\end{array}\right.$	5 ten-thousandths	14 ten-thousandths

Finally, set up the following proportion, and solve for x:

$$\frac{x}{10} = \frac{5}{14}$$

$$x = \frac{5}{14}(10)$$

$$x \approx 3.57$$

$$x \approx 4$$

A good approximation for N is found by adding 4 thousandths to 2.970. Hence, log 2.974 $\approx$ 0.4733 and $N \approx$ 2.974.   ∎

Example 10   Use logarithms to calculate $\dfrac{(0.01320)(50.80)}{8.724}$ to four-digit accuracy.

Solution   First, calculate log 0.01320, log 50.80, and log 8.724:

log 0.01320 = log(1.32 × 10^{-2})	Write 0.01320 in scientific notation.
= log 1.32 + log 10^{-2}	Use Property 5 of logarithms.
= log 1.32 − 2 log 10	Use Property 7 of logarithms.
$\approx$ 0.1206 − 2	Find log 1.32 and use Property 2 of logarithms.
$\approx$ −1.8794	Simplify.
log 50.80 = log(5.08 × 10^1)	Write 50.80 in scientific notation.
= log 5.08 + log 10^1	Use Property 5 of logarithms.
$\approx$ 0.7059 + 1	Find log 5.08 and use Properties 7 and 2 of logarithms.
$\approx$ 1.7059	

Finding log 8.724 requires linear interpolation:

$$10 \left[4 \begin{bmatrix} \log 8.720 \approx 0.9405 \\ \log 8.724 \approx ? \\ \log 8.730 \approx 0.9410 \end{bmatrix} x \right] 5$$

Set up a proportion and solve for x:

$$\frac{4}{10} = \frac{x}{5}$$

$$5\left(\frac{4}{10}\right) = x$$

$$2 = x$$

Hence, you have

$$\log 8.724 \approx 0.9405 + 0.0002$$
$$\approx 0.9407$$

Now, form the equation

$$N = \frac{(0.01320)(50.80)}{8.724}$$

Take the common logarithm of both sides, and find log N as follows:

$$N = \frac{(0.01320)(50.80)}{8.724}$$

$$\log N = \log \frac{(0.01320)(50.80)}{8.724}$$

Take the common logarithm of both sides.

$$= \log 0.01320 + \log 50.80 - \log 8.724$$

Use Properties 5 and 6 of logarithms.

$$\approx -1.8794 + 1.7059 - 0.9407$$

Substitute the values of the logarithms.

$$\approx -1.1142$$

Because -1.1142 means $-1 - 0.1142$ and the decimal -0.1142 is negative, the antilogarithm of -1.1142 cannot be found directly from the table. However, if you add and subtract 2 from -1.1142, you can establish a positive mantissa:

$$\log N \approx -1.1142 \approx (-1.1142 + 2) - 2$$
$$\approx 0.8858 - 2$$

Write this logarithmic equation in exponential form, and solve as follows:

$$\log N \approx 0.8858 - 2$$
$$N \approx 10^{0.8858-2}$$
$$\approx 10^{0.8858} \cdot 10^{-2}$$

You can evaluate the factor $10^{0.8858}$ by forming the exponential equation $M = 10^{0.8858}$, and using linear interpolation on its equivalent logarithmic form, $\log M = 0.8858$.

$$10 \left[x \left[\begin{matrix} \log 7.680 \approx 0.8854 \\ \log M \approx 0.8858 \end{matrix} \right] 4 \\ \log 7.690 \approx 0.8859 \end{matrix} \right] 5$$

Thus,

$$\frac{x}{10} = \frac{4}{5}$$
$$x = 8$$

The antilogarithm of 0.8858 is approximately $7.680 + 0.008$, so $M \approx 10^{0.8858} \approx 7.688$. Therefore,

$$N \approx 10^{0.8858} \cdot 10^{-2}$$
$$\approx 7.688 \cdot 10^{-2}$$
$$\approx 0.07688$$

and

$$\frac{(0.01320)(50.80)}{8.724} \approx 0.07688$$

∎

Before calculators, mathematicians also relied on tables to find base-e logarithms. The next example uses a table of base-e logarithmic values.

Example 11   Use Table C in Appendix II to find the value of ln 2.34.

Solution   Look up 2.3 in the left column of Table C (reproduced in Figure 10-11), and follow that row to the column headed by 4. From that position in the table, you can read that ln 2.34 ≈ 0.8502.

N	0	1	2	3	4	5	6	7	8	9
2.2	.7885	.7930	.7975	.8020	.8065	.8109	.8154	.8198	.8242	.8286
2.3	.8329	.8372	.8416	.8459	.8502	.8544	.8587	.8629	.8671	.8713
2.4	.8755	.8796	.8838	.8879	.8920	.8961	.9002	.9042	.9083	.9123

Figure 10-11   ■

■ EXERCISE 10.5

In Exercises 1–8, use Table B to find each indicated logarithm.

1. log 5.97   **2.** log 3.15   **3.** log 4.23   **4.** log 9.83

5. log 432,000   **6.** log 57,900,000 **7.** log 0.00137   **8.** log 0.0838

In Exercises 9–16, use Table B to find the value of N.

9. $\log N = 0.4969$ **10.** $\log N = 0.8785$ **11.** $\log N = 0.0334$ **12.** $\log N = 0.7427$

13. $\log N = 3.9232$ **14.** $\log N = 4.6149$ **15.** $\log N = -2.5467$ **16.** $\log N = -4.4377$

In Exercises 17–20, use linear interpolation and Table B to find each indicated logarithm.

17. log 6.894   **18.** log 37.43   **19.** log 0.003456   **20.** log 0.04376

In Exercises 21–26, use linear interpolation and Table B to find the value of N to four significant digits.

21. $\log N = 0.6315$ **22.** $\log N = 0.0437$

23. $\log N = 3.2036$ **24.** $\log N = 0.8508 - 4$

25. $\log N = -2.1134$ **26.** $\log N = -0.4467$

In Exercises 27–30, use Table C to find each indicated logarithm. Use linear interpolation, if needed.

27. ln 4.65   **28.** ln 2.93   **29.** ln 8.325   **30.** ln 9.993

In Exercises 31–40, use Table B to calculate the approximate value of each indicated number. Do not use linear interpolation.

31. $(2.3)^{1/4}$ **32.** $\sqrt[3]{0.007}$ **33.** $(0.012)^{-0.03}$ **34.** $(1.05)^{25}$

35. $10^{5.4942}$ **36.** $(1.73)^{-1.73}$ **37.** $\sqrt{0.071}$ **38.** $4.3^{-5.2} + 3.1^{1.3}$

39. $(\log 4.1)^{2.4}$ **40.** $(2.3 + 1.79)^{-0.157}$

In Exercises 41–48, use Table B to calculate each indicated value. Use linear interpolation.

41. $(34.41)(0.4455)$ **42.** $\dfrac{648.1}{2.798}$ **43.** $(0.0004519)^{2.5}$ **44.** $(0.04263)^{-0.2}$

45. $5,258,000^{7/8}$ **46.** $\sqrt[15]{38,670}$ **47.** $\dfrac{(8.034)(32.68)}{\sqrt{3.869}}$ **48.** $\dfrac{(0.0004629)^{0.3}}{\sqrt[3]{0.004567}}$

In Exercises 49–52, use Table C and the fact that $\ln 10 \approx 2.3026$ *to calculate each value.*

49. $\ln 29.4$ **50.** $\ln 751$ **51.** $\ln 0.00823$ **52.** $\ln 0.436$

CHAPTER SUMMARY

Key Words

annual growth rate   (10.1)
antilogarithms   (10.5)
characteristic   (10.5)
common logarithms   (10.3)
decibel   (10.3)
decreasing functions   (10.1)
exponential equation   (10.4)
exponential function   (10.1)
half-life of radioactive
    material   (10.4)
increasing functions   (10.1)

linear interpolation   (10.5)
logarithmic equation   (10.4)
logarithmic function   (10.2)
Malthusian model for population
    growth   (10.1)
mantissa   (10.5)
model of an event   (10.1)
natural logarithms   (10.3)
pH scale   (10.3)
Richter scale   (10.3)

Key Ideas

(10.1)   The **exponential function** $y = b^x$, where $b > 0$, $b \neq 1$, and x is a real number, is a one-to-one function. Its domain is the set of real numbers, and its range is the set of positive real numbers.

If $b > 0$, $b \neq 1$, and $b^m = b^n$, then $m = n$.

Formula for compound interest: $A = A_0 \left(1 + \dfrac{r}{k}\right)^{kt}$

$e \approx 2.718281828 \ldots$

Malthusian model for population growth: $P = P_0 e^{kt}$

(10.2)   The **logarithmic function** $y = \log_b x$, where $b > 0$, $b \neq 1$, and x is a positive real number, is a one-to-one function. Its domain is the set of positive real numbers and its range is the set of all real numbers.

The equation $y = \log_b x$ is equivalent to the equation $x = b^y$.

Logarithms of negative numbers do not exist.

The functions defined by $y = \log_b x$ and $y = b^x$ are inverse functions.

Properties of logarithms:

1. $\log_b 1 = 0$ **2.** $\log_b b = 1$

3. $\log_b b^x = x$ **4.** $b^{\log_b x} = x$

5. $\log_b MN = \log_b M + \log_b N$ **6.** $\log_b \dfrac{M}{N} = \log_b M - \log_b N$

7. $\log_b M^p = p \log_b M$ **8.** If $\log_b x = \log_b y$, then $x = y$.

(10.3)   Common logarithms are base-10 logarithms.

Natural logarithms are base-e logarithms.

pH scale: $\text{pH} = -\log[\text{H}^+]$ **Decibel-voltage gain** $= 20\,\dfrac{\log E_O}{\log E_I}$

Richter scale: $R = \log \dfrac{A}{P}$ **Weber–Fechner law:** $L = k \ln I$

(10.4)   **Change-of-base formula:** $\log_b y = \dfrac{\log_a y}{\log_a b}$

Formula for radioactive decay: $A = A_0 2^{-t/h}$

(10.5)   **Linear interpolation** can be used to extend the accuracy of a table of logarithms.
(Optional)

━━━━━━━━━━   **REVIEW EXERCISES**   ━━━━━━━━━━

In Review Exercises 1–4, graph both the exponential function defined by each equation and its inverse on one set of coordinate axes.

1. $y = \left(\dfrac{6}{5}\right)^x$ **2.** $y = \left(\dfrac{3}{4}\right)^x$ **3.** $y = (0.8)^x$ **4.** $y = (0.4)^x$

In Review Exercises 5–8, graph each pair of equations on one set of coordinate axes.

5. $y = \left(\dfrac{1}{3}\right)^x$ and $y = \log_{1/3} x$ **6.** $y = \left(\dfrac{2}{5}\right)^x$ and $\log_{2/5} x = y$

7. $y = 4^x$ and $y = \log_4 x$ **8.** $y = 3^x$ and $y = \log_3 x$

In Review Exercises 9–26, find the value of each indicated variable.

9. $\log_2 M = 3;\ \ M$ **10.** $\log_3 N = -2;\ \ N$

11. $\log_b 9 = 2;\ \ b$ **12.** $\log_c 0.125 = -3;\ \ c$

13. $\log_7 7 = P;\ \ P$ **14.** $\log_3 \sqrt{3} = Q;\ \ Q$

15. $\log_8 \sqrt{2} = R;\ \ R$ **16.** $\log_6 36 = S;\ \ S$

17. $\log_{1/3} 9 = T;\ \ T$ **18.** $\log_{1/2} 1 = U;\ \ U$

19. $\log_v 3 = \dfrac{1}{3};\ \ v$ **20.** $\log_w 25 = -2;\ \ w$

21. $\log_2 N = 5;\quad N$

22. $\log_{\sqrt{3}} K = 4;\quad K$

23. $\log_{\sqrt{3}} M = 6;\quad M$

24. $\log_{0.1} 10 = M;\quad M$

25. $\log_b 2 = -\dfrac{1}{3};\quad b$

26. $\log_b 32 = 5;\quad b$

In Review Exercises 27–28, write each expression in terms of the logarithms of x, y and z.

27. $\log_b \dfrac{\sqrt{xz}}{y}$

28. $\log_b \sqrt{\dfrac{x}{yz}}$

In Review Exercises 29–30, write each expression as the logarithm of a single quantity.

29. $2 \log_b x - 3 \log_b y - 2 \log_b z$

30. $\log_b(2x - 4z) - \log_b 2y + \log_b y$

In Review Exercises 31–34, use a calculator to find the value of the variable, if possible. Express all answers to four decimal places.

31. $\log 735.4 = x$

32. $\log(-0.002345) = y$

33. $\ln \dfrac{2}{15} = z$

34. $\ln M = 5.345$

In Review Exercises 35–44, solve for x.

35. $3^x = 7$

36. $1.2 = (3.4)^{5.6x}$

37. $2^x = 3^{x-1}$

38. $\log x + \log(29 - x) = 2$

39. $\log_2 x + \log_2(x - 2) = 3$

40. $\log_2(x + 2) + \log_2(x - 1) = 2$

41. $e^{x \ln 2} = 9$

42. $\ln x = \log x$

43. $\ln x = \ln(x - 1) + 1$

44. $\ln x = \ln(x - 1)$

45. A wooden statue excavated from the sands of Egypt has a carbon-14 content that is one-third of that found in living wood. The half-life of carbon-14 is 5700 years. How old is the statue?

46. The pH of grapefruit juice is approximately 3.1. What is its hydrogen ion concentration?

47. Some chemistry texts define the pH of a solution as the common logarithm of the reciprocal of the hydrogen ion concentration:

$$\text{pH} = \log_{10} \dfrac{1}{[\text{H}^+]}$$

Show that this definition is equivalent to the one given in the text.

48. What is the half-life of a radioactive material if one-third of it decays in 20 years?

In Review Exercises 49–52, use Table B and perform the indicated calculations. Use linear interpolation, if necessary. These exercises are from an optional section.

49. $(34.5)(0.236)$

50. $\sqrt[5]{456{,}000}$

51. $\dfrac{(0.00235)^3}{(0.00896)^2}$

52. $\dfrac{(3.476)(0.003456)}{3.45}$

CHAPTER TEN TEST

1. Graph the equation $y = 2(3^x)$.

2. A deposit of $5000 earns 7% annual interest, compounded semiannually. How much will be on deposit after 6 years? $\left[Hint: \quad A = A_0 \left(1 + \dfrac{r}{k} \right)^{kt}. \right]$

In Problems 3–8, find each value of x.

3. $\log_4 16 = x$ 4. $\log_x 81 = 4$ 5. $\log_{16} x = \dfrac{1}{2}$ 6. $\log 100 = x$

7. $\ln e^2 = x$ 8. $\log(\log 10) = x$

9. Graph the equation $y = -\log_3 x$.   10. Graph the equation $y = \ln x$.

In Problems 11–12, assume that $\log a = 0.8$, $\log b = 3.2$, and $\log c = 1.6$. Find the value of each expression.

11. $\log(ab)$ 12. $\log \dfrac{c}{a}$

In Problems 13–16, assume that $\log a = 1.2$, $\log b = 0.6$, and $\log c = 2.4$. Find the value of each expression.

13. $\log b^2$ 14. $\dfrac{\log a}{\log b}$ 15. $\log(abc)$ 16. $\log \dfrac{ac}{b}$

In Problems 17–20, solve each equation for x.

17. $5^x = 3$ 18. $3^{x-1} = 100^x$

19. $\log(5x + 2) = \log(2x + 5)$ 20. $\log x + \log(x - 9) = 1$

21. Given that $\log 3 = 0.48$ and $\log 8 = 0.90$, find the value of $\log_3 8$.

22. Recall the formula

$$\text{db voltage gain} = 20 \log \frac{E_O}{E_I}$$

Find the db voltage gain of an amplifier if E_O is 60 volts and E_I is 0.6 volt.

23. Recall the formula

$$pH = -\log_{10}[H^+]$$

Find the pH of a solution with a hydrogen ion concentration of 3.7×10^{-6} gram ions per liter.

24. What is the hydrogen ion concentration of a solution with pH of 8.2?

CAREERS AND MATHEMATICS

Medical laboratory workers, also called **clinical laboratory** workers, include three levels of personnel: medical technologists, technicians, and assistants. They perform laboratory tests on specimens taken from patients by other health professionals, such as physicians.

Qualifications

The minimum educational requirement for a beginning job as a medical technologist is 4 years of college including completion of a training program in medical technology.

Medical laboratory technicians acquire their training by enrolling in accredited, 2-year programs, offered by community and junior colleges, colleges, and universities. Some are trained in the armed forces.

Most lab assistants are either trained on the job or complete a one-year training program in a hospital or community college.

Job outlook

Employment of medical laboratory workers is expected to expand at a rate about average for all occupations through the mid-1990s as physicians continue to use laboratory tests in routine physical checkups and in the diagnosis and treatment of disease.

**Example
application**

During bacterial reproduction, the time required for a population to double is called the **generation time**. If B bacteria are inoculated into a medium, and after the generation time of the organism has elapsed, then there are $2B$ cells. After another generation, there are $2(2B)$ or $4B$ cells, and so on. After n generations, the number of cells present will be

$$b = B \cdot 2^n$$

Solving this equation for n, we have

$$\log b = \log(B \cdot 2^n) \qquad \text{Take the common logarithm of both sides.}$$
$$\log b = \log B + n \log 2 \qquad \text{Apply Properties 5 and 7 of logarithms.}$$
$$\log b - \log B = n \log 2 \qquad \text{Add } -\log B \text{ to both sides.}$$
$$n = \frac{\log b - \log B}{\log 2} \qquad \text{Divide both sides by } \log 2.$$

If we substitute 0.301 for $\log 2$ and simplify, we obtain the formula that gives the number of generations that have passed while the population grew from B bacteria to b bacteria.

$$n = \frac{\log b - \log B}{0.301}$$

$$n = \frac{1}{0.301} \log \frac{b}{B}$$

1. $\quad n = 3.3 \log \dfrac{b}{B}$

The generation time G is given by the formula

$$G = \frac{t}{n} = \frac{t}{3.3 \log(b/B)}$$

where t is the length of the time of growth.

If a medium is inoculated with a bacterial culture that contains 1000 cells per milliliter, how many generations will have passed by the time the culture has grown to a population of 1 million cells per milliliter? What is the generation time if the culture is 10 hours old?

Solution   To find the number of generations that have passed, substitute 1000 for B, and 1,000,000 for b into Equation 1, and solve for n.

$$n = 3.3 \log \frac{1,000,000}{1000}$$

$n = 3.3 \log 1000$ Simplify.

$n = 3.3(3)$ $\log 1000 = 3$.

$n = 9.9$ Simplify.

Approximately 10 generations have passed.

To find the generation time, divide 10 hours by 9.9.

$G = 10/9.9$

$G \approx 1$

The generation time is approximately 1 hour.

Exercises

1. If a medium is inoculated with a bacterial culture containing 500 cells per milliliter, how many generations have passed by the time the culture contains 5×10^6 cells per milliliter?

2. What is the generation time in Exercise 1 if the time period is 26.5 hours?

3. It takes 12 hours for a culture containing 300 cells per milliliter to grow to contain 7×10^6 cells per milliliter. What is the generation time?

4. Show that in any bacterial culture, the per milliliter cell count will grow by a factor of k in $3.3 \log k$ generations.

(Answers: **1.**   13.2   **2.**   2 hours   **3.**   0.83 hours)

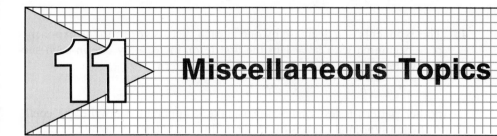

Miscellaneous Topics

In this chapter, we introduce several topics that have applications in advanced mathematics and in certain occupational areas. The binomial theorem, permutations, and combinations are used in statistics. Arithmetic and geometric progressions are used in mathematics of finance.

11.1 THE BINOMIAL THEOREM

We have discussed how to raise binomials to positive integral powers. For example, we know that

$$(a + b)^2 = a^2 + 2ab + b^2$$

and that

$$(a + b)^3 = (a + b)(a + b)^2$$
$$= (a + b)(a^2 + 2ab + b^2)$$
$$= a^3 + 2a^2b + ab^2 + ba^2 + 2ab^2 + b^3$$
$$= a^3 + 3a^2b + 3ab^2 + b^3$$

In this section, we will show how to raise binomials to positive integral powers without performing the actual multiplications. To this end, we consider the following binomial expansions:

$$(a + b)^0 = 1$$
$$(a + b)^1 = a + b$$
$$(a + b)^2 = a^2 + 2ab + b^2$$
$$(a + b)^3 = a^3 + 3a^2b + 3ab^2 + b^3$$
$$(a + b)^4 = a^4 + 4a^3b + 6a^2b^2 + 4ab^3 + b^4$$
$$(a + b)^5 = a^5 + 5a^4b + 10a^3b^2 + 10a^2b^3 + 5ab^4 + b^5$$
$$(a + b)^6 = a^6 + 6a^5b + 15a^4b^2 + 20a^3b^3 + 15a^2b^4 + 6ab^5 + b^6$$

Several patterns appear in these expansions:

1.   Each expansion has one more term than the power of the binomial.
2.   The degree of each term in each expansion is equal to the exponent of the binomial that is being expanded.
3.   The first term in each expansion is a, raised to the power of the binomial.

4. The exponents of a decrease by one in each successive term. The exponents of b, beginning with $b^0 = 1$ in the first term, increase by one in each successive term. Thus, the variables have the pattern

$$a^n, a^{n-1}b, a^{n-2}b^2, \ldots, ab^{n-1}, b^n$$

Another pattern is less obvious. To make it apparent, we write the coefficients of each binomial expansion in the following triangular array:

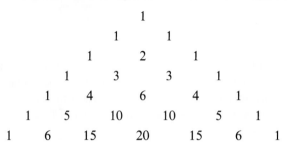

In this triangular array, called **Pascal's triangle**, each entry between the 1's is the sum of the closest pair of numbers in the line immediately above it.

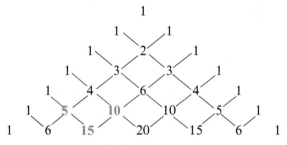

For example, the first 15 in the bottom row is the sum of the 5 and 10 immediately above it. Pascal's triangle continues with the same pattern forever. The next two lines are:

$$1 \quad 7 \quad 21 \quad 35 \quad 35 \quad 21 \quad 7 \quad 1$$
$$1 \quad 8 \quad 28 \quad 56 \quad 70 \quad 56 \quad 28 \quad 8 \quad 1$$

Example 1   Expand $(x + y)^5$.

Solution   The first term in the expansion is x^5, and the exponents of x decrease by one in each successive term. The y first appears in the second term, and the exponents of y increase by one in each successive term, concluding when the term y^5 is reached. Hence, the variables in the expansion are

$$x^5, x^4y, x^3y^2, x^2y^3, xy^4, y^5$$

The coefficients of these variables are given in Pascal's triangle in the row whose second entry is 5:

$$1 \quad 5 \quad 10 \quad 10 \quad 5 \quad 1$$

Putting these two pieces of information together gives the required expansion:

$$(x + y)^5 = x^5 + 5x^4y + 10x^3y^2 + 10x^2y^3 + 5xy^4 + y^5$$

■

Example 2 Expand $(u - v)^4$.

Solution Note that the expression $(u - v)^4$ can be written in the form $[u + (-v)]^4$. The variables in this expansion are

$$u^4, u^3(-v), u^2(-v)^2, u(-v)^3, (-v)^4$$

and the coefficients are given in Pascal's triangle in the row whose second entry is 4:

$$1 \quad 4 \quad 6 \quad 4 \quad 1$$

Hence, the required expansion is

$$(u - v)^4 = u^4 + 4u^3(-v) + 6u^2(-v)^2 + 4u(-v)^3 + (-v)^4$$
$$= u^4 - 4u^3v + 6u^2v^2 - 4uv^3 + v^4$$

■

Although Pascal's triangle gives the coefficients of the terms in a binomial expansion, it is not always the most convenient way to expand a binomial. A more formal method, called the **binomial theorem**, makes use of **factorial** notation.

Definition.   The symbol $n!$ (read as "**n factorial**" or as "**factorial n**") is defined as

$$n! = n(n - 1)(n - 2)(n - 3) \cdots (3)(2)(1)$$

where n is a natural number.

Example 3 Find   **a.** $2!$,   **b.** $5!$,   **c.** $-9!$, and   **d.** $(n - 2)!$.

Solution **a.** $2! = 2 \cdot 1 = 2$

b. $5! = 5 \cdot 4 \cdot 3 \cdot 2 \cdot 1 = 120$

c. $-9! = -9 \cdot 8 \cdot 7 \cdot 6 \cdot 5 \cdot 4 \cdot 3 \cdot 2 \cdot 1 = -362,880$

d. $(n - 2)! = (n - 2)(n - 3)(n - 4) \cdots 3 \cdot 2 \cdot 1$

According to the previous definition, part **d** is meaningful only if $n - 2$ is a natural number.

Mathematicians have found it convenient to define zero factorial as follows.

Definition. $0! = 1$

We note that

$$5 \cdot 4! = 5 \cdot 4 \cdot 3 \cdot 2 \cdot 1 = 5!$$
$$7 \cdot 6! = 7 \cdot 6 \cdot 5 \cdot 4 \cdot 3 \cdot 2 \cdot 1 = 7!$$

and

$$10 \cdot 9! = 10 \cdot 9 \cdot 8 \cdot 7 \cdot 6 \cdot 5 \cdot 4 \cdot 3 \cdot 2 \cdot 1 = 10!$$

We generalize this idea in the following theorem.

Theorem.   If n is a positive integer, then

$$n(n - 1)! = n!$$

We now state the binomial theorem.

The Binomial Theorem.   If n is any positive integer, then

$$(a + b)^n = a^n + \frac{n!}{1!(n - 1)!} a^{n-1}b + \frac{n!}{2!(n - 2)!} a^{n-2}b^2$$

$$+ \frac{n!}{3!(n - 3)!} a^{n-3}b^3 + \cdots + \frac{n!}{r!(n - r)!} a^{n-r}b^r + \cdots + b^n$$

In the binomial theorem, the exponents of the variables follow the familiar pattern: The sum of the exponents of a and b in each term is n, the exponents of a decrease, and the exponents of b increase. Only the method of finding the coefficients is different. Except for the first and last term, the numerator of each coefficient is $n!$. If the exponent of b in a particular term is r, the denominator of the coefficient of that term is $r!(n - r)!$.

Example 4   Use the binomial theorem to expand $(a + b)^3$.

Solution   Substitute directly into the binomial theorem, and simplify:

$$(a + b)^3 = a^3 + \frac{3!}{1!(3 - 1)!} a^2b + \frac{3!}{2!(3 - 2)!} ab^2 + b^3$$

$$= a^3 + \frac{3 \cdot 2 \cdot 1}{1 \cdot 2 \cdot 1} a^2b + \frac{3 \cdot 2 \cdot 1}{2 \cdot 1 \cdot 1} ab^2 + b^3$$

$$= a^3 + 3a^2b + 3ab^2 + b^3$$

∎

Example 5   Use the binomial theorem to expand $(x - y)^4$.

Solution   Write $(x - y)^4$ in the form $[x + (-y)]^4$, substitute directly into the binomial theorem, and simplify:

$$(x - y)^4 = [x + (-y)]^4$$

$$= x^4 + \frac{4!}{1!(4-1)!} x^3(-y) + \frac{4!}{2!(4-2)!} x^2(-y)^2 + \frac{4!}{3!(4-3)!} x(-y)^3 + (-y)^4$$

$$= x^4 - \frac{4 \cdot 3!}{3!} x^3 y + \frac{4 \cdot 3 \cdot 2!}{2!2!} x^2 y^2 - \frac{4 \cdot 3!}{3!} xy^3 + y^4$$

$$= x^4 - 4x^3 y + 6x^2 y^2 - 4xy^3 + y^4$$ ∎

Example 6   Use the binomial theorem to expand $(3u - 2v)^4$.

Solution   Write $(3u - 2v)^4$ in the form $[3u + (-2v)]^4$ and let $a = 3u$ and $b = -2v$. Then use the binomial theorem to expand $(a + b)^4$.

$$(a + b)^4 = a^4 + \frac{4!}{1!(4-1)!} a^3 b + \frac{4!}{2!(4-2)!} a^2 b^2 + \frac{4!}{3!(4-3)!} ab^3 + b^4$$

$$= a^4 + 4a^3 b + 6a^2 b^2 + 4ab^3 + b^4$$

Now substitute $3u$ for a and $-2v$ for b, and simplify:

$$(3u - 2v)^4 = (3u)^4 + 4(3u)^3(-2v) + 6(3u)^2(-2v)^2 + 4(3u)(-2v)^3 + (-2v)^4$$

$$= 81u^4 - 216u^3 v + 216u^2 v^2 - 96uv^3 + 16v^4$$ ∎

■ EXERCISE 11.1

In Exercises 1–18, evaluate each expression.

1. $3!$

2. $7!$

3. $-5!$

4. $-6!$

5. $3! + 4!$

6. $2!(3!)$

7. $3!(4!)$

8. $4! + 4!$

9. $8(7!)$

10. $4!(5)$

11. $\dfrac{9!}{11!}$

12. $\dfrac{13!}{10!}$

13. $\dfrac{49!}{47!}$

14. $\dfrac{101!}{100!}$

15. $\dfrac{5!}{3!(5-3)!}$

16. $\dfrac{6!}{4!(6-4)!}$

17. $\dfrac{7!}{5!(7-5)!}$

18. $\dfrac{8!}{6!(8-6)!}$

In Exercises 19–32, use the binomial theorem to expand each expression.

19. $(x + y)^2$

20. $(x + y)^4$

21. $(x - y)^3$

22. $(x - y)^2$

23. $(2x + y)^3$

24. $(x + 2y)^3$

25. $(x - 2y)^3$

26. $(2x - y)^3$

27. $(2x + 3y)^4$

28. $(3x - 2y)^4$

29. $\left(\dfrac{x}{2} - \dfrac{y}{3}\right)^4$

30. $\left(\dfrac{x}{3} + \dfrac{y}{2}\right)^4$

31. $(3 + 2y)^5$

32. $(2x + 3)^5$

33.   Without referring to the text, write the first ten rows of Pascal's triangle.

34.   Find the sum of the numbers in each row of the first ten rows of Pascal's triangle. What is the pattern?

35. Find the sum of the numbers in the designated diagonal rows of Pascal's triangle shown in Illustration 1. What is the pattern?

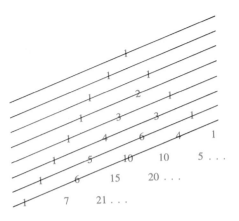

Illustration 1

11.2 THE nTH TERM OF A BINOMIAL EXPANSION

Suppose we want to find the fourth term of the expansion of $(a + b)^9$. We could raise the binomial $a + b$ to the ninth power, and then look at the fourth term. This task would be very tedious. By using the binomial theorem, we can construct the fourth term without finding the complete expansion of $(a + b)^9$. The following example shows how.

Example 1   Find the fourth term in the expansion of $(a + b)^9$.

Solution   Because b^1 appears in the second term, b^2 appears in the third term, and so on, the exponent of b in the fourth term is 3. Because the exponent of b added to the exponent of a must equal 9, the exponent of a must be 6. Hence, the variables of the fourth term are

$$a^6b^3$$

Because of the binomial theorem, the coefficient of the variables must be

$$\frac{n!}{r!(n - r)!} = \frac{9!}{3!(9 - 3)!}$$

Hence, the complete fourth term is

$$\frac{9!}{3!(9 - 3)!} a^6b^3 = \frac{9 \cdot 8 \cdot 7 \cdot 6!}{3 \cdot 2 \cdot 1 \cdot 6!} a^6b^3$$

$$= 84a^6b^3 \qquad\blacksquare$$

Example 2   Find the sixth term in the expansion of $(x - y)^7$.

Solution   Find the sixth term of $[x + (-y)]^7$. In the sixth term, the exponent of $(-y)$ is 5. Hence, the variables in the sixth term are

$$x^2(-y)^5 \qquad \text{The sum of the exponents must be 7.}$$

The coefficient of these variables is

$$\frac{n!}{r!(n-r)!} = \frac{7!}{5!(7-5)!}$$

The complete sixth term is

$$\frac{7!}{5!(7-5)!} x^2(-y)^5 = -\frac{7 \cdot 6 \cdot 5!}{5! \cdot 2 \cdot 1} x^2 y^5$$

$$= -21x^2 y^5 \qquad \blacksquare$$

Example 3   Find the fourth term of the expansion of $(2x - 3y)^6$.

Solution   Let $a = 2x$ and $b = -3y$ Then, find the fourth term of the expansion of $(a + b)^6$. The fourth term is

$$\frac{6!}{3!(6-3)!} a^3 b^3 = \frac{6 \cdot 5 \cdot 4 \cdot 3!}{3! \cdot 3 \cdot 2 \cdot 1} a^3 b^3$$

$$= 20 a^3 b^3$$

Now substitute $2x$ for a and $-3y$ for b, and simplify:

$$20 a^3 b^3 = 20(2x)^3(-3y)^3$$

$$= -4320 x^3 y^3$$

The fourth term in the expansion of $(2x - 3y)^6$ is $-4320x^3 y^3$. $\qquad \blacksquare$

The binomial theorem can be used to approximate powers of certain decimals to any required degree of accuracy. To do so, we use the binomial expansion

$$(1 + b)^n = 1^n + \frac{n!}{1!(n-1)!} (1)^{n-1}b + \frac{n!}{2!(n-2)!} (1)^{n-2}b^2$$

$$+ \frac{n!}{3!(n-3)!} (1)^{n-3}b^3 + \cdots$$

Example 4   Find the value of $(1.01)^{12}$ to three decimal places.

Solution   Write $(1.01)^{12}$ as $(1 + 0.01)^{12}$, and use the binomial theorem with $b = 0.01$:

$$(1 + 0.01)^{12} = (1)^{12} + \frac{12!}{1!11!} (1)^{11}(0.01) + \frac{12!}{2!10!} (1)^{10}(0.01)^2$$

$$+ \frac{12!}{3!9!} (1)^9 (0.01)^3 + \cdots$$

$$\approx 1 + 12(0.01) + \frac{12(11)}{2} (0.0001) + \frac{12(11)(10)}{6} (0.000001)$$

$$\approx 1 + 0.12 + 0.0066 + 0.00022$$

Because the fourth term 0.00022 does not effect the third decimal place of the sum, you only need the sum of the first three terms. Thus, you have

$$(1.01)^{12} \approx 1.1266$$

$$\approx 1.127 \qquad \text{Round to three decimal places.} \qquad ■$$

■ EXERCISE 11.2

In Exercises 1–30, use the binomial theorem to find the required term of each expansion.

1. $(a + b)^3$;   second term

2. $(a + b)^3$;   third term

3. $(x - y)^4$;   fourth term

4. $(x - y)^5$;   second term

5. $(x + y)^6$;   fifth term

6. $(x + y)^7$;   fifth term

7. $(x - y)^8$;   third term

8. $(x - y)^9$;   seventh term

9. $(x + 3)^5$;   third term

10. $(x - 2)^4$;   second term

11. $(4x + y)^5$;   third term

12. $(x + 4y)^5$;   fourth term

13. $(x - 3y)^4$;   second term

14. $(3x - y)^5$;   third term

15. $(2x - 5)^7$;   fourth term

16. $(2x + 3)^6$;   sixth term

17. $(2x - 3y)^5$;   fifth term

18. $(3x - 2y)^4$;   second term

19. $(\sqrt{2}x + \sqrt{3}y)^6$;   third term

20. $(\sqrt{3}x + \sqrt{2}y)^5$;   second term

21. $\left(\dfrac{x}{2} - \dfrac{y}{3}\right)^4$;   second term

22. $\left(\dfrac{x}{3} + \dfrac{y}{2}\right)^5$;   fourth term

23. $(a + b)^n$;   fourth term

24. $(a + b)^n$;   third term

25. $(a - b)^n$;   fifth term

26. $(a - b)^n$;   sixth term

27. $(a + b)^n$; rth term

28. $(a + b)^n$; $(r + 1)$th term

29. $(2a - 3b)^n$;   fifth term

30. $(3a - 2b)^n$;   fourth term

In Exercises 31–34, use the binomial theorem to evaluate the following powers correctly to two decimal places.

31. $(1.02)^9$

32. $(1.01)^{10}$

33. $(0.99)^{12}$ [*Hint:*   Use $(1 - .01)^{12}$.]

34. $(1.98)^8$

11.3   PERMUTATIONS

Steven goes to the school cafeteria for lunch. He has a choice of three different sandwiches (hamburger, hot dog, or ham and cheese) and four different beverages (cola, root beer, orange, or milk). How many lunches can Steven choose? There are three choices of sandwich and, for any one of these choices, there are four choices of drink. The different options are shown in the "tree diagram" in Figure 11-1.

The tree diagram shows that Steven can choose from 12 different lunches. One possibility is a hamburger with a cola. Another possibility is a hot dog with milk.

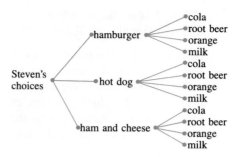

Figure 11-1

Any situation that can have several outcomes—such as choosing a sandwich—is called an **event**. Choosing a sandwich and choosing a beverage can be thought of as two events. The preceding example illustrates the **multiplication principle for events**.

The Multiplication Principle for Events.   Let E_1 and E_2 be two events. If E_1 can be done in a_1 ways, and if—after E_1 has occurred—E_2 can be done in a_2 ways, then the event "E_1 followed by E_2" can be done in $a_1 \cdot a_2$ ways.

The multiplication principle for events can be extended to cover any number of events.

Example 1   Before studying for an examination, Heidi plans to watch the early evening news and then a situation comedy on television. If she has a choice of four news broadcasts and two situation comedies, in how many ways can she choose to watch television?

Solution   Let E_1 be the event "watching the news" and E_2 be the event "watching the situation comedy." Because there are four ways to accomplish E_1, and two ways to accomplish E_2, the number of choices that Heidi has is $4 \cdot 2 = 8$.   ■

We can use the multiplication principle for events to compute the number of ways that we can arrange a certain number of objects in a row.

Example 2   In how many ways can you arrange five books on a shelf?

Solution   You can fill the first space with any of the five books, the second space with any of the remaining four books, the third space with any of the remaining three books, the fourth space with any of the remaining two books, and the fifth space with the last book. According to the multiplication principle for events, the number of ways that the books can be arranged is $5 \cdot 4 \cdot 3 \cdot 2 \cdot 1 = 120$.   ■

Example 3   If Sally has six flags, each of a different color, to hang on a flagpole, how many different signals can she send by using four flags?

Solution   Sally must find the number of arrangements of four flags when she has six flags to choose from. She can hang any one of the six flags in the top position, any one of the remaining five flags in the second position, any one of the remaining four flags in the third position, and any one of the remaining three flags in the lowest position. According to the multiplication principle for events, the total number of signals that Sally can send is $6 \cdot 5 \cdot 4 \cdot 3 = 360$.   ∎

When computing the number of possible arrangements of objects such as books on a shelf or flags on a pole, we are determining the number of **permutations** of those objects. In Example 2, you determined that the number of permutations of five books, using all five books, is 120. In Example 3, you determined that the number of permutations of six flags, using four of them, is 360.

The symbol $_nP_r$, read as "the number of permutations of n things r at a time," often is used to express permutation problems. In Example 2, you found that $_5P_5 = 120$. In Example 3, you found that $_6P_4 = 360$.

Example 4   If Sarah has seven flags, each of a different color, to hang on a flagpole, how many different signals can she send by using three flags?

Solution   Sarah must evaluate $_7P_3$ (the number of permutations of seven things three at a time). In the top position, Sarah can hang any one of seven flags, in the middle position any one of the remaining six, and in the bottom position any one of the remaining five. According to the multiplication principle for events,

$$_7P_3 = 7 \cdot 6 \cdot 5 = 210$$

Sarah can send 210 different signals using three of the available seven flags.   ∎

Although it is proper to write $_7P_3 = 7 \cdot 6 \cdot 5$, there is an advantage in changing the form of this answer to obtain a formula for computing $_7P_3$:

$$_7P_3 = 7 \cdot 6 \cdot 5 = \frac{7 \cdot 6 \cdot 5 \cdot 4 \cdot 3 \cdot 2 \cdot 1}{4 \cdot 3 \cdot 2 \cdot 1} = \frac{7!}{4!} = \frac{7!}{(7-3)!}$$

The generalization of this idea gives the following formula.

The Formula for Computing $_nP_r$. The number of permutations of n things r at a time is given by the formula

$$_nP_r = \frac{n!}{(n-r)!}$$

Example 5   Compute   **a.** $_8P_2$,   **b.** $_7P_5$, and   **c.** $_nP_n$.

Solution   **a.** $_8P_2 = \dfrac{8!}{(8-2)!} = \dfrac{8 \cdot 7 \cdot 6!}{6!} = 8 \cdot 7 = 56$

b. $_7P_5 = \dfrac{7!}{(7-5)!} = \dfrac{7 \cdot 6 \cdot 5 \cdot 4 \cdot 3 \cdot 2!}{2!} = 7 \cdot 6 \cdot 5 \cdot 4 \cdot 3 = 2520$

c. $_nP_n = \dfrac{n!}{(n-n)!} = \dfrac{n!}{0!} = \dfrac{n!}{1} = n!$ ■

Part **c** of Example 5 establishes the following formula.

The Formula for Computing $_nP_n$.   The number of permutations of n things n at a time is given by the formula

$$_nP_n = n!$$

Example 6   **a.**   In how many ways can a television executive arrange the Saturday night lineup of 6 programs, if there are 15 programs to choose from?

b.   If there are only 6 programs to choose from?

Solution   **a.**   To find the number of permutations of 15 programs 6 at a time, use the formula $_nP_r = \dfrac{n!}{(n-r)!}$ with $n = 15$ and $r = 6$.

$$_{15}P_6 = \frac{15!}{(15-6)!} = \frac{15 \cdot 14 \cdot 13 \cdot 12 \cdot 11 \cdot 10 \cdot 9!}{9!}$$

$$= 15 \cdot 14 \cdot 13 \cdot 12 \cdot 11 \cdot 10$$

$$= 3{,}603{,}600$$

b.   To find the number of permutations of 6 programs 6 at a time, use the formula $_nP_n = n!$ with $n = 6$.

$$_6P_6 = 6! = 720$$ ■

■ **EXERCISE 11.3**

1.   Kristy intends to go out to dinner and see a movie. In how many ways can she arrange her evening if she has a choice of five movies and seven restaurants?

2.   Paula has five ways to travel from New York to Chicago, three ways to travel from Chicago to Denver, and four ways to travel from Denver to Los Angeles. How many choices are available to Paula if she travels from New York to Los Angeles?

3.   For lunch, Bill has a choice of five sandwiches, seven drinks, and six desserts. If a lunch consists of one sandwich, one drink, and one dessert, how many choices does Bill have?

4. Sarah has 11 blouses, 7 skirts, 9 pairs of socks, and 3 pairs of shoes. How many choices does she have every time she gets dressed?

5. How many six-digit license plates can be manufactured? Note there are ten choices—0, 1, 2, 3, 4, 5, 6, 7, 8, 9—for each digit.

6. How many seven-digit license plates can be manufactured?

7. How many six-digit license plates can be manufactured if no digit can be repeated?

8. How many seven-digit license plates can be manufactured if no digit can be repeated?

9. How many six-digit license plates can be manufactured if no license can begin with 0 and if no digit can be repeated?

10. How many seven-digit license plates can be manufactured if no license can begin with 0 or 1?

11. How many license plates can be manufactured with two letters followed by four digits?

12. How many license plates can be manufactured with two letters followed by four digits if the letter O (oh) is not used and if no digit can be repeated?

13. How many seven-digit phone numbers are available in area code 815 if no phone number can begin with 0 or 1?

14. How many ten-digit phone numbers (three for area code and seven for local number) are available if area codes of 000, 800, and 900 cannot be used and if no local number can begin with 0 or 1?

In Exercises 15–28, evaluate each expression.

15. $_3P_3$

16. $_4P_4$

17. $_5P_3$

18. $_3P_2$

19. $_{15}P_{14}$

20. $_{20}P_{18}$

21. $_2P_2 \cdot {}_3P_3$

22. $_3P_2 \cdot {}_3P_3$

23. $_5P_3 \cdot {}_6P_2$

24. $_7P_5 \cdot {}_6P_3$

25. $\dfrac{_5P_3}{_4P_2}$

26. $\dfrac{_6P_2}{_5P_4}$

27. $\dfrac{_8P_6}{_6P_5}$

28. $\dfrac{_{10}P_9}{_8P_5}$

29. In how many ways can six girls be placed in a line?

30. In how many ways can seven boys be placed in a line?

31. In how many ways can four girls and five boys be arranged in a line if all of the girls are placed first?

32. In how many ways can six girls and four boys be arranged in a line if all of the boys are placed first?

33. In how many ways can four girls and five boys be arranged in a line if the boys and girls must alternate?

34. In how many ways can four girls and four boys be arranged in a line if the boys and girls must alternate?

35. How many permutations does a combination lock have if each combination has three numbers, no two numbers of any combination are equal, and the lock has 25 numbers?

36. How many permutations does a combination lock have if each combination has three numbers, no two numbers of any combination are equal, and the lock has 50 numbers?

37. Ten Ping-Pong balls are labeled 1 through 10, placed in a paper bag, mixed thoroughly, and then removed and placed in a row. In how many ways can they be placed so they are *not* in the order 1, 2, 3, . . . , 10?

38. The 25 sophomore members of Auburn High School wish to elect class officers: president, vice-president, secretary, and treasurer. How many different administrations can be elected?

39. The frazzled receptionist at Dr. Gasem's dental office has only three appointment times available before next Tuesday, and ten patients with a toothache. In how many ways can the receptionist fill those appointments?

40. In some computers, a *word* consists of 32 *bits*—a string of 32 1s and 0s. How many distinct words are possible?

41. A byte consists of 8 bits. (Refer to Exercise 40.) How many distinct bytes are possible?

11.4 COMBINATIONS

Suppose that Jerry must read four books from a reading list of ten books. The order in which Jerry reads the books is not important; all that matters is that he read four books. For the moment, however, we will assume that order is important and compute the number of permutations of 10 things 4 at a time:

$$_{10}P_4 = \frac{10!}{(10 - 4)!} = \frac{10 \cdot 9 \cdot 8 \cdot 7 \cdot 6!}{6!} = 5040$$

This result indicates the number of ways of arranging four books if there are ten books from which to choose. However, in the original setting of this problem, the order of the books does not matter. Because there are 24 or 4! ways of ordering the four books that are chosen, the calculation of $_{10}P_4 = 5040$ provides an answer that is 4! or 24 times too big. Actually, the number of possible choices that Jerry has is the number of permutations of 10 things 4 at a time, divided by 24:

$$\frac{_{10}P_4}{24} = \frac{5040}{24} = 210$$

Jerry has 210 options for choosing four books to read from the list of ten books.

In situations where order is *not* important, we are interested in **combinations**, not permutations. The symbols $_nC_r$ and $\binom{n}{r}$ both mean the number of combinations of n things r at a time.

If a set of r books is chosen from a total of n books, the number of possible selections is $_nC_r$, and there are $r!$ arrangements of the r books in the chosen set. If we consider the chosen set of r books taken from a set of n books as an ordered grouping, the number of orderings of the r books is $_nP_r$. Hence, the number of combinations of n things r at a time multiplied by $r!$ equals the number of permutations of n things r at a time. This relationship is shown by the equation

$$r! \cdot {_nC_r} = {_nP_r}$$

This equation can be rewritten to obtain the standard formula for computing $_nC_r$:

$$r! \cdot {_nC_r} = {_nP_r}$$

$${_nC_r} = \frac{1}{r!} \cdot {_nP_r} \qquad \text{Divide both sides by } r!.$$

$$= \frac{1}{r!} \cdot \frac{n!}{(n - r)!} \qquad \text{Substitute } \frac{n!}{(n - r)!} \text{ for } {_nP_r}.$$

$$= \frac{n!}{r!(n - r)!}$$

> **The Formula for Computing $_nC_r$.** The number of combinations of n things r at a time is given by the formula
>
> $$_nC_r = \binom{n}{r} = \frac{n!}{r!(n-r)!}$$

Example 1 Compute   **a.** $_8C_5$,   **b.** $\binom{7}{2}$,   **c.** $_nC_n$, and   **d.** $_nC_0$.

Solution   **a.** $_8C_5 = \dfrac{8!}{5!(8-5)!} = \dfrac{8 \cdot 7 \cdot 6 \cdot 5!}{5!3!} = 8 \cdot 7 = 56$

b. $\binom{7}{2} = \dfrac{7!}{2!(7-2)!} = \dfrac{7 \cdot 6 \cdot 5!}{2 \cdot 1 \cdot 5!} = 21$

c. $_nC_n = \dfrac{n!}{n!(n-n)!} = \dfrac{1}{0!} = \dfrac{1}{1} = 1$

d. $_nC_0 = \dfrac{n!}{0!(n-0)!} = \dfrac{n!}{0!n!} = \dfrac{1}{0!} = \dfrac{1}{1} = 1$

The symbol $_nC_0$ indicates that you chose 0 things from the available n things. ∎

Parts **c** and **d** of Example 1 establish the following formulas.

> **The Formulas for Computing $_nC_n$ and $_nC_0$.** The number of combinations of n things n at a time is 1. The number of combinations of n things 0 at a time is 1.
>
> $$_nC_n = 1 \quad \text{and} \quad _nC_0 = 1$$

Example 2 A group of 15 students wants to pick a committee of 4 students to plan a party. How many committees are possible?

Solution The ordering of the people on each possible committee is unimportant, so calculate the number of combinations of 15 people 4 at a time:

$$_{15}C_4 = \frac{15!}{4!(15-4)!} = \frac{15 \cdot 14 \cdot 13 \cdot 12 \cdot 11!}{4 \cdot 3 \cdot 2 \cdot 1 \cdot 11!}$$

$$= \frac{15 \cdot 14 \cdot 13 \cdot 12}{4 \cdot 3 \cdot 2 \cdot 1}$$

$$= 1365$$

There are 1365 possible committees. ∎

Example 3 A committee in Congress consists of ten Democrats and eight Republicans. In how many ways can a subcommittee be chosen if it is to contain five Democrats and four Republicans?

Solution There are $_{10}C_5$ ways of choosing the five Democrats, and $_8C_4$ ways of choosing the four Republicans. By the multiplication principle for events, there are $_{10}C_5 \cdot {}_8C_4$ ways of choosing the subcommittee:

$$_{10}C_5 \cdot {}_8C_4 = \frac{10!}{5!(10-5)!} \cdot \frac{8!}{4!(8-4)!}$$

$$= \frac{10 \cdot 9 \cdot 8 \cdot 7 \cdot 6 \cdot 5!}{120 \cdot 5!} \cdot \frac{8 \cdot 7 \cdot 6 \cdot 5 \cdot 4!}{24 \cdot 4!}$$

$$= \frac{10 \cdot 9 \cdot 8 \cdot 7 \cdot 6}{120} \cdot \frac{8 \cdot 7 \cdot 6 \cdot 5}{24}$$

$$= 17,640$$

There are 17,640 possible subcommittees. ∎

We note that the expansion of $(x + y)^3$ is

$$(x + y)^3 = 1x^3 + 3x^2y + 3xy^2 + 1y^3$$

and that

$$\binom{3}{0} = 1, \qquad \binom{3}{1} = 3, \qquad \binom{3}{2} = 3, \qquad \text{and} \qquad \binom{3}{3} = 1$$

Putting these facts together gives the following alternative way of writing the expansion of $(x + y)^3$:

$$(x + y)^3 = \binom{3}{0}x^3 + \binom{3}{1}x^2y + \binom{3}{2}xy^2 + \binom{3}{3}y^3$$

Similarly, we have

$$(x + y)^4 = 1x^4 \quad + 4x^3y \quad + 6x^2y^2 \quad + 4xy^3 \quad + 1y^4$$

$$= \binom{4}{0}x^4 + \binom{4}{1}x^3y + \binom{4}{2}x^2y^2 + \binom{4}{3}xy^3 + \binom{4}{4}y^4$$

The generalization of this idea allows us to state the binomial theorem in an alternative form using combination notation.

Alternative Form of the Binomial Theorem. If n is any positive integer, then

$$(a + b)^n = \binom{n}{0}a^n + \binom{n}{1}a^{n-1}b + \binom{n}{2}a^{n-2}b^2 + \cdots$$

$$+ \binom{n}{r}a^{n-r}b^r + \cdots + \binom{n}{n}b^n$$

Example 4   Use the alternative form of the binomial theorem to expand $(x + y)^6$.

Solution
$$(x + y)^6 = \binom{6}{0}x^6 + \binom{6}{1}x^5y + \binom{6}{2}x^4y^2 + \binom{6}{3}x^3y^3$$
$$+ \binom{6}{4}x^2y^4 + \binom{6}{5}xy^5 + \binom{6}{6}y^6$$
$$= x^6 + 6x^5y + 15x^4y^2 + 20x^3y^3 + 15x^2y^4 + 6xy^5 + y^6 \quad \blacksquare$$

Example 5   Use the alternative form of the binomial theorem to expand $(2x - y)^3$.

Solution
$$(2x - y)^3 = [2x + (-y)]^3$$
$$= \binom{3}{0}(2x)^3 + \binom{3}{1}(2x)^2(-y) + \binom{3}{2}(2x)(-y)^2 + \binom{3}{3}(-y)^3$$
$$= (2x)^3 + 3(4x^2)(-y) + 3(2x)(y^2) + (-y)^3$$
$$= 8x^3 - 12x^2y + 6xy^2 - y^3 \quad \blacksquare$$

■ **EXERCISE 11.4** ━━━━━━━━━━━━━━━━━━━━━━━━━━━━━━

In Exercises 1–20, evaluate each expression.

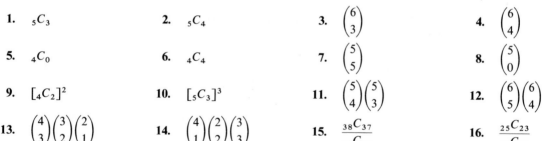

1. $_5C_3$

2. $_5C_4$

3. $\binom{6}{3}$

4. $\binom{6}{4}$

5. $_4C_0$

6. $_4C_4$

7. $\binom{5}{5}$

8. $\binom{5}{0}$

9. $[_4C_2]^2$

10. $[_5C_3]^3$

11. $\binom{5}{4}\binom{5}{3}$

12. $\binom{6}{5}\binom{6}{4}$

13. $\binom{4}{3}\binom{3}{2}\binom{2}{1}$

14. $\binom{4}{1}\binom{2}{2}\binom{3}{3}$

15. $\dfrac{_{38}C_{37}}{_{19}C_{18}}$

16. $\dfrac{_{25}C_{23}}{_{40}C_{39}}$

17. $_nC_2$

18. $_nC_3$

19. $_nC_{n-3}$

20. $_nC_{n-2}$

21. A class of 14 students wants to pick a committee of 3 students to plan a picnic. How many committees are possible?

22. Jeffrey must read 3 books from a reading list of 15 books. How many choices does he have?

23. In how many ways can you pick five books from a group of nine books?

24. In how many ways can you pick 5 persons from a group of 15 persons?

25. The number of three-person committees that can be formed from a group of persons is ten. How many persons are in the group?

26. The number of three-person committees that can be formed from a group of persons is 20. How many persons are in the group?

27. In how many ways can you select a committee of two boys and two girls from a group containing three boys and four girls?

28. In how many ways can you select a committee of three boys and two girls from a group containing five boys and three girls?

29. In how many ways can you select 5 cards from a deck of 12 cards?

30. In how many ways can you select seven cubes from a box containing ten cubes each of a different color?

31. In how many ways can you select 2 shirts and 3 neckties from a group of 12 shirts and 10 neckties?

32. In how many ways can you select five dresses and two coats from a wardrobe containing nine dresses and three coats?

In Exercises 33–46, use the alternative form of the binomial theorem to expand each expression.

33. $(x + y)^4$

34. $(x + y)^2$

35. $(x - y)^2$

36. $(x - y)^4$

37. $(2x + y)^3$

38. $(x + 2y)^3$

39. $(x + 1)^5$

40. $(x - 1)^5$

41. $(2x + 1)^4$

42. $(3x - 2)^4$

43. $(3 - x^2)^3$

44. $(2 - x^2)^5$

45. $(2x + 3y)^4$

46. $(3x + 2y)^4$

In Exercises 47–50, find the indicated term of the binomial expansion.

47. $(x - 5y)^5$; fourth term

48. $(2x - y)^5$; third term

49. $(x^2 - y^3)^4$; second term

50. $(x^3 - y^2)^4$; fourth term

11.5 ARITHMETIC PROGRESSIONS

A **sequence** is a function whose domain is the set of natural numbers. For example, the function $f(n) = 3n + 2$, where n is a natural number, is a sequence. Because a sequence is a function whose domain is the set of natural numbers, it is easy to write its values as a list. If the natural numbers are substituted for n, the function $f(n) = 3n + 2$ generates the list

$$5, 8, 11, 14, 17, \ldots$$

It is common to call the list, as well as the function, a sequence. Each number in the list is called a **term** of the sequence. Other examples of sequences are

a. $1^3, 2^3, 3^3, 4^3, \ldots$

b. $4, 8, 12, 16, \ldots$

c. $2, 3, 5, 7, 11, \ldots$

d. $1, 1, 2, 3, 5, 8, 13, 21, \ldots$

The sequence in part a is the ordered list of the cubes of the natural numbers. The sequence in part b is the ordered list of the multiples of 4 that are positive integers. The sequence in part c is the ordered list of the prime numbers. The sequence in part d is called the **Fibonacci sequence**, after the 12th-century mathematician Leonardo of Pisa—known to his friends as Fibonacci. Beginning with the 2, each term of the Fibonacci sequence is the sum of the two preceding terms.

One important type of sequence is called an **arithmetic progression**.

> **Definition.**   An **arithmetic progression** is a sequence of the form
>
> $$a, a + d, a + 2d, a + 3d, \ldots, a + (n - 1)d, \ldots$$
>
> where a is the **first term**, $a + (n - 1)d$ is the **nth term**, and d is the **common difference**.

Note that the second term of an arithmetic progression has an addend of $1d$, the third term has an addend of $2d$, the fourth term has an addend of $3d$, and the nth term has an addend of $(n - 1)d$. Also note that the difference between any two consecutive terms is d.

Example 1   An arithmetic progression has a first term of 5 and a common difference of 4.

a.   Write the first six terms of the progression.

b.   Write the 25th term of the progression.

Solution   a.   Because the first term is $a = 5$ and the common difference is $d = 4$, the first six terms are

$$5, \quad 5 + 4, \quad 5 + 2(4), \quad 5 + 3(4), \quad 5 + 4(4), \quad 5 + 5(4)$$

or

$$5, 9, 13, 17, 21, 25$$

b.   The nth term is $a + (n - 1)d$. Because you want the 25th term, let $n = 25$:

$$n\text{th term} = a + (n - 1)d$$
$$25\text{th term} = 5 + (25 - 1)4 \qquad \text{Remember } a = 5 \text{ and } d = 4.$$
$$= 5 + 24(4)$$
$$= 5 + 96$$
$$= 101 \qquad\qquad\qquad\qquad\qquad \blacksquare$$

Example 2   The first three terms of an arithmetic progression are 3, 8, and 13. Find
a.   the 67th term, and   b.   the 100th term.

Solution   First find d, the common difference. It is the difference between successive terms: $d = 8 - 3 = 13 - 8 = 5$.

a.   Substitute 3 for a, 67 for n, and 5 for d in the formula for the nth term, and simplify:

$$n\text{th term} = a + (n - 1)d$$
$$67\text{th term} = 3 + (67 - 1)5$$
$$= 3 + (66)5$$
$$= 333$$

b.   Substitute 3 for a, 100 for n, and 5 for d in the formula for the nth term, and simplify:

$$n\text{th term} = a + (n - 1)d$$
$$100\text{th term} = 3 + (100 - 1)5$$
$$= 3 + (99)5$$
$$= 498 \qquad \blacksquare$$

Example 3   The first term of an arithmetic progression is 12 and the 50th term is 3099. Write the first six terms of the progression.

Solution   The key is to find the common difference. Because the 50th term of this progression is 3099, you can let $n = 50$ and solve the following equation for d:

$$n\text{th term} = a + (n - 1)d$$
$$3099 = 12 + (50 - 1)d$$
$$3099 = 12 + 49d \qquad \text{Simplify.}$$
$$3087 = 49d \qquad \text{Add } -12 \text{ to both sides.}$$
$$63 = d \qquad \text{Divide both sides by 49.}$$

The first term of the required progression is 12, and the common difference is 63. Hence, the first six terms of the progression are

$$12, 75, 138, 201, 264, 327 \qquad \blacksquare$$

If numbers are inserted between two given numbers a and b to form an arithmetic progression, the inserted numbers are called **arithmetic means** between a and b. If a single number is inserted between the numbers a and b, that number is called **the arithmetic mean** between a and b.

Example 4   Insert two arithmetic means between 6 and 27.

Solution   In this example, the first term is $a = 6$, and the fourth term (or the last term) is $l = 27$. You must find the common difference d so that the terms

$$6, 6 + d, 6 + 2d, 27$$

form an arithmetic progression. Substitute 6 for a and 4 for n into the formula for the nth term, and solve for d:

$$n\text{th term} = a + (n - 1)d$$
$$4\text{th term} = 6 + (4 - 1)d$$
$$27 = 6 + 3d \qquad \text{Simplify.}$$
$$21 = 3d \qquad \text{Add } -6 \text{ to both sides.}$$
$$7 = d \qquad \text{Divide both sides by 3.}$$

The two arithmetic means between 6 and 27 are

$$6 + d = 6 + 7 = 13$$

and

$$6 + 2d = 6 + 2(7) = 6 + 14 = 20$$

Note that 6, 13, 20, and 27 are the first four terms of an arithmetic progression.

∎

There is a formula that gives the sum of the first n terms of an arithmetic progression. To develop this formula, we let S_n represent the sum of the first n terms of an arithmetic progression:

$$S_n = \quad a \quad + \quad [a + d] \quad + \quad [a + 2d] \quad + \cdots + [a + (n - 1)d]$$

We write the same sum again, but in reverse order:

$$S_n = [a + (n - 1)d] + [a + (n - 2)d] + [a + (n - 3)d] + \cdots + \quad a$$

We add these two equations together, term by term, to get

$$2S_n = [2a + (n - 1)d] + [2a + (n - 1)d] + [2a + (n - 1)d] + \cdots + [2a + (n - 1)d]$$

Because there are n equal terms on the right-hand side of the preceding equation, we can write

$$2S_n = n[2a + (n - 1)d]$$
$$2S_n = n[a + a + (n - 1)d]$$
$$2S_n = n[a + l]$$
$$S_n = \frac{n(a + l)}{2}$$

Substitute l for $a + (n - 1)d$, because $a + (n - 1)d$ is the last term of the progression.

This reasoning establishes the following theorem.

Theorem. The sum of the first n terms of an arithmetic progression is given by the formula

$$S_n = \frac{n(a + l)}{2} \qquad \text{with } l = a + (n - 1)d$$

where a is the first term, l is the last (or nth) term, and n is the number of terms in the progression.

Example 5   Find the sum of the first 40 terms of the arithmetic progression 4, 10, 16, ...

Solution   In this example, let $a = 4$, $n = 40$, $d = 6$ and $l = 4 + (40 - 1)6 = 238$. Substitute these values into the formula $S_n = \frac{n(a + l)}{2}$, and solve for S_n:

$$S_n = \frac{n(a + l)}{2}$$

$$= \frac{40(4 + 238)}{2}$$

$$= 20(242)$$

$$= 4840$$

The sum of the first 40 terms is 4840. ∎

Summation Notation

There is a shorthand notation for indicating the sum of a finite (ending) number of consecutive terms in a sequence. This notation, called **summation notation**, involves the Greek letter Σ (sigma). The expression

$$\sum_{k=2}^{5} 3k \qquad \text{Read as "the summation of } 3k \text{ as } k \text{ runs from 2 to 5."}$$

designates the sum of all terms obtained if we successively substitute the numbers 2, 3, 4, and 5 for k, called the **index of the summation**. Thus, we have

$$\sum_{k=2}^{5} 3k = 3(2) + 3(3) + 3(4) + 3(5)$$

$$= 6 + 9 + 12 + 15$$

$$= 42$$

Example 6 Find each sum: **a.** $\displaystyle\sum_{k=3}^{5} (2k + 1)$, **b.** $\displaystyle\sum_{k=2}^{5} k^2$, and **c.** $\displaystyle\sum_{k=1}^{3} (3k^2 + 3)$.

Solution **a.** $\displaystyle\sum_{k=3}^{5} (2k + 1) = [2(3) + 1] + [2(4) + 1] + [2(5) + 1]$

$$= 7 + 9 + 11$$

$$= 27$$

b. $\displaystyle\sum_{k=2}^{5} k^2 = 2^2 + 3^2 + 4^2 + 5^2$

$$= 4 + 9 + 16 + 25$$

$$= 54$$

c. $\displaystyle\sum_{k=1}^{3} (3k^2 + 3) = [3(1^2) + 3] + [3(2^2) + 3] + [3(3^2) + 3]$

$$= 6 + 15 + 30$$

$$= 51$$ ∎

▪ EXERCISE 11.5 ▬▬▬▬▬▬▬▬▬▬▬▬▬▬▬▬▬▬▬▬▬

In Exercises 1–14, write the first five terms of each arithmetic progression with the given properties.

1. $a = 3, \quad d = 2$

2. $a = -2, \quad d = 3$

3. $a = -5, \quad d = -3$

4. $a = 8, \quad d = -5$

5. $a = 5,$ fifth term is 29

6. $a = 4,$ sixth term is 39

7. $a = -4,$ sixth term is -39

8. $a = -5,$ fifth term is -37

9. $d = 7,$ sixth term is -83

10. $d = 3,$ seventh term is 12

11. $d = -3,$ seventh term is 16

12. $d = -5,$ seventh term is -12

13. The 19th term is 131, and the 20th term is 138.

14. The 16th term is 70, and the 18th term is 78.

15. Find the 30th term of the arithmetic progression with $a = 7$ and $d = 12$.

16. Find the 55th term of the arithmetic progression with $a = -5$ and $d = 4$.

17. Find the 37th term of the arithmetic progression with a 2nd term of -4 and a 3rd term of -9.

18. Find the 40th term of the arithmetic progression with a 2nd term of 6 and a 4th term of 16.

19. Find the 1st term of the arithmetic progression with a common difference of 11 and whose 27th term is 263.

20. Find the common difference of the arithmetic progression with a 1st term of -164 if its 36th term is -24.

21. Find the common difference of the arithmetic progression with a 1st term of 40 if its 44th term is 556.

22. Find the 1st term of the arithmetic progression with a common difference of -5 and whose 23rd term is -625.

23. Insert three arithmetic means between 2 and 11.

24. Insert four arithmetic means between 5 and 25.

25. Insert four arithmetic means between 10 and 20.

26. Insert three arithmetic means between 20 and 30.

27. Find the arithmetic mean between 10 and 19.

28. Find the arithmetic mean between 5 and 23.

29. Find the arithmetic mean between -4.5 and 7.

30. Find the arithmetic mean between -6.3 and -5.2.

In Exercises 31–38, find the sum of the first n terms of each arithmetic progression.

31. $1, 4, 7, \ldots ; \quad n = 30$

32. $2, 6, 10, \ldots ; \quad n = 28$

33. $-5, -1, 3, \ldots ; \quad n = 17$

34. $-7, -1, 5, \ldots ; \quad n = 15$

35. Second term is 7, third term is 12; $n = 12$

36. Second term is 5, fourth term is 9; $n = 16$

37. $f(n) = 2n + 1,$ nth term is 31; n is a natural number

38. $f(n) = 4n + 3,$ nth term is 23; n is a natural number

39. Find the sum of the first 50 natural numbers.

40. Find the sum of the first 100 natural numbers.

41. Find the sum of the first 50 positive odd natural numbers.

42. Find the sum of the first 50 positive even natural numbers.

43. Fred puts $60 in a shoebox. After each succeeding month, he puts an additional $50 in the shoebox. Write the first six terms of an arithmetic progression that gives the monthly amounts in the box, and find the sum in the box after 10 years.

44. Freda borrowed $10,000, interest free, from her mother. Freda agrees to pay back the loan in monthly installments of $275. Write the first six terms of an arithmetic progression that shows the balance due after each month, and calculate the balance due after 17 months.

45. The equation $s = 16t^2$ represents the distance s in feet that an object will fall in t seconds. After 1 second, the object has fallen 16 feet. After 2 seconds, the object has fallen 64 feet, and so on. Find the distance that the object will fall during the second and third seconds.

46. Refer to Exercise 45. How far will the object fall during the 12th second? Ignore any wind resistance.

47. Show that the arithmetic mean between a and b is the average of a and b:

$$\frac{a + b}{2}$$

48. Show that the sum of the two arithmetic means between a and b is $a + b$.

In Exercises 49–54, find each sum.

49. $\sum_{k=1}^{4} 6k$

50. $\sum_{k=2}^{5} 3k$

51. $\sum_{k=3}^{4} (k^2 + 3)$

52. $\sum_{k=2}^{6} (k^2 + 1)$

53. $\sum_{k=4}^{4} (2k + 4)$

54. $\sum_{k=3}^{5} (3k^2 - 7)$

55. Show that $\sum_{k=1}^{5} 5k = 5 \sum_{k=1}^{5} k$.

56. Show that $\sum_{k=3}^{6} (k^2 + 3k) = \sum_{k=3}^{6} k^2 + \sum_{k=3}^{6} 3k$.

57. Show that $\sum_{k=1}^{n} 3 = 3n$. (*Hint:* Consider 3 to be $3k^0$.)

11.6 GEOMETRIC PROGRESSIONS

Another important type of sequence is called a **geometric progression**.

Definition. A **geometric progression** is a sequence of the form

$$a, ar, ar^2, ar^3, \ldots, ar^{n-1}, \ldots$$

where a is the **first term**, ar^{n-1} is the **nth term**, and r is the **common ratio**.

Note that the second term of a geometric progression has a factor of r^1, the third term has a factor of r^2, the fourth term has a factor of r^3, and the nth term has a factor of r^{n-1}. Also note that the quotient obtained when any term is divided by the previous term is r.

Example 1 A geometric progression has a first term of 5 and a common ratio of 3.

a. Write the first five terms of the progression.

b. Write the ninth term of the progression.

Solution **a.** Because the first term is $a = 5$ and the common ratio is $r = 3$, the first five terms are

$$5, \quad 5(3), \quad 5(3)^2, \quad 5(3)^3, \quad 5(3)^4$$

or

$$5, 15, 45, 135, 405$$

b.   The nth term is ar^{n-1} where $a = 5$ and $r = 3$. Because you want the ninth term, let $n = 9$:

$$n\text{th term} = ar^{n-1}$$
$$9\text{th term} = 5(3)^{9-1}$$
$$= 5(3)^8$$
$$= 5(6561)$$
$$= 32{,}805$$ ∎

Example 2   The first three terms of a geometric progression are 16, 4, and 1. Find the seventh term of the progression.

Solution   In the given progression, $a = 16$, $r = \frac{1}{4}$, and $n = 7$. Substitute these values into the expression for the nth term, and simplify:

$$n\text{th term} = ar^{n-1}$$
$$7\text{th term} = 16\left(\frac{1}{4}\right)^{7-1}$$
$$= 16\left(\frac{1}{4}\right)^{6}$$
$$= 16\left(\frac{1}{4096}\right)$$
$$= \frac{1}{256}$$ ∎

If numbers are inserted between two given numbers a and b to form a geometric progression, the inserted numbers are called **geometric means** between a and b. If a single number is inserted between the numbers a and b, that number is called a **geometric mean** between a and b.

Example 3   Insert two geometric means between 7 and 1512.

Solution   In this example, the first term is $a = 7$, and the fourth term (or last term) is $l = 1512$. You must find the common ratio r so that the terms

$$7, \quad 7r, \quad 7r^2, \quad 1512$$

form a geometric progression. Substitute 4 for n and 7 for a into the formula for the nth term of a geometric progression, and solve for r:

$$n\text{th term} = ar^{n-1}$$
$$4\text{th term} = 7r^{4-1}$$
$$1512 = 7r^3$$
$$216 = r^3 \qquad \text{Divide both sides by 7.}$$
$$6 = r \qquad \text{Take the cube root of both sides.}$$

The two geometric means between 7 and 1512 are

$$7r = 7(6) = 42$$

and

$$7r^2 = 7(6)^2 = 7(36) = 252$$

Note that 7, 42, 252, and 1512 are the first four terms of a geometric progression.   ■

Example 4   Find a geometric mean between 2 and 20.

Solution   You wish to find the middle term of the three-termed geometric progression

$$2,   2r,   20$$

with $a = 2$, $l = 20$, and $n = 3$. Substitute into the formula for the nth term of a geometric progression, and solve for r:

$$n\text{th term} = ar^{n-1}$$
$$20\text{th term} = 2r^{3-1}$$
$$20 = 2r^2$$
$$10 = r^2 \qquad \text{Divide both sides by 2.}$$
$$\pm\sqrt{10} = r \qquad \text{Take the square root of both sides.}$$

Because r can be either $\sqrt{10}$ or $-\sqrt{10}$, there are two values for a geometric mean between 2 and 20. They are

$$2r = 2\sqrt{10}$$

and

$$2r = -2\sqrt{10}$$

Note that 2, $2\sqrt{10}$, 20 and 2, $-2\sqrt{10}$, 20 are both geometric progressions. The common ratio of the first is $\sqrt{10}$, and the common ratio of the second is $-\sqrt{10}$.   ■

There is a formula that gives the sum of the first n terms of a geometric progression. To develop this formula, we let S_n represent the sum of the first n terms of a geometric progression.

1. $S_n = a + ar + ar^2 + ar^3 + \cdots + ar^{n-1}$

We multiply both sides of Equation 1 by r to get

2. $S_n r = \qquad ar + ar^2 + ar^3 + \cdots + ar^{n-1} + ar^n$

We now subtract Equation 2 from Equation 1, and solve for S_n:

$$S_n - S_n r = a - ar^n$$
$$S_n(1 - r) = a - ar^n \qquad \text{Factor out } S_n \text{ from the left side.}$$
$$S_n = \frac{a - ar^n}{1 - r} \qquad \text{Divide both sides by } 1 - r.$$

This reasoning establishes the following theorem.

Theorem.   The sum of the first n terms of a geometric progression is given by the formula

$$S_n = \frac{a - ar^n}{1 - r} \qquad (r \neq 1)$$

where S_n is the sum, a is the first term, r is the common ratio, and n is the number of terms.

Example 5   Find the sum of the first six terms of the geometric progression $250, 50, 10, \ldots$.

Solution   In this geometric progression, $a = 250$, $r = \frac{1}{5}$, and $n = 6$. Substitute these values into the formula for the sum of the first n terms of a geometric progression, and simplify:

$$S_n = \frac{a - ar^n}{1 - r}$$

$$= \frac{250 - 250\left(\frac{1}{5}\right)^6}{1 - \frac{1}{5}}$$

$$= \frac{250 - 250\left(\frac{1}{15{,}625}\right)}{\frac{4}{5}}$$

$$= \frac{5}{4}\left(250 - \frac{250}{15{,}625}\right)$$

$$= \frac{5}{4}\left(\frac{3{,}906{,}000}{15{,}625}\right)$$

$$= 312.48$$

The sum of the first six terms is 312.48.   ∎

Example 6   Big City, California, has a population of 250 people. The mayor predicts a growth rate of 4% each year for the next 10 years. If this is true, what will be the population of Big City 10 years from now?

Solution   Let P_0 be the initial population of Big City. After 1 year, there will be a different population, P_1. The initial population (P_0) plus the growth (the product of P_0 and the rate of growth, r) will equal this new population, P_1:

$$P_1 = P_0 + P_0 r = P_0(1 + r)$$

The population after 2 years will be P_2, and

$$P_2 = P_1 + P_1 r$$
$$= P_1(1 + r)$$
$$= P_0(1 + r)(1 + r) \qquad \text{Remember that } P_1 = P_0(1 + r).$$
$$= P_0(1 + r)^2$$

Similarly, the population after 3 years will be P_3, and

$$P_3 = P_2 + P_2 r$$
$$= P_2(1 + r)$$
$$= P_0(1 + r)^2(1 + r) \qquad \text{Remember that } P_2 = P_0(1 + r)^2.$$
$$= P_0(1 + r)^3$$

Note that the yearly population figures

$$P_0, \quad P_1, \quad P_2, \quad P_3, \ldots$$

or

$$P_0, \quad P_0(1 + r), \quad P_0(1 + r)^2, \quad P_0(1 + r)^3, \ldots$$

form a geometric progression with a first term of P_0 and a common ratio of $1 + r$. The population of Big City after 10 years is P_{10}, which is the 11th term of this progression:

$$n\text{th term} = ar^{n-1}$$
$$P_{10} = 11\text{th term} = P_0(1 + r)^{10}$$
$$= 250(1 + 0.04)^{10}$$
$$= 250(1.04)^{10}$$
$$\approx 250(1.48)$$
$$\approx 370$$

The estimated population of Big City 10 years from now is 370 people. ∎

■ EXERCISE 11.6

In Exercises 1–14, write the first five terms of each geometric progression with the given properties.

1. $a = 3, \quad r = 2$

2. $a = -2, \quad r = 2$

3. $a = -5, \quad r = \dfrac{1}{5}$

4. $a = 8, \quad r = \dfrac{1}{2}$

5. $a = 2, r > 0$, and third term is 32

6. $a = 3$, fourth term is 24

7. $a = -3$, fourth term is -192

8. $a = 2, r < 0$, and third term is 50

9. $a = -64, r < 0$, and fifth term is -4

10. $a = -64, r > 0$, and fifth term is -4

11. $a = -64, r > 0$, and sixth term is -2

12. $a = -81, r < 0$, and sixth term is $\dfrac{1}{3}$

13. The second term is 10, and the third term is 50.

14. The third term is -27, and the fourth term is 81.

15. Find the tenth term of the geometric progression with $a = 7$ and $r = 2$.

16. Find the 12th term of the geometric progression with $a = 64$ and $r = \frac{1}{2}$.

17. Find the first term of the geometric progression with a common ratio of -3 and an eighth term of -81.

18. Find the first term of the geometric progression with a common ratio of 2 and a tenth term of 384.

19. Find the common ratio of the geometric progression with a first term of -8 and a sixth term of -1944.

20. Find the common ratio of the geometric progression with a first term of 12 and a sixth term of $\frac{3}{8}$.

21. Insert three positive geometric means between 2 and 162.

22. Insert four geometric means between 3 and 96.

23. Insert four geometric means between -4 and $-12,500$.

24. Insert three geometric means (two positive and one negative) between -64 and -1024.

25. Find the negative geometric mean between 2 and 128.

26. Find the positive geometric mean between 3 and 243.

27. Find the positive geometric mean between 10 and 20.

28. Find the negative geometric mean between 5 and 15.

29. Find a geometric mean, if possible, between -50 and 10.

30. Find a negative geometric mean, if possible, between -25 and -5.

In Exercises 31–42, find the sum of the first n terms of each geometric progression.

31. $2, 6, 18, \ldots$; $n = 6$

32. $2, -6, 18, \ldots$; $n = 6$

33. $2, -6, 18, \ldots$; $n = 5$

34. $2, 6, 18, \ldots$; $n = 5$

35. $3, -6, 12, \ldots$; $n = 8$

36. $3, 6, 12, \ldots$; $n = 8$

37. $3, 6, 12, \ldots$; $n = 7$

38. $3, -6, 12, \ldots$; $n = 7$

39. The second term is 1, and the third term is $\frac{1}{5}$; $n = 4$.

40. The second term is 1, and the third term is 4; $n = 5$.

41. The third term is -2, and the fourth term is 1; $n = 6$.

42. The third term is -3, and the fourth term is 1; $n = 5$.

In Exercises 43–48, use a calculator to solve each problem.

43. The population of Mud City is predicted to increase by 6% each year. What will be the population of Mud City 5 years from now if its current population is 500?

44. The population of Hicksville is decreasing by 10% each year. If its current population is 98, what will be the population 8 years from now?

45. John has $10,000 in a safety deposit box. Each year he spends 12% of what is left in the box. How much will be in the box after 15 years?

46. Sally has $5000 in a savings account earning 12% annual interest. How much will be in her account 10 years from now? (Assume that Sally makes no deposits or withdrawals.)

47. A house appreciates by 6% per year. If that house is worth $70,000 today, how much will it be worth 12 years from now?

48. A motorboat that originally cost $5000 depreciates at a rate of 9% per year. How much will the boat be worth in 5 years?

49. Show that the formula for the sum of the first n terms of a geometric progression can be written in the form

$$S_n = \frac{a - lr}{1 - r} \qquad \text{where} \quad l = ar^{n-1}$$

50. Show that the formula for the sum of the first n terms of a geometric progression can be written in the form

$$S_n = \frac{a(1 - r^n)}{1 - r}$$

11.7   INFINITE GEOMETRIC PROGRESSIONS

Under certain conditions, it is possible to find the sum of the terms of a geometric progression, even though the progression is unending.

We have established that the formula

$$S_n = \frac{a - ar^n}{1 - r}$$

gives the sum of the first n terms of a finite (ending) geometric progression. If $|r| < 1$, then the term ar^n in the above formula approaches 0 as n becomes very large. For example,

$$a\left(\frac{1}{4}\right)^1 = \frac{1}{4}a, \qquad a\left(\frac{1}{4}\right)^2 = \frac{1}{16}a, \qquad a\left(\frac{1}{4}\right)^3 = \frac{1}{64}a,$$

and so on. Thus, when n is very large, the value of ar^n is negligible, and the term ar^n in the above formula can be ignored. This reasoning justifies the following theorem.

Theorem.   If a is the first term and r is the common ratio of an infinite (unending) geometric progression, and if $|r| < 1$, then the sum of the terms of the progression is given by the formula

$$S = \frac{a}{1 - r}$$

Example 1   Find the sum of the terms of the infinite geometric progression $125, 25, 5, \ldots$.

Solution   In this geometric progression, $a = 125$ and $r = \frac{1}{5}$. Because $|r| = \left|\frac{1}{5}\right| = \frac{1}{5} < 1$, you can find the sum of all the terms of the progression. Do this by substituting

125 for a and $\frac{1}{5}$ for r in the formula $S = \dfrac{a}{1-r}$ and simplifying:

$$S = \frac{a}{1-r} = \frac{125}{1 - \dfrac{1}{5}} = \frac{125}{\dfrac{4}{5}} = \frac{5}{4}(125) = \frac{625}{4}$$

The sum of all the terms of the progression 125, 25, 5, . . . is $\dfrac{625}{4}$. ∎

Example 2 Find the sum of the infinite geometric progression $64, -4, \dfrac{1}{4}, \dots$.

Solution In this geometric progression, $a = 64$ and $r = -\frac{1}{16}$. Because $|r| = \left|-\frac{1}{16}\right| = \frac{1}{16} < 1$, you can find the sum of all the terms of the progression. Substitute 64 for a and $-\frac{1}{16}$ for r in the formula

$$S = \frac{a}{1-r}$$

and simplify:

$$S = \frac{a}{1-r} = \frac{64}{1 - \left(-\dfrac{1}{16}\right)} = \frac{64}{\dfrac{17}{16}} = \frac{16}{17}(64) = \frac{1024}{17}$$

The sum of all the terms of the progression $64, -4, \dfrac{1}{4}, \dots$ is $\dfrac{1024}{17}$. ∎

Example 3 Change $0.888\dots$ to a common fraction.

Solution Note that the decimal $0.888\dots$ can be written as the infinite geometric progression

$$\frac{8}{10} + \frac{8}{100} + \frac{8}{1000} + \cdots$$

where $a = \frac{8}{10}$ and $r = \frac{1}{10}$. Because $|r| = \left|\frac{1}{10}\right| = \frac{1}{10} < 1$, you can find the sum as follows:

$$S = \frac{a}{1-r} = \frac{\dfrac{8}{10}}{1 - \dfrac{1}{10}} = \frac{\dfrac{8}{10}}{\dfrac{9}{10}} = \frac{8}{9}$$

Hence, $0.888\dots$ is equal to $\frac{8}{9}$. Long division will verify that $\frac{8}{9}$ is indeed equal to $0.888\dots$. ∎

Example 4 Change $0.252525\dots$ to a common fraction.

Solution   Note that the decimal 0.252525 . . . can be written as the infinite geometric progression

$$\frac{25}{100} + \frac{25}{10,000} + \frac{25}{1,000,000} + \cdots$$

where $a = 25/100$ and $r = 1/100$ Because $|r| = \left|\frac{1}{100}\right| = \frac{1}{100} < 1$, you can find the sum as follows:

$$S = \frac{a}{1-r} = \frac{\dfrac{25}{100}}{1 - \dfrac{1}{100}} = \frac{\dfrac{25}{100}}{\dfrac{99}{100}} = \frac{25}{99}$$

Hence, 0.252525 . . . is equal to 25/99. Long division will verify that this is true.   ■

■ EXERCISE 11.7

In Exercises 1–12, find the sum of each infinite geometric progression, if possible.

1.	8, 4, 2, . . .	**2.**	12, 6, 3, . . .	**3.**	54, 18, 6, . . .	**4.**	45, 15, 5, . . .
5.	12, −6, 3, . . .	**6.**	8, −4, 2, . . .	**7.**	−45, 15, −5, . . .	**8.**	−54, 18, −6, . . .
9.	$\frac{9}{2}$, 6, 8, . . .	**10.**	−112, −28, −7, . . .	**11.**	$-\frac{27}{2}$, −9, −6, . . .	**12.**	$\frac{18}{25}$, $\frac{6}{5}$, 2, . . .

In Exercises 13–20, change each decimal to a common fraction. Then check your answer by performing a long division.

13.	0.111 . . .	**14.**	0.222 . . .	**15.**	−0.333 . . .	**16.**	−0.444 . . .
17.	0.121212 . . .	**18.**	0.212121 . . .	**19.**	0.757575 . . .	**20.**	0.575757 . . .

21. A rubber ball is dropped from a height of 10 meters. On each bounce, it returns to a height one-half of that from which it fell. What is the total distance the ball travels?

22. A golf ball is dropped from a height of 12 feet. On each bounce, it returns to a height two-thirds of that from which it fell. What is the total distance the ball travels?

23. Show that 0.999 . . . is equal to 1.   **24.** Show that 1.999 . . . is equal to 2.

25. Does 0.999999 = 1? Explain.

26. Suppose that $f(x) = 1 + x + x^2 + x^3 + x^4 + \cdots$. Calculate $f(\frac{1}{2})$ and $f(-\frac{1}{2})$.

11.8   MATHEMATICAL INDUCTION

Suppose we ask a theatergoer whether everyone in line for a movie gained admittance. The theatergoer answers that "the first person in line was admitted." Does this response answer our question? Certainly more than one person was admitted, but this does not mean that everyone was admitted. As a matter of fact, we know little more now than before we asked the question.

Meeting a second theatergoer, we ask the question again. This time the answer is "they promised that, if anyone was admitted, the person next in line also would be admitted." On the basis of this response we know that, if anyone was admitted, the person behind was admitted also. But a promise that begins with "if someone is admitted" does not guarantee that anyone actually was admitted.

However, when we consider both the first and second answers, we know that everyone in line gained admittance. The first theatergoer said that the first person got in; the second theatergoer said that, if anyone got in, the next person in line got in also. Because the first person was admitted, the second person was admitted also. And, if the second person was admitted, then so was the third. This pattern would have continued until everyone was admitted to the theater.

This situation is very similar to a children's game played with dominoes. The dominoes are placed in a row, fairly close together. When the first domino is pushed over, it falls against the second, knocking it down. The second domino, in turn, knocks down the third, which topples the fourth, and so on until all the dominoes fall. Two things must happen to guarantee that all the dominoes fall: (1) the first domino must be knocked over, and (2) every domino that falls must topple the next one. When both of these conditions are met, it is certain that all the dominoes will fall.

The preceding examples illustrate the basic idea that underlies the principle of **mathematical induction**.

The Principle of Mathematical Induction.   If a statement involving the natural number n has the two properties that

1.   the statement is true for $n = 1$ and
2.   the statement is true for $n = k + 1$ whenever it is true for $n = k$, then the statement is true for all natural numbers.

The principle of mathematical induction provides a method for proving many theorems. Note that any such induction proof involves two parts: First show that the formula holds true for the natural number 1, and then show that, *if* the formula holds true for any natural number k, then it also holds true for the natural number $k + 1$. A proof by mathematical induction is complete only if both the required properties are established.

There is a formula for finding the sum of the first n natural numbers:

$$1 + 2 + 3 + 4 + \cdots + n = \frac{n(n + 1)}{2}$$

To show that this formula is correct, we will prove it by using mathematical induction.

Example 1   Use mathematical induction to prove that the formula

$$1 + 2 + 3 + \cdots + n = \frac{n(n + 1)}{2}$$

is true for every natural number n.

Solution   The proof has two parts.

Part 1

Verify that the formula holds true for the value $n = 1$. Substituting $n = 1$ into the term n on the left side of the equation yields a single term, the number 1. Substituting the number 1 for n on the right side, the formula becomes

$$1 = \frac{n(n + 1)}{2}$$

$$1 = \frac{(1)(1 + 1)}{2}$$

$$1 = 1$$

Thus, the formula is true for $n = 1$, and Part 1 of the proof is complete.

Part 2

Assume that the given formula is true when n is replaced by *some* natural number k. By this assumption, called the **induction hypothesis**, you accept that

1. $1 + 2 + 3 + \cdots + k = \dfrac{k(k + 1)}{2}$

is a true statement. The plan is to show that the given formula is true for the next natural number, $k + 1$. Do this by verifying the equation

2. $1 + 2 + 3 + \cdots + k + (k + 1) = \dfrac{(k + 1)[(k + 1) + 1]}{2}$

obtained from the given formula by replacing n with $k + 1$.

Compare the left sides of Equations 1 and 2, and note that the left side of Equation 2 contains an extra term of $k + 1$. Hence, add $k + 1$ to both sides of Equation 1 (which was assumed to be true) to obtain the equation

$$1 + 2 + 3 + \cdots + k + (k + 1) = \frac{k(k + 1)}{2} + (k + 1)$$

Because both terms on the right side of this equation have a common factor of $k + 1$, the right side factors and the equation can be rewritten as follows:

$$1 + 2 + 3 + \cdots + k + (k + 1) = (k + 1)\left(\frac{k}{2} + 1\right)$$

$$= (k + 1)\left(\frac{k + 2}{2}\right)$$

$$= \frac{(k + 1)(k + 2)}{2}$$

$$= \frac{(k + 1)[(k + 1) + 1]}{2}$$

This final result has the same form as Equation 2. Because the truth of Equation 1 implies the truth of Equation 2, Part 2 of the proof is complete. Parts 1 and 2 together establish that the formula

$$1 + 2 + 3 + \cdots + n = \frac{n(n + 1)}{2}$$

is true for any natural number n.   ∎

Here is a brief overview of Example 1:

1. Did the first domino fall? That is, is the formula

$$1 + 2 + 3 + \cdots + n = \frac{n(n + 1)}{2}$$

true for $n = 1$? Yes, Part 1 verified this.

2. Will toppling any domino knock over the next domino? If the given formula is true for the value $n = k$, is it also true for the value $n = k + 1$? Yes, Part 2 of the proof verified this.

Because both of the induction requirements were verified, the formula is true for all natural numbers n.

Example 2   Use mathematical induction to prove the formula

$$1 + 5 + 9 + 13 + \cdots + (4n - 3) = n(2n - 1)$$

for all natural numbers n.

Solution   The proof has two parts.

Part 1

First verify the formula for the value $n = 1$. Substituting the value $n = \mathbf{1}$ into the term $4n - 3$ on the left side of the formula gives the single term 1. After substituting the same value into the right side, the equation becomes

$$1 = \mathbf{1}[2(\mathbf{1}) - 1]$$
$$1 = 1$$

Thus, the formula holds true for $n = 1$, and Part 1 of the proof is complete.

Part 2

The induction hypothesis is the assumption that the formula holds true for $n = k$. Hence, you assume that

$$1 + 5 + 9 + 13 + \cdots + (4k - 3) = k(2k - 1)$$

is a true statement. Because the truth of this assumption must guarantee the truth of the formula for $k + 1$ terms, add the $(k + 1)$th term to both sides of the induction hypothesis formula. In this example, the terms on the left side increase by 4, so the $(k + 1)$th term is $(4k - 3) + 4$, or $4k + 1$. Adding $4k + 1$ to both sides of the induction hypothesis formula gives

$$1 + 5 + 9 + 13 + \cdots + (4k - 3) + (4k + 1) = k(2k - 1) + (4k + 1)$$

Simplify the right side and rewrite the equation as follows:

$$1 + 5 + 9 + 13 + \cdots + (4k - 3) + [4(k + 1) - 3] = 2k^2 + 3k + 1$$
$$= (k + 1)(2k + 1)$$
$$= (k + 1)[2(k + 1) - 1]$$

Because the above equation has the same form as the given formula, except that $k + 1$ appears in place of n, the truth of the formula for $n = k$ implies the truth of the formula for $n = k + 1$. Part 2 of the proof is complete.

Because both of the induction requirements have been verified, the given formula is proved for all natural numbers.   ∎

Example 3   Prove that $\dfrac{1}{2} + \dfrac{1}{4} + \dfrac{1}{8} + \cdots + \dfrac{1}{2^n} < 1.$

Solution   The proof is by induction.

Part 1

Verify the formula for $n = 1$. Substituting 1 for n on the left side of the inequality gives $\frac{1}{2} < 1$. The formula holds true for $n = 1$, and Part 1 of the proof is complete.

Part 2

The induction hypothesis is the assumption that the inequality holds true for $n = k$. Thus, you assume that

$$\frac{1}{2} + \frac{1}{4} + \frac{1}{8} + \cdots + \frac{1}{2^k} < 1$$

Multiply both sides of the above inequality by $\frac{1}{2}$ to obtain the inequality

$$\frac{1}{2}\left(\frac{1}{2} + \frac{1}{4} + \frac{1}{8} + \cdots + \frac{1}{2^k}\right) < 1\left(\frac{1}{2}\right)$$

or

$$\frac{1}{4} + \frac{1}{8} + \frac{1}{16} + \cdots + \frac{1}{2^{k+1}} < \frac{1}{2}$$

Now add $\frac{1}{2}$ to both sides of this inequality to get

$$\frac{1}{2} + \frac{1}{4} + \frac{1}{8} + \frac{1}{16} + \cdots + \frac{1}{2^{k+1}} < \frac{1}{2} + \frac{1}{2}$$

or

$$\frac{1}{2} + \frac{1}{4} + \frac{1}{8} + \frac{1}{16} + \cdots + \frac{1}{2^{k+1}} < 1$$

The resulting inequality is the same as the original except that $k + 1$ appears in place of n. The truth of the inequality for $n = k$ implies the truth of the inequality for $n = k + 1$. Part 2 of the proof is complete.

Because both of the induction requirements have been verified, this inequality is true for all natural numbers.   ∎

There are statements that are not true when $n = 1$, but that are true for all natural numbers equal to or greater than some given natural number, say q. In these cases, verify the given statements for $n = q$ in Part 1 of the induction proof. After establishing Part 2 of the induction proof, the given statement is proved for all natural numbers that are greater than or equal to q.

EXERCISE 11.8

In Exercises 1–4, verify each given formula for $n = 1, 2, 3,$ and 4.

1. $5 + 10 + 15 + \cdots + 5n = \dfrac{5n(n + 1)}{2}$

2. $1^2 + 2^2 + 3^2 + \cdots + n^2 = \dfrac{n(n + 1)(2n + 1)}{6}$

3. $7 + 10 + 13 + \cdots + (3n + 4) = \dfrac{n(3n + 11)}{2}$

4. $1(3) + 2(4) + 3(5) + \cdots + n(n + 2) = \dfrac{n}{6}(n + 1)(2n + 7)$

In Exercises 5–20, prove each of the following formulas by mathematical induction, if possible.

5. $2 + 4 + 6 + \cdots + 2n = n(n + 1)$

6. $1 + 3 + 5 + \cdots + (2n - 1) = n^2$

7. $3 + 7 + 11 + \cdots + (4n - 1) = n(2n + 1)$

8. $4 + 8 + 12 + \cdots + 4n = 2n(n + 1)$

9. $10 + 6 + 2 + \cdots + (14 - 4n) = 12n - 2n^2$

10. $8 + 6 + 4 + \cdots + (10 - 2n) = 9n - n^2$

11. $2 + 5 + 8 + \cdots + (3n - 1) = \dfrac{n(3n + 1)}{2}$

12. $3 + 6 + 9 + \cdots + 3n = \dfrac{3n(n + 1)}{2}$

13. $1^2 + 2^2 + 3^2 + \cdots + n^2 = \dfrac{n(n + 1)(2n + 1)}{6}$

14. $1 + 2 + 3 + \cdots + (n - 1) + n + (n - 1) + \cdots + 3 + 2 + 1 = n^2$

15. $\dfrac{1}{3} + 2 + \dfrac{11}{3} + \cdots + \left(\dfrac{5}{3}n - \dfrac{4}{3}\right) = n\left(\dfrac{5}{6}n - \dfrac{1}{2}\right)$

16. $\dfrac{1}{1 \cdot 2} + \dfrac{1}{2 \cdot 3} + \dfrac{1}{3 \cdot 4} + \cdots + \dfrac{1}{n(n + 1)} = \dfrac{n}{n + 1}$

17. $\dfrac{1}{2} + \dfrac{1}{4} + \dfrac{1}{8} + \cdots + \left(\dfrac{1}{2}\right)^n = 1 - \left(\dfrac{1}{2}\right)^n$

18. $\dfrac{1}{3} + \dfrac{2}{9} + \dfrac{4}{27} + \cdots + \dfrac{1}{3}\left(\dfrac{2}{3}\right)^{n-1} = 1 - \left(\dfrac{2}{3}\right)^n$

19. $2^0 + 2^1 + 2^2 + 2^3 + \cdots + 2^{n-1} = 2^n - 1$

20. $1^3 + 2^3 + 3^3 + \cdots + n^3 = \left[\dfrac{n(n + 1)}{2}\right]^2$

21. Prove that $x - y$ is a factor of $x^n - y^n$. (*Hint:* Consider subtracting and adding xy^k to the binomial $x^{k+1} - y^{k+1}$.)

22. Prove that $n < 2^n$.

23. The sum of the angles of any triangle is 180°. Prove by induction that $(n - 2)180°$ gives the sum of the angles of any simple polygon when n is the number of sides of that polygon. (*Hint:* If a polygon has $k + 1$ sides, it has $k - 2$ sides plus three more sides.)

24. If $1 + 2 + 3 + \cdots + n = \dfrac{n}{2}(n + 1) + 1$ were true for $n = k$, show that it would be true for $n = k + 1$. Is it true for $n = 1$?

25. Prove that $n + 1 = 1 + n$ for each natural number n.

26. If n is any natural number, prove that $7^n - 1$ is divisible by 6.

27. Prove that $1 + 2n < 3^n$ for $n > 1$.

28. Prove that, if r is a real number and $r \neq 1$, then

$$1 + r + r^2 + \cdots + r^n = \frac{1 - r^{n+1}}{1 - r}$$

29. The expression a^m where m is a natural number was defined in Section 2.1. An alternative definition of a^m, useful in proofs by induction, is (Part 1) $a^1 = a$ and (Part 2) $a^{m+1} = a^m \cdot a$. Use mathematical induction on n to prove the familiar law of exponents $a^m a^n = a^{m+n}$.

CHAPTER SUMMARY

Key Words

arithmetic mean   (11.5)

arithmetic progression   (11.5)

binomial theorem   (11.1)

combinations   (11.4)

event   (11.3)

factorial n   (11.1)

Fibonacci sequence   (11.5)

geometric means   (11.6)

geometric progression   (11.6)

index of summation   (11.5)

induction hypothesis   (11.8)

infinite geometric progression   (11.7)

mathematical induction   (11.8)

multiplication principle for events   (11.3)

Pascal's triangle   (11.1)

permutations   (11.3)

sequence   (11.5)

summation notation   (11.5)

Key Ideas

(11.1) The symbol $n!$ (**n factorial**) is defined as

$$n! = n(n - 1)(n - 2)(n - 3) \cdots \cdot 3 \cdot 2 \cdot 1$$

where n is a natural number.

$0! = 1$

$n(n - 1)! = n!$, provided that n is a natural number.

The binomial theorem:

$$(a + b)^n = a^n + \frac{n!}{1!(n - 1)!} a^{n-1}b + \frac{n!}{2!(n - 2)!} a^{n-2}b^2 + \frac{n!}{3!(n - 3)!} a^{n-3}b^3 + \cdots + b^n$$

(11.2) The binomial theorem can be used to find the nth term of a binomial expansion.

(11.3)   **The multiplication principle for events:**   If E_1 and E_2 are two events, and if E_1 can be done in a_1 ways and E_2 can be done in a_2 ways, then the event "E_1 followed by E_2" can be done in $a_1 \cdot a_2$ ways.

The formula for computing the number of permutations of n things r at a time is

$$_nP_r = \frac{n!}{(n-r)!}$$

The number of permutations of n things n at a time is

$$_nP_n = n!$$

(11.4)   The formula for computing the number of combinations of n things r at a time is

$$_nC_r = \binom{n}{r} = \frac{n!}{r!(n-r)!}$$

The number of combinations of n things n at a time is

$$_nC_n = \binom{n}{n} = 1$$

The number of combinations of n things 0 at a time is

$$_nC_0 = \binom{n}{0} = 1$$

(11.5)   An **arithmetic progression** is a sequence of the form

$$a, a + d, a + 2d, a + 3d, \ldots, a + (n-1)d$$

where a is the first term, $a + (n-1)d$ is the nth term, and d is the common difference.

If numbers are inserted between two given numbers a and b to form an arithmetic progression, the inserted numbers are **arithmetic means** between a and b.

The sum of the first n terms of an arithmetic progression is given by the formula

$$S_n = \frac{n(a + l)}{2} \quad \text{with} \quad l = a + (n-1)d$$

where a is the first term, l is the last (or nth) term, and n is the number of terms in the progression.

$$\sum_{k=1}^{n} f(k) = f(1) + f(2) + f(3) + \cdots + f(n)$$

(11.6)   A **geometric progression** is a sequence of the form

$$a, ar, ar^2, ar^3, \ldots, ar^{n-1}$$

where a is the first term, ar^{n-1} is the nth term, and r is the common ratio.

If numbers are inserted between two given numbers a and b to form a geometric progression, the inserted numbers are **geometric means** between a and b.

The sum of the first n terms of a geometric progression is given by the formula

$$S_n = \frac{a - ar^n}{1 - r} \qquad (r \neq 1)$$

where S_n is the sum, a is the first term, r is the common ratio, and n is the number of terms in the progression.

(11.7) If r is the common ratio of an infinite geometric progression, and if $|r| < 1$, then the sum of the terms of the infinite geometric progression is given by the formula

$$S = \frac{a}{1 - r}$$

where a is the first term and r is the common ratio.

(11.8) **The principle of mathematical induction.**   If a formula involving the natural number n has the two properties that

1.   the formula is true when $n = 1$, and
2.   the formula is true when $n = k + 1$ whenever it is true for $n = k$,

then the formula is true for all natural numbers n.

REVIEW EXERCISES

In Review Exercises 1–4, evaluate each expression.

1. $(4!)(3!)$

2. $\dfrac{5!}{3!}$

3. $\dfrac{6!}{2!(6-2)!}$

4. $\dfrac{12!}{3!(12-3)!}$

In Review Exercises 5–8, use the binomial theorem to find each expansion.

5. $(x + y)^5$

6. $(x - y)^4$

7. $(4x - y)^3$

8. $(x + 4y)^3$

In Review Exercises 9–12, find the required term in each expansion.

9. $(x + y)^4$;   third term

10. $(x - y)^5$;   fourth term

11. $(3x - 4y)^3$;   second term

12. $(4x + 3y)^4$;   third term

13.   In how many ways can five persons be arranged in a line?

14.   In how many ways can three boys and five girls be arranged in a line if the girls are placed ahead of the boys?

In Review Exercises 15–28, evaluate each expression.

15. $_7P_7$

16. $_7P_0$

17. $_8P_6$

18. $\dfrac{_9P_6}{_{10}P_7}$

19. $_7C_7$

20. $_7C_0$

21. $\dbinom{8}{6}$

22. $\dbinom{9}{6}$

23. $_6C_3 \cdot {_7C_3}$

24. $\dfrac{_7C_3}{_6C_3}$

25. $\sum\limits_{k=4}^{6} \dfrac{1}{2}k$

26. $\sum\limits_{k=2}^{5} 7k^2$

27. $\sum\limits_{k=1}^{4} (3k-4)$

28. $\sum\limits_{k=10}^{10} 36k$

29. In how many ways can you pick three persons from a group of ten persons?

30. In how many ways can you pick a committee of two Democrats and two Republicans from a group containing five Democrats and six Republicans?

31. Write the first five terms of the arithmetic progression whose ninth term is 242 and whose seventh term is 212.

32. Find two arithmetic means between 8 and 25.

33. Find the sum of the first 20 terms of the progression 11, 18, 25, ...

34. Find the sum of the first ten terms of the progression 9, $6\frac{1}{2}$, 4, ...

35. Write the first five terms of the geometric progression whose fourth term is 3 and whose fifth term is $\frac{3}{2}$.

36. Find two geometric means between -6 and 384.

37. Find the sum of the first eight terms of the progression $\frac{1}{8}$, $-\frac{1}{4}$, $\frac{1}{2}$, ...

38. Find the sum of the first seven terms of the progression 162, 54, 18, ...

39. Find the sum of the infinite geometric progression 25, 20, 16, ...

40. Change the decimal fraction 0.050505 ... to a common fraction.

41. A \$5000 car depreciates at the rate of 20% per year. How much is the car worth after 5 years?

42. The value of Mary's stock portfolio is expected to appreciate at the rate of 18% per year. How much will the portfolio be worth in 10 years if its current value is \$25,700?

43. Harold planted 300 acres in corn this year. He intends to plant an additional 75 acres in corn in each successive year until he has 1200 acres in corn. In how many years will that be?

44. If an object is in free fall, the sequence 16, 48, 80, ... represents the distance in feet that object falls during the first second, during the second second, during the third second, and so on. How far will the object fall during the first ten seconds?

45. Prove that
$$6 + 16 + 26 + \cdots + (10n - 4) = n(5n + 1)$$
by using mathematical induction.

CHAPTER ELEVEN TEST

1. Evaluate $\dfrac{7!}{4!}$

2. Evaluate 0!

3. Find the second term in the expansion of $(x - y)^5$.

4. Find the third term in the expansion of $(x + 2y)^4$.

In Problems 5–10, find the value of each expression.

5. $_5P_4$

6. $_7P_7$

7. $_6C_3$

8. $_5C_5$

9. $_6C_0 \cdot {_6P_4}$

10. $\dfrac{_6P_4}{_6C_4}$

11. In how many ways can you pick 3 persons from a group of 7 persons?

12. Find the tenth term of an arithmetic progression with the first three terms of 3, 10, and 17.

13. Find the sum of the first 12 terms of the progression $-2, 3, 8, \ldots$.

14. Find two arithmetic means between 2 and 98.

15. Evaluate $\displaystyle\sum_{k=1}^{3} (2k - 3)$.

16. Find the seventh term of the geometric progression with the first three terms of $-\frac{1}{9}$, $-\frac{1}{3}$, and -1.

17. Find the sum of the first six terms of the progression $\frac{1}{27}, \frac{1}{9}, \frac{1}{3}, \ldots$.

18. Find two geometric means between 3 and 648.

19. Find the sum of all the terms of the infinite geometric progression $9, 3, 1, \ldots$.

20. Is the formula $1 + 2 + 3 + \cdots + n = 4n$ true when $n = 6$?

21. Is the formula of Problem 20 true for any natural number n?

CUMULATIVE REVIEW EXERCISES (CHAPTERS 10–11)

1. Graph the function determined by $y = \left(\dfrac{1}{2}\right)^{x}$.

2. Write $y = \log_2 x$ as an exponential equation.

In Exercises 3–6, find x.

3. $\log_x 25 = 2$

4. $\log_5 125 = x$

5. $\log_3 x = -3$

6. $\log_5 x = 0$

7. Find the inverse of $y = \log_2 x$.

8. If $\log_{10} 10^x = y$, then y must equal what quantity?

In Exercises 9–12, assume that $\log 7 = 0.8451$ and $\log 14 = 1.1461$. Evaluate each expression without using a calculator or tables.

9. $\log 98$

10. $\log 2$

11. $\log 49$

12. $\log \dfrac{7}{5}$ (*Hint:* $\log 10 = 1$.)

13. Solve the equation $2^{x+2} = 3^x$.

14. Solve the equation $2 \log 5 + \log x - \log 4 = 2$.

In Exercises 15–16, use a calculator.

15. Suppose a boat depreciates 12% per year. How much will a $9000 boat be worth after 9 years?

16. Find $\log_6 8$.

17. Evaluate $\dfrac{6!7!}{5!}$.

18. Use the binomial theorem to expand $(3a - b)^4$.

19. Find the seventh term in the expansion of $(2x - y)^8$.

20. In how many ways can seven persons stand in a line?

21. Evaluate $_6P_3$.

22. Evaluate $_6C_3$.

23. In how many ways can a committee of three persons be chosen from a group containing nine persons?

24. Choose the smaller: $_nP_n$ or $_nC_n$ $(n > 1)$.

25. Find the 20th term of an arithmetic progression with a first term of -11 and a common difference of 6.

26. Find the sum of the first 20 terms of an arithmetic progression with a first term of 6 and a common difference of 3.

27. Insert two arithmetic means between -3 and 30.

28. Evaluate $\sum_{k=1}^{3} 3k^2$.

29. Evaluate $\sum_{k=3}^{5} (2k + 1)$.

30. Find the seventh term of a geometric progression with a first term of $\frac{1}{27}$ and a common ratio of 3.

31. Find the sum of the first ten terms of the progression

$$\frac{1}{64}, \frac{1}{32}, \frac{1}{16}, \ldots$$

32. Insert two geometric means between -3 and 192.

33. Find the sum of all the terms of the progression $9, 3, 1, \ldots$.

34. Use mathematical induction to prove that

$$1 + 4 + 7 + \cdots + (3n - 2) = \frac{n(3n - 1)}{2}$$

for all natural numbers n.

SAMPLE FINAL EXAMINATION

1. How many prime numbers are there in the interval from 40 to 50?
  a. 3   **b.** 4   **c.** 5   **d.** none of the above

2. The commutative property of multiplication is written symbolically as
  a. $ab = ba$ **b.** $(ab)c = a(bc)$ **c.** If $a = b$, then $b = a$ **d.** none of the above

3. If $a = 3$, $b = -2$, and $c = 6$, the value of $\dfrac{c - ab}{bc}$ is
  a. 1   **b.** 0   **c.** -1 **d.** none of the above

4. Evaluate $\dfrac{1}{2} + \dfrac{3}{4} \div \dfrac{5}{6}$
  a. $\dfrac{3}{2}$ **b.** $\dfrac{7}{2}$ **c.** $\dfrac{9}{8}$ **d.** none of the above

5. The expression $\left(\dfrac{a^2}{a^5}\right)^{-5}$ can be written as
  a. 15   **b.** a^{-15} **c.** a^{15} **d.** none of the above

6. Write 0.0000234 in scientific notation.
  a. 2.34×10^{-5} **b.** 2.34×10^5 **c.** $234. \times 10^{-7}$ **d.** none of the above

7. If $P(x) = 2x^2 - x - 1$, find $P(-1)$.
    a. 4   **b.** 0   **c.** 2   **d.** none of the above

8. Simplify: $(3x + 2) - (2x - 1) + (x - 3)$
    a. $2x - 2$ **b.** $2x$ **c.** 2   **d.** none of the above

9. Multiply: $(3x - 2)(2x + 3)$
    a. $6x^2 + 5x - 6$ **b.** $6x^2 - 5x - 6$ **c.** $6x^2 + 5x + 6$ **d.** none of the above

10. Divide: $2x + 1 \overline{)\, 2x^2 - 3x - 2}$
    a. $x + 2$ **b.** $x - 1$ **c.** $x - 2$ **d.** none of the above

11. Factor completely: $3ax^2 + 6a^2x$
    a. $3(ax^2 + 2a^2x)$ **b.** $3a(x^2 + 2ax)$ **c.** $3x(ax + 2a^2)$ **d.** none of the above

12. The sum of the prime factors of $x^4 - 16$ is
    a. $2x^2$ **b.** $x^2 + 2x + 4$ **c.** $4x + 4$ **d.** none of the above

13. The sum of the factors of $8x^2 - 2x - 3$ is
    a. $6x - 2$ **b.** $8x - 3$ **c.** $6x + 2$ **d.** none of the above

14. One of the factors of $27a^3 + 8$ is
    a. $3a - 2$ **b.** $9a^2 + 12a + 4$ **c.** $9a^2 + 6a + 4$ **d.** none of the above

15. Solve for x: $5x - 3 = -2x + 10$
    a. $\dfrac{11}{7}$ **b.** $\dfrac{13}{3}$ **c.** 1   **d.** none of the above

16. The sum of two consecutive odd integers is 44. The product of the integers is
    a. an even integer   **b.** 463   **c.** 483   **d.** none of the above

17. Solve for x: $2ax - a = b + x$
    a. $\dfrac{a + b}{2a}$ **b.** $\dfrac{a + b}{2a - 1}$ **c.** $a + b - 2a$ **d.** $-a + b + 1$

18. The smallest solution of the equation $6x^2 - 5x - 6 = 0$ is
    a. $-\dfrac{3}{2}$ **b.** $\dfrac{2}{3}$ **c.** $-\dfrac{2}{3}$ **d.** none of the above

19. The sum of the solutions of $|2x + 5| = 13$ is
    a. 4   **b.** 8   **c.** 12   **d.** none of the above

20. Solve for x: $-2x + 5 > 9$
    a. $x > 7$ **b.** $x < 7$ **c.** $x < -2$ **d.** $x > -2$

21. Solve for x: $|2x - 5| \le 9$
    a. $2 \le x \le 7$ **b.** $-2 \ge x$ and $x \le 7$ **c.** $x \le 7$ **d.** $-2 \le x \le 7$

22. Simplify: $\dfrac{x^2 + 5x + 6}{x^2 - 9}$
    a. $-\dfrac{2}{3}$ **b.** $\dfrac{x + 2}{x - 3}$ **c.** $\dfrac{-5x + 6}{9}$ **d.** none of the above

23. Simplify: $\dfrac{3x + 6}{x + 3} - \dfrac{x^2 - 4}{x^2 + x - 6}$
    a. $\dfrac{3(x + 2)(x + 2)}{(x + 3)(x + 3)}$ **b.** 3   **c.** $\dfrac{1}{3}$ **d.** none of these

24. The numerator of the sum $\dfrac{y}{x+y} + \dfrac{x}{x-y}$ is

    a. $x^2 + 2xy - y^2$ **b.** $y + x$ **c.** $y - x$ **d.** none of the above

25. Simplify: $\dfrac{\dfrac{1}{x} + \dfrac{1}{y}}{\dfrac{1}{y}}$

    a. $\dfrac{1}{xy}$ **b.** 1 **c.** $\dfrac{y+x}{x}$ **d.** none of the above

26. Solve for y: $\dfrac{2}{y+1} = \dfrac{1}{y+1} - \dfrac{1}{3}$

    a. 4 **b.** 3 **c.** -4 **d.** none of the above

27. The expression $x^{a/2}x^{a/5}$ can be expressed as

    a. $x^{a/7}$ **b.** $x^{7a/10}$ **c.** $x^{a/10}$ **d.** $x^{a/5}$

28. The product of $(x^{1/2} + 2)$ and $(x^{-1/2} - 2)$ is

    a. -3 **b.** $-3 + 2x$ **c.** $-3 - 2x^{1/2} + 2x^{-1/2}$ **d.** none of the above

29. Completely simplify $\sqrt{112a^3}$

    a. $a\sqrt{112a}$ **b.** $4a^2\sqrt{7a}$ **c.** $4\sqrt{7a^2}$ **d.** none of the above

30. Simplify and combine terms. $\sqrt{50} - \sqrt{98} + \sqrt{128}$

    a. $-2\sqrt{2} + \sqrt{128}$ **b.** $6\sqrt{2}$ **c.** $20\sqrt{2}$ **d.** $-4\sqrt{2}$

31. Rationalize the denominator and simplify $3/(2 - \sqrt{3})$

    a. $-6 - \sqrt{3}$ **b.** $6 + 3\sqrt{3}$ **c.** $2(2 + \sqrt{3})$ **d.** $6(2 + \sqrt{3})$

32. Write as a single radical $\sqrt{5}\sqrt[3]{2}$

    a. $\sqrt[5]{10}$ **b.** $\sqrt[6]{7}$ **c.** $\sqrt[6]{10}$ **d.** $\sqrt[6]{500}$

33. Solve for x: $\sqrt{x+7} - 2x = -1$

    a. 2 **b.** $-\dfrac{3}{4}$ **c.** solutions are extraneous   **d.** $2, -\dfrac{3}{4}$

34. The sum of the x and y intercepts of the graph of $2x + 3y = 6$ is

    a. $-\dfrac{2}{3}$ **b.** 0 **c.** 5 **d.** none of the above

35. The distance between the points $(-2, 3)$ and $(6, -8)$ is

    a. $\sqrt{185}$ **b.** $\sqrt{41}$ **c.** $\sqrt{57}$ **d.** none of the above

36. The slope of the line passing through $(3, -2)$ and $(5, -1)$ is

    a. $-\dfrac{1}{2}$ **b.** 2 **c.** -2 **d.** $\dfrac{1}{2}$

37. The graphs of the equations $\begin{cases} 2x - 3y = 4 \\ 3x + 2y = 1 \end{cases}$

    a. are parallel   **b.** are perpendicular   **c.** do not intersect   **d.** are the same line

38. The equation of the line passing through $(-2, 5)$ and $(6, 7)$ is

a. $y = -\dfrac{1}{4}x - \dfrac{11}{2}$ **b.** $y = \dfrac{1}{4}x + \dfrac{11}{2}$ **c.** $y = \dfrac{1}{4}x - \dfrac{11}{2}$ **d.** $y = -\dfrac{1}{4}x + \dfrac{11}{2}$

39. If $g(x) = x^2 - 3$, then $g(t + 1)$ is
    a. $t^2 - 2$ **b.** -2 **c.** $t - 2$ **d.** $t^2 + 2t - 2$

40. The inverse function of $y = 3x + 2$ is

    a. $y = \dfrac{x + 2}{3}$ **b.** $x = 3y - 2$ **c.** $y = \dfrac{x - 2}{3}$ **d.** $x = \dfrac{y - 2}{3}$

41. Assume that d varies directly with t. Find the constant of variation if $d = 12$ when $t = 3$.
    a. 36   **b.** 4   **c.** -36 **d.** none of the above

42. The graph of $y > 3x + 2$ contains no points in the
    a. 1st Quadrant   **b.** 2nd Quadrant   **c.** 3rd Quadrant   **d.** 4th Quadrant

43. The quadratic formula is

    a. $x = \dfrac{b \pm \sqrt{b^2 - 4ac}}{2a}$ **b.** $x = \dfrac{-b \pm \sqrt{b^2 - 4ac}}{a}$

    c. $x = \dfrac{-b \pm \sqrt{b - 4ac}}{a}$ **d.** $x = \dfrac{-b \pm \sqrt{b^2 - 4ac}}{2a}$

44. Write the complex number $(2 + 3i)^2$ in $a + bi$ form.
    a. $-5 + 12i$ **b.** -5 **c.** $13 + 12i$ **d.** 13

45. Write the complex number $\dfrac{i}{3 + i}$ in $a + bi$ form

    a. $\dfrac{1}{3} + 0i$ **b.** $\dfrac{1}{10} + \dfrac{3}{10}i$ **c.** $\dfrac{1}{8} + \dfrac{3}{8}i$ **d.** none of the above

46. The vertex of the parabola determined by $y = 2x^2 + 4x - 3$ is at point
    a. $(0, -3)$ **b.** $(2, 13)$ **c.** $(-1, -5)$ **d.** $(1, 3)$

47. Solve for x. $\dfrac{2}{x} < 3$

    a. $0 < x < \dfrac{2}{3}$ **b.** $x > \dfrac{2}{3}$ **c.** $x < 0$ or $x > \dfrac{2}{3}$ **d.** $x < 0$ and $x > \dfrac{2}{3}$

48. The equation of a circle with center at $(-2, 4)$ and radius of 4 units is
    a. $(x + 2)^2 + (y - 4)^2 = 16$ **b.** $(x + 2)^2 + (y - 4)^2 = 4$
    c. $(x + 2)^2 + (y - 4)^2 = 2$ **d.** $(x - 2)^2 + (y + 4)^2 = 16$

49. The sum of the solutions of the system $\begin{cases} \dfrac{4}{x} + \dfrac{2}{y} = 2 \\ \dfrac{2}{x} - \dfrac{3}{y} = -1 \end{cases}$ is

    a. 6   **b.** -6 **c.** 2   **d.** -4

50. The y value of the solution of the system $\begin{cases} 4x + 6y = 5 \\ 8x - 9y = 3 \end{cases}$ is

    a. $\dfrac{3}{4}$ **b.** $\dfrac{1}{3}$ **c.** $\dfrac{1}{2}$ **d.** $\dfrac{2}{3}$

51. The value of the determinant $\begin{vmatrix} 2 & -3 \\ 4 & 4 \end{vmatrix}$ is

    a. 0   **b.** 20   **c.** -4 **d.** -20

52. The value of z in the system $\begin{cases} x + y + z = 4 \\ 2x + y + z = 6 \\ 3x + y + 2z = 8 \end{cases}$ is

    a. 0   **b.** 1   **c.** 2   **d.** 3

53. $\log_a N = x$ means

    a. $a^x = N$ **b.** $a^N = x$ **c.** $x^N = a$ **d.** $N^a = x$

54. $\log_2 \dfrac{1}{32} =$

    a. 5   **b.** $\dfrac{1}{5}$ **c.** -5 **d.** $-\dfrac{1}{5}$

55. $\log 7 + \log 5 =$

    a. $\log 12$ **b.** $\log \dfrac{7}{12}$ **c.** $\log 2$ **d.** $\log 35$

56. $b^{\log_b x} =$

    a. b **b.** x **c.** 10   **d.** 0

57. Solve for y: $\log y + \log(y + 3) = 1$

    a. $-5, 2$ **b.** 2   **c.** 5   **d.** none of the above

58. The coefficient of the 3rd term in the expansion of $(a + b)^6$ is

    a. 30   **b.** $2!$ **c.** 15   **d.** 120

59. Compute $_7P_3$

    a. 35   **b.** 210   **c.** 21   **d.** none of the above

60. Compute $_7C_3$

    a. 35   **b.** 210   **c.** $7! \cdot 3!$ **d.** none of the above

61. Find the one-hundredth term of 2, 5, 8, 11, ...

    a. 301   **b.** 299   **c.** 297   **d.** 295

62. Find the sum of all the terms of $1, \frac{1}{3}, \frac{1}{9}, \frac{1}{27}, \ldots$

    a. 2   **b.** $\dfrac{2}{3}$ **c.** $\dfrac{3}{2}$ **d.** $\dfrac{5}{3}$

CAREERS AND MATHEMATICS

ACTUARY Why do young persons pay more for automobile insurance than older persons? How much should an insurance policy cost? How much should an organization contribute each year to its pension fund? Answers to these and similar questions are provided by actuaries who design insurance and pension plans and follow their experience to make sure that their plans are maintained on a sound financial basis.

Qualifications A good educational background for a beginning job in a large life or casualty company is a bachelor's degree in mathematics or statistics; a degree in actuarial science is even better. Companies prefer well-rounded individuals with a liberal arts background, including social science and communication courses, as well as a good technical background.

Job Outlook Shifts in the age distribution of the population will result in large increases in the number of people with established careers and family responsibilities. This group will account for increases in insurance and pension plan sales. The outlook for employment is very good through the mid-1990s. Job opportunities will be best for new college graduates who have passed at least two actuarial examinations while still in school and have a strong mathematical and statistical background.

Example application 1 An **annuity** is a sequence of equal payments made periodically over a length of time. The sum of the payments and the interest earned during the **term** of the annuity is called the **amount** of the annuity.

After a sales clerk works 6 months, her employer will begin an annuity for her, and will contribute $500 semiannually to a fund that pays 8% annual interest. After she has been employed 2 years, what will be the amount of her annuity?

Solution Because the payments are to be made semiannually, there will be four payments of $500, each earning a rate of 4% per six-month period. These payments will occur at the end of 6 months, 12 months, 18 months, and 24 months. The first payment that is to be made after 6 months will earn interest for three interest periods. Thus, the amount of the first payment after 2 years is $500(1.04)^3$. The amounts of each of the four payments after 2 years are shown in the following table.

Payment (at the end of period)	Amount of payment at the end of 2 years
1	$500(1.04)^3 = \$ 562.43$
2	$500(1.04)^2 = \$ 540.80$
3	$500(1.04)^1 = \$ 520.00$
4	$\$500 = \$ 500.00$
	$A_n = \$2123.23$

The amount of the annuity is the sum of the amounts of the individual payments. This sum is $2123.23.

Example application 2   The **present value** of the above annuity is the one lump principal that must be invested on day one of her employment at 8% annual interest, compounded semiannually, to grow to $2123.23 at the end of 2 years. Find the present value of the annuity.

Solution   A sum of $500(1.04)$^{-1}$ invested at 4% interest per six-month period will grow to be the first payment of $500 to be made after 6 months. A sum of $500(1.04)$^{-2}$ will grow to be the second payment of $500 to be made after 12 months. The present values of each of the four payments are shown in the following table.

Payment	Present value of each payment
1	$500(1.04)^{-1} = \$\ 480.77$
2	$500(1.04)^{-2} = \$\ 462.28$
3	$500(1.04)^{-3} = \$\ 444.50$
4	$500(1.04)^{-4} = \$\ 427.40$
	Present value $= \$1814.95$

The present value of the annuity is $1814.95. Verify that $1814.95 will grow to $2132.23 in 2 years at 8% annual interest compounded semiannually.

Exercises

1. Find the amount of an annuity if $1000 is paid semiannually for 2 years at the rate of 6% annual interest. Assume the first of the four payments is made immediately.

2. Find the present value of the annuity of Exercise 1.

3. Note that the amounts for the payments in the first table form a geometric progression. Verify the answer for the first example by using the formula for the sum of a geometric progression.

4. Note that the present values for the payments in the second table form a geometric progression. Verify the answer for the second example by using the formula for the sum of a geometric progression.

(Answers: **1.** $4309.14   **2.** $3717.10)

Exercise 1.1 (page 6)

1. $\subseteq$ **3.** $\in$ **5.** $\subseteq$ **7.** 2, 3, 5, 7 **9.** No such numbers
11. {2, 3, 4, 5, 6, 7, 8, 10} **13.** $\varnothing$ **15.** $\varnothing$ **17.** {3, 5, 7} **19.** infinite **21.** finite
23. infinite **25.** natural number, whole number, integer, rational number, real number
27. whole number, integer, rational number, real number **29.** rational number, real number
31. rational number, real number **33.** natural number, whole number, integer, rational number, real number
35. irrational number, real number **37.** even integer, composite number **39.** odd integer
41. even integer, prime number **43.** even integer **45.** odd integer, composite number
47. even integer **49.** true; $4 \times 5 = 20$ **51.** False; 0 is a whole number but not a natural number.
53. False; $3 + 5 = 8$ and 8 is a composite. **55.** true; $6 + 8 = 14$
57. false; $5 + 9 = 14$ and 14 is even. **59.** false; $4 + 9 = 13$ and 13 is prime.
61. true; $5 \times 11 = 55$ **63.** true; $0 + 8 = 8$
65. true; all other even natural numbers are divisible by 2. **67.** true; $7 = \frac{7}{1} = \frac{14}{2} = \frac{21}{3}$, and so on
69. 0.875; terminating decimal **71.** $-0.733\ldots$; repeating decimal **73.** $\frac{1}{2}$ **75.** $\frac{3}{4}$

Exercise 1.2 (page 13)

1. $=$ **3.** $\neq$ **5.** symmetric property of equality **7.** reflexive property of equality
9. transitive property of equality **11.** substitution property of equality
13. $5 + (2 + x) = (5 + 2) + x = 7 + x$ **15.** $5(3b) = (5 \cdot 3)b = 15b$
17. $(3 + b) + 7 = 7 + (3 + b) = (7 + 3) + b = 10 + b$ **19.** $(3y)2 = 2(3y) = (2 \cdot 3)y = 6y$
21. $3(x + 2) = 3x + 3 \cdot 2 = 3x + 6$ **23.** $5(x + y + 4) = 5x + 5y + 5 \cdot 4 = 5x + 5y + 20$
25. closure property of multiplication **27.** commutative property of addition
29. distributive property **31.** commutative property of addition **33.** additive identity property
35. multiplicative inverse property **37.** associative property of addition
39. commutative property of multiplication **41.** The additive inverse of 1 is -1.
43. The additive inverse of -8 is 8. **45.** The additive inverse of 0 is 0.
47. The additive inverse of π is $-\pi$. **49.** The additive inverse of 10 is -10.
51. The additive inverse of -3 is 3. **53.** The multiplicative inverse of 1 is 1.
55. The multiplicative inverse of $\frac{1}{2}$ is 2. **57.** The multiplicative inverse of -0.25 (or $-\frac{1}{4}$) is -4.
59. 0 does not have a multiplicative inverse.
61. $(a + b) + c = a + (b + c)$ associative property of addition
$\qquad = a + (c + b)$ commutative property of addition
63. $(b + c)a = a(b + c)$ commutative property of multiplication
$\qquad = ab + ac$ distributive property
$\qquad = ba + ac$ commutative property of multiplication
$\qquad = ba + ca$ commutative property of multiplication

Exercise 1.3 (page 19)

1. (number line with points at 2, 3, 5, 7; marks 0 1 2 3 4 5 6 7 8 9 10)

3. (number line with points at 11, 13, 15, 17, 19; marks 10 11 12 13 14 15 16 17 18 19 20)   **5.** <   **7.** >   **9.** <

11. $12 < 19$ **13.** $-5 \geq -6$ **15.** $-3 \leq 5$ **17.** $x > 3$ **19.** $z < 4$ **21.** $x \geq 7$

23. $x \leq -3$ **25.** (number line, open circle at 3, arrow left)   **27.** (number line, open circles at -3 and 2)   **29.** (number line, open circles at 0 and 5)   **31.** (number line, open circle at -2, arrow right)

33. (number line, closed points at -6 and 9)   **35.** (number line, open circle at 2, closed at 4)   **37.** (number line, closed points at 2 and 4)   **39.** (number line, closed at -4, open at 8)

41. (number line, closed points at 2 and 4)   **43.** 20   **45.** -6 **47.** 5   **49.** 7   **51.** 3   **53.** 99

55. all positive numbers and 0   **57.** Both are 0.   **59.** yes   **61.** 3 or -3

Exercise 1.4 (page 29)

1. -8 **3.** -9 **5.** -6 **7.** -5 **9.** 6   **11.** 17   **13.** -93 **15.** -2
17. -7 **19.** 0   **21.** -12 **23.** 21   **25.** -2 **27.** 4   **29.** -1 **31.** 0
33. 2   **35.** -13 **37.** 1   **39.** 4   **41.** -1 **43.** -4 **45.** -12
47. -20 **49.** 1   **51.** -8 **53.** -9 **55.** $-\frac{5}{4}$ **57.** $-\frac{1}{8}$ **59.** 4   **61.** 14
63. $\frac{20}{21}$ **65.** $22.17 + 39.56 = 61.73$; \$61.73   **67.** $7 + (-3) = 4$; $+4$ degrees
69. $437 + 25 + 37 + 45 + (-17) + (-83) + (-22) = 422$;   \$422
71. $15(-30) = -450$;   Harry would lose \$450.   **73.** $-120(-12) = 1440$;   1440 gal   **75.** equal
77. $\frac{4}{5}$ **79.** $-\frac{2}{3}$ **81.** $\frac{3}{20}$ **83.** $\frac{14}{9}$ **85.** $\frac{5}{27}$ **87.** 2   **89.** $-\frac{17}{45}$ **91.** $-\frac{17}{35}$
93. 16   **95.** 26 lbs

REVIEW EXERCISES (page 34)

1. true   **3.** false; $B \subseteq A$ **5.** true   **7.** integer, rational number, real number
9. irrational number, real number   **11.** rational number, real number
13. $5 - 5 = 0$, whole number, integer, rational number, real number   **15.** even number, composite number
17. odd number   **19.** odd number   **21.** odd number   **23.** distributive property
25. commutative property of addition   **27.** additive identity property
29. reflexive property of equality   **31.** associative property of multiplication
33. multiplicative inverse property   **35.** -1; 1   **37.** 0;   no multiplicative inverse exists.
39. $a < 4$ **41.** $10 \geq 3$ **43.** (number line, closed points at 25 and 29; marks 20 21 22 23 24 25 26 27 28 29 30)

45. (number line, closed point at -4, arrow right)   **47.** (number line, open circles at -2 and 3)   **49.** (number line, open circle at 2, arrow right)   **51.** 0   **53.** -8 **55.** -2

57. 3   **59.** -5 **61.** -12 **63.** 12   **65.** -4 **67.** 4   **69.** -1
71. $-\frac{19}{6}$ **73.** $-\frac{4}{3}$ **75.** $-\frac{3}{4}$ **77.** $-\frac{27}{8}$

CHAPTER ONE TEST (page 35)

1. $\in$ **3.** $\subseteq$ **5.** $\{1, 2, 3, 4, 5, 6\}$ **7.** $\varnothing$ **9.** 1, 2, 5   **11.** $-2, 0, 1, 2, \frac{6}{5}, 5$
13. $0.77\overline{7}$ **15.** commutative property of addition   **17.** associative property of addition
19. (number line, closed points at -3, -1, 1, 3, 5; marks -5 -4 -3 -2 -1 0 1 2 3 4 5 6)   **21.** (number line, closed at -2, open at 4)   **23.** -3

25. -6 **27.** 6   **29.** 1

Exercise 2.1   (page 46)

1. 9   **3.** -9 **5.** 9   **7.** $-32x^5$ **9.** $-128x^7$ **11.** $64x^6$ **13.** $\frac{1}{25}$
15. $-\frac{1}{25}$ **17.** $\frac{1}{25}$ **19.** 1   **21.** -1 **23.** 1   **25.** x^5 **27.** k^7 **29.** x^{10}

31. p^{10} **33.** a^4b^5 **35.** x^5y^4 **37.** x^{28} **39.** $\frac{1}{b^{72}}$ **41.** $x^{12}y^8$ **43.** $\frac{s^3}{r^9}$

45. a^{20} **47.** $-\frac{1}{d^3}$ **49.** x^2y^{10} **51.** $\frac{a^{15}}{b^{10}}$ **53.** $\frac{a^6}{b^4}$ **55.** a^5 **57.** c^7

59. m **61.** a^4 **63.** m **65.** $\frac{64b^{12}}{27a^9}$ **67.** 1   **69.** $\frac{-b^3}{8a^{21}}$ **71.** $\frac{27}{8a^{18}}$

73. $\frac{1}{9x^3}$ **75.** $\frac{-3y^2}{x^2}$ **77.** $\frac{-n^3}{48}$ **79.** 108   **81.** -8 **83.** $-\frac{1}{216}$ **85.** $\frac{1}{324}$

87. $\frac{27}{2}$ **89.** $\frac{27}{8}$ **91.** a^{n-1} **93.** b^{3n-9} **95.** a^{-n-1} or $\frac{1}{a^{n+1}}$ **97.** a^{-n+2} or $\frac{1}{a^{n-2}}$

107. $2^2 + 2^3 \neq 2^{2+3}$

Exercise 2.2   (page 50)

1. 3.9×10^3 **3.** 7.8×10^{-3} **5.** 1.76×10^7 **7.** 9.6×10^{-6} **9.** 3.23×10^7
11. 6.0×10^{-4} **13.** 5.27×10^3 **15.** 3.17×10^{-4} **17.** 7.31×10^6 **19.** 9.137×10^1
21. 270   **23.** 0.00323   **25.** 796,000   **27.** 0.00037   **29.** 5.23   **31.** 23,650,000,000
33. 0.04   **35.** 14,400,000   **37.** 6000   **39.** 0.64   **41.** 1.1916×10^8 cm/hr
43. 1.67248×10^{-18} g   **45.** 1.48896×10^{10} in.   **47.** 5.58×10^7 mi to 1.674×10^9 mi
49. about 2.49×10^{13} mi   **51.** about 1.29×10^{13} **53.** about 5.67×10^{10}
55. about 3.62×10^{25}

Exercise 2.3   (page 56)

1. monomial   **3.** trinomial   **5.** binomial   **7.** monomial   **9.** 2   **11.** 8
13. 10   **15.** 1   **17.** 0   **19.** 2   **21.** $3x^3 - 4x^2 + 2x + 7$
23. $-x^4 + 2x^3y - 3x^2y^2 + 5xy^3 + y^4$ **25.** $-4x^6y + 3x^3z - 4x + 3$ **27.** $P(0) = 2$
29. $P(4) = 38$ **31.** $P(1) = -2$ **33.** $P(-2) = -23$ **35.** $P(t) = -3t^2 + 4t - 3$
37. $P(-x) = -3x^2 - 4x - 3$ **39.** $P(2x) = -12x^2 + 8x - 3$ **41.** $P(P(0)) = P(-3) = -42$
43. $P(P(-1)) = P(-10) = -343$ **45.** -2 **47.** 13   **49.** 35   **51.** -90 **53.** 84
55. $-\frac{6}{19}$ **57.** $\frac{78}{77}$ **59.** about -34.225 **61.** about 0.1728   **63.** about -0.1227
65. like terms; $10x$ **67.** unlike terms   **69.** like terms; 12   **71.** like terms; $12x^2$
73. $12x$ **75.** $2x^3y^2z$ **77.** $-7x^2y^3 + 3xy^4$ **79.** $10x^4y^2$ **81.** $x^2 - 5x + 6$
83. $-5a^2 + 4a + 4$ **85.** $-y^3 + 4y^2 + 6$ **87.** $4x^2 - 11$ **89.** $3x^3 - x + 13$
91. $4y^2 - 9y + 3$ **93.** $6x^3 - 6x^2 + 14x - 17$ **95.** $-9y^4 + 3y^3 - 15y^2 + 20y + 12$
97. $x^2 - 8x + 22$ **99.** $-3y^3 + 18y^2 - 28y + 35$ **101.** $8x^3 - x^2$ **103.** $-16x^3 - 27x^2 - 12x$
105. $-8z^2 - 40z + 54$ **107.** $-2x + 9$ **109.** $-8x^2 - 2x + 2$

Exercise 2.4   (page 62)

1. $-6a^3b$ **3.** $-15a^2b^2c^3$ **5.** $-120a^9b^3$ **7.** $12x^5y^9$ **9.** $25x^8y^8$ **11.** $-405x^7y^4$
13. $100x^{10}y^6$ **15.** $3x + 6$ **17.** $-a^2 + ab$ **19.** $3x^3 + 9x^2$ **21.** $-6x^3 + 6x^2 - 4x$
23. $10a^6b^4 - 25a^2b^6$ **25.** $7r^3st + 7rs^3t - 7rst^3$ **27.** $-12x^4y^3 + 16x^3y^4 - 4x^2y^5$
29. $-12m^4n^2 - 12m^3n^3$ **31.** $x^2 + 4x + 4$ **33.** $a^2 - 8a + 16$ **35.** $a^2 + 2ab + b^2$
37. $4x^2 - 4xy + y^2$ **39.** $x^2 - 4$ **41.** $a^2 - b^2$ **43.** $x^2 + 5x + 6$ **45.** $z^2 - 9z + 14$
47. $2a^2 - 3a - 2$ **49.** $6y^2 - 5yz + z^2$ **51.** $2x^2 + xy - 6y^2$ **53.** $9 + 6x - 8x^2$

55. $9x^2 - 6xy - 3y^2$ **57.** $8a^2 + 14a - 15$ **59.** $u^2 - 2uv + v^2$ **61.** $4x^2 + 4x + 1$
63. $9x^2 + 12xy + 4y^2$ **65.** $6y^3 + 11y^2 + 9y + 2$ **67.** $6a^3 - 7a^2b + 6ab^2 - 2b^3$
69. $2a^2 + ab - b^2 - 3bc - 2c^2$ **71.** $x^2 + 4xy + 6xz + 4y^2 + 12yz + 9z^2$ **73.** $r^4 - 2r^2s^2 + s^4$
75. $4x^3 - 8x^2 - 9x + 6$ **77.** $-27x^6 + 135x^5 - 225x^4 + 125x^3$ **79.** $a^3 - 3a^2b - ab^2 + 3b^3$
81. $2x^5 + x$ **83.** $\dfrac{3x}{y^4z} - \dfrac{x^6}{y^{10}z^2}$ **85.** $\dfrac{1}{x^2} - y^2$ **87.** $2x^3y^3 - \dfrac{2}{x^3y^3} + 3$

89. $8x^6 + 24x^4y^2 + 18x^2y^4 - \dfrac{12x^4}{y^2} - 36x^2 - 27y^2$ **91.** $x^{3n} - x^{2n}$ **93.** $x^{2n} - 1$

95. $x^{2n} - x^ny^{-n} - x^ny^n + 1$ **97.** $x^{4n} - y^{4n}$ **99.** $-3x^{-n}y^{2n} + 2x^ny^{-2n} + 5$
101. $x^{2n} - y^{2n} + 2y^n - 1$ **103.** $5x^2 - 36x + 7$ **105.** $21x^2 - 6xy + 29y^2$
107. $24y^2 - 4yz + 21z^2$ **113.** $9.2127x^2 - 7.7956x - 36.0315$ **115.** $299.29y^2 - 150.51y + 18.9225$

Exercise 2.5   (page 70)

1. $\dfrac{y}{2x^3}$ **3.** $\dfrac{3b^4}{4a^4}$ **5.** $\dfrac{-5}{7xy^7t^2}$ **7.** $\dfrac{13a^nb^{2n}c^{3n-1}}{3}$ **9.** $\dfrac{2x}{3} - \dfrac{x^2}{6}$ **11.** $\dfrac{2xy^2}{3} + \dfrac{x^2y}{3}$

13. $\dfrac{x^4y^4}{2} - \dfrac{x^3y^9}{4} + \dfrac{3}{4xy^2}$ **15.** $\dfrac{b^2}{4a} - \dfrac{a^3}{2b^4} + \dfrac{3}{4ab}$ **17.** $1 - 3x^ny^n + 6x^{2n}y^{2n}$ **19.** $x + 2$

21. $x + 7$ **23.** $3x - 5 + \dfrac{3}{2x + 3}$ **25.** $3x^2 + x + 2 + \dfrac{8}{x - 1}$ **27.** $2x^2 + 5x + 3 + \dfrac{4}{3x - 2}$

29. $3x^2 + 4x + 3$ **31.** $a + 1$ **33.** $2y + 2$ **35.** $6x - 12$ **37.** $3x^2 - x + 2$

39. $4x^3 - 3x^2 + 3x + 1$ **41.** $a^2 + a + 1 + \dfrac{2}{a - 1}$ **43.** $5a^2 - 3a - 4$ **45.** $6y - 12$

47. $16x^4 - 8x^3y + 4x^2y^2 - 2xy^3 + y^4$ **49.** $x^4 + x^2 + 4$ **51.** $x^2 + x + 1$ **53.** $x^2 + 3$

55. $9.8x + 16.4 + \dfrac{-36.5}{x - 2}$

Exercise 2.6   (page 75)

1. $x + 2$ **3.** $x - 3$ **5.** $x + 2$ **7.** $x - 7 + \dfrac{28}{x + 2}$ **9.** $3x^2 - x + 2$

11. $2x^2 + 4x + 3$ **13.** $6x^2 - x + 1 + \dfrac{3}{x + 1}$ **15.** $7.2x - 0.66 + \dfrac{0.368}{x - 0.2}$

17. $2.7x - 3.59 + \dfrac{0.903}{x + 1.7}$ **19.** $9x^2 - 513x + 29{,}241 + \dfrac{-1{,}666{,}762}{x + 57}$ **21.** $P(1) = -1$

23. $P(-2) = -37$ **25.** $P(3) = 23$ **27.** $P(0) = -1$ **29.** $Q(-1) = 2$ **31.** $Q(2) = -1$

33. $Q(3) = 18$ **35.** $Q(-3) = 174$ **37.** -8 **39.** 59 **41.** 44 **43.** $\dfrac{29}{32}$ **45.** 64

REVIEW EXERCISES   (page 77)

1. 729 **3.** -64 **5.** $-6x^6$ **7.** $\dfrac{1}{x}$ **9.** $27x^6$ **11.** $-32x^{10}$ **13.** $\dfrac{1}{x^{10}}$

15. $\dfrac{x^6}{9}$ **17.** x^2 **19.** $\dfrac{1}{a^5}$ **21.** $\dfrac{1}{y^7}$ **23.** $\dfrac{1}{x}$ **25.** $9x^4y^6$ **27.** $\dfrac{64y^9}{27x^6}$

29. 1.93×10^{10} **31.**   72,000,000   **33.**   6   **35.** $-t^2 - 4t + 6$ **37.**   5   **39.** $5x^2 + 3x$
41. $4x^2 - 9x + 19$ **43.** $-16a^3b^3c$ **45.** $2x^4y^3 - 8x^2y^7$ **47.** $16x^2 + 14x - 15$

49. $15x^4 - 22x^3 + 73x^2 - 50x + 50$ **51.** $3x^2 - 13x + 42 + \dfrac{-124}{x + 3}$ **53.** $x^4 + x^3 + x^2 + x + 1$

55. $2x^2 + 3x + 1 + \dfrac{10}{3x - 2}$ **57.** $x^2 - 2x + \dfrac{6x - 3}{x^2 + 2x + 3}$ **59.** $x^2 + 4x + 3$ **61.** $P(2) = 11$

CHAPTER TWO TEST   (page 79)

1. x^8 **3.** $\dfrac{1}{m^8}$ **5.**   3   **7.** 4.7×10^6 **9.**   653,000   **11.** 3.76×10^5 km

13. $P(2) = -9$ **15.**   5   **17.** $5y^2 + y - 1$ **19.** $10a^2 + 22$ **21.** $-6x^4yz^4$

23. $z^2 - 16$ **25.** $2x^3 - x^2 - 7x - 3$ **27.** $\dfrac{-6x}{y} + \dfrac{4x^2}{y^2} - \dfrac{3}{y^3}$ **29.** -7

Exercise 3.1   (page 85)

1. $2 \cdot 3$ **3.** $3^3 \cdot 5$ **5.** 2^7 **7.** $5^2 \cdot 13$ **9.**   12   **11.**   2   **13.** $4a^2$
15. $6xy^2z^2$ **17.** $2(x + 4)$ **19.** $2x(x - 3)$ **21.** $5xy(1 + 2y)$ **23.** $5x^2y(3 - 2y)$
25. $3rt^4(4rs^3 + 5t^2)$ **27.** $6rs(4rs^2 - 2r^2st + t^2)$ **29.** $9x^7y^3(5x^3 - 7y^4 + 9x^3y^7)$ **31.**   prime
33. $8a^3b^3(3b^2 + 4a^2 - 8a^2b^2c^5)$ **35.** $(x + y)(4 + t)$ **37.** $(a - b)(r - s)$
39. $(m + n + p)(3 + x)$ **41.** $b(x - a)$ **43.** $(x + y)(x + y + z)$ **45.** $(u + v)(1 - u - v)$
47. $-(x - y)$ **49.** $-6ab(3a + 2b)$ **51.** $-7u^2v^3z^2(9uv^3z^7 - 4v^4 + 3uz^2)$ **53.** $-(x + y)(a - b)$
55. $-8x^2(m - n + p)(4x + 5 - 2x^2)$ **57.** $x^2(x^n + x^{n+1})$ **59.** $y^n(y^2 - y^3)$ **61.** $x^{-2}(x^6 - 5x^8)$
63. $y^{-2n}(y^{4n} + y^{2n} + 1)$ **69.**   relatively prime   **71.**   not relatively prime
73.   relatively prime   **75.**   relatively prime

Exercise 3.2   (page 90)

1. $(x + 2)(x - 2)$ **3.** $(3y + 8)(3y - 8)$ **5.**   prime   **7.** $(25a + 13b^2)(25a - 13b^2)$
9. $(9a^2 + 7b)(9a^2 - 7b)$ **11.** $(6x^2y + 7z^2)(6x^2y - 7z^2)$ **13.** $(x + y + z)(x + y - z)$
15. $(a - b + c + d)(a - b - c - d)$ **17.** $(x^2 + y^2)(x + y)(x - y)$
19. $(16x^2y^2 + z^4)(4xy + z^2)(4xy - z^2)$ **21.** $2(x + 12)(x - 12)$ **23.** $2x(x + 4)(x - 4)$
25. $5x(x + 5)(x - 5)$ **27.** $t^2(rs + x^2y)(rs - x^2y)$ **29.** $2(c - d)(x + 3)(x - 3)$
31. $(r + s)(r^2 - rs + s^2)$ **33.** $(x - 2y)(x^2 + 2xy + 4y^2)$ **35.** $(4a - 5b^2)(16a^2 + 20ab^2 + 25b^4)$
37. $27(xy^2 + 2z^3)(x^2y^4 - 2xy^2z^3 + 4z^6)$ **39.** $[3a + (x + y)^2][9a^2 - 3a(x + y)^2 + (x + y)^4]$
41. $(x^2 + y^2)(x^4 - x^2y^2 + y^4)$ **43.** $5(x + 5)(x^2 - 5x + 25)$ **45.** $4x^2(x - 4)(x^2 + 4x + 16)$
47. $2u^2(4v - t)(16v^2 + 4vt + t^2)$ **49.** $(a + b)(x + 3)(x^2 - 3x + 9)$
51. $6[(a + b) - z][(a + b)^2 + z(a + b) + z^2]$ **53.** $(x^m + y^{2n})(x^m - y^{2n})$
55. $(10a^{2m} + 9b^n)(10a^{2m} - 9b^n)$ **57.** $(x^n - 2)(x^{2n} + 2x^n + 4)$ **59.** $(a^m + b^n)(a^{2m} - a^mb^n + b^{2n})$
61. $2(x^{2m} + 2y^m)(x^{4m} - 2x^{2m}y^m + 4y^{2m})$

Exercise 3.3   (page 100)

1. $(x + 1)(x + 1)$ **3.** $(a - 9)(a - 9)$ **5.** $(2y + 1)(2y + 1)$ **7.** $(3b - 2)(3b - 2)$
9. $(3z + 4)(3z + 4)$ **11.** $(x + 3)(x + 2)$ **13.** $(x - 2)(x - 2)$ **15.**   prime
17. $(x - 6)(x + 5)$ **19.** $(a + 10)(a - 5)$ **21.** $(y - 7)(y + 3)$ **23.** $3(x + 7)(x - 3)$
25. $b^2(a - 11)(a - 2)$ **27.** $x^2(b - 7)(b - 5)$ **29.** $-(a - 8)(a + 4)$ **31.** $-3(x - 3)(x - 2)$
33. $-4(x - 5)(x + 4)$ **35.** $(3y + 2)(2y + 1)$ **37.** $(4a - 3)(2a + 3)$ **39.** $(3x - 4)(2x + 1)$

41. prime   **43.** $(4x - 3)(2x - 1)$ **45.** $(3z - 4)(2z + 5)$ **47.** $(a + b)(a - 4b)$
49. $(2y - 3t)(y + 2t)$ **51.** $x(3x - 1)(x - 3)$ **53.** $-(3a + 2b)(a - b)$ **55.** $(3t + 2)(3t - 1)$
57. $(3x - 2)(3x - 2)$ **59.** $-(2x - 3)(2x - 3)$ **61.** $(5x - 1)(3x - 2)$ **63.** $5(a - 3b)(a - 3b)$
65. $z(8x^2 + 6xy + 9y^2)$ **67.** $(15x - 1)(x + 5)$ **69.** $x^2(7x - 8)(3x + 2)$
71. $(3xy - 4z)(2xy - 3z)$ **73.** $(x^2 + 5)(x^2 + 3)$ **75.** $(y^2 - 10)(y^2 - 3)$
77. $(a + 3)(a - 3)(a + 2)(a - 2)$ **79.** $(z^2 + 3)(z + 2)(z - 2)$ **81.** $(x^3 + 3)(x + 1)(x^2 - x + 1)$
83. $(y - 2)(y^2 + 2y + 4)(y - 1)(y^2 + y + 1)$ **85.** $(x + 2)(x + 2)$ **87.** $(a + b + 4)(a + b - 6)$
89. $(3x + 3y + 4)(2x + 2y - 5)$ **91.** $(5x - 9)(x - 3)$ **93.** $(4x - 5)(x - 2)$ **95.** $(x^n + 1)(x^n + 1)$
97. $(2a^{3n} + 1)(a^{3n} - 2)$ **99.** $(x^{2n} + y^{2n})(x^{2n} + y^{2n})$ **101.** $(3x^n - 1)(2x^n + 3)$

Exercise 3.4   (page 104)

1. $(x + y)(a + b)$ **3.** $(x + 2)(x + y)$ **5.** $(3 - c)(c + d)$ **7.** $(a + b)(a - 4)$
9. $(a + b)(x - 1)$ **11.** $(x + y)(x + y + z)$ **13.** $y(x + y)(x + y + 2z)$
15. $n(2n - p + 2m)(n^2p - 1)$ **17.** $(x - 1)(x^3 - x^2 + 1)$ **19.** $(a - b)(a + b + 2)$
21. $(y + 4)(2 + a)$ **23.** $(x + 1)(x - 1 + 3x^4 + 3x^3)$ **25.** $(x + 2 + y)(x + 2 - y)$
27. $(x + 1 + 3z)(x + 1 - 3z)$ **29.** $(2a - b + c)(2a - b - c)$ **31.** $(a + 4 + b)(a + 4 - b)$
33. $(2x + y + z)(2x + y - z)$ **35.** $(x^2 + x + 3)(x^2 - x + 3)$ **37.** $(2a^2 + a + 1)(2a^2 - a + 1)$
39. $(a^2 + 2ab + 2b^2)(a^2 - 2ab + 2b^2)$ **41.** $(x^2 + 3x + 1)(x^2 - 3x + 1)$
43. $(x + 2)(x^2 - 2x + 4)(x - 2)(x^2 + 2x + 4)$ **45.** $(a - 16)(a - 1)$ **47.** $(2u + 3)(u + 1)$
49. $(4r - 3s)(5r + 2s)$

Exercise 3.5   (page 107)

1. $(x + 4)(x + 4)$ **3.** $(2xy - 3)(4x^2y^2 + 6xy + 9)$ **5.** $(x - t)(y + s)$ **7.** $(5x + 4y)(5x - 4y)$
9. $(6x + 5)(2x + 7)$ **11.** $2(3x - 4)(x - 1)$ **13.** $(8x - 1)(7x - 1)$ **15.** $y^2(2x + 1)(2x + 1)$
17. $(x + a^2y)(x^2 - a^2xy + a^4y^2)$ **19.** $2(x - 3)(x^2 + 3x + 9)$ **21.** $(a + b)(f + e)$
23. $(2x + 2y + 3)(x + y - 1)$ **25.** $(25x^2 + 16y^2)(5x + 4y)(5x - 4y)$ **27.** $36(x^2 + 1)(x + 1)(x - 1)$
29. $2(x^2 + y^2)(x^4 - x^2y^2 + y^4)$ **31.** $(a + 3)(a - 3)(a + 2)(a - 2)$ **33.** $(x + 3 + y)(x + 3 - y)$
35. $(2x + 1 + 2y)(2x + 1 - 2y)$ **37.** $(z^2 + 4 + z)(z^2 + 4 - z)$ **39.** $(x + 1)(x^2 - x + 1)(x + 1)(x - 1)$
41. $(x + 3)(x - 3)(x + 2)(x^2 - 2x + 4)$ **43.** $2z(x + y)(x - y)(x - y)(x^2 + xy + y^2)$

REVIEW EXERCISES   (page 108)

1. $4(x + 2)$ **3.** $5xy^2(xy - 2)$ **5.** $-4x^2y^3z^2(2z^2 + 3x^2)$ **7.** $9x^2y^3z^2(3xz + 9x^2y^2 - 10z^5)$
9. $5x^2(x + y)^3(1 - 3x^2 - 3xy)$ **11.** $(z + 4)(z - 4)$ **13.** $(xy^2 + 8z^3)(xy^2 - 8z^3)$
15. $(x + z + t)(x + z - t)$ **17.** $2(x^2 + 7)(x^2 - 7)$ **19.** $(x + 7)(x^2 - 7x + 49)$
21. $8(y - 4)(y^2 + 4y + 16)$ **23.** $(y + 20)(y + 1)$ **25.** $-(x + 7)(x - 4)$ **27.** $(4a - 1)(a - 1)$
29. prime   **31.** $y(y + 2)(y - 1)$ **33.** $-3(x + 2)(x + 1)$ **35.** $3(5x + y)(x - 4y)$
37. $(8x + 3y)(3x - 4y)$ **39.** $(x^2 + x + 7)(x^2 - x + 7)$ **41.** $(x + 2)(y + 4)$
43. $(x^2 + 4)(x^2 + y)$ **45.** $(z - 2)(z + x + 2)$ **47.** $(x + 2 + 2p^2)(x + 2 - 2p^2)$
49. $(a^2 + 4 + 2a)(a^2 + 4 - 2a)$

CHAPTER THREE TEST   (page 109)

1. $2^2 \cdot 3 \cdot 19$ **3.** $3xy(y + 2x)$ **5.** $(u - v)(r + s)$ **7.** $y^n(x^2y^2 + 1)$ **9.** $(x + 7)(x - 7)$
11. $4(y^2 + 4)(y + 2)(y - 2)$ **13.** $(b - 3)(b^2 + 3b + 9)$ **15.** $(3z + 4t^2)(3z - 4t^2)$
17. $(x + 5)(x + 3)$ **19.** $(2a + 3)(a - 4)$ **21.** $3(2u - 1)(u + 2)$ **23.** $(3x - y)(2x + y)$
25. $(x^n + 1)(x^n + 1)$ **27.** $(a - y)(x + y)$ **29.** $(x + 3 + y)(x + 3 - y)$
31. $(x + 2)(x - 2)(x + 4)(x - 4)$

CUMULATIVE REVIEW EXERCISES (CHAPTERS 1–3)   (page 109)

1. rational number, real number, positive number   **3.** odd integer, prime number   **5.** -2
7. $\frac{7}{8}$ **9.** transitive property of equality   **11.** commutative property of addition   **13.** 5
15. 22   **17.** -4 **19.** $-\frac{1}{8}$ **21.** $-\frac{169}{72}$ **23.** $x^8 y^{12}$ **25.** $-b^3/a^2$
27. 4.97×10^{-6} **29.** trinomial   **31.** 18   **33.** $4x^2 - 4x + 14$ **35.** $6x^2 - 7x - 20$
37. $x + 4$ **39.** $3rs^3(r - 2s)$ **41.** $(x + y)(u + v)$ **43.** $(2x - 3y^2)(4x^2 + 6xy^2 + 9y^4)$
45. $(3x - 5)(3x - 5)$ **47.** $(3a + 2b)(9a^2 - 6ab + 4b^2)$ **49.** $(x + 5 + y^2)(x + 5 - y^2)$

Exercise 4.1   (page 122)

1. all real numbers   **3.** all real numbers except 2   **5.** all real numbers except 2 and -3
7. all real numbers except 0, 10, and -10 **9.** $x = 2$ **11.** $z = 14$ **13.** $u = 3$
15. $x = 28$ **17.** $x = \frac{2}{3}$ **19.** $x = 6$ **21.** $x = -8$ **23.** $r = 2$ **25.** $a = -2$
27. $a = 3$ **29.** $x = 4$ **31.** $a = -4$ **33.** $y = -6$ **35.** $y = 6$ **37.** $y = 13$
39. $x = -8$ **41.** no solution   **43.** $x = 24$ **45.** $x = 6$ **47.** $x = 4$ **49.** $x = 3$
51. $a = 0$ **53.** $a = 6$ **55.** an identity   **57.** $x = 2000$ **59.** $x = 5$ **61.** $\frac{1}{3}$
63. $-\frac{691}{1980}$ **65.** 37 and 38   **67.** 32   **69.** 29   **71.** 3 ft, 5 ft, and 9 ft   **73.** 32 ft
75. 12 cm by 24 cm   **77.** 315 sq ft   **79.** 20 ft by 45 ft

Exercise 4.2   (page 131)

1. 6 cm   **3.** 6 in. by 4 in.   **5.** 50 m   **7.** 5 ft from the fulcrum   **9.** 90 lb
11. 2 ft in front of Jim   **13.** 10 nickels, 20 dimes, 40 pennies   **15.** 13 nickels and 11 quarters
17. 3 nickels and 5 dimes   **19.** Heidi invested $2000 at 8% and $10,000 at 9%   **21.** 10.8%
23. $4200   **25.** 25 student and 175 adult tickets   **27.** 3 hr   **29.** 1.5 hr
31. 4 hr at each rate   **33.** 20 lb of the $0.95 candy and 10 lb of the $1.10 candy   **35.** 10 oz
37. 2 gal of cream

Exercise 4.3   (page 135)

1. $w = \dfrac{A}{l}$ **3.** $r^2 = \dfrac{A}{\pi}$ **5.** $B = \dfrac{3V}{h}$ **7.** $t = \dfrac{I}{pr}$ **9.** $w = \dfrac{p - 2l}{2}$ **11.** $b_1 = \dfrac{2A}{h} - b_2$

13. $x = z\sigma + \mu$ **15.** $x = \dfrac{y - b}{m}$ **17.** $n = \dfrac{l - a + d}{d}$ **19.** $r_1 = \dfrac{rr_2}{r_2 - r}$

21. $\Sigma x^2 = n(\sigma^2 + \mu^2)$ **23.** $r = \dfrac{S - a}{S - l}$ **25.** $n = \dfrac{-360}{a - 180} = \dfrac{360}{180 - a}$ **27.** $s = \dfrac{f(P - L)}{i}$

29. $r_2 = \dfrac{rr_1}{r_1 - r}$ **31.** $x = \dfrac{y - y_1 + mx_1}{m}$ **33.** $a = \dfrac{Hb}{2b - H}$ **35.** $a^2 = \dfrac{x^2 b^2}{b^2 - y^2}$

37. $h = x^2 - \dfrac{y - k}{a} = \dfrac{ax^2 - y + k}{a}$ **39.** $h = \dfrac{3V}{B_1 + B_2 + \sqrt{B_1 B_2}}$ **41.** $C = \dfrac{5F - 160}{9}$; $0°$, $21.1°$, $100°$

43. $n = \dfrac{C - 6.50}{0.07}$; 620 kwh, 980 kwh, 1700 kwh   **45.** 60 mi; 160 mi   **47.** $600

Exercise 4.4   (page 140)

1. $0, -2$ **3.** $4, -4$ **5.** $0, -1$ **7.** $0, 5$ **9.** $-3, -5$ **11.** $1, 6$ **13.** $3, 4$
15. $-2, -4$ **17.** $-\frac{1}{3}, -3$ **19.** $\frac{1}{2}, 2$ **21.** $1, -\frac{1}{2}$ **23.** $2, -\frac{1}{3}$ **25.** $3, 3$
27. $\frac{1}{4}, -\frac{3}{2}$ **29.** $2, -\frac{5}{6}$ **31.** $\frac{1}{3}, -1$ **33.** $2, \frac{1}{2}$ **35.** $\frac{1}{5}, -\frac{5}{3}$ **37.** $0, 0, -1$
39. $0, 7, -7$ **41.** $0, 7, -3$ **43.** $3, -3, 2, -2$ **45.** $0, -2, -3$ **47.** $0, \frac{5}{6}, -7$

49. $1, -1, -3$ **51.** $3, -3, -\frac{3}{2}$ **53.** $2, -2, -\frac{1}{3}$ **55.** 16 and 18, or -18 and -16
57. 6 and 7   **59.** 40 m   **61.** 4 units   **63.** 5 ft by 12 ft   **65.** 20 ft by 40 ft
67. $x^2 - 8x + 15 = 0$ **69.** $x^2 + 5x = 0$

Exercise 4.5   (page 144)

1. 8   **3.** 12   **5.** -2 **7.** -30 **9.** 50   **11.** $4 - \pi$ **13.** $|2|$ **15.** $|5|$
17. $|-2|$ **19.** $-|-4|$ **21.** $-|-7|$ **23.** $|x + 1|$ **25.** $8, -8$ **27.** $9, -3$
29. $4, -1$ **31.** $\frac{14}{3}, -6$ **33.** no solutions   **35.** $8, -4$ **37.** -8 **39.** $-4, -28$
41. $-2, -\frac{4}{5}$ **43.** $3, -1$ **45.** 0   **47.** $\frac{4}{3}$ **49.** no solutions

Exercise 4.6   (page 150)

1. [number line: open circle at 1, ray left]
3. [number line: closed circle at -2, ray right]
5. [number line: open circle at 2, ray left]
7. [number line: closed circle at 1, ray right]
9. [number line: open circle at $-\frac{8}{5}$, ray right]

11. [number line: closed circle at 20, ray left]
13. [number line: open circle at 10, ray right]
15. [number line: closed circle at -36, ray right]
17. [number line: open circle at -100, ray right]
19. [number line: closed circle at $\frac{45}{7}$, ray left]

21. [number line: open circles at -2 and 5, segment between]
23. [number line: closed at 8, open at 11, segment between]
25. [number line: closed at -4, open at 6, segment between]
27. no solutions

29. [number line: closed circles at -2 and 4, segment between]
31. [number line: open circles at 2 and 3, segment between]
33. [number line: closed at 1, open at $\frac{9}{4}$, segment between]
35. [number line: open circle at -15, ray left]

37. [number line: open circles at 2 and 7, segment between]
39. [number line: open circle at 1, ray left]
41. no solutions   **43.** $|7| + |-2| \geq |7 + (-2)|$

45. x and y must differ in sign.   **47.** no   **49.** no   **51.** profit $\leq$ \$15
53. 185 in. $\leq p \leq$ 260 in.   **55.** score ≥ 88

Exercise 4.7   (page 153)

1. [number line: open circles at -4 and 4, segment between]
3. [number line: closed circles at -21 and 3, segment between]
5. no solutions
7. [number line: closed circles at $-\frac{3}{2}$ and 2, segment between]

9. [number line: open circles at -1 and 1, rays out both sides]
11. [number line: closed circles at -12 and 36, segment between]
13. [number line: open circles at $-\frac{16}{3}$ and 4, segment between]
15. [number line: closed circle at 0, rays out]

17. [number line: open circles at -2 and 5, segment between]
19. [number line: open circle at $\frac{3}{8}$, rays out]
21. [number line: closed circles at -10 and 14, segment between]
23. [number line: open circles at $-\frac{5}{3}$ and 1, segment between]

25. [number line: closed circles at -4 and -1, rays out]
27. no solutions
29. [number line: open circles at -24 and -18, segment between]
31. [number line: open circle at 25, rays out]

33. [number line: open circles at $-\frac{65}{9}$ and $-\frac{5}{9}$, segment between]
35. [number line: open circle at -4, rays out]
37. [number line: closed circle at -7, ray right]
39. [number line: closed circle at 5, ray left]

REVIEW EXERCISES   (page 155)

1. $x = 5$ **3.** $y = 8$ **5.** $x = 19$ **7.** $y = 12$ **9.** 88, 90, 92

11. \$18,000 at 10% and \$7000 at 9%   **13.** $r^3 = \dfrac{3V}{4\pi}$ **15.** $S_0 = \dfrac{6V}{H} - 4S_1 - S_2$ **17.** $\frac{1}{2}, -\frac{5}{6}$

19. $0, -\frac{2}{3}, \frac{4}{5}$ **21.** $3, -\frac{11}{3}$ **23.** $\frac{1}{5}, -5$ **25.** [number line: open circle at -24, ray right]   **27.** [number line: open circles at $-\frac{1}{3}$ and 2, segment between]

29. [number line: open circles at -5 and -2, segment between]   **31.** [number line: all numbers x]

CHAPTER FOUR TEST   (page 156)

1.   all real numbers except 3 and -2 **3.** $y = 6$ **5.**   36 sq cm   **7.**   330 mi

9. $i = \dfrac{f(P - L)}{s}$ **11.** $6, -1$ **13.**   48 units   **15.**   3   **17.** $4, -7$

19. $\xleftarrow{\quad\bullet\quad}_{-5}$ **21.** $\xleftarrow{\;\bullet\;\bullet\;}_{-7\quad 1}$

Exercise 5.1   (page 163)

1. $\dfrac{2}{3}$ **3.** $-\dfrac{28}{9}$ **5.** $\dfrac{12}{13}$ **7.** $-\dfrac{122}{37}$ **9.** $4x^2$ **11.** $-\dfrac{4y}{3x}$ **13.** x^2 **15.** $-\dfrac{x}{2}$

17. $\dfrac{3y}{7(y - z)}$ **19.** $\dfrac{1}{x - y}$ **21.** $\dfrac{5}{x - 2}$ **23.** $\dfrac{-3(x + 2)}{x + 1}$ **25.**   3   **27.** $x + 2$

29. $\dfrac{x + 1}{x + 3}$ **31.** $\dfrac{m - 2n}{n - 2m}$ **33.** $\dfrac{x + 4}{2(2x - 3)}$ **35.** $\dfrac{3(x - y)}{x + 2}$ **37.** $\dfrac{2x + 1}{2 - x}$

39. $\dfrac{a^2 - 3a + 9}{4(a - 3)}$ **41.**   in lowest terms   **43.** $m + n$ **45.** $\dfrac{x - y}{x + y}$ **47.** $\dfrac{2a - 3b}{a - 2b}$

49. $\dfrac{1}{x^2 + xy + y^2 - 1}$ **51.** $x^2 - y^2$ **53.** $x^2 + 8 + x$ **55.** $\dfrac{(x + 1)^2}{(x - 1)^3}$ **57.** $\dfrac{3a + b}{y + b}$

Exercise 5.2   (page 168)

1. $\dfrac{10}{7}$ **3.** $-\dfrac{5}{6}$ **5.** $\dfrac{xy^2 d}{c^3}$ **7.** $-\dfrac{x^{10}}{y^2}$ **9.** $x + 1$ **11.**   1   **13.** $\dfrac{x - 4}{x + 5}$

15. $\dfrac{(a + 7)^2(a - 5)(a - 3)}{12x^2}$ **17.** $\dfrac{1}{x + 1}$ **19.** $(x + 1)^2$ **21.** $x - 5$ **23.** $\dfrac{x + y}{x - y}$

25. $-\dfrac{x + 3}{x + 2}$ **27.** $\dfrac{a + b}{(x - 3)(c + d)}$ **29.** $-\dfrac{x^7}{18y^4}$ **31.** $x^2(x + 3)$ **33.** $\dfrac{3x}{2}$ **35.** $\dfrac{t + 1}{t}$

37. $\dfrac{x^2 + 4 + x}{x^3}$ **39.** $\dfrac{x + 2}{x - 2}$ **41.** $\dfrac{x - 1}{3x + 2}$ **43.** $\dfrac{x - 7}{x + 7}$ **45.**   1

Exercise 5.3   (page 173)

1. $\dfrac{5}{2}$ **3.** $-\dfrac{1}{3}$ **5.** $\dfrac{11}{4y}$ **7.** $\dfrac{3 - a}{a + b}$ **9.** $\dfrac{2(x + 2)}{x + 1}$ **11.**   3   **13.**   3

15. $\dfrac{6x}{(x - 3)(x - 2)}$ **17.**   72   **19.** $x(x + 3)(x - 3)$ **21.** $(x + 3)^2(x^2 - 3x + 9)$

23. $(2x + 3)^2(x + 1)^2$ **25.** $\dfrac{5}{6}$ **27.** $-\dfrac{16}{75}$ **29.** $\dfrac{9a}{10}$ **31.** $\dfrac{21a - 8b}{14}$ **33.** $\dfrac{17}{12x}$

35. $\dfrac{9a^2 - 4b^2}{6ab}$ **37.** $\dfrac{10a + 4b}{21}$ **39.** $\dfrac{8x - 2}{(x + 2)(x - 4)}$ **41.** $\dfrac{7x + 29}{(x + 5)(x + 7)}$ **43.** $\dfrac{x^2 + 1}{x}$

45. $\dfrac{2x^2 + x}{(x + 3)(x + 2)(x - 2)}$ **47.** $\dfrac{-x^2 + 11x + 8}{(3x + 2)(x + 1)(x - 3)}$ **49.** $\dfrac{-4x^2 + 14x + 54}{x(x + 3)(x - 3)}$

51. $\dfrac{2x^2 + 5x + 4}{x + 1}$ **53.** $\dfrac{x^2 - 5x - 5}{x - 5}$ **55.** $\dfrac{-x^3 - x^2 + 5x}{x - 1}$ **57.** $\dfrac{-y^2 + 48y + 161}{(y + 4)(y + 3)}$

59. $\dfrac{2}{x + 1}$ **61.** $\dfrac{3x + 1}{x(x + 3)}$ **63.** $\dfrac{2x^3 + x^2 - 43x - 35}{(x + 5)(x - 5)(2x + 1)}$ **65.** $\dfrac{3x^2 - 2x - 17}{(x - 3)(x - 2)}$ **67.** $\dfrac{2a}{a - 1}$

69. $\dfrac{x^2 - 6x - 1}{2(x + 1)(x - 1)}$ **71.** $\dfrac{2b}{a + b}$ **73.** $\dfrac{-7a^2 + 8ab + 2a - 2b - 3b^2}{(2a - b)(2a - b)}$ **75.** $\dfrac{x}{3}$

77. $\dfrac{2}{x}$ **79.** $\dfrac{x^2 - 6x + 9}{x^2}$

Exercise 5.4   (page 179)

1. $\dfrac{2}{3}$ **3.** -1 **5.** $\dfrac{10}{3}$ **7.** $-\dfrac{5}{6}$ **9.** $\dfrac{2y}{3z}$ **11.** $125b$ **13.** $-\dfrac{1}{y}$ **15.** $\dfrac{y - x}{x^2 y^2}$

17. $\dfrac{b + a}{b}$ **19.** $\dfrac{y + x}{y - x}$ **21.** $x^2 + x - 6$ **23.** $\dfrac{5x^2 y^2}{xy + 1}$ **25.** -1 **27.** $\dfrac{x + 2}{x - 3}$

29. $\dfrac{a - 1}{a + 1}$ **31.** $\dfrac{y + x}{x^2 y}$ **33.** $\dfrac{xy^2}{y - x}$ **35.** $\dfrac{y + x}{y - x}$ **37.** xy **39.** $\dfrac{x^3 y^2 - x^2}{x^2 y^3 - y^2}$

41. $\dfrac{(b + a)(b - a)}{b(b - a - ab)}$ **43.** $\dfrac{x - 1}{x}$ **45.** $\dfrac{5b}{5b + 4}$ **47.** $\dfrac{3a^2 + 2a}{2a + 1}$

49. $\dfrac{(-x^2 + 2x - 2)(3x + 2)}{(2 - x)(-3x^2 - 2x + 9)}$ or $\dfrac{-3x^3 + 4x^2 - 2x - 4}{3x^3 - 4x^2 - 13x + 18}$

Exercise 5.5   (page 185)

1. $x = 12$ **3.** $x = 40$ **5.** $y = \frac{1}{2}$ **7.** no solution   **9.** $x = \frac{17}{25}$ **11.** $y = 0$
13. $a = 1$ **15.** $x = 2$ **17.** $x = 2$ **19.** $x = \frac{1}{3}$ **21.** $a = 0$ **23.** $2, -5$
25. $-4, 3$ **27.** $6, \frac{17}{3}$ **29.** $1, -11$ **31.** $1\frac{7}{8}$ days   **33.** $4\frac{1}{2}$ hr   **35.** $3\frac{3}{7}$ hr
37. 3 mph   **39.** 60 mph and 40 mph   **41.** 2   **43.** 7 motors

REVIEW EXERCISES   (page 188)

1. $\dfrac{31x}{72y}$ **3.** $\dfrac{x - 7}{x + 7}$ **5.** $\dfrac{1}{2(x + 2)}$ **7.** 1   **9.** $\dfrac{5y - 3}{x - y}$ **11.** $\dfrac{5x + 13}{(x + 2)(x + 3)}$

13. $\dfrac{3x(x - 1)}{(x - 3)(x + 1)}$ **15.** $\dfrac{5x^2 + 11x}{(x + 1)(x + 2)}$ **17.** $\dfrac{5x^2 + 23x + 4}{(x + 1)(x - 1)(x - 1)}$ **19.** $\dfrac{2x^3 + 13x^2 + 31x + 31}{(x + 3)(x + 2)(x - 3)}$

21. $\dfrac{3y - 2x}{x^2 y^2}$ **23.** $\dfrac{2x + 1}{x + 1}$ **25.** $\dfrac{1}{x}$ **27.** $\dfrac{x^2 y^2}{(x - y)^2(y^2 - x^2)}$ **29.** $\dfrac{y(1 + xy)}{x(xy - 1)}$

31. $x = 5$ **33.** no solution   **35.** 50 mph

CHAPTER FIVE TEST   (page 189)

1. $\dfrac{-2}{3xy}$ **3.** -3 **5.** $\dfrac{xz}{y^4}$ **7.** 1   **9.** $\dfrac{2}{x + 1}$ **11.** 3   **13.** $\dfrac{2s + r^2}{rs}$

15. $\dfrac{u^2}{2vw}$ **17.** $x = \dfrac{5}{2}$ **19.** 10 days

Exercise 6.1   (page 196)

1. 4   **3.** 3   **5.** 3   **7.** 2   **9.** $\dfrac{1}{2}$ **11.** $\dfrac{1}{2}$ **13.** -2 **15.** -3

17. -4 **19.** 0   **21.** 216   **23.** 27   **25.** 1728   **27.** $\dfrac{1}{4}$ **29.** $\dfrac{4}{9}$ **31.** $\dfrac{1}{8}$

33. $\dfrac{1}{64}$ **35.** $\dfrac{1}{9}$ **37.** $\dfrac{1}{4}$ **39.** 8   **41.** $\dfrac{16}{81}$ **43.** $-\dfrac{3}{2}$ **45.** $5^{5/7}$ **47.** $4^{3/5}$

49. $9^{1/5}$ **51.** $7^{1/2}$ **53.** $\dfrac{1}{36}$ **55.** $2^{2/3}$ **57.** $2x^{1/3}$ **59.** $9x^{4/3}y^{2/3}$

61. $125x^3y^{3/2}$ **63.** $\dfrac{1}{2x^{1/2}y^{3/2}}$ **65.** $\dfrac{2x}{3}$ **67.** $4y$ **69.** $x^{5a/6}$ **71.** $\dfrac{1}{x^{a/12}}$

73. $\dfrac{1}{x^{a/5}}$ **75.** x^{5b-a} **77.** $y+y^2$ **79.** a^2+1 **81.** $x^2-x+x^{3/5}$

83. $y^{1/2}+z^{1/3}y^{1/2}-z^{1/3}$ **85.** $x-4$ **87.** $x^{4/3}-x^2$ **89.** $2+x^{1/2}y^{1/2}+\dfrac{1}{x^{1/2}y^{1/2}}$

91. $x^{4/3}+2x^{2/3}y^{2/3}+y^{4/3}$ **93.** $a^3-2a^{3/2}b^{3/2}+b^3$ **95.** $x+1$ **97.** $729-54x+x^2$

Exercise 6.2   (page 205)

1. 11   **3.** -8 **5.** 1   **7.** -5 **9.** 3   **11.** -3 **13.** -2 **15.** 27
17. 8   **19.** -4 **21.** $-\frac{1}{4}$ **23.** $\frac{1}{9}$ **25.** $2|x|$ **27.** $2a$ **29.** $|x|$ **31.** $|x^3|$
33. $-x$ **35.** $-3a^2$ **37.** 5   **39.** 2   **41.** t **43.** $2\sqrt{5}$ **45.** $2\sqrt{6}$
47. $10\sqrt{2}$ **49.** $-5x\sqrt{2}$ **51.** $4\sqrt{2b}$ **53.** $-4a\sqrt{7a}$ **55.** $5ab\sqrt{7b}$ **57.** $-10\sqrt{3xy}$
59. $3\sqrt[3]{3}$ **61.** $-2\sqrt[3]{10}$ **63.** $-3x^2\sqrt[3]{2}$ **65.** $-2x^4y\sqrt[3]{2}$ **67.** $2x^3y\sqrt[4]{2}$
69. $-2x^2y\sqrt[5]{2}$ **71.** -10 **73.** 7   **75.** $6b$ **77.** $4x^2\sqrt{x}$ **79.** 2   **81.** $3a$
83. $\dfrac{\sqrt{7}}{3}$ **85.** $\dfrac{1}{4x}$ **87.** $\dfrac{c\sqrt{5}}{7ab^2}$ **89.** $\dfrac{x\sqrt{11}}{11}$ **91.** $\dfrac{\sqrt[3]{4}}{3a}$ **93.** $\dfrac{a^3\sqrt[3]{6}}{2}$ **95.** $\dfrac{\sqrt{7}}{7}$

97. $\dfrac{\sqrt{6}}{3}$ **99.** $\dfrac{\sqrt{10}}{4}$ **101.** 2   **103.** $\dfrac{\sqrt[3]{4}}{2}$ **105.** $\sqrt[3]{3}$ **107.** $\dfrac{\sqrt[3]{18}}{3}$ **109.** $2\sqrt{2x}$

111. $\dfrac{\sqrt{5y}}{y}$ **113.** $\dfrac{\sqrt[3]{2ab^2}}{b}$ **115.** $xz^2\sqrt[4]{xy^3}$ **117.** $-2ab\sqrt[3]{2a^2c^2}$ **119.** $-2abc^2\sqrt[5]{2abc}$

121. $x+1$ **123.** $x+4$

Exercise 6.3   (page 208)

1. $10\sqrt{2}$ **3.** $\sqrt[5]{7}$ **5.** $14\sqrt{x}$ **7.** $4\sqrt{3}$ **9.** $-\sqrt{2}$ **11.** $2\sqrt{2}$ **13.** $9\sqrt{6}$
15. $3\sqrt[3]{3}$ **17.** $-\sqrt[3]{4}$ **19.** -10 **21.** $-41\sqrt[3]{2}$ **23.** $-17\sqrt[4]{2}$ **25.** $-4\sqrt{2}$
27. $3\sqrt{2}+\sqrt{3}$ **29.** $-11\sqrt[3]{2}$ **31.** $y\sqrt{z}$ **33.** $(x^2y-x^2y^2-x^2y^3)\sqrt{y}$
35. $(3x+x^2)\sqrt[5]{xy^2}$ **37.** $2x+3$ **39.** $(2x+1)\sqrt{3}$ **41.** 0   **43.** $3\sqrt{x}$
45. $\dfrac{5}{2}\sqrt{2}-\dfrac{5}{3}\sqrt{3}$ **47.** $\dfrac{\sqrt{3a}}{a}$ **49.** $\dfrac{3\sqrt{3x}}{x}$ **51.** $\dfrac{5x\sqrt{7xy}}{6}$

Exercise 6.4   (page 212)

1. 4   **3.** $5\sqrt{2}$ **5.** $3\sqrt{2}$ **7.** 5   **9.** 3   **11.** $2\sqrt[3]{3}$ **13.** ab^2 **15.** $5a\sqrt{b}$
17. $r\sqrt[3]{10s}$ **19.** $2a^2b^2\sqrt[3]{2c^2}$ **21.** $x^2(x+3)$ **23.** $3x(y+z)\sqrt[3]{4}$ **25.** $12\sqrt{5}-15$

27. $12\sqrt{6} + 6\sqrt{14}$ **29.** $-8\sqrt{10} + 6\sqrt{15}$ **31.** $-1 - 2\sqrt{2}$ **33.** $9 - 14\sqrt{3}$

35. $8 + 2\sqrt{15}$ **37.** 1 **39.** $-9 + 5\sqrt{15}$ **41.** $22 - 12\sqrt{2}$ **43.** $-20\sqrt{3} - 12\sqrt{7}$

45. $\sqrt{2} + 1$ **47.** $\dfrac{6(\sqrt{5} - 4)}{11}$ **49.** $\sqrt{3} - 1$ **51.** $5(\sqrt{6} - 1)$ **53.** $\dfrac{3\sqrt{2} - \sqrt{10}}{4}$

55. $-2\sqrt{7} - \sqrt{35}$ **57.** $\sqrt{7} + \sqrt{5}$ **59.** $10(\sqrt{3} - 1)$ **61.** $2 + \sqrt{3}$

63. $\dfrac{7 + 2\sqrt{2} - 2\sqrt{7} - \sqrt{14}}{5}$ **65.** $\dfrac{2(\sqrt{x} - 1)}{x - 1}$ **67.** $\dfrac{x(\sqrt{x} + 4)}{x - 16}$ **69.** $\sqrt{2z} + 1$

71. $\dfrac{x - 2\sqrt{xy} + y}{x - y}$

Exercise 6.5   (page 215)

1. $\sqrt{3}$ **3.** $\sqrt{x}$ **5.** x^2 **7.** y^3 **9.** $2x^3$ **11.** $\sqrt{3x}$ **13.** $-x\sqrt[3]{2}$

15. $-4a^4$ **17.** $3\sqrt{3}$ **19.** $-2\sqrt{2}$ **21.** 0 **23.** $3x\sqrt[3]{y}$ **25.** $17xy^2z^3\sqrt[5]{xyz}$

27. $\sqrt[6]{27}$ **29.** $\sqrt[8]{625}$ **31.** $\sqrt[6]{225}$ **33.** $\sqrt[9]{64}$ **35.** $\sqrt{3}$ **37.** $\sqrt[3]{5}$ **39.** $\sqrt[6]{8x^3y^3}$

41. $\sqrt[6]{16x^4y^2}$ **43.** $\sqrt[6]{6125}$ **45.** $\sqrt[8]{1125}$ **47.** $\sqrt[6]{32}$ **49.** $\sqrt[15]{16{,}384}$ **51.** $\sqrt[6]{125x^5y^2}$

53. $\sqrt[15]{5^{10} \cdot 3^3x^5y^6}$ **55.** $\sqrt[35]{x^{26}y^{43}} = y\sqrt[35]{x^{26}y^8}$ **57.** $\dfrac{\sqrt[6]{72}}{2}$ **59.** $\dfrac{\sqrt[15]{2^5 \cdot 3^3}}{2}$ or $\dfrac{\sqrt[15]{864}}{2}$

61. $\dfrac{\sqrt[10]{3^7}}{3}$ or $\dfrac{\sqrt[10]{2187}}{3}$ **63.** $\dfrac{\sqrt[14]{(4x)^2(2x)^7}}{2x} = \dfrac{\sqrt[14]{2^{11}x^9}}{2x} = \dfrac{\sqrt[14]{2048x^9}}{2x}$ **65.** $\dfrac{\sqrt[15]{(3x)^3 \cdot 9^5}}{3} = \dfrac{\sqrt[15]{3^{13}x^3}}{3}$

Exercise 6.6   (page 219)

1. $x = 2$ **3.** $x = 4$ **5.** $x = 0$ **7.** $n = 4$ **9.** $p = 8$ **11.** $x = \frac{5}{2}, x = \frac{1}{2}$

13. $r = 6, r = -5$ **15.** $x = 4, x = 3$ **17.** $y = 2, y = 7$ **19.** $x = 9, x = -25$

21. $x = 2, x = -1$ **23.** $x = 1, x = -1$ **25.** no solution   **27.** $y = 0, y = 4$

29. no solution   **31.** $v = 0$ **33.** $u = 1, u = 9$ **35.** $t = 4, t = 0$ **37.** $x = 2, x = 142$

39. $z = 2$ **41.** no solution   **43.** $x = -\frac{3}{2}$ **45.** $x = 0, x = -\frac{12}{11}$ **47.** $a = 1$

49. $x = 4, x = -9$

REVIEW EXERCISES   (page 221)

1. 5 **3.** 27 **5.** -2 **7.** $\frac{1}{4}$ **9.** $-16{,}807$ **11.** 8 **13.** $3xy^{1/3}$

15. $125x^9y^6$ **17.** $u - 1$ **19.** $x + 2x^{1/2}y^{1/2} + y$ **21.** 7 **23.** -6 **25.** -3

27. 5 **29.** $4\sqrt{15}$ **31.** $2\sqrt[4]{2}$ **33.** $2|x|\sqrt{2x}$ **35.** $2xyz\sqrt[3]{2x^2y}$ **37.** 5

39. $3x$ **41.** $-2x$ **43.** $\dfrac{\sqrt{3}}{3}$ **45.** $\dfrac{\sqrt{xy}}{y}$ **47.** $x + 3$ **49.** $3\sqrt{2}$ **51.** 0

53. $29\sqrt{2}$ **55.** $13\sqrt[3]{2}$ **57.** 4 **59.** $\sqrt{10} - \sqrt{5}$ **61.** 1 **63.** $x - y$

65. $2(\sqrt{2} + 1)$ **67.** $2(\sqrt{x} - 4)$ **69.** $\sqrt{5}$ **71.** $\sqrt{2}$ **73.** x^2 **75.** $x^2y^2\sqrt{x}$

77. $\sqrt[6]{5^3 \cdot 2^2} = \sqrt[6]{500}$ **79.** $\sqrt[6]{3^2 \cdot 2^3} = \sqrt[6]{72}$ **81.** $\dfrac{\sqrt[6]{2000}}{2}$ **83.** $\dfrac{\sqrt[10]{6075}}{3}$ **85.** $y = 22$

87. $r = 3, r = 9$ **89.** $x = 2$

CHAPTER SIX TEST   (page 223)

1. 2 **3.** $\dfrac{1}{216}$ **5.** $2^{4/3}$ **7.** $4\sqrt{3}$ **9.** $2x^5y\sqrt[3]{3}$ **11.** $\dfrac{\sqrt{5}}{5}$ **13.** $2\sqrt[3]{3}$

15. $2\sqrt{3}$ **17.** $2y^2\sqrt{3y}$ **19.** $-6x\sqrt{y} - 2xy^2$ **21.** $\dfrac{\sqrt{2}(\sqrt{5} - 3)}{4}$ **23.** $2\sqrt{2}$

25. $4\sqrt{6}$ **27.** $\sqrt[6]{5^2 \cdot 7^3} = \sqrt[6]{8575}$ **29.** $n = 10$

CUMULATIVE REVIEW EXERCISES (CHAPTERS 4–6)   (page 224)

1. $x = 8$ **3.** $y = -1$ **5.** $28, 30, 32$ **7.** $a = \dfrac{2S}{n} - l$ **9.** $x = -\frac{1}{3}, x = -\frac{7}{2}$

11. $x = -5, x = -\frac{3}{5}$ **13.** $-3 < x < 3$ **15.** $-\frac{2}{3} \le x \le 2$ **17.** $\dfrac{2x - 3}{3x - 1}$ **19.** $\dfrac{4}{x - y}$

21. $x = 0$ **23.** 16 **25.** $x^{17/12}$ **27.** $-3x$ **29.** $4x$ **31.** $7\sqrt{2}$ **33.** $-18\sqrt{6}$

35. $\dfrac{x + 3\sqrt{x} + 2}{x - 1}$ **37.** $x = 2, x = 7$

Exercise 7.1   (page 235)

1.

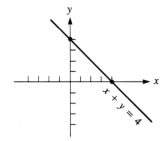

3.

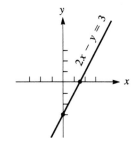

5.

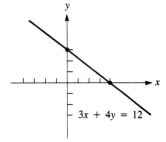

7.

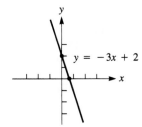

9.

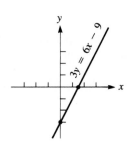

11.

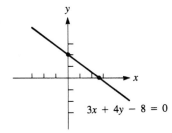

13. 5 **15.** 5 **17.** 13 **19.** 10 **21.** $\sqrt{104}$ or $2\sqrt{26}$ **23.** $(\frac{7}{2}, 6)$ **25.** $(\frac{1}{2}, -2)$

27. $(9, 12)$ **29.** $(-4, 0)$ **31.** $(\frac{3}{2}, -2)$ **33.** $\left(\dfrac{a + c}{2}, \dfrac{b + d}{2}\right)$ **35.** $\left(\dfrac{\sqrt{2} + \sqrt{3}}{2}, \dfrac{\sqrt{3} + \sqrt{2}}{2}\right)$

39. $(4, 1)$ **41.** $x = 7$ **43.** $(7, 0), (3, 0)$

Exercise 7.2   (page 242)

1. 3 **3.** -1 **5.** $-\frac{1}{3}$ **7.** 0 **9.** no defined slope   **11.** $-\frac{3}{2}$ **13.** $\frac{3}{4}$
15. $\frac{1}{2}$ **17.** 0 **19.** $-\frac{2}{3}$ **21.** negative   **23.** positive   **25.** no defined slope
27. perpendicular   **29.** parallel   **31.** perpendicular   **33.** neither   **35.** parallel
37. perpendicular   **39.** neither   **41.** on same line   **43.** not on same line
45. on same line   **47.** $y = 0$; slope is 0 **55.** $a = 4$

Exercise 7.3   (page 250)

1. $5x - y = -7$ **3.** $3x + y = 6$ **5.** $3x - 2y = -4$ **7.** $y = x$ **9.** $y = \frac{7}{3}x - 3$
11. $y = -\frac{9}{5}x + \frac{2}{5}$ **13.** $y = 3x + 17$ **15.** $y = -7x + 54$ **17.** $y = -4$ **19.** $y = -\frac{1}{2}x + 11$
21. $m = 1;\ \ b = -1$ **23.** $m = \frac{2}{3};\ \ b = 2$

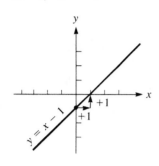

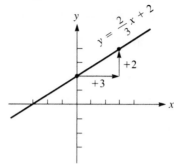

25. $m = -\frac{2}{3};\ \ b = 6$ **27.** $m = \frac{3}{2};\ \ b = -4$

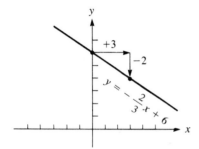

29. $m = -\frac{1}{3};\ \ b = -\frac{5}{6}$ **31.** $m = \frac{7}{2};\ \ b = 2$ **33.** parallel   **35.** perpendicular
37. parallel   **39.** perpendicular   **41.** perpendicular   **43.** perpendicular   **45.** $y = 4x$
47. $y = 4x - 3$ **49.** $y = \frac{4}{5}x - \frac{26}{5}$ **51.** $y = -\frac{1}{4}x$ **53.** $y = -\frac{1}{4}x + \frac{11}{2}$ **55.** $y = -\frac{5}{4}x + 3$
57. perpendicular   **59.** parallel   **61.** $x = -2$ **63.** $x = 5$

Exercise 7.4   (page 256)

1. a function   **3.** a function   **5.** not a function   **7.** not a function   **9.** a function
11. Both the domain and range are the set of real numbers.
13. The domain is the set of all real numbers except 2; the range is the set of all real numbers except 0.
15. The domain is the set of all real numbers; the range is the set of all nonnegative real numbers.
17. The domain is the set of all real numbers; the range is $\{3\}$.
19. The domain is the set of all real numbers greater than or equal to 2; the range is the set of all nonnegative real numbers.
21. 7   **23.** -8 **25.** 0   **27.** $5x + 12$ **29.** 4   **31.** 5   **33.** $\frac{40}{9}$
35. $x^4 + 4$ **37.** 7   **39.** 24   **41.** $2x^2 - 1$ **43.** 58   **45.** 110
47. $9x^2 - 9x + 2$ **53.** $2x + h$ **55.** $3x - 2$; The domain is the set of real numbers.
57. $2x^2 - 5x - 3$; The domain is the set of real numbers.
59. $-2x^2 + 3x - 3$; The domain is the set of real numbers.

61. $\dfrac{3x - 2}{2x^2 + 1}$; The domain is the set of real numbers.   **63.**   3; The domain is set of real numbers.

65. $\dfrac{x^2 - 4}{x^2 - 1}$; The domain is the set of real numbers except 1 and -1.

Exercise 7.5   (page 261)

1.   a linear function   **3.**   a linear function   **5.**   not a linear function   **7.**   a linear function
9.   not a linear function   **11.**   The domain is $\{1, 3, 5\}$; the range is $\{2, 4, 9, 12\}$; not a function
13.   The domain is $\{1, -1, 4, -4\}$; the range is $\{2, 3, 4, 5\}$; a function
15.   The domain is $\{5, 6\}$; the range is $\{8, 9, 10\}$; not a function   **17.** $\{(2, 3), (1, 2), (0, 1)\}$;   a function
19. $\{(2, 1), (3, 2), (3, 1), (5, 1)\}$;   not a function   **21.** $\{(1, 1), (4, 2), (9, 3), (16, 4)\}$;   a function
23. $f^{-1}(x) = \frac{1}{3}x - \frac{1}{3}$;   a linear function   **25.** $f^{-1}(x) = 5x - 4$;   a linear function
27. $f^{-1}(x) = 5x + 4$;   a linear function   **29.** $y = \frac{5}{4}x^2 + 5$;   a function but not a linear function

31. $f^{-1}(x) = \dfrac{x + 3}{2}$ **33.** $f^{-1}(x) = \dfrac{x - 2}{3}$

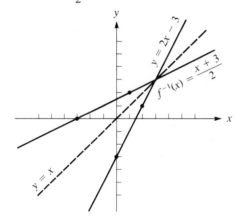

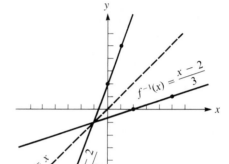

The axis of symmetry is the line $y = x$.   The axis of symmetry is the line $y = x$.

35. $f^{-1}(x) = \dfrac{x + 5}{3}$ **37.** $f^{-1}(x) = -3x + 4$

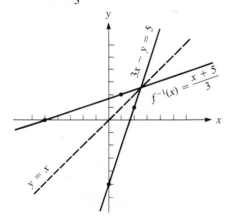

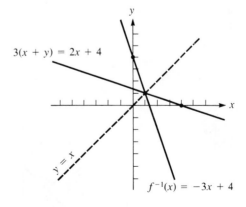

The axis of symmetry is the line $y = x$.   The axis of symmetry is the line $y = x$.

39. $f^{-1}(x) = -\frac{2}{3}x + 4$ **41.** $g = 5l$ **43.** $c = 10 + n$

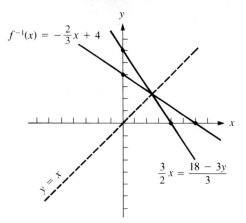

The axis of symmetry is the line $y = x$.

45. $C = \frac{5}{9}F - \frac{160}{9}$ or $C = \frac{5}{9}(F - 32)$

Exercise 7.6   (page 267)

1. $x = 3$ **3.** $r = 5$ **5.** $y = 4$ or $y = -4$ **7.** $n = -3$

9. $x = 5$ **11.** $t = -3$ is extraneous   **13.** $A = kp^2$ **15.** $v = \dfrac{k}{\sqrt[4]{r}}$

17. $B = kmn$ **19.** $P = \dfrac{ka^2}{j^3}$ **21.** $F = \dfrac{km_1 m_2}{d^2}$

23. L varies directly with the product of m and n, or L varies jointly with m and n.
25. E varies directly with the product ab^2, or E varies jointly with a and the square of b.
27. X varies directly with the square of x and inversely with the square of y.
29. R varies directly with L and inversely with the square of d.
31. 36π sq in.   **33.** 432 mi   **35.** 25 days   **37.** 12 cu in.   **39.** approximately 85.3
41. The volume is multiplied by 12.   **43.** 26, 437.5 gallons   **45.** 3 ohms
47. 4.4 in. (on side);   0.275 in. (on edge)

Exercise 7.7   (page 272)

1.
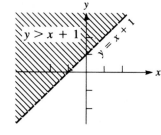

3.
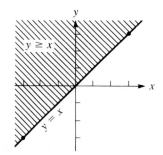

5.
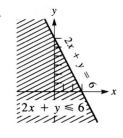

7.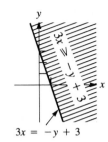

$3x = -y + 3$

9.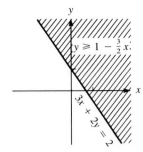

11.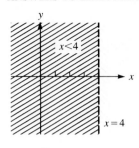

13.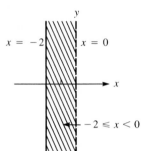

15.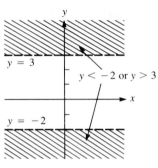

17.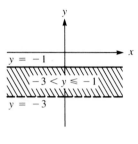

19. $3x + 2y > 6$ **21.** $x \leq 3$ **23.** $y \leq x$ **25.** $-2 \leq x \leq 3$ **27.** $y > -1$ or $y \leq -3$

REVIEW EXERCISES   (page 275)

1.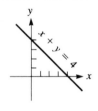

3.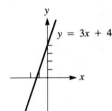

5. 5   **7.** $\sqrt{233}$ **9.** $(2, 9)$ **11.** $(\frac{7}{2}, 2)$

13. 1   **15.** 5

17. $y - 5 = -\frac{3}{2}(x + 2)$ or $y = -\frac{3}{2}x + 2$ **19.** $3x - y = -29$ **21.** $3x - 2y = 1$
23. The domain is the set of real numbers; the range is the set of real numbers; a function
25. The domain is the set of real numbers; the range is the set of real numbers greater than or equal to 1; a function
27. The domain is the set of real numbers; the range is the set of real numbers; a function
29. -7 **31.** 12   **33.** $3x^2 - 10$ **35.** $5x - 1$; the domain is the set of real numbers.

37. $6x^2 - x - 2$; the domain is the set of real numbers.   **39.** $f^{-1}(x) = \dfrac{x - 2}{7}$; a function

41. $x = 5$ **43.** 72   **45.** $k = 2$ **47.**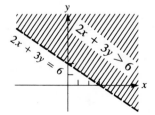

49.

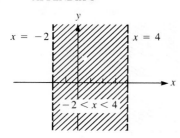

CHAPTER SEVEN TEST   (page 276)

1.

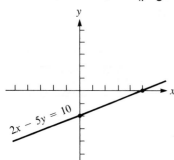

3. $\sqrt{116} = 2\sqrt{29}$

5. $\frac{2}{5}$ **7.** $y = \frac{2}{3}x - \frac{23}{3}$ **9.** $m = -\frac{1}{3}$; $b = -\frac{3}{2}$ **11.** perpendicular   **13.** $y = \frac{3}{2}x + \frac{21}{2}$
15. The domain is the set of all real numbers except 2; the range is the set of all real numbers except 0.
17. $g(0) = -2$ **19.** $g(f(x)) = 9x^2 + 6x - 1$ **21.** $(f \cdot g)x = 3x^3 + x^2 - 6x - 2$ **23.** yes
25. $x = 6$ or $x = -1$ **27.** $\frac{135}{2}$ **29.**

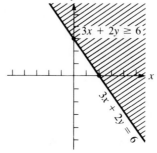

Exercise 8.1   (page 287)

1. $x = 0$; $x = -2$ **3.** $y = 5$; $y = -5$ **5.** $r = -2$; $r = -4$ **7.** $x = 6$; $x = 1$
9. $z = 2$; $z = \frac{1}{2}$ **11.** $s = \frac{2}{3}$; $s = -\frac{5}{2}$ **13.** $x = 6$; $x = -6$ **15.** $z = \sqrt{5}$; $z = -\sqrt{5}$
17. $x = \frac{4\sqrt{3}}{3}$; $x = -\frac{4\sqrt{3}}{3}$ **19.** $x = 0$; $x = -2$ **21.** $s = 4$ or $s = 10$
23. $x = -5 + \sqrt{3}$; $x = -5 - \sqrt{3}$ **25.** $x = 2$; $x = -4$ **27.** $x = 2$; $x = 4$
29. $x = -1$; $x = -4$ **31.** $x = 1$; $x = -\frac{1}{2}$ **33.** $x = -\frac{1}{3}$; $x = -\frac{3}{2}$ **35.** $r = \frac{3}{4}$; $r = -\frac{3}{2}$
37. $x = \frac{-7 + \sqrt{29}}{10}$; $x = \frac{-7 - \sqrt{29}}{10}$ **39.** $x = -1$; $x = -2$ **41.** $x = -6$; $x = -6$
43. $x = \frac{-5 + \sqrt{5}}{10}$; $x = \frac{-5 - \sqrt{5}}{10}$ **45.** $u = -\frac{3}{2}$; $u = -\frac{1}{2}$ **47.** $y = \frac{1}{4}$; $y = -\frac{3}{4}$

49. $x = \dfrac{-5 + \sqrt{17}}{2};\ \ x = \dfrac{-5 - \sqrt{17}}{2}$ **51.** either 16 and 18 or -18 and -16 **53.** 6 and 7

55. 8 ft by 12 ft   **57.** 4 units   **59.** 10 ft by 5 ft   **61.** $b = \frac{4}{3}$ cm   **63.** $r = 30$ mph

65. either \$4.80 or \$5.20   **67.** 4000 subscribers   **69.** about 2.26 in.   **71.** $x^2 - 2x - 24 = 0$

Exercise 8.2   (page 296)

1. $3i, -3i$ **3.** $\dfrac{4\sqrt{3}}{3}\,i, -\dfrac{4\sqrt{3}}{3}\,i$ **5.** $-1 + i, -1 - i$ **7.** $-\dfrac{1}{4} + \dfrac{\sqrt{7}}{4}\,i, -\dfrac{1}{4} - \dfrac{\sqrt{7}}{4}\,i$

9. $\dfrac{2}{3} + \dfrac{\sqrt{2}}{3}\,i, \dfrac{2}{3} - \dfrac{\sqrt{2}}{3}\,i$ **11.** i **13.** $-i$ **15.** 1 **17.** i **19.** $8 - 2i$ **21.** $3 - 5i$

23. $15 + 7i$ **25.** $-15 + 2\sqrt{3}\,i$ **27.** $3 + 6i$ **29.** $7 + i$ **31.** $14 - 8i$ **33.** $8 + \sqrt{2}\,i$

35. $6 - 8i$ **37.** $6 + \sqrt{6} + (3\sqrt{3} - 2\sqrt{2})i$ **39.** $-16 - \sqrt{35} + (2\sqrt{5} - 8\sqrt{7})i$ **41.** $0 - i$

43. $0 + \frac{4}{5}i$ **45.** $\frac{1}{8} + 0i$ **47.** $0 + \frac{3}{5}i$ **49.** $0 + \dfrac{3\sqrt{2}}{4}\,i$ **51.** $\frac{15}{26} - \frac{3}{26}i$ **53.** $-\frac{42}{25} - \frac{6}{25}i$

55. $\frac{1}{4} + \frac{3}{4}i$ **57.** $\frac{5}{13} - \frac{12}{13}i$ **59.** $\dfrac{6 + \sqrt{10}}{9} + \dfrac{2\sqrt{2} - 3\sqrt{5}}{9}\,i$ **61.** 10 **63.** 13

65. $\sqrt{74}$ **67.** $3\sqrt{2}$ **69.** $\sqrt{69}$

Exercise 8.3   (page 302)

1. rational and equal   **3.** nonreal and complex conjugates   **5.** irrational and unequal

7. rational and unequal   **9.** 6 and -6 **11.** 12 and -12 **13.** 5   **15.** 12 and -3

17. The solutions are real numbers.   **19.** $k < -\frac{4}{3}$ **21.** $1, -1, 4, -4$ **23.** $1, -1, \sqrt{2}, -\sqrt{2}$

25. $1, -1, \sqrt{5}, -\sqrt{5}$ **27.** $1, -1, 2, -2$ **29.** 1   **31.** no solution   **33.** $-8, -27$

35. $-1, 27$ **37.** $-1, -4$ **39.** $4, -5$ **41.** $0, 2$ **43.** $-1, -\frac{27}{13}$ **45.** $1, 1, -1, -1$

47. $x = \pm\sqrt{r^2 - y^2}$ **49.** $d = \pm\sqrt{\dfrac{k}{I}} = \dfrac{\pm\sqrt{kI}}{I}$ **51.** $y = \dfrac{-3x \pm \sqrt{9x^2 - 28x}}{2x}$

53. $\mu^2 = \dfrac{\Sigma x^2}{N} - \sigma^2$ **55.** $\frac{2}{3}, -\frac{1}{4}$ **57.** $x = \dfrac{-5 + \sqrt{17}}{4};\ x = \dfrac{-5 - \sqrt{17}}{4}$

59. $x = \dfrac{1 + i\sqrt{11}}{3};\ x = \dfrac{1 - i\sqrt{11}}{3}$

Exercise 8.4   (page 307)

1.

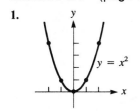

$y = x^2$

3.

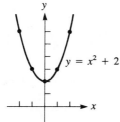

$y = x^2 + 2$

5.

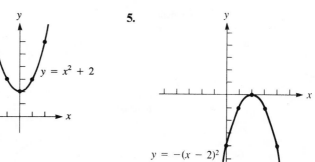

$y = -(x - 2)^2$

7.

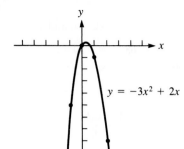

$y = -3x^2 + 2x$

9.

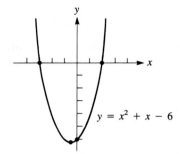

$y = x^2 + x - 6$

11.

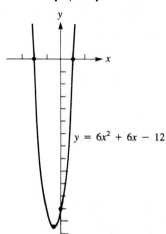

$y = 6x^2 + 6x - 12$

13. vertex at $(1, 2)$;   axis of symmetry is $x = 1$

15. vertex at $(-3, -4)$; axis of symmetry is $x = -3$ **17.** vertex at $(0, 0)$; axis of symmetry is $x = 0$
19. vertex at $(1, -2)$;   axis of symmetry is $x = 1$ **21.** vertex at $(2, 21)$; axis of symmetry is $x = 2$
23. vertex at $\left(\frac{5}{12}, \frac{143}{24}\right)$; axis of symmetry is $x = \frac{5}{12}$ **25.** vertex at (h, k)
27. Both numbers are 25.   **29.**   36 ft;   3 sec   **31.**   50 ft by 50 ft;   2500 sq. ft   **33.**   \$35

Exercise 8.5   (page 314)

1.

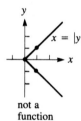

$x = |y|$

not a
function

3.

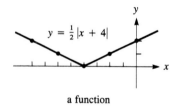

$y = \frac{1}{2}|x + 4|$

a function

5.

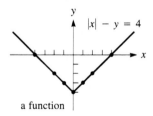

$|x| - y = 4$

a function

7.

$x = y^2 + 4$

not a
function

9.

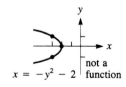

$x = -y^2 - 2$ not a
function

11.

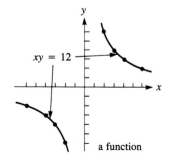

$xy = 12$

a function

13.

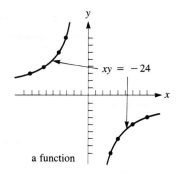

$xy = -24$

a function

15.

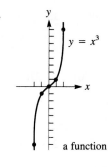

$y = x^3$

a function

17.

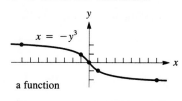

$x = -y^3$

a function

19.

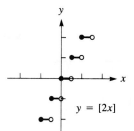

$y = [2x]$

21.

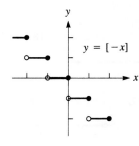

$y = [-x]$

23.

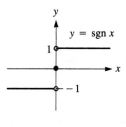

$y = \operatorname{sgn} x$

25. $y = \pm\sqrt{x - 4}$; not a function   **27.** $y = x^2 - 4$; a function   **29.** $y = \sqrt[3]{x}$; a function

31. $y = x^2,\ x \ge 0$; a function   **33.** $y = \sqrt{x} - 1$; a function

35. $f^{-1}(x) = \sqrt[3]{\dfrac{x + 3}{2}}$ does determine a function.

37.

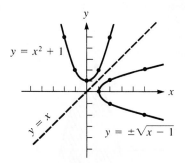

$y = x^2 + 1$

$y = x$

$y = \pm\sqrt{x - 1}$

The axis of symmetry is the
line $y = x$.

39.

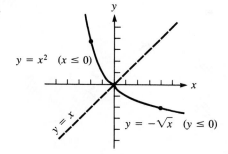

$y = x^2 \ (x \le 0)$

$y = x$

$y = -\sqrt{x} \ (y \le 0)$

The axis of symmetry is the
line $y = x$.

41. a one-to-one function   **43.** a one-to-one function   **45.** not a one-to-one function
47. a one-to-one function   **49.** not a one-to-one function

Exercise 8.6   (page 319)

1. $\{x : 1 < x < 4\}$ **3.** $\{x : x < 3 \text{ or } x > 5\}$ **5.** $\{x : -4 \le x \le 3\}$ **7.** $\{x : x \le -5 \text{ or } x \ge 3\}$

9. no solutions   **11.** $\{x : x \le -3 \text{ or } x \ge 3\}$ **13.** $\{x : -5 < x < 5\}$ **15.** $\{x : x < 0 \text{ or } x > \frac{1}{2}\}$

17. $\{x : 0 < x \le 2\}$ **19.** $\{x : x < -\frac{5}{3} \text{ or } x > 0\}$ **21.** $\{x : x < -3 \text{ or } 1 < x < 4\}$

23. $\{x : -5 \le x < -2 \text{ or } x \ge 4\}$ **25.** $\{x : x \le -2 \text{ or } \frac{3}{2} \le x < 3\}$ **27.** $\{x : -\frac{1}{2} < x < \frac{1}{3} \text{ or } x > \frac{1}{2}\}$

29. $\{x : 0 < x < 2 \text{ or } x > 8\}$ **31.** $\{x : x < -2 \text{ or } 2 < x \le 18\}$ **33.** $\{x : -\frac{34}{5} \le x < -4 \text{ or } x > 3\}$

35. $\{x : -4 < x \le -2 \text{ or } -1 < x \le 2\}$ **37.** $\{x : x < -16 \text{ or } -4 < x < -1 \text{ or } x > 4\}$

39. $\{x : x < -2 \text{ or } x > -2\}$ or $\{x : x \ne -2\}$ **41.**

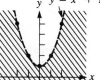

43.

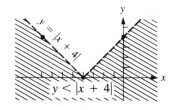

45.

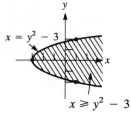

47.

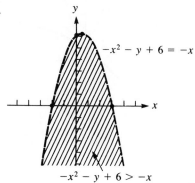

49.

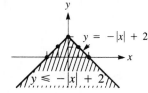

51.

Exercise 8.7   (page 325)

1.

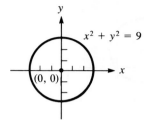

$x^2 + y^2 = 9$

$(0, 0)$

3.

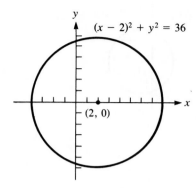

$(x - 2)^2 + y^2 = 36$

$(2, 0)$

5.

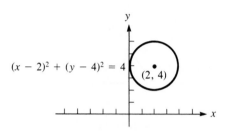

$(x - 2)^2 + (y - 4)^2 = 4$

$(2, 4)$

7.

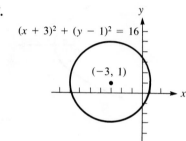

$(x + 3)^2 + (y - 1)^2 = 16$

$(-3, 1)$

9.

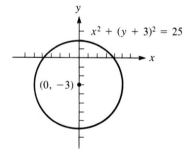

$x^2 + (y + 3)^2 = 25$

$(0, -3)$

11. $x^2 + y^2 = 1$ **13.** $(x - 6)^2 + (y - 8)^2 = 25$

15. $(x + 2)^2 + (y - 6)^2 = 144$ **17.** $x^2 + y^2 = 2$ **19.**

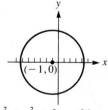

$(-1, 0)$

$x^2 + y^2 + 2x - 26 = 0$

21.

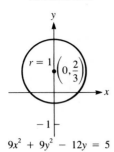

$9x^2 + 9y^2 - 12y = 5$

23.

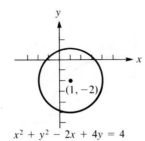

$x^2 + y^2 - 2x + 4y = 4$

25.

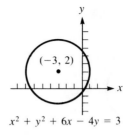

$x^2 + y^2 + 6x - 4y = 3$

27.

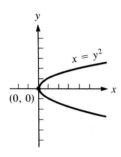

29.

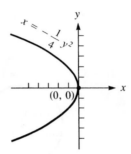

31.

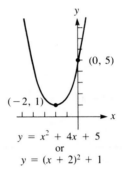

$y = x^2 + 4x + 5$
or
$y = (x + 2)^2 + 1$

33.

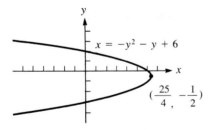

35.

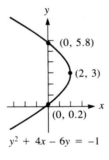

$y^2 + 4x - 6y = -1$

37.

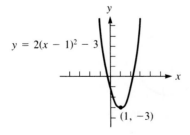

39.

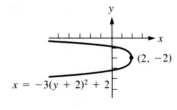

Exercise 8.8   (page 331)

1.

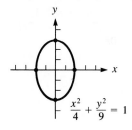

$$\frac{x^2}{4} + \frac{y^2}{9} = 1$$

3.

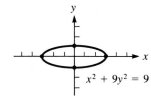

$x^2 + 9y^2 = 9$

5.

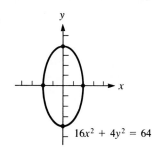

$16x^2 + 4y^2 = 64$

7.

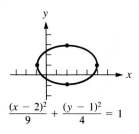

$$\frac{(x - 2)^2}{9} + \frac{(y - 1)^2}{4} = 1$$

9.

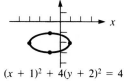

$(x + 1)^2 + 4(y + 2)^2 = 4$

11.

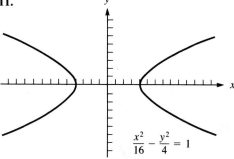

$$\frac{x^2}{16} - \frac{y^2}{4} = 1$$

13.

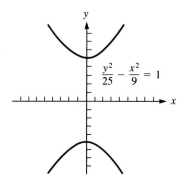

$$\frac{y^2}{25} - \frac{x^2}{9} = 1$$

15.

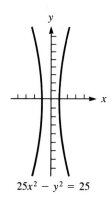

$25x^2 - y^2 = 25$

17.

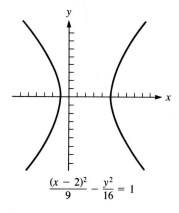

$$\frac{(x - 2)^2}{9} - \frac{y^2}{16} = 1$$

19.

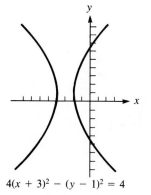

$4(x + 3)^2 - (y - 1)^2 = 4$

21.

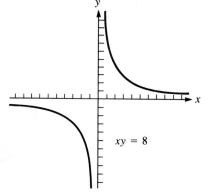

$xy = 8$

498   APPENDIX I

REVIEW EXERCISES   (page 333)

1. $x = \frac{2}{3}$; $x = -\frac{3}{4}$ **3.** $x = \frac{2}{3}$; $x = -\frac{4}{5}$ **5.** $x = -4$; $x = -2$ **7.** $x = 9$; $x = -1$
9. $x = \frac{1}{2}$; $x = -7$ **11.** $12 - 8i$ **13.** $-96 + 3i$ **15.** $-28 - 21i$ **17.** $-24 + 28i$
19. $0 - \frac{3}{4}i$ **21.** $\frac{12}{5} - \frac{6}{5}i$ **23.** $\frac{15}{17} + \frac{8}{17}i$ **25.** $\frac{15}{29} - \frac{6}{29}i$ **27.** $15 + 0i$
29. real and unequal **31.** $k = 12$ or $k = 152$ **33.** 4 cm by 6 cm **35.** $x = 1$; $x = 144$
37. $x = 1$ **39.** $\frac{14}{3}$ **41.**

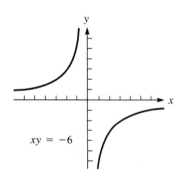

43.

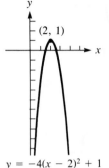

45.

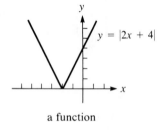

a function

47.

a function

49. $y = f^{-1}(x) = \dfrac{x + 3}{6}$ **51.** $y = f^{-1}(x) = \sqrt{\dfrac{x + 1}{2}} = \dfrac{\sqrt{2x + 2}}{2}$ **53.** one-to-one function

55.

57.

59.

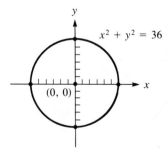

61.

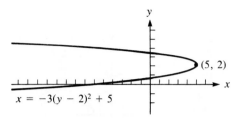

$x = -3(y - 2)^2 + 5$

63.

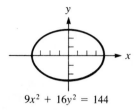

$9x^2 + 16y^2 = 144$

65.

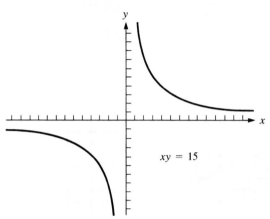

$xy = 15$

67.

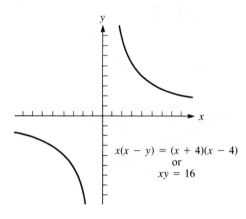

$x(x - y) = (x + 4)(x - 4)$
or
$xy = 16$

CHAPTER EIGHT TEST   (page 335)

1. $x = 3$; $x = -6$ **3.**   144   **5.** $x = -2 \pm \sqrt{3}$ **7.** $4 - 7i$ **9.** $-10 - 11i$

11. $\dfrac{1}{2} + \dfrac{1}{2}i$ **13.** $k = 2$ **15.** $y = 1$; $y = \frac{1}{4}$

17.

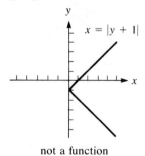

$x = |y + 1|$

not a function

19.

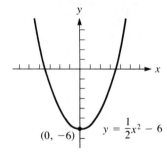

$(0, -6)$ $y = \dfrac{1}{2}x^2 - 6$

not one-to-one

21. $\{x: -3 < x \le 2\}$

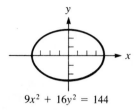

23.

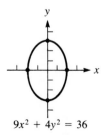

$9x^2 + 4y^2 = 36$

25.

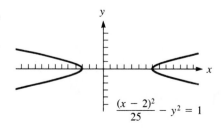

$\dfrac{(x - 2)^2}{25} - y^2 = 1$

Exercise 9.1   (page 341)

1.
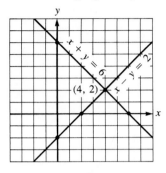

3.
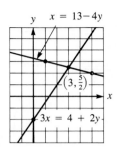

5.
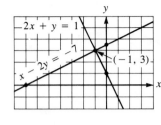

7.
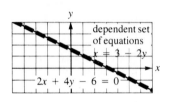

9.
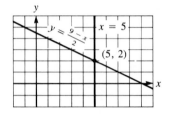

11.
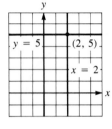

13.
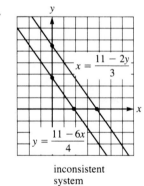

inconsistent
system

15.
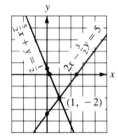

17.
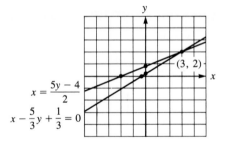

19.
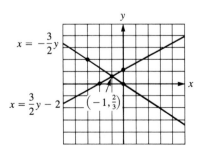

21.   one possible answer is $\begin{cases} x + y = -3 \\ x + y = -7 \end{cases}$

Exercise 9.2   (page 348)

1. $(2, 2)$ **3.** $(5, 3)$ **5.** $(-2, 4)$ **7.** inconsistent system; no solution   **9.** $(5, \frac{3}{2})$
11. $(5, 2)$ **13.** $(-4, -2)$ **15.** $(1, 2)$ **17.** $(\frac{1}{2}, \frac{2}{3})$ **19.** dependent equations
21. $(4, 8)$ **23.** $(20, -12)$ **25.** $(\frac{2}{3}, \frac{3}{2})$ **27.** $(\frac{1}{2}, -3)$ **29.** $(2, 3)$ **31.** $(-\frac{1}{3}, 1)$
33. 28 and 21   **35.** 16 in. by 20 in.   **37.** $40°$ **39.** 40 oz of 8% solution; 60 oz of 15% solution
41. 55 mph   **43.** $270

Exercise 9.3   (page 354)

1. 8   **3.** -2 **5.** $x^2 - y^2$ **7.** $-2x^2$ **9.** $-2x - 2$ **11.** $(4, 2)$ **13.** $(-1, 3)$
15. $(-\frac{1}{2}, \frac{1}{3})$ **17.** $(2, -1)$ **19.** inconsistent system; no solution   **21.** $(5, \frac{14}{5})$

23. $x = \dfrac{kd - q}{ad - 1}, \quad y = \dfrac{aq - k}{ad - 1}$ **25.** $x = 2$ **27.** $x = 2$

Exercise 9.4   (page 358)

1. $(1, 1, 2)$ **3.** $(0, 2, 2)$ **5.** $(3, 2, 1)$ **7.** inconsistent system; no solution
9. dependent equations; one possible solution is $(\frac{1}{2}, \frac{5}{3}, 0)$ **11.** $(2, 6, 9)$ **13.** $-2, 4,$ and 16
15. 9, 10, and 11   **17.** 10 nickels, 5 dimes, 2 quarters
19. 50 expensive footballs, 75 middle-priced footballs, 1000 cheap footballs
21. 250 five-dollar tickets, 375 three-dollar tickets, 125 two-dollar tickets

Exercise 9.5   (page 364)

1. 0   **3.** -13 **5.** 26   **7.** 0   **9.** $10a$ **11.** $(1, 1, 2)$ **13.** $(3, 2, 1)$
15. Cramer's Rule fails.   **17.** $(3, -2, 1)$ **19.** Cramer's Rule fails.   **21.** $(-2, 3, 1)$
23. Cramer's Rule fails.

Exercise 9.6   (page 370)

1. $(1, 1)$ **3.** $(2, -3)$ **5.** $(0, -3)$ **7.** $(1, 2, 3)$ **9.** $(-1, -1, 2)$ **11.** $(2, 1, 0)$
13. $(1, 2)$ **15.** $(2, 0)$ **17.** no solution   **19.** $x = -6 - z, \quad y = 2 - z, \quad z$ can be any number
21. $x = 2 - z, \quad y = 1 - z, \quad z$ can be any number

Exercise 9.7   (page 373)

1. $y = -2x + 3$

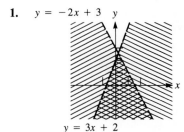

$y = 3x + 2$

3.

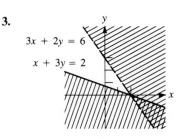

$3x + 2y = 6$

$x + 3y = 2$

5.

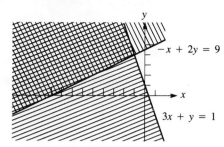

$-x + 2y = 9$

$3x + y = 1$

7.

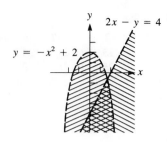

$2x - y = 4$

$y = -x^2 + 2$

9.

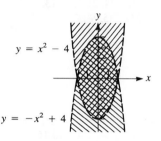

$y = x^2 - 4$

$y = -x^2 + 4$

11.

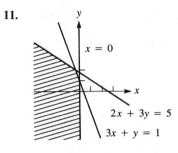

$x = 0$

$2x + 3y = 5$

$3x + y = 1$

13.

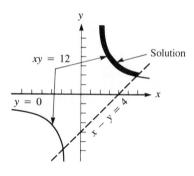

$xy = 12$

$y = 0$

Solution

$x - y = 4$

15.

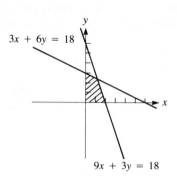

$3x + 6y = 18$

$9x + 3y = 18$

REVIEW EXERCISES   (page 375)

1. $(3, 5)$ **3.** dependent equations   **5.** $(-1, 3)$ **7.** $(3, 4)$ **9.** $(-3, 1)$ **11.** $(9, -4)$
13. $(1, 2, 3)$ **15.** 18   **17.** -3 **19.** $(2, 1)$ **21.** $(1, -2, 3)$ **23.** $(2, 1)$
25.

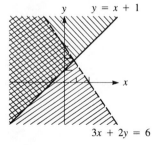

$y = x + 1$

$3x + 2y = 6$

CHAPTER NINE TEST   (page 376)

1.

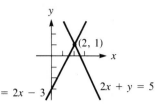

3. consistent   **5.** $y = 4$ **7.** -17 **9.** $\begin{vmatrix} 1 & -1 \\ 3 & 1 \end{vmatrix}$

11. $x = 3$ **13.** $z = -1$ **15.** $\begin{bmatrix} 1 & 1 & 1 & \vdots & 4 \\ 1 & 1 & -1 & \vdots & 6 \\ 2 & -3 & 1 & \vdots & -1 \end{bmatrix}$

CUMULATIVE REVIEW EXERCISES (CHAPTER 7–9)   (page 377)

1.

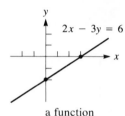

a function

3. $(3, -2)$ **5.** $y = -\frac{7}{5}x + \frac{11}{5}$ **7.** $f(-1) = 5$

9. $(f \circ g)(x) = 12x^2 - 12x + 5$ **11.** $y = f^{-1}(x) = 3x - 3;$ a function   **13.** $y = \dfrac{kxz}{rs}$

15.

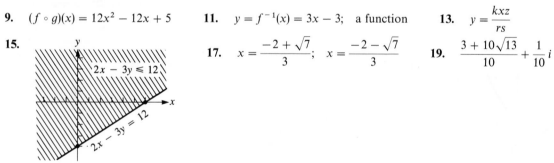

17. $x = \dfrac{-2 + \sqrt{7}}{3};\quad x = \dfrac{-2 - \sqrt{7}}{3}$ **19.** $\dfrac{3 + 10\sqrt{13}}{10} + \dfrac{1}{10}i$

21. 16   **23.**

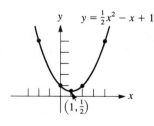

25. yes   **27.**

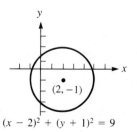

29. $(1, 1)$ **31.** $(3, 1)$ **33.** $y = -1$ **35.** $z = 1$

504   APPENDIX I

37.

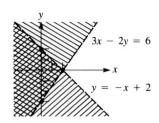

$3x - 2y = 6$

$y = -x + 2$

Exercise 10.1   (page 387)

1.

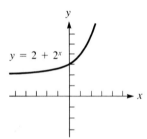

$y = 3^x$

3.

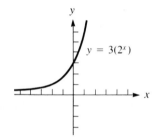

$y = \left(\dfrac{1}{5}\right)^x$

5.

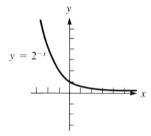

$y = \left(\dfrac{3}{4}\right)^x$

7.

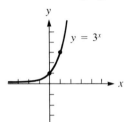

$y = 2 + 2^x$

9.

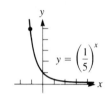

$y = 3(2^x)$

11.

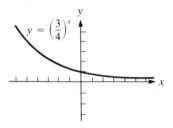

$y = 2^{-x}$

13.

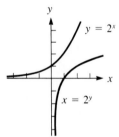

$y = 2^x$

$x = 2^y$

15.

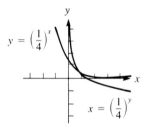

$y = \left(\dfrac{1}{4}\right)^x$

$x = \left(\dfrac{1}{4}\right)^y$

17. $b = 5$

19.   no value of b **21.** $b = \frac{1}{2}$ **23.**   \$22,080.40   **25.**   \$114,508.36   **27.**   \$14,576.95
29.   3645 fish   **31.**   about 8.58 billion   **33.** 1.679046×10^8 **35.** 5.0421×10^{-5}

Exercise 10.2   (page 395)

1. $3^4 = 81$ **3.** $\left(\dfrac{1}{2}\right)^3 = \dfrac{1}{8}$ **5.** $4^{-3} = \dfrac{1}{64}$ **7.** $x^z = y$ **9.** $\log_8 64 = 2$

11. $\log_4 \dfrac{1}{16} = -2$ **13.** $\log_{1/2} 32 = -5$ **15.** $\log_x z = y$ **17.**   0   **19.**   2   **21.**   1

23.   3   **25.**   3   **27.**   3   **29.**   16   **31.**   81   **33.**   2   **35.**   2   **37.**   2

39.   3   **41.**   81   **43.**   25   **45.** -2 **47.** $\frac{1}{3}$ **49.**   4

51. 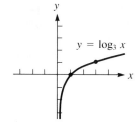   **53.** 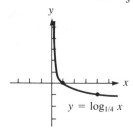   **55.**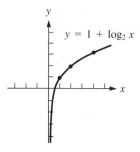

57. 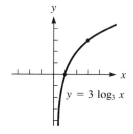   **59.** 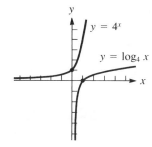   **61.**   no value of b **63.** $b = 3$

65.   no value of b

67. $\log_b x + \log_b y + \log_b z$ **69.** $2 \log_b x - 2 \log_b y$ **71.** $\log_b x + \frac{1}{2} \log_b(x - z)$

73. $\frac{3}{4} \log_b x + \frac{1}{2} \log_b y - \log_b z$ **75.** $\log_b \dfrac{x + 1}{x}$ **77.** $\log_b x^2 \sqrt[3]{y}$ **79.** $\log_b \dfrac{\sqrt{z}}{x^3 y^2}$

81. $\log_b \dfrac{x}{y}$ **83.**   false   **85.**   true   **87.**   false   **89.**   true   **91.**   true

93.   false   **95.**   false   **97.**   false   **99.**   1.4472   **101.**   0.3521   **103.**   1.1972

105.   2.4014   **107.**   0.3011   **109.**   0.7782   **111.**   1.6990

Exercise 10.3   (page 402)

1.   0.5119   **3.**   0.3854   **5.**   1949.8446   **7.**   0.0001   **9.**   1.3610   **11.** -3.6109

13.   0.0002   **15.**   16.9455   **17.**   2.3282   **19.**   4.77   **21.** 6.31×10^{-14} gram ions per L

23.   from 2.5119×10^{-8} to 1.585×10^{-7} **25.**   0.71 V   **27.**   49.54 db   **29.**   4.4

31.   2500 microns   **33.**   12.2 years   **35.**   23.1 years

37.   The original loudness L_0 is increased by $k \ln 2$ because $L = L_0 + k \ln 2$,   where $L_0 = k \ln I$

39.   The original intensity must be cubed.

Exercise 10.4   (page 409)

1. $\dfrac{\log 5}{\log 4}$ **3.** $\dfrac{1}{\log 2} + 1$ **5.** -2 **7.** $\pm\sqrt{2}$ **9.**   0   **11.** $\dfrac{\log 3}{\log 4 - \log 3}$ **13.**   9

15.   8   **17.**   2   **19.**   3   **21.**   7   **23.**   4   **25.**   10, -10 **27.**   50   **29.**   20

31.   10   **33.**   10   **35.**   2   **37.** 10^{10} **39.**   1.771   **41.** -1 **43.** $\frac{1}{6}$

45.   0.3028   **51.**   5.146 years   **53.**   about 42.7 days   **55.**   about 4200 years old

57.   It will be 2.828 times larger.   **59.**   about 9.46 years   **61.**   about 9.694 years

Exercise 10.5   (page 418)

1. 0.7760   **3.** 0.6263   **5.** 5.6355   **7.** -2.8633 **9.** 3.14   **11.** 1.08   **13.** 8380
15. 0.00284   **17.** 0.8384   **19.** -2.4614 **21.** 4.281   **23.** 1598   **25.** 0.007702
27. 1.5369   **29.** 2.1193   **31.** 1.23   **33.** 1.14   **35.** 312,000   **37.** 0.266
39. 0.309   **41.** 15.33   **43.** 0.00000000434   **45.** 759,800   **47.** 133.5   **49.** 3.3810
51. -4.8000

REVIEW EXERCISES   (page 420)

1.
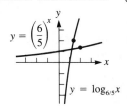

3.
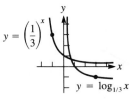

5.
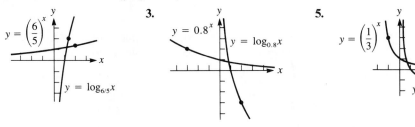

7.
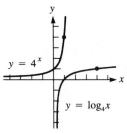

9. $M = 8$ **11.** $b = 3$ **13.** $P = 1$ **15.** $R = \frac{1}{6}$

17. $T = -2$ **19.** $v = 27$ **21.** $N = 32$ **23.** $M = 27$ **25.** $b = \frac{1}{8}$

27. $\frac{1}{2}\log_b x + \frac{1}{2}\log_b z - \log_b y$ **29.** $\log_b \frac{x^2}{y^3 z^2}$ **31.** $x = 2.8665$ **33.** $z = -2.0149$

35. $x = \dfrac{\log 7}{\log 3}$ **37.** $x = \dfrac{\log 3}{\log 3 - \log 2}$ **39.** $x = 4$ **41.** $x = \dfrac{\ln 9}{\ln 2}$ **43.** $x = \dfrac{e}{e-1}$

45. approximately 9034 years old   **47.** $\text{pH} = \log_{10} \dfrac{1}{[\text{H}^+]} = \log_{10}[\text{H}^+]^{-1} = -\log_{10}[\text{H}^+]$

49. 8.141   **51.** 0.0001617

CHAPTER TEN TEST   (page 422)

1.
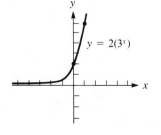

3. 2   **5.** 4   **7.** 2   **9.**
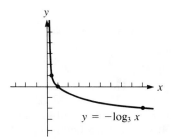

11. 4.0   **13.** 1.2   **15.** 4.2   **17.** $x = \dfrac{\log 3}{\log 5}$ **19.** $x = 1$

21. $\log_3 8 = \dfrac{\log 8}{\log 3} = \dfrac{0.90}{0.48} \approx 1.9$ **23.** 5.4

Exercise 11.1   (page 429)

1. 6   **3.** -120 **5.** 30   **7.** 144   **9.** $8! = 40,320$ **11.** $\frac{1}{110}$ **13.** 2352
15. 10   **17.** 21   **19.** $x^2 + 2xy + y^2$ **21.** $x^3 - 3x^2y + 3xy^2 - y^3$
23. $8x^3 + 12x^2y + 6xy^2 + y^3$ **25.** $x^3 - 6x^2y + 12xy^2 - 8y^3$

27. $16x^4 + 96x^3y + 216x^2y^2 + 216xy^3 + 81y^4$ **29.** $\dfrac{x^4}{16} - \dfrac{1}{6}x^3y + \dfrac{1}{6}x^2y^2 - \dfrac{2}{27}xy^3 + \dfrac{1}{81}y^4$

31. $243 + 810y + 1080y^2 + 720y^3 + 240y^4 + 32y^5$
35. $1, 1, 2, 3, 5, 8, 13, \ldots$;   beginning with 2, each number is the sum of the previous two numbers.

Exercise 11.2   (page 432)

1. $3a^2b$ **3.** $-4xy^3$ **5.** $15x^2y^4$ **7.** $28x^6y^2$ **9.** $90x^3$ **11.** $640x^3y^2$

13. $-12x^3y$ **15.** $-70,000x^4$ **17.** $810xy^4$ **19.** $180x^4y^2$ **21.** $-\dfrac{1}{6}x^3y$

23. $\dfrac{n!}{3!(n-3)!}a^{n-3}b^3$ **25.** $\dfrac{n!}{4!(n-4)!}a^{n-4}b^4$ **27.** $\dfrac{n!}{(r-1)!(n-r+1)!}a^{n-r+1}b^{r-1}$

29. $\dfrac{81(2^{n-4})n!}{4!(n-4)!}a^{n-4}b^4$ **31.** 1.20   **33.** 0.89

Exercise 11.3   (page 435)

1. 35   **3.** 210   **5.** 1,000,000   **7.** 151,200   **9.** 136,080   **11.** 6,760,000
13. 8,000,000   **15.** 6   **17.** 60   **19.** about 1.3×10^{12} **21.** 12   **23.** 1800
25. 5   **27.** 28   **29.** 720   **31.** 2880   **33.** 2880   **35.** 13,800   **37.** 3,628,799
39. 720   **41.** 256

Exercise 11.4   (page 440)

1. 10   **3.** 20   **5.** 1   **7.** 1   **9.** 36   **11.** 50   **13.** 24   **15.** 2

17. $\dfrac{n!}{2!(n-2)!}$ **19.** $\dfrac{n!}{(n-3)!3!}$ **21.** 364   **23.** 126   **25.** 5   **27.** 18   **29.** 792

31. 7920   **33.** $\dbinom{4}{0}x^4 + \dbinom{4}{1}x^3y + \dbinom{4}{2}x^2y^2 + \dbinom{4}{3}xy^3 + \dbinom{4}{4}y^4 = x^4 + 4x^3y + 6x^2y^2 + 4xy^3 + y^4$

35. $\dbinom{2}{0}x^2 + \dbinom{2}{1}x(-y) + \dbinom{2}{2}(-y)^2 = x^2 - 2xy + y^2$

37. $\dbinom{3}{0}(2x)^3 + \dbinom{3}{1}(2x)^2y + \dbinom{3}{2}(2x)y^2 + \dbinom{3}{3}y^3 = 8x^3 + 12x^2y + 6xy^2 + y^3$

39. $\dbinom{5}{0}x^5 + \dbinom{5}{1}x^4 + \dbinom{5}{2}x^3 + \dbinom{5}{3}x^2 + \dbinom{5}{4}x + \dbinom{5}{5} = x^5 + 5x^4 + 10x^3 + 10x^2 + 5x + 1$

41. $\dbinom{4}{0}(2x)^4 + \dbinom{4}{1}(2x)^3 + \dbinom{4}{2}(2x)^2 + \dbinom{4}{3}(2x) + \dbinom{4}{4} = 16x^4 + 32x^3 + 24x^2 + 8x + 1$

43. $\binom{3}{0}(3)^3 + \binom{3}{1}(3)^2(-x^2) + \binom{3}{2}(3)(-x^2)^2 + \binom{3}{3}(-x^2)^3 = 27 - 27x^2 + 9x^4 - x^6$

45. $\binom{4}{0}(2x)^4 + \binom{4}{1}(2x)^3(3y) + \binom{4}{2}(2x)^2(3y)^2 + \binom{4}{3}(2x)(3y)^3 + \binom{4}{4}(3y)^4 = 16x^4 + 96x^3y + 216x^2y^2 + 216xy^3 + 81y^4$

47. $\binom{5}{3}x^2(-5y)^3 = -1250x^2y^3$ **49.** $\binom{4}{1}(x^2)^3(-y^3) = -4x^6y^3$

Exercise 11.5   (page 446)

1. 3, 5, 7, 9, 11   **3.** $-5, -8, -11, -14, -17$ **5.** 5, 11, 17, 23, 29
7. $-4, -11, -18, -25, -32$ **9.** $-118, -111, -104, -97, -90$ **11.** 34, 31, 28, 25, 22
13. 5, 12, 19, 26, 33   **15.** 355   **17.** -179 **19.** -23 **21.** 12   **23.** $\frac{17}{4}, \frac{13}{2}, \frac{35}{4}$
25. 12, 14, 16, 18   **27.** $\frac{29}{2}$ **29.** $\frac{5}{4}$ **31.** 1335   **33.** 459   **35.** 354   **37.** 255
39. 1275   **41.** 2500   **43.** 60, 110, 160, 210, 260, 310; $6060   **45.** 48 ft; 80 ft   **49.** 60
51. 31   **53.** 12

Exercise 11.6   (page 451)

1. 3, 6, 12, 24, 48   **3.** $-5, -1, -\frac{1}{5}, -\frac{1}{25}, -\frac{1}{125}$ **5.** 2, 8, 32, 128, 512
7. $-3, -12, -48, -192, -768$ **9.** $-64, 32, -16, 8, -4$ **11.** $-64, -32, -16, -8, -4$
13. 2, 10, 50, 250, 1250   **15.** 3584   **17.** $\frac{1}{27}$ **19.** 3   **21.** 6, 18, 54
23. $-20, -100, -500, -2500$ **25.** -16 **27.** $10\sqrt{2}$ **29.** No geometric mean exists.
31. 728   **33.** 122   **35.** -255 **37.** 381   **39.** $\frac{156}{25}$ **41.** $-\frac{21}{4}$
43. about 669 people   **45.** $1469.74   **47.** $140,853.75

Exercise 11.7   (page 455)

1. 16   **3.** 81   **5.** 8   **7.** $-\frac{135}{4}$ **9.** no sum   **11.** $-\frac{81}{2}$ **13.** $\frac{1}{9}$ **15.** $-\frac{1}{3}$

17. $\frac{4}{33}$ **19.** $\frac{25}{33}$ **21.** 30 m   **25.** no; $0.999999 = \dfrac{999999}{1,000,000} < 1$

REVIEW EXERCISES   (page 463)

1. 144   **3.** 15   **5.** $x^5 + 5x^4y + 10x^3y^2 + 10x^2y^3 + 5xy^4 + y^5$ **7.** $64x^3 - 48x^2y + 12xy^2 - y^3$
9. $6x^2y^2$ **11.** $-108x^2y$ **13.** 120   **15.** 5040   **17.** 20,160   **19.** 1   **21.** 28
23. 700   **25.** $\frac{15}{2}$ **27.** 14   **29.** 120   **31.** 122, 137, 152, 167, 182   **33.** 1550
35. 24, 12, 6, 3, $\frac{3}{2}$ **37.** $-\frac{85}{8}$ **39.** 125   **41.** $1638.40   **43.** 13 yrs

CHAPTER ELEVEN TEST   (page 464)

1. 210   **3.** $-5x^4y$ **5.** 120   **7.** 20   **9.** 360   **11.** 35   **13.** 306   **15.** 3
17. $\frac{364}{27}$ **19.** $\frac{27}{2}$ **21.** Yes; it is true for $n = 7$.

CUMULATIVE REVIEW EXERCISES   (CHAPTERS 10–11)   (page 465)

1.

$y = \left(\frac{1}{2}\right)^x$ **3.** 5   **5.** $\frac{1}{27}$ **7.** $y = 2^x$ **9.** 1.9912   **11.** 1.6902

13. $x = \dfrac{2 \log 2}{\log 3 - \log 2}$ **15.** $2848.31 **17.** 30,240 **19.** $112x^2y^6$ **21.** 120 **23.** 84

25. 103 **27.** 8 and 19 **29.** 27 **31.** $\frac{1023}{64}$ **33.** $\frac{27}{2}$

ANSWERS TO SAMPLE FINAL EXAMINATION (page 466)

1. a	**12.** b	**23.** d	**34.** c	**45.** b	**56.** b
2. a	**13.** a	**24.** a	**35.** a	**46.** c	**57.** b
3. c	**14.** d	**25.** c	**36.** d	**47.** c	**58.** c
4. d	**15.** d	**26.** c	**37.** b	**48.** a	**59.** b
5. c	**16.** c	**27.** b	**38.** b	**49.** a	**60.** a
6. a	**17.** b	**28.** c	**39.** d	**50.** b	**61.** b
7. c	**18.** c	**29.** d	**40.** c	**51.** b	**62.** c
8. b	**19.** d	**30.** b	**41.** b	**52.** a	
9. a	**20.** c	**31.** b	**42.** d	**53.** a	
10. c	**21.** d	**32.** d	**43.** d	**54.** c	
11. d	**22.** b	**33.** a	**44.** a	**55.** d	

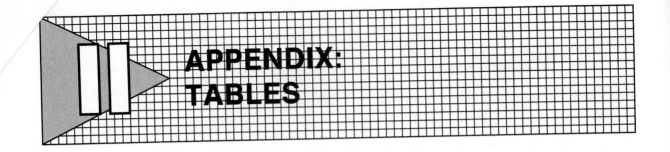

APPENDIX: TABLES

Table A Powers and Roots

n	n^2	$\sqrt{n}$	n^3	$\sqrt[3]{n}$	n	n^2	$\sqrt{n}$	n^3	$\sqrt[3]{n}$
1	1	1.000	1	1.000	51	2,601	7.141	132,651	3.708
2	4	1.414	8	1.260	52	2,704	7.211	140,608	3.733
3	9	1.732	27	1.442	53	2,809	7.280	148,877	3.756
4	16	2.000	64	1.587	54	2,916	7.348	157,464	3.780
5	25	2.236	125	1.710	55	3,025	7.416	166,375	3.803
6	36	2.449	216	1.817	56	3,136	7.483	175,616	3.826
7	49	2.646	343	1.913	57	3,249	7.550	185,193	3.849
8	64	2.828	512	2.000	58	3,364	7.616	195,112	3.871
9	81	3.000	729	2.080	59	3,481	7.681	205,379	3.893
10	100	3.162	1,000	2.154	60	3,600	7.746	216,000	3.915
11	121	3.317	1,331	2.224	61	3,721	7.810	226,981	3.936
12	144	3.464	1,728	2.289	62	3,844	7.874	238,328	3.958
13	169	3.606	2,197	2.351	63	3,969	7.937	250,047	3.979
14	196	3.742	2,744	2.410	64	4,096	8.000	262,144	4.000
15	225	3.873	3,375	2.466	65	4,225	8.062	274,625	4.021
16	256	4.000	4,096	2.520	66	4,356	8.124	287,496	4.041
17	289	4.123	4,913	2.571	67	4,489	8.185	300,763	4.062
18	324	4.243	5,832	2.621	68	4,624	8.246	314,432	4.082
19	361	4.359	6,859	2.668	69	4,761	8.307	328,509	4.102
20	400	4.472	8,000	2.714	70	4,900	8.367	343,000	4.121
21	441	4.583	9,261	2.759	71	5,041	8.426	357,911	4.141
22	484	4.690	10,648	2.802	72	5,184	8.485	373,248	4.160
23	529	4.796	12,167	2.844	73	5,329	8.544	389,017	4.179
24	576	4.899	13,824	2.884	74	5,476	8.602	405,224	4.198
25	625	5.000	15,625	2.924	75	5,625	8.660	421,875	4.217
26	676	5.099	17,576	2.962	76	5,776	8.718	438,976	4.236
27	729	5.196	19,683	3.000	77	5,929	8.775	456,533	4.254
28	784	5.292	21,952	3.037	78	6,084	8.832	474,552	4.273
29	841	5.385	24,389	3.072	79	6,241	8.888	493,039	4.291
30	900	5.477	27,000	3.107	80	6,400	8.944	512,000	4.309
31	961	5.568	29,791	3.141	81	6,561	9.000	531,441	4.327
32	1,024	5.657	32,768	3.175	82	6,724	9.055	551,368	4.344
33	1,089	5.745	35,937	3.208	83	6,889	9.110	571,787	4.362
34	1,156	5.831	39,304	3.240	84	7,056	9.165	592,704	4.380
35	1,225	5.916	42,875	3.271	85	7,225	9.220	614,125	4.397
36	1,296	6.000	46,656	3.302	86	7,396	9.274	636,056	4.414
37	1,369	6.083	50,653	3.332	87	7,569	9.327	658,503	4.431
38	1,444	6.164	54,872	3.362	88	7,744	9.381	681,472	4.448
39	1,521	6.245	59,319	3.391	89	7,921	9.434	704,969	4.465
40	1,600	6.325	64,000	3.420	90	8,100	9.487	729,000	4.481
41	1,681	6.403	68,921	3.448	91	8,281	9.539	753,571	4.498
42	1,764	6.481	74,088	3.476	92	8,464	9.592	778,688	4.514
43	1,849	6.557	79,507	3.503	93	8,649	9.644	804,357	4.531
44	1,936	6.633	85,184	3.530	94	8,836	9.695	830,584	4.547
45	2,025	6.708	91,125	3.557	95	9,025	9.747	857,375	4.563
46	2,116	6.782	97,336	3.583	96	9,216	9.798	884,736	4.579
47	2,209	6.856	103,823	3.609	97	9,409	9.849	912,673	4.595
48	2,304	6.928	110,592	3.634	98	9,604	9.899	941,192	4.610
49	2,401	7.000	117,649	3.659	99	9,801	9.950	970,299	4.626
50	2,500	7.071	125,000	3.684	100	10,000	10.000	1,000,000	4.642

Table B Base-10 Logarithms

N	0	1	2	3	4	5	6	7	8	9
1.0	.0000	.0043	.0086	.0128	.0170	.0212	.0253	.0294	.0334	.0374
1.1	.0414	.0453	.0492	.0531	.0569	.0607	.0645	.0682	.0719	.0755
1.2	.0792	.0828	.0864	.0899	.0934	.0969	.1004	.1038	.1072	.1106
1.3	.1139	.1173	.1206	.1239	.1271	.1303	.1335	.1367	.1399	.1430
1.4	.1461	.1492	.1523	.1553	.1584	.1614	.1644	.1673	.1703	.1732
1.5	.1761	.1790	.1818	.1847	.1875	.1903	.1931	.1959	.1987	.2014
1.6	.2041	.2068	.2095	.2122	.2148	.2175	.2201	.2227	.2253	.2279
1.7	.2304	.2330	.2355	.2380	.2405	.2430	.2455	.2480	.2504	.2529
1.8	.2553	.2577	.2601	.2625	.2648	.2672	.2695	.2718	.2742	.2765
1.9	.2788	.2810	.2833	.2856	.2878	.2900	.2923	.2945	.2967	.2989
2.0	.3010	.3032	.3054	.3075	.3096	.3118	.3139	.3160	.3181	.3201
2.1	.3222	.3243	.3263	.3284	.3304	.3324	.3345	.3365	.3385	.3404
2.2	.3424	.3444	.3464	.3483	.3502	.3522	.3541	.3560	.3579	.3598
2.3	.3617	.3636	.3655	.3674	.3692	.3711	.3729	.3747	.3766	.3784
2.4	.3802	.3820	.3838	.3856	.3874	.3892	.3909	.3927	.3945	.3962
2.5	.3979	.3997	.4014	.4031	.4048	.4065	.4082	.4099	.4116	.4133
2.6	.4150	.4166	.4183	.4200	.4216	.4232	.4249	.4265	.4281	.4298
2.7	.4314	.4330	.4346	.4362	.4378	.4393	.4409	.4425	.4440	.4456
2.8	.4472	.4487	.4502	.4518	.4533	.4548	.4564	.4579	.4594	.4609
2.9	.4624	.4639	.4654	.4669	.4683	.4698	.4713	.4728	.4742	.4757
3.0	.4771	.4786	.4800	.4814	.4829	.4843	.4857	.4871	.4886	.4900
3.1	.4914	.4928	.4942	.4955	.4969	.4983	.4997	.5011	.5024	.5038
3.2	.5051	.5065	.5079	.5092	.5105	.5119	.5132	.5145	.5159	.5172
3.3	.5185	.5198	.5211	.5224	.5237	.5250	.5263	.5276	.5289	.5302
3.4	.5315	.5328	.5340	.5353	.5366	.5378	.5391	.5403	.5416	.5428
3.5	.5441	.5453	.5465	.5478	.5490	.5502	.5514	.5527	.5539	.5551
3.6	.5563	.5575	.5587	.5599	.5611	.5623	.5635	.5647	.5658	.5670
3.7	.5682	.5694	.5705	.5717	.5729	.5740	.5752	.5763	.5775	.5786
3.8	.5798	.5809	.5821	.5832	.5843	.5855	.5866	.5877	.5888	.5899
3.9	.5911	.5922	.5933	.5944	.5955	.5966	.5977	.5988	.5999	.6010
4.0	.6021	.6031	.6042	.6053	.6064	.6075	.6085	.6096	.6107	.6117
4.1	.6128	.6138	.6149	.6160	.6170	.6180	.6191	.6201	.6212	.6222
4.2	.6232	.6243	.6253	.6263	.6274	.6284	.6294	.6304	.6314	.6325
4.3	.6335	.6345	.6355	.6365	.6375	.6385	.6395	.6405	.6415	.6425
4.4	.6435	.6444	.6454	.6464	.6474	.6484	.6493	.6503	.6513	.6522
4.5	.6532	.6542	.6551	.6561	.6571	.6580	.6590	.6599	.6609	.6618
4.6	.6628	.6637	.6646	.6656	.6665	.6675	.6684	.6693	.6702	.6712
4.7	.6721	.6730	.6739	.6749	.6758	.6767	.6776	.6785	.6794	.6803
4.8	.6812	.6821	.6830	.6839	.6848	.6857	.6866	.6875	.6884	.6893
4.9	.6902	.6911	.6920	.6928	.6937	.6946	.6955	.6964	.6972	.6981
5.0	.6990	.6998	.7007	.7016	.7024	.7033	.7042	.7050	.7059	.7067
5.1	.7076	.7084	.7093	.7101	.7110	.7118	.7126	.7135	.7143	.7152
5.2	.7160	.7168	.7177	.7185	.7193	.7202	.7210	.7218	.7226	.7235
5.3	.7243	.7251	.7259	.7267	.7275	.7284	.7292	.7300	.7308	.7316
5.4	.7324	.7332	.7340	.7348	.7356	.7364	.7372	.7380	.7388	.7396

Table B (*Continued*)

N	0	1	2	3	4	5	6	7	8	9
5.5	.7404	.7412	.7419	.7427	.7435	.7443	.7451	.7459	.7466	.7474
5.6	.7482	.7490	.7497	.7505	.7513	.7520	.7528	.7536	.7543	.7551
5.7	.7559	.7566	.7574	.7582	.7589	.7597	.7604	.7612	.7619	.7627
5.8	.7634	.7642	.7649	.7657	.7664	.7672	.7679	.7686	.7694	.7701
5.9	.7709	.7716	.7723	.7731	.7738	.7745	.7752	.7760	.7767	.7774
6.0	.7782	.7789	.7796	.7803	.7810	.7818	.7825	.7832	.7839	.7846
6.1	.7853	.7860	.7868	.7875	.7882	.7889	.7896	.7903	.7910	.7917
6.2	.7924	.7931	.7938	.7945	.7952	.7959	.7966	.7973	.7980	.7987
6.3	.7993	.8000	.8007	.8014	.8021	.8028	.8035	.8041	.8048	.8055
6.4	.8062	.8069	.8075	.8082	.8089	.8096	.8102	.8109	.8116	.8122
6.5	.8129	.8136	.8142	.8149	.8156	.8162	.8169	.8176	.8182	.8189
6.6	.8195	.8202	.8209	.8215	.8222	.8228	.8235	.8241	.8248	.8254
6.7	.8261	.8267	.8274	.8280	.8287	.8293	.8299	.8306	.8312	.8319
6.8	.8325	.8331	.8338	.8344	.8351	.8357	.8363	.8370	.8376	.8382
6.9	.8388	.8395	.8401	.8407	.8414	.8420	.8426	.8432	.8439	.8445
7.0	.8451	.8457	.8463	.8470	.8476	.8482	.8488	.8494	.8500	.8506
7.1	.8513	.8519	.8525	.8531	.8537	.8543	.8549	.8555	.8561	.8567
7.2	.8573	.8579	.8585	.8591	.8597	.8603	.8609	.8615	.8621	.8627
7.3	.8633	.8639	.8645	.8651	.8657	.8663	.8669	.8675	.8681	.8686
7.4	.8692	.8698	.8704	.8710	.8716	.8722	.8727	.8733	.8739	.8745
7.5	.8751	.8756	.8762	.8768	.8774	.8779	.8785	.8791	.8797	.8802
7.6	.8808	.8814	.8820	.8825	.8831	.8837	.8842	.8848	.8854	.8859
7.7	.8865	.8871	.8876	.8882	.8887	.8893	.8899	.8904	.8910	.8915
7.8	.8921	.8927	.8932	.8938	.8943	.8949	.8954	.8960	.8965	.8971
7.9	.8976	.8982	.8987	.8993	.8998	.9004	.9009	.9015	.9020	.9025
8.0	.9031	.9036	.9042	.9047	.9053	.9058	.9063	.9069	.9074	.9079
8.1	.9085	.9090	.9096	.9101	.9106	.9112	.9117	.9122	.9128	.9133
8.2	.9138	.9143	.9149	.9154	.9159	.9165	.9170	.9175	.9180	.9186
8.3	.9191	.9196	.9201	.9206	.9212	.9217	.9222	.9227	.9232	.9238
8.4	.9243	.9248	.9253	.9258	.9263	.9269	.9274	.9279	.9284	.9289
8.5	.9294	.9299	.9304	.9309	.9315	.9320	.9325	.9330	.9335	.9340
8.6	.9345	.9350	.9355	.9360	.9365	.9370	.9375	.9380	.9385	.9390
8.7	.9395	.9400	.9405	.9410	.9415	.9420	.9425	.9430	.9435	.9440
8.8	.9445	.9450	.9455	.9460	.9465	.9469	.9474	.9479	.9484	.9489
8.9	.9494	.9499	.9504	.9509	.9513	.9518	.9523	.9528	.9533	.9538
9.0	.9542	.9547	.9552	.9557	.9562	.9566	.9571	.9576	.9581	.9586
9.1	.9590	.9595	.9600	.9605	.9609	.9614	.9619	.9624	.9628	.9633
9.2	.9638	.9643	.9647	.9652	.9657	.9661	.9666	.9671	.9675	.9680
9.3	.9685	.9689	.9694	.9699	.9703	.9708	.9713	.9717	.9722	.9727
9.4	.9731	.9736	.9741	.9745	.9750	.9754	.9759	.9763	.9768	.9773
9.5	.9777	.9782	.9786	.9791	.9795	.9800	.9805	.9809	.9814	.9818
9.6	.9823	.9827	.9832	.9836	.9841	.9845	.9850	.9854	.9859	.9863
9.7	.9868	.9872	.9877	.9881	.9886	.9890	.9894	.9899	.9903	.9908
9.8	.9912	.9917	.9921	.9926	.9930	.9934	.9939	.9943	.9948	.9952
9.9	.9956	.9961	.9965	.9969	.9974	.9978	.9983	.9987	.9991	.9996

Table C (Continued)

N	0	1	2	3	4	5	6	7	8	9
5.5	1.7047	.7066	.7084	.7102	.7120	.7138	.7156	.7174	.7192	.7210
5.6	.7228	.7246	.7263	.7281	.7299	.7317	.7334	.7352	.7370	.7387
5.7	.7405	.7422	.7440	.7457	.7475	.7492	.7509	.7527	.7544	.7561
5.8	.7579	.7596	.7613	.7630	.7647	.7664	.7681	.7699	.7716	.7733
5.9	.7750	.7766	.7783	.7800	.7817	.7834	.7851	.7867	.7884	.7901
6.0	1.7918	.7934	.7951	.7967	.7984	.8001	.8017	.8034	.8050	.8066
6.1	.8083	.8099	.8116	.8132	.8148	.8165	.8181	.8197	.8213	.8229
6.2	.8245	.8262	.8278	.8294	.8310	.8326	.8342	.8358	.8374	.8390
6.3	.8405	.8421	.8437	.8453	.8469	.8485	.8500	.8516	.8532	.8547
6.4	.8563	.8579	.8594	.8610	.8625	.8641	.8656	.8672	.8687	.8703
6.5	1.8718	.8733	.8749	.8764	.8779	.8795	.8810	.8825	.8840	.8856
6.6	.8871	.8886	.8901	.8916	.8931	.8946	.8961	.8976	.8991	.9006
6.7	.9021	.9036	.9051	.9066	.9081	.9095	.9110	.9125	.9140	.9155
6.8	.9169	.9184	.9199	.9213	.9228	.9242	.9257	.9272	.9286	.9301
6.9	.9315	.9330	.9344	.9359	.9373	.9387	.9402	.9416	.9430	.9445
7.0	1.9459	.9473	.9488	.9502	.9516	.9530	.9544	.9559	.9573	.9587
7.1	.9601	.9615	.9629	.9643	.9657	.9671	.9685	.9699	.9713	.9727
7.2	.9741	.9755	.9769	.9782	.9796	.9810	.9824	.9838	.9851	.9865
7.3	.9879	.9892	.9906	.9920	.9933	.9947	.9961	.9974	.9988	2.0001
7.4	2.0015	.0028	.0042	.0055	.0069	.0082	.0096	.0109	.0122	.0136
7.5	2.0149	.0162	.0176	.0189	.0202	.0215	.0229	.0242	.0255	.0268
7.6	.0281	.0295	.0308	.0321	.0334	.0347	.0360	.0373	.0386	.0399
7.7	.0412	.0425	.0438	.0451	.0464	.0477	.0490	.0503	.0516	.0528
7.8	.0541	.0554	.0567	.0580	.0592	.0605	.0618	.0631	.0643	.0656
7.9	.0669	.0681	.0694	.0707	.0719	.0732	.0744	.0757	.0769	.0782
8.0	2.0794	.0807	.0819	.0832	.0844	.0857	.0869	.0882	.0894	.0906
8.1	.0919	.0931	.0943	.0956	.0968	.0980	.0992	.1005	.1017	.1029
8.2	.1041	.1054	.1066	.1078	.1090	.1102	.1114	.1126	.1138	.1150
8.3	.1163	.1175	.1187	.1199	.1211	.1223	.1235	.1247	.1258	.1270
8.4	.1282	.1294	.1306	.1318	.1330	.1342	.1353	.1365	.1377	.1389
8.5	2.1401	.1412	.1424	.1436	.1448	.1459	.1471	.1483	.1494	.1506
8.6	.1518	.1529	.1541	.1552	.1564	.1576	.1587	.1599	.1610	.1622
8.7	.1633	.1645	.1656	.1668	.1679	.1691	.1702	.1713	.1725	.1736
8.8	.1748	.1759	.1770	.1782	.1793	.1804	.1815	.1827	.1838	.1849
8.9	.1861	.1872	.1883	.1894	.1905	.1917	.1928	.1939	.1950	.1961
9.0	2.1972	.1983	.1994	.2006	.2017	.2028	.2039	.2050	.2061	.2072
9.1	.2083	.2094	.2105	.2116	.2127	.2138	.2148	.2159	.2170	.2181
9.2	.2192	.2203	.2214	.2225	.2235	.2246	.2257	.2268	.2279	.2289
9.3	.2300	.2311	.2322	.2332	.2343	.2354	.2364	.2375	.2386	.2396
9.4	.2407	.2418	.2428	.2439	.2450	.2460	.2471	.2481	.2492	.2502
9.5	2.2513	.2523	.2534	.2544	.2555	.2565	.2576	.2586	.2597	.2607
9.6	.2618	.2628	.2638	.2649	.2659	.2670	.2680	.2690	.2701	.2711
9.7	.2721	.2732	.2742	.2752	.2762	.2773	.2783	.2793	.2803	.2814
9.8	.2824	.2834	.2844	.2854	.2865	.2875	.2885	.2895	.2905	.2915
9.9	.2925	.2935	.2946	.2956	.2966	.2976	.2986	.2996	.3006	.3016

Use the properties of logarithms and $\ln 10 \approx 2.3026$ to find logarithms of numbers less than 1 or greater than 10.

Table C   Base e Logarithms

N	0	1	2	3	4	5	6	7	8	9
1.0	.0000	.0100	.0198	.0296	.0392	.0488	.0583	.0677	.0770	.0862
1.1	.0953	.1044	.1133	.1222	.1310	.1398	.1484	.1570	.1655	.1740
1.2	.1823	.1906	.1989	.2070	.2151	.2231	.2311	.2390	.2469	.2546
1.3	.2624	.2700	.2776	.2852	.2927	.3001	.3075	.3148	.3221	.3293
1.4	.3365	.3436	.3507	.3577	.3646	.3716	.3784	.3853	.3920	.3988
1.5	.4055	.4121	.4187	.4253	.4318	.4383	.4447	.4511	.4574	.4637
1.6	.4700	.4762	.4824	.4886	.4947	.5008	.5068	.5128	.5188	.5247
1.7	.5306	.5365	.5423	.5481	.5539	.5596	.5653	.5710	.5766	.5822
1.8	.5878	.5933	.5988	.6043	.6098	.6152	.6206	.6259	.6313	.6366
1.9	.6419	.6471	.6523	.6575	.6627	.6678	.6729	.6780	.6831	.6881
2.0	.6931	.6981	.7031	.7080	.7129	.7178	.7227	.7275	.7324	.7372
2.1	.7419	.7467	.7514	.7561	.7608	.7655	.7701	.7747	.7793	.7839
2.2	.7885	.7930	.7975	.8020	.8065	.8109	.8154	.8198	.8242	.8286
2.3	.8329	.8372	.8416	.8459	.8502	.8544	.8587	.8629	.8671	.8713
2.4	.8755	.8796	.8838	.8879	.8920	.8961	.9002	.9042	.9083	.9123
2.5	.9163	.9203	.9243	.9282	.9322	.9361	.9400	.9439	.9478	.9517
2.6	.9555	.9594	.9632	.9670	.9708	.9746	.9783	.9821	.9858	.9895
2.7	.9933	.9969	1.0006	.0043	.0080	.0116	.0152	.0188	.0225	.0260
2.8	1.0296	.0332	.0367	.0403	.0438	.0473	.0508	.0543	.0578	.0613
2.9	.0647	.0682	.0716	.0750	.0784	.0818	.0852	.0886	.0919	.0953
3.0	1.0986	.1019	.1053	.1086	.1119	.1151	.1184	.1217	.1249	.1282
3.1	.1314	.1346	.1378	.1410	.1442	.1474	.1506	.1537	.1569	.1600
3.2	.1632	.1663	.1694	.1725	.1756	.1787	.1817	.1848	.1878	.1909
3.3	.1939	.1969	.2000	.2030	.2060	.2090	.2119	.2149	.2179	.2208
3.4	.2238	.2267	.2296	.2326	.2355	.2384	.2413	.2442	.2470	.2499
3.5	1.2528	.2556	.2585	.2613	.2641	.2669	.2698	.2726	.2754	.2782
3.6	.2809	.2837	.2865	.2892	.2920	.2947	.2975	.3002	.3029	.3056
3.7	.3083	.3110	.3137	.3164	.3191	.3218	.3244	.3271	.3297	.3324
3.8	.3350	.3376	.3403	.3429	.3455	.3481	.3507	.3533	.3558	.3584
3.9	.3610	.3635	.3661	.3686	.3712	.3737	.3762	.3788	.3813	.3838
4.0	1.3863	.3888	.3913	.3938	.3962	.3987	.4012	.4036	.4061	.4085
4.1	.4110	.4134	.4159	.4183	.4207	.4231	.4255	.4279	.4303	.4327
4.2	.4351	.4375	.4398	.4422	.4446	.4469	.4493	.4516	.4540	.4563
4.3	.4586	.4609	.4633	.4656	.4679	.4702	.4725	.4748	.4770	.4793
4.4	.4816	.4839	.4861	.4884	.4907	.4929	.4951	.4974	.4996	.5019
4.5	1.5041	.5063	.5085	.5107	.5129	.5151	.5173	.5195	.5217	.5239
4.6	.5261	.5282	.5304	.5326	.5347	.5369	.5390	.5412	.5433	.5454
4.7	.5476	.5497	.5518	.5539	.5560	.5581	.5602	.5623	.5644	.5665
4.8	.5686	.5707	.5728	.5748	.5769	.5790	.5810	.5831	.5851	.5872
4.9	.5892	.5913	.5933	.5953	.5974	.5994	.6014	.6034	.6054	.6074
5.0	1.6094	.6114	.6134	.6154	.6174	.6194	.6214	.6233	.6253	.6273
5.1	.6292	.6312	.6332	.6351	.6371	.6390	.6409	.6429	.6448	.6467
5.2	.6487	.6506	.6525	.6544	.6563	.6582	.6601	.6620	.6639	.6658
5.3	.6677	.6696	.6715	.6734	.6752	.6771	.6790	.6808	.6827	.6845
5.4	.6864	.6882	.6901	.6919	.6938	.6956	.6974	.6993	.7011	.7029

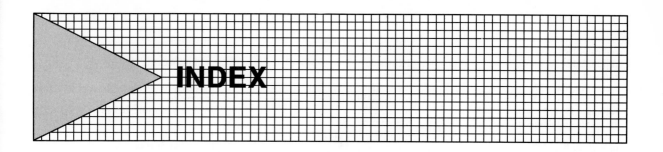

INDEX

FUNCTIONS AND FUNCTION NOTATION

$(f \circ g)(x) = f(g(x))$

$(f + g)(x) = f(x) + g(x)$

$(f - g)(x) = f(x) - g(x)$

$(f \cdot g)(x) = f(x)g(x)$

$(f/g)(x) = \dfrac{f(x)}{g(x)}$

7.6 PROPORTION AND VARIATION

$y = kx$ y varies directly as x

$y = \dfrac{k}{x}$ y varies inversely as x

$y = kxz$ y varies jointly with x and z

$y = \dfrac{kx}{z}$ y varies directly as x, but inversely as z

8.1 COMPLETING THE SQUARE AND THE QUADRATIC FORMULA

The solutions of the equation $x^2 = c$ are

$x = \sqrt{c}$ and $x = -\sqrt{c}$

$x = \dfrac{-b \pm \sqrt{b^2 - 4ac}}{2a}$ $(a \neq 0)$ quadratic formula

8.2 COMPLEX NUMBERS

$i^2 = -1$

$a + bi = c + di$ if and only if $a = c$ and $b = d$

$(a + bi) + (c + di) = (a + c) + (b + d)i$

$(a + bi)(c + di) = (ac - bd) + (ad + bc)i$

$a + bi$ and $a - bi$ are complex conjugates

$|a + bi| = \sqrt{a^2 + b^2}$

8.3 MORE ON QUADRATIC EQUATIONS

If $\begin{cases} b^2 - 4ac > 0 \\ b^2 - 4ac = 0 \\ b^2 - 4ac < 0 \end{cases}$, then the solutions

of $ax^2 + bx + c = 0$ are $\begin{cases} \text{real and unequal} \\ \text{real and equal} \\ \text{nonreal} \end{cases}$.

If x_1 and x_2 are roots of the equation $ax^2 + bx + c = 0$, then

$x_1 + x_2 = -\dfrac{b}{a}$

$x_1 x_2 = \dfrac{c}{a}$

8.4 GRAPHS OF QUADRATIC FUNCTIONS

The graph of $y = a(x - h)^2 + k$ is a parabola with vertex at (h, k).

The vertex of the parabola determined by $y = ax^2 + bx + c$ has coordinates of

$$\left(-\frac{b}{2a}, \frac{4ac - b^2}{4a} \right)$$

8.7 THE CIRCLE AND THE PARABOLA

$(x - h)^2 + (y - k)^2 = r^2$ circle centered at (h, k) with radius $|r|$

$x^2 + y^2 = r^2$ circle centered at $(0, 0)$ with radius $|r|$

Parabola opening	Vertex at origin	Vertex at (h, k)
Up	$y = ax^2$	$y = a(x - h)^2 + k$
Down	$y = -ax^2$	$y = -a(x - h)^2 + k$
Right	$x = ay^2$	$x = a(y - k)^2 + h$
Left	$x = -ay^2$	$x = -a(y - k)^2 + h$

8.8 THE ELLIPSE AND THE HYPERBOLA

$\dfrac{x^2}{a^2} + \dfrac{y^2}{b^2} = 1$ ellipse centered at $(0, 0)$; x-intercepts at $(a, 0)$ and $(-a, 0)$; y-intercepts at $(0, b)$ and $(0, -b)$

$\dfrac{(x - h)^2}{a^2} + \dfrac{(y - k)^2}{b^2} = 1$ ellipse centered at (h, k)

$\dfrac{x^2}{a^2} - \dfrac{y^2}{b^2} = 1$ hyperbola centered at $(0, 0)$; x-intercepts at $(a, 0)$ and $(-a, 0)$

$\dfrac{y^2}{b^2} - \dfrac{x^2}{a^2} = 1$ hyperbola centered at $(0, 0)$; y-intercepts at $(0, b)$ and $(0, -b)$

$\left. \begin{array}{l} \dfrac{(x - h)^2}{a^2} - \dfrac{(y - k)^2}{b^2} = 1 \\[2mm] \dfrac{(y - k)^2}{b^2} - \dfrac{(x - h)^2}{a^2} = 1 \end{array} \right\}$ hyperbolas centered at (h, k)

$xy = k$ hyperbola not intersecting either axis

9.3 SOLUTION BY DETERMINANTS

$\begin{vmatrix} a & b \\ c & d \end{vmatrix} = ad - bc$

The solutions of $\begin{cases} ax + by = e \\ cx + dy = f \end{cases}$ are

$x = \dfrac{D_x}{D}$ and $y = \dfrac{D_y}{D}$ where $D = \begin{vmatrix} a & b \\ c & d \end{vmatrix}$,

$D_x = \begin{vmatrix} e & b \\ f & d \end{vmatrix}$,   and $D_y = \begin{vmatrix} a & e \\ c & f \end{vmatrix}$